T0188924

Communications
in Computer and Information Science　　　1967

Rationale

The CCIS series is devoted to the publication of proceedings of computer science conferences. Its aim is to efficiently disseminate original research results in informatics in printed and electronic form. While the focus is on publication of peer-reviewed full papers presenting mature work, inclusion of reviewed short papers reporting on work in progress is welcome, too. Besides globally relevant meetings with internationally representative program committees guaranteeing a strict peer-reviewing and paper selection process, conferences run by societies or of high regional or national relevance are also considered for publication.

Topics

The topical scope of CCIS spans the entire spectrum of informatics ranging from foundational topics in the theory of computing to information and communications science and technology and a broad variety of interdisciplinary application fields.

Information for Volume Editors and Authors

Publication in CCIS is free of charge. No royalties are paid, however, we offer registered conference participants temporary free access to the online version of the conference proceedings on SpringerLink (http://link.springer.com) by means of an http referrer from the conference website and/or a number of complimentary printed copies, as specified in the official acceptance email of the event.

CCIS proceedings can be published in time for distribution at conferences or as post-proceedings, and delivered in the form of printed books and/or electronically as USBs and/or e-content licenses for accessing proceedings at SpringerLink. Furthermore, CCIS proceedings are included in the CCIS electronic book series hosted in the SpringerLink digital library at http://link.springer.com/bookseries/7899. Conferences publishing in CCIS are allowed to use Online Conference Service (OCS) for managing the whole proceedings lifecycle (from submission and reviewing to preparing for publication) free of charge.

Publication process

The language of publication is exclusively English. Authors publishing in CCIS have to sign the Springer CCIS copyright transfer form, however, they are free to use their material published in CCIS for substantially changed, more elaborate subsequent publications elsewhere. For the preparation of the camera-ready papers/files, authors have to strictly adhere to the Springer CCIS Authors' Instructions and are strongly encouraged to use the CCIS LaTeX style files or templates.

Abstracting/Indexing

CCIS is abstracted/indexed in DBLP, Google Scholar, EI-Compendex, Mathematical Reviews, SCImago, Scopus. CCIS volumes are also submitted for the inclusion in ISI Proceedings.

How to start

To start the evaluation of your proposal for inclusion in the CCIS series, please send an e-mail to ccis@springer.com.

Biao Luo · Long Cheng · Zheng-Guang Wu ·
Hongyi Li · Chaojie Li
Editors

Neural Information Processing

30th International Conference, ICONIP 2023
Changsha, China, November 20–23, 2023
Proceedings, Part XIII

Springer

Editors
Biao Luo ⓘ
School of Automation
Central South University
Changsha, China

Long Cheng ⓘ
Institute of Automation
Chinese Academy of Sciences
Beijing, China

Zheng-Guang Wu ⓘ
Institute of Cyber-Systems and Control
Zhejiang University
Hangzhou, China

Hongyi Li ⓘ
School of Automation
Guangdong University of Technology
Guangzhou, China

Chaojie Li ⓘ
School of Electrical Engineering
and Telecommunications
UNSW Sydney
Sydney, NSW, Australia

ISSN 1865-0929 ISSN 1865-0937 (electronic)
Communications in Computer and Information Science
ISBN 978-981-99-8177-9 ISBN 978-981-99-8178-6 (eBook)
https://doi.org/10.1007/978-981-99-8178-6

This Springer imprint is published by the registered company Springer Nature Singapore Pte Ltd.
The registered company address is: 152 Beach Road, #21-01/04 Gateway East, Singapore 189721, Singapore

Paper in this product is recyclable.

Preface

Welcome to the 30th International Conference on Neural Information Processing (ICONIP2023) of the Asia-Pacific Neural Network Society (APNNS), held in Changsha, China, November 20–23, 2023.

The mission of the Asia-Pacific Neural Network Society is to promote active interactions among researchers, scientists, and industry professionals who are working in neural networks and related fields in the Asia-Pacific region. APNNS has Governing Board Members from 13 countries/regions – Australia, China, Hong Kong, India, Japan, Malaysia, New Zealand, Singapore, South Korea, Qatar, Taiwan, Thailand, and Turkey. The society's flagship annual conference is the International Conference of Neural Information Processing (ICONIP). The ICONIP conference aims to provide a leading international forum for researchers, scientists, and industry professionals who are working in neuroscience, neural networks, deep learning, and related fields to share their new ideas, progress, and achievements.

ICONIP2023 received 1274 papers, of which 394 papers were accepted for publication in Communications in Computer and Information Science (CCIS), representing an acceptance rate of 30.93% and reflecting the increasingly high quality of research in neural networks and related areas. The conference focused on four main areas, i.e., "Theory and Algorithms", "Cognitive Neurosciences", "Human-Centered Computing", and "Applications". All the submissions were rigorously reviewed by the conference Program Committee (PC), comprising 258 PC members, and they ensured that every paper had at least two high-quality single-blind reviews. In fact, 5270 reviews were provided by 2145 reviewers. On average, each paper received 4.14 reviews.

We would like to take this opportunity to thank all the authors for submitting their papers to our conference, and our great appreciation goes to the Program Committee members and the reviewers who devoted their time and effort to our rigorous peer-review process; their insightful reviews and timely feedback ensured the high quality of the papers accepted for publication. We hope you enjoyed the research program at the conference.

October 2023

Biao Luo
Long Cheng
Zheng-Guang Wu
Hongyi Li
Chaojie Li

Organization

Honorary Chair

Weihua Gui   Central South University, China

Advisory Chairs

Jonathan Chan   King Mongkut's University of Technology
                                    Thonburi, Thailand

Zeng-Guang Hou   Chinese Academy of Sciences, China

Nikola Kasabov   Auckland University of Technology, New Zealand

Derong Liu   Southern University of Science and Technology,
                                    China

Seiichi Ozawa   Kobe University, Japan

Kevin Wong   Murdoch University, Australia

General Chairs

Tingwen Huang   Texas A&M University at Qatar, Qatar

Chunhua Yang   Central South University, China

Program Chairs

Biao Luo   Central South University, China

Long Cheng   Chinese Academy of Sciences, China

Zheng-Guang Wu   Zhejiang University, China

Hongyi Li   Guangdong University of Technology, China

Chaojie Li   University of New South Wales, Australia

Technical Chairs

Xing He   Southwest University, China

Keke Huang   Central South University, China

Huaqing Li   Southwest University, China

Qi Zhou   Guangdong University of Technology, China

Local Arrangement Chairs

Wenfeng Hu   Central South University, China
Bei Sun   Central South University, China

Finance Chairs

Fanbiao Li   Central South University, China
Hayaru Shouno   University of Electro-Communications, Japan
Xiaojun Zhou   Central South University, China

Special Session Chairs

Hongjing Liang   University of Electronic Science and Technology,
                                  China
Paul S. Pang   Federation University, Australia
Qiankun Song   Chongqing Jiaotong University, China
Lin Xiao   Hunan Normal University, China

Tutorial Chairs

Min Liu   Hunan University, China
M. Tanveer   Indian Institute of Technology Indore, India
Guanghui Wen   Southeast University, China

Publicity Chairs

Sabri Arik   Istanbul University-Cerrahpaşa, Turkey
Sung-Bae Cho   Yonsei University, South Korea
Maryam Doborjeh   Auckland University of Technology, New Zealand
El-Sayed M. El-Alfy   King Fahd University of Petroleum and Minerals,
                                  Saudi Arabia
Ashish Ghosh   Indian Statistical Institute, India
Chuandong Li   Southwest University, China
Weng Kin Lai   Tunku Abdul Rahman University of
                                  Management & Technology, Malaysia
Chu Kiong Loo   University of Malaya, Malaysia
Qinmin Yang   Zhejiang University, China
Zhigang Zeng   Huazhong University of Science and Technology,
                                  China

Publication Chairs

Zhiwen Chen   Central South University, China
Andrew Chi-Sing Leung   City University of Hong Kong, China
Xin Wang   Southwest University, China
Xiaofeng Yuan   Central South University, China

Secretaries

Yun Feng   Hunan University, China
Bingchuan Wang   Central South University, China

Webmasters

Tianmeng Hu   Central South University, China
Xianzhe Liu   Xiangtan University, China

Program Committee

Rohit Agarwal   UiT The Arctic University of Norway, Norway
Hasin Ahmed   Gauhati University, India
Harith Al-Sahaf   Victoria University of Wellington, New Zealand
Brad Alexander   University of Adelaide, Australia
Mashaan Alshammari   Independent Researcher, Saudi Arabia
Sabri Arik   Istanbul University, Turkey
Ravneet Singh Arora   Block Inc., USA
Zeyar Aung   Khalifa University of Science and Technology, UAE
Monowar Bhuyan   Umeå University, Sweden
Jingguo Bi   Beijing University of Posts and Telecommunications, China
Xu Bin   Northwestern Polytechnical University, China
Marcin Blachnik   Silesian University of Technology, Poland
Paul Black   Federation University, Australia
Anoop C. S.   Govt. Engineering College, India
Ning Cai   Beijing University of Posts and Telecommunications, China
Siripinyo Chantamunee   Walailak University, Thailand
Hangjun Che   City University of Hong Kong, China

Wei-Wei Che	Qingdao University, China
Huabin Chen	Nanchang University, China
Jinpeng Chen	Beijing University of Posts & Telecommunications, China
Ke-Jia Chen	Nanjing University of Posts and Telecommunications, China
Lv Chen	Shandong Normal University, China
Qiuyuan Chen	Tencent Technology, China
Wei-Neng Chen	South China University of Technology, China
Yufei Chen	Tongji University, China
Long Cheng	Institute of Automation, China
Yongli Cheng	Fuzhou University, China
Sung-Bae Cho	Yonsei University, South Korea
Ruikai Cui	Australian National University, Australia
Jianhua Dai	Hunan Normal University, China
Tao Dai	Tsinghua University, China
Yuxin Ding	Harbin Institute of Technology, China
Bo Dong	Xi'an Jiaotong University, China
Shanling Dong	Zhejiang University, China
Sidong Feng	Monash University, Australia
Yuming Feng	Chongqing Three Gorges University, China
Yun Feng	Hunan University, China
Junjie Fu	Southeast University, China
Yanggeng Fu	Fuzhou University, China
Ninnart Fuengfusin	Kyushu Institute of Technology, Japan
Thippa Reddy Gadekallu	VIT University, India
Ruobin Gao	Nanyang Technological University, Singapore
Tom Gedeon	Curtin University, Australia
Kam Meng Goh	Tunku Abdul Rahman University of Management and Technology, Malaysia
Zbigniew Gomolka	University of Rzeszow, Poland
Shengrong Gong	Changshu Institute of Technology, China
Xiaodong Gu	Fudan University, China
Zhihao Gu	Shanghai Jiao Tong University, China
Changlu Guo	Budapest University of Technology and Economics, Hungary
Weixin Han	Northwestern Polytechnical University, China
Xing He	Southwest University, China
Akira Hirose	University of Tokyo, Japan
Yin Hongwei	Huzhou Normal University, China
Md Zakir Hossain	Curtin University, Australia
Zengguang Hou	Chinese Academy of Sciences, China

Lu Hu   Jiangsu University, China
Zeke Zexi Hu   University of Sydney, Australia
He Huang   Soochow University, China
Junjian Huang   Chongqing University of Education, China
Kaizhu Huang   Duke Kunshan University, China
David Iclanzan   Sapientia University, Romania
Radu Tudor Ionescu   University of Bucharest, Romania
Asim Iqbal   Cornell University, USA
Syed Islam   Edith Cowan University, Australia
Kazunori Iwata   Hiroshima City University, Japan
Junkai Ji   Shenzhen University, China
Yi Ji   Soochow University, China
Canghong Jin   Zhejiang University, China
Xiaoyang Kang   Fudan University, China
Mutsumi Kimura   Ryukoku University, Japan
Masahiro Kohjima   NTT, Japan
Damian Kordos   Rzeszow University of Technology, Poland
Marek Kraft   Poznań University of Technology, Poland
Lov Kumar   NIT Kurukshetra, India
Weng Kin Lai   Tunku Abdul Rahman University of
                                Management & Technology, Malaysia
Xinyi Le   Shanghai Jiao Tong University, China
Bin Li   University of Science and Technology of China,
                                China
Hongfei Li   Xinjiang University, China
Houcheng Li   Chinese Academy of Sciences, China
Huaqing Li   Southwest University, China
Jianfeng Li   Southwest University, China
Jun Li   Nanjing Normal University, China
Kan Li   Beijing Institute of Technology, China
Peifeng Li   Soochow University, China
Wenye Li   Chinese University of Hong Kong, China
Xiangyu Li   Beijing Jiaotong University, China
Yantao Li   Chongqing University, China
Yaoman Li   Chinese University of Hong Kong, China
Yinlin Li   Chinese Academy of Sciences, China
Yuan Li   Academy of Military Science, China
Yun Li   Nanjing University of Posts and
                                Telecommunications, China
Zhidong Li   University of Technology Sydney, Australia
Zhixin Li   Guangxi Normal University, China
Zhongyi Li   Beihang University, China

Ziqiang Li	University of Tokyo, Japan
Xianghong Lin	Northwest Normal University, China
Yang Lin	University of Sydney, Australia
Huawen Liu	Zhejiang Normal University, China
Jian-Wei Liu	China University of Petroleum, China
Jun Liu	Chengdu University of Information Technology, China
Junxiu Liu	Guangxi Normal University, China
Tommy Liu	Australian National University, Australia
Wen Liu	Chinese University of Hong Kong, China
Yan Liu	Taikang Insurance Group, China
Yang Liu	Guangdong University of Technology, China
Yaozhong Liu	Australian National University, Australia
Yong Liu	Heilongjiang University, China
Yubao Liu	Sun Yat-sen University, China
Yunlong Liu	Xiamen University, China
Zhe Liu	Jiangsu University, China
Zhen Liu	Chinese Academy of Sciences, China
Zhi-Yong Liu	Chinese Academy of Sciences, China
Ma Lizhuang	Shanghai Jiao Tong University, China
Chu-Kiong Loo	University of Malaya, Malaysia
Vasco Lopes	Universidade da Beira Interior, Portugal
Hongtao Lu	Shanghai Jiao Tong University, China
Wenpeng Lu	Qilu University of Technology, China
Biao Luo	Central South University, China
Ye Luo	Tongji University, China
Jiancheng Lv	Sichuan University, China
Yuezu Lv	Beijing Institute of Technology, China
Huifang Ma	Northwest Normal University, China
Jinwen Ma	Peking University, China
Jyoti Maggu	Thapar Institute of Engineering and Technology Patiala, India
Adnan Mahmood	Macquarie University, Australia
Mufti Mahmud	University of Padova, Italy
Krishanu Maity	Indian Institute of Technology Patna, India
Srimanta Mandal	DA-IICT, India
Wang Manning	Fudan University, China
Piotr Milczarski	Lodz University of Technology, Poland
Malek Mouhoub	University of Regina, Canada
Nankun Mu	Chongqing University, China
Wenlong Ni	Jiangxi Normal University, China
Anupiya Nugaliyadde	Murdoch University, Australia

Toshiaki Omori	Kobe University, Japan
Babatunde Onasanya	University of Ibadan, Nigeria
Manisha Padala	Indian Institute of Science, India
Sarbani Palit	Indian Statistical Institute, India
Paul Pang	Federation University, Australia
Rasmita Panigrahi	Giet University, India
Kitsuchart Pasupa	King Mongkut's Institute of Technology Ladkrabang, Thailand
Dipanjyoti Paul	Ohio State University, USA
Hu Peng	Jiujiang University, China
Kebin Peng	University of Texas at San Antonio, USA
Dawid Połap	Silesian University of Technology, Poland
Zhong Qian	Soochow University, China
Sitian Qin	Harbin Institute of Technology at Weihai, China
Toshimichi Saito	Hosei University, Japan
Fumiaki Saitoh	Chiba Institute of Technology, Japan
Naoyuki Sato	Future University Hakodate, Japan
Chandni Saxena	Chinese University of Hong Kong, China
Jiaxing Shang	Chongqing University, China
Lin Shang	Nanjing University, China
Jie Shao	University of Science and Technology of China, China
Yin Sheng	Huazhong University of Science and Technology, China
Liu Sheng-Lan	Dalian University of Technology, China
Hayaru Shouno	University of Electro-Communications, Japan
Gautam Srivastava	Brandon University, Canada
Jianbo Su	Shanghai Jiao Tong University, China
Jianhua Su	Institute of Automation, China
Xiangdong Su	Inner Mongolia University, China
Daiki Suehiro	Kyushu University, Japan
Basem Suleiman	University of New South Wales, Australia
Ning Sun	Shandong Normal University, China
Shiliang Sun	East China Normal University, China
Chunyu Tan	Anhui University, China
Gouhei Tanaka	University of Tokyo, Japan
Maolin Tang	Queensland University of Technology, Australia
Shu Tian	University of Science and Technology Beijing, China
Shikui Tu	Shanghai Jiao Tong University, China
Nancy Victor	Vellore Institute of Technology, India
Petra Vidnerová	Institute of Computer Science, Czech Republic

Shanchuan Wan	University of Tokyo, Japan
Tao Wan	Beihang University, China
Ying Wan	Southeast University, China
Bangjun Wang	Soochow University, China
Hao Wang	Shanghai University, China
Huamin Wang	Southwest University, China
Hui Wang	Nanchang Institute of Technology, China
Huiwei Wang	Southwest University, China
Jianzong Wang	Ping An Technology, China
Lei Wang	National University of Defense Technology, China
Lin Wang	University of Jinan, China
Shi Lin Wang	Shanghai Jiao Tong University, China
Wei Wang	Shenzhen MSU-BIT University, China
Weiqun Wang	Chinese Academy of Sciences, China
Xiaoyu Wang	Tokyo Institute of Technology, Japan
Xin Wang	Southwest University, China
Xin Wang	Southwest University, China
Yan Wang	Chinese Academy of Sciences, China
Yan Wang	Sichuan University, China
Yonghua Wang	Guangdong University of Technology, China
Yongyu Wang	JD Logistics, China
Zhenhua Wang	Northwest A&F University, China
Zi-Peng Wang	Beijing University of Technology, China
Hongxi Wei	Inner Mongolia University, China
Guanghui Wen	Southeast University, China
Guoguang Wen	Beijing Jiaotong University, China
Ka-Chun Wong	City University of Hong Kong, China
Anna Wróblewska	Warsaw University of Technology, Poland
Fengge Wu	Institute of Software, Chinese Academy of Sciences, China
Ji Wu	Tsinghua University, China
Wei Wu	Inner Mongolia University, China
Yue Wu	Shanghai Jiao Tong University, China
Likun Xia	Capital Normal University, China
Lin Xiao	Hunan Normal University, China
Qiang Xiao	Huazhong University of Science and Technology, China
Hao Xiong	Macquarie University, Australia
Dongpo Xu	Northeast Normal University, China
Hua Xu	Tsinghua University, China
Jianhua Xu	Nanjing Normal University, China

Xinyue Xu	Hong Kong University of Science and Technology, China
Yong Xu	Beijing Institute of Technology, China
Ngo Xuan Bach	Posts and Telecommunications Institute of Technology, Vietnam
Hao Xue	University of New South Wales, Australia
Yang Xujun	Chongqing Jiaotong University, China
Haitian Yang	Chinese Academy of Sciences, China
Jie Yang	Shanghai Jiao Tong University, China
Minghao Yang	Chinese Academy of Sciences, China
Peipei Yang	Chinese Academy of Science, China
Zhiyuan Yang	City University of Hong Kong, China
Wangshu Yao	Soochow University, China
Ming Yin	Guangdong University of Technology, China
Qiang Yu	Tianjin University, China
Wenxin Yu	Southwest University of Science and Technology, China
Yun-Hao Yuan	Yangzhou University, China
Xiaodong Yue	Shanghai University, China
Paweł Zawistowski	Warsaw University of Technology, Poland
Hui Zeng	Southwest University of Science and Technology, China
Wang Zengyunwang	Hunan First Normal University, China
Daren Zha	Institute of Information Engineering, China
Zhi-Hui Zhan	South China University of Technology, China
Baojie Zhang	Chongqing Three Gorges University, China
Canlong Zhang	Guangxi Normal University, China
Guixuan Zhang	Chinese Academy of Science, China
Jianming Zhang	Changsha University of Science and Technology, China
Li Zhang	Soochow University, China
Wei Zhang	Southwest University, China
Wenbing Zhang	Yangzhou University, China
Xiang Zhang	National University of Defense Technology, China
Xiaofang Zhang	Soochow University, China
Xiaowang Zhang	Tianjin University, China
Xinglong Zhang	National University of Defense Technology, China
Dongdong Zhao	Wuhan University of Technology, China
Xiang Zhao	National University of Defense Technology, China
Xu Zhao	Shanghai Jiao Tong University, China

Liping Zheng	Hefei University of Technology, China
Yan Zheng	Kyushu University, Japan
Baojiang Zhong	Soochow University, China
Guoqiang Zhong	Ocean University of China, China
Jialing Zhou	Nanjing University of Science and Technology, China
Wenan Zhou	PCN&CAD Center, China
Xiao-Hu Zhou	Institute of Automation, China
Xinyu Zhou	Jiangxi Normal University, China
Quanxin Zhu	Nanjing Normal University, China
Yuanheng Zhu	Chinese Academy of Sciences, China
Xiaotian Zhuang	JD Logistics, China
Dongsheng Zou	Chongqing University, China

Contents – Part XIII

Applications

xx   Contents – Part XIII

Applications

Improve Conversational Search
with Multi-document Information

Shiyulong He, Sai Zhang, Xiaowang Zhang$^{(\boxtimes)}$, and Zhiyong Feng

College of Intelligence and Computing, Tianjin University, Tianjin 300350, China
{hsyl1104,zhang_sai,xiaowangzhang,zyfeng}@tju.edu.cn

Abstract. Conversational Search (CS) aims to satisfy complex information needs via multi-turn user-agent interactions. During this process, multiple documents need to be retrieved based on the conversation history to respond to the user. However, existing approaches still make it difficult to distinguish irrelevant information from the user's question at the semantic level. When a conversation involves multiple documents, it sometimes affects the retrieval performance negatively. In order to enhance the model's ability to comprehend conversations and distinguish passages in irrelevant documents via multi-document information, we propose an unsupervised multi-document conversation segmentation method and a zero-shot Large Language Model (LLM)-based document summarization method to extract multi-document information from conversation history and documents respectively for amending the lack of training data for extracting multiple document information. We further present the Passage-Segment-Document (PSD) post-training method to train the reranker using the extracted multi-document information in combination with a multi-task learning method. The results on the MultiDoc2Dial dataset verifies the improvement of our method on retrieval performance. Extensive experiments show the strong performance of our method for dealing with conversation histories that contain multi-document information.

Keywords: conversational search · conversational dense retrieval · multi-document · post-training

1 Introduction

Conversational Search (CS) has recently become a growing research frontier in the Information Retrieval (IR) community [17]. It aims to better meet the user's information needs by retrieving the most relevant passages to the current user's question from multiple documents based on the conversation history of the multi-turn of natural language interactions between the user and the agent. Compared with single-turn question answering retrieval or machine reading comprehension tasks given only one single document (passage), CS assumes that users ask questions and get responses step by step in a conversation. Therefore, the

© The Author(s), under exclusive license to Springer Nature Singapore Pte Ltd. 2024
B. Luo et al. (Eds.): ICONIP 2023, CCIS 1967, pp. 3–15, 2024.
https://doi.org/10.1007/978-981-99-8178-6_1

conversation history is crucial to understanding the user's current question and performing relevant passage retrieval [4, 19].

During the conversation, the user may query passages in the same document or different documents. Therefore, the conversation history will include information related to multiple documents. However, it will also contain information about documents that are irrelevant to understanding the user's current question (interfering documents), and this information does not provide a positive effect and could even have a negative impact [14]. Therefore, it is necessary but challenging for the model to distinguish information from the semantic-level irrelevant to the user's current question during retrieval.

Most of the current approaches employ sparse or dense retrievers based on the entire conversation history and apply lexical or vector similarity matching to make full use of information in the conversation history for retrieval [1, 9, 12, 21]. However, these approaches cannot address the negative impact brought by the information of interfering documents, resulting in a decrease in retrieval performance when multiple documents are involved in the conversation history. Some approaches remove utterances irrelevant to the current question from the conversation history and directly use the remaining history for retrieval to avoid the influence of information of interfering documents [16]. However, the structure of the conversation history is destroyed because of directly removing utterances, resulting in the loss of semantic information and making it difficult to improve retrieval performance. In addition, these approaches directly model the similarity between a single passage and a conversation, which would lose the context information of the passage, making it difficult for the model to distinguish similar passages in different contexts.

To better deal with this challenge, our work focuses on improving retrieval performance when the conversation history contains multi-document information. Since multi-document information is implicit in the conversation history and passage context and considering the lack of training data for extracting the information, we propose an unsupervised multi-document conversation segmentation method and a zero-shot LLM-based document summarization method. Meanwhile, we propose the Passage-Segment-Document (PSD) post-training method, including the passage-segment and passage-document tasks. Combined with the extracted multi-document information, we use multi-task learning to joint train the two tasks to allow the model to learn the correlation between passages and conversation segments containing different document information and enhance the model's ability to distinguish similar passages in different contexts through supplementing the passages' context information.

Our contributions can be summarized as follows:

- We propose the multi-document conversation segmentation method for unsupervised multi-turn conversation segmentation and extract conversation segments containing different document information.
- We propose the zero-shot LLM-based document summarization method and design a prompt more conducive to information retrieval, to extract the multi-document information implicit in the passage context.

- We propose the Passage-Segment-Document (PSD) post-training method, including the passage-segment task and the passage-document task, to fully use extracted multi-document information and let the model learn the semantic associations between conversation history and passages from multiple documents.

2 Related Works

2.1 Conversational Search

There are two main types of methods for this task recently. One is based on the conversational query rewriting [18], which rewrites the conversation into an independent question and then retrieves passages using it. However, this method will cause information loss and bring a long time delay [11], and some research has shown that rewriting does not actually improve performance [6]. The other focuses on information retrieval, [10] employs a bi-encoder as the retriever in single-stage retrieval. [3] further applies BM25 to construct negative samples, and [7] explores strategies such as data augmentation using synonym enhancers. To better achieve recall and rerank performance, CPII-NLP [12] expands the retrieval into two phases, using DPR as the retriever, and the reranker is an ensemble of three cross-encoder models. UGent-T2K [9] divides the retrieval into document retrieval and passage retrieval, and uses the LambdaMART algorithm combined with TF-IDF and term-matching techniques to synthesize passage scoring and ranking. R3 [1] also uses a two-stage retrieval method. Besides, [16] proposes the Passage Checking Model to denoise the conversation history, and use passage information to assist in removing the utterance irrelevant to the current question in the conversation history during retrieval.

Our work focuses on the latter method. Although we adopt a two-stage retrieval framework, we focus on leveraging multiple document information from conversation history and passage context to improve the model's retrieval ability when a conversation involves varying numbers of documents.

2.2 LLM for Information Retrieval

Large Language Models (LLMs) such as GPT-series have shown excellent performance in various NLP tasks, including text summarization, and have strong zero-shot capabilities [22]. Despite the excellent performance of LLMs, billion-parameter models are rarely used in IR. One reason is that IR tasks are computationally intensive, and the straightforward use of LLMs can lead to very expensive expenses [2]. Therefore, how to use LLMs cost-effectively in IR tasks is still an open question. [2] introduces a method for efficiently using LLMs in information retrieval tasks by guiding the LLMs to generate relevant queries for documents with a small number of examples and then using these synthesized query-document pairs to train the retriever. Based on [2,8] further replaces GPT-3 with the open source GPT-J (6B), and applies the correlation score calculated by monoT5 fine-tuned on MS MARCO to screen the top 10,000 pairs. In addition, [15] uses LLMs for query enhancement, [20] proposes the generate-then-read, that is, directly use LLMs to generate a contextual document according to a given question, and then read the document to generate an answer to the question.

These approaches aim to use LLMs to improve the performance of single-turn question answering retrieval. Unlike these approaches, we focus on improving the retrieval performance in CS. Instead of directly generating data for retrieval, we summarize documents to provide auxiliary information via LLMs, aiming to explore a new perspective of using LLMs to promote retrieval in CS.

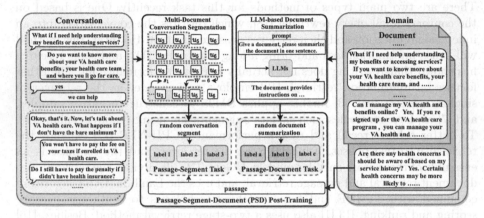

Fig. 1. The overview framework.

3   Method

As shown in Fig. 1, our method mainly includes three parts: (1) Multi-Document Conversation Segmentation, which segments multi-turn conversations unsupervised and extracts conversation segments containing information about different documents. (2) LLM-based Document Summarization. We adopt a zero-shot approach and carefully design the prompt for document summaries based on LLMs. (3) Reranker based on PSD Post-training, which uses multiple training tasks to enhance the model's ability to model conversation history and distinguish passages of interfering documents based on the multi-task learning method.

3.1   Preliminaries

Given a conversation $C = \{u_1, \ldots, u_{t-1}, u_t\}$ and a collection of multiple documents $D = \{D_1, ..., D_m\}$, where t represents the number of utterances, u_t is the user's current question, $\{u_1, \ldots, u_{t-1}\}$ is the conversation history, and m is the number of documents. Each document is divided into passages based on the structure of the document or a fixed-size sliding window. That is, for each document $Di \in D$, the segment set $P_i = \{p_{i0}, ..., p_{ij}\}$ is obtained, where j is segmentation number of passages of document i . The goal is to retrieve the most relevant passage to u_t from the collection of passage sets $P = P_1 \cup \cdots \cup P_m$ obtained from all document segmentations based on $\{u_1, \ldots, u_{t-1}\}$ and u_t.

3.2 Multi-document Conversation Segmentation

To extract the multi-document information contained in the conversation, we propose the Multi-Document Conversation Segmentation method to segment multi-turns of conversations unsupervised and extract conversation segments containing different document information.

Algorithm 1. Multi-Document Conversation Segmentation Algorithm

---

Input: Conversation $C = \{u_1, \ldots, u_{t-1}, u_t\}$; Maximum window size W
Output: Conversation segment list $S = \{s_1, \ldots, s_n\}$
 1: set $i \leftarrow 1$, $S \leftarrow [\,]$;
 2: **while** $i \leq t$ **do**
 3:     $j \leftarrow 1$; $c_0 \leftarrow$ ""
 4:     **while** $i + j \leq t + 1$ and $j < W$ **do**
 5:         $c_j \leftarrow concatenate(c_{j-1}, u_{i+j-1})$
 6:         $cost_{c_j} \leftarrow max(similarity(c_j, u_{i-1}), similarity(c_j, u_{i+j}))$
 7:         $j \leftarrow j + 1$
 8:     **end while**
 9:     $segIndex \leftarrow min_j cost_{c_j}$
10:     $S.append(c_{segIndex})$; $i \leftarrow segIndex + 1$
11: **end while**

As shown in Algorithm 1. Given a conversation C, we select the current utterance segment c through a dynamically sized window and expand the range of c by gradually increasing the window size. During this process, we compute the similarity score between the c and the preceding and following utterances, and keep the maximum value of the two. By limiting the size to W, when the window increases to W, we segment the conversation from the position where the calculated similarity value is the smallest and construct a new window from the next utterance at the segmented position until we reach the end of the conversation. Finally, for the conversation C, we get the segment list $\{s_1, \ldots, s_n\}$.

3.3 LLM-Based Document Summarization

Considering the effectiveness of prompt in guiding LLMs to play a role, and the stable performance of LLMs in scenarios lacking training data, we adopt a zero-shot approach to carefully construct the prompt to guide LLMs to summarize for a given document.

We consider that the document summarization information that contributes to information retrieval should meet the following conditions: (1) The description should be concise and avoid redundant vocabulary; (2) The summarization should emphasize the content or aspects of the core description of the article, so as to provide differentiated contextual information effectively. Therefore, we propose the prompt schema: "*Give a document, please summarize the document in one sentence. \n document:* $\{D_i\}$". Specifically, given a document $D_i \in D$, we fill its content into the placeholder of prompt and denote the document summarization information generated by LLMs as $LLM(D_i)$.

For documents that exceed the length limit, we apply a multi-stage approach to summarize the document. Specifically, we first split the document into multiple short document segments that can satisfy the length constraint according to its passage structure. Then we fill them into the prompt to generate a summarization through LLMs. We concatenate multiple summarizations together to form a new document, fill it into the prompt for a new round of summary generation, and use the final generated result as the summarization of the original document.

3.4   Reranker Based on Passage-Segment-Document Post-training

To utilize the extracted multi-document information, we introduce the passage-segment and the passage-document tasks and use multi-task learning to integrate the feature information learned by each task.

Passage-Segment Task. In order to learn the correlation between passages and conversation segments containing different document information, and enhance the model's ability to distinguish the passages of interfering documents, we propose the passage-segment training task. Specifically, for each passage $p \in P$, we randomly sample conversation segments s with three categories of labels and let the model classify the correlation between p and s. The three categories are 1. conversation segments mentioning the same document (label 1); 2. conversation segments mentioning other documents in the same domain (label 2); 3. conversation segments mentioning documents in different domains (label 3). The conversation segments are obtained from the multi-document conversation segmentation method; the domain represents a larger document category that includes multiple documents. We train this task using the cross-entropy loss function L_{pseg}:

$$L_{pseg} = -\sum \sum_{i=1}^{3} x_i log(MLP(h_s, h_p)_i), \tag{1}$$

where h_s and h_p represent the embedding vectors of conversation segment and passage, respectively; x_i is a symbolic function; MLP represents a multi-layer perceptron.

Passage-Document Task. In order to learn the correlation between document passages and their contextual information, so that the model can better learn the representation of passages, we propose the passage-document task. Similar to the passage-segment task, for each passage $p \in P$, we randomly sample the summarization information $LLM(D)$ of three categories of labels and let the model classify the correlation between p and $LLM(D)$. The three categories are a. the summarization information of the document to which the passage belongs (label a); b. the summarization information of other documents in the same domain (label b); c. the summarization information of other documents in a different domain (label c). The summarization information of each document is obtained from the LLM-based document summarization method. We train this task using the cross-entropy loss function L_{pdoc}:

$$L_{pdoc} = -\sum\sum_{i=1}^{3} y_i log(MLP(h_{LLM(D)}, h_p)_i), \tag{2}$$

where $h_{LLM(D)}$ and h_p represent the embedding vectors of conversation segment and passage respectively; y_i is a symbolic function.

Multi-task Post-training. By utilizing the potential correlation between different tasks, we adopt a multi-task learning strategy to jointly optimize the above training tasks and integrate the feature information. We apply the following loss function to optimize the model:

$$L_{PSD} = w_1 Lpseg + w_2 Lpdoc, \tag{3}$$

where $w_i, i \in \{1, 2\}$ denote the weights of the loss functions of the two tasks. We use Dynamic Weight Averaging [13] to adaptively and dynamically adjust the weights to balance the learning speed of different tasks. Specifically, using Eq. (4) [13] to calculate the change of the continuous loss of each task, which is used as the learning speed of the task, and the weight of each task is obtained after normalization using Equation (5) [13].

$$r_i(z-1) = L(z-1)/L(z-2), \tag{4}$$

$$w_i(z) = Nexp(r_i(z-1)/T)/\sum_j exp(r_j(z-1)/T), \tag{5}$$

where $L \in \{Lpseg, Lpdoc\}$. z is an iteration index. N represents the number of tasks, and T represents a temperature that controls the softness of task weighting. The larger T is, the more uniform the weight between tasks is. When T is large enough, $w_i \approx 1$, at this time, each task is weighted equally.

4 Experiments

4.1 Experimental Settings

Dataset. We consider that other relevant datasets lack fluent conversations or are based only on single documents or web pages. In order to better verify the retrieval performance of our method in conversations containing multi-document information, we choose the MultiDoc2Dial [3] as the evaluation dataset. The dataset contains a total of 488 documents and 4796 conversations from four domains, each conversation contains an average of 14 turns, and each document contains an average of about 800 words. We follow MultiDoc2Dial's method of dividing and constructing the dataset, and follow CPII-NLP [12] to preprocess the document data to ensure the fairness of the evaluation results.

Evaluation Metrics. To evaluate the effect of the multi-document conversation segmentation method, we use **F1** as the evaluation metric. F1 is the harmonic mean of the recall and precision of the segmentation point. To quantify the retrieval performance, we use **R@k** ($k = 1/5/10$), and **MRR@k** ($k = 5/10$) as the evaluation metrics. R@k is a measure to evaluate how many correct passages are recalled at top K results. MRR@k is a measure to evaluate the position of the most relevant passage in the top K ranking result.

Parameter Settings. We set the maximum window parameter $W = 6$, employ *gpt-3.5-turbo2* as the LLM, and use the *ms-marco-MiniLM-L-12-v2* to implement retriever and reranker. We adopt a representation-based approach to construct the retriever, applying in-batch samples as negative samples and setting batch size as 32, the gradient accumulation steps as 1, the maximum sequence length as 256, the learning rate as 2e-05 using Adam, linear scheduling with warmup, and dropout rate as 0.1. In the post-training, we set the learning rate as 1e-05 using Adam, linear scheduling with warmup, the dropout rate as 0.1, the batch size as 4, the gradient accumulation steps as 4, the maximum sequence length as 512, and the temperature T as 2.0. Additionally, we apply contrastive learning to fine-tune the model on downstream tasks, using 7 hard negative samples sampled from the top 50 retrieved by the retriever.

4.2 Performance Comparison

We compare the evaluation results with the following models on this dataset:
   RAG [3] applies the bi-encoder model DPR pre-trained on the NQ dataset as the retriever, and **G4** [21] uses another dense retrieval bi-encoder model ANCE; **CPII-NLP** [12] follows the baseline setting, employs DPR as the retriever, and uses a collection of three cross-encoder models as the reranker; **PC** [16] builds a classification model to retain the utterances related to the current question in the conversation history for retrieval, and reuses the probability score of the classification model as a reranking metric; **UtMP** [5] employs a post-training

approach consisting of three tasks to train the reranker and fine-tune using contrastive learning, achieving the SOTA results on the dataset; **R3** [1] replaces the dense retriever with a sparse retriever based on DistilSplade, and adds a cross-encoder passage reranker based on RoBERTa; **UGent-T2K** [9] divides the retrieval part into document retrieval and passage retrieval, using LambdaMART algorithm combined with TF-IDF and other methods to score comprehensively.

4.3   Experimental Results

Conversational Search. We conduct the experiment on the validation set of MultiDoc2Dial to verify the retrieval effect of our model the results are shown in Table 1.

Table 1. Results on MultiDoc2Dial validation set.

Models	R@1	R@5	R@10
G4	0.395	0.685	0.773
RAG	0.490	0.723	0.800
PC	0.525	0.754	0.823
R3*	0.558	0.767	0.847
UGent-T2K	0.570	0.821	0.883
CPII-NLP*	0.614	0.821	0.881
UtMP	0.625	0.837	0.892
PSD(Ours)	**0.644**	**0.857**	**0.911**

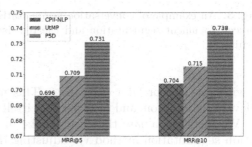

Fig. 2. Comparison with the SOTA models at MRR@k (k = 5/10).

The results show that our model achieves a certain improvement in all metrics, demonstrating the effectiveness of our method. Among them, R@1 and R@5 have been greatly improved by 1.9% and 2.0%, respectively, which shows that our model makes full use of the multi-document information, better modeling the interactive relationship between conversation segments and passages with contextual information and improving retrieval capabilities.

To verify the improvement of our post-training method on the effect of reranking, we further compared it with the SOTA models on the MRR@k metric, and the results are shown in Fig. 2. The results show that our model outperforms the SOTA models, which means our method can rank more relevant passages at higher positions.

Effectiveness of Multi-Document Conversation Segmentation. Figure 3 shows a case of conversation segmentation. It can be seen that our method can effectively divide utterance fragments involving different documents. But when there are too few utterances involved in a document, our method tends to group

Role	Utterance
user	Is it possible for getting more than one VA education benefit at the same time?
agent	It is possible to qualify for more than one at once.
agent	You should know, however, that if you qualify for more than one, you cannot use both at the same time. They must be used separately.
user	I did find out that I qualify for two different ones. How do I decide what to do now?
agent	We can help you make the right decision for you to help maximize the benefits in your situation.
user	Ok thank you, and about accessing and managing the VA benefits and health care, what benefits can someone who is a service member find?
agent	On our page you can find out what benefits you may be eligible for during service and which time-sensitive benefits to consider when separating or retiring.
user	Could you give me some info about my qualifications for more than one VA education benefit? Is there any need for me to make a phone call? Deafness is a condition that I suffer from. Is there a phone number for people who are deaf or hard of hearing?
agent	Yes, we have a regular phone number as well as a TTY phone number.
user	Even, what are the benefits of VA education, and are there any exceptions?

Fig. 3. An example of conversation segmentation, where the solid line represents the correct document segmentation and the red dashed line represents our segmentation results.

it with previous or later utterances. This may be because there is insufficient relevant information, and the utterance itself has semantic overlap in the context.

We further evaluate the effect of the unsupervised multi-document conversation segmentation method via adjusting the maximum window parameter W from 3 to 7. Results are shown in Fig. 4(left). It can be seen that the score is the highest when the window size is 6. The drop in F1 is more pronounced when the window is too small, which may be due to the lack of utterances contained in the segment, resulting in semantic incoherence.

Improvements for CS with Different Numbers of Documents. To verify the improvement of our method when the conversation history contains different numbers of documents, we divide the MutiDoc2Dial dataset into several subsets according to the number of documents involved in the conversation. The number of documents involved only indicates how many documents have been used in the conversation history, and the content retrieved during the entire conversation process will be repeatedly switched among these documents. We use the model that has not been post-trained (miniLM), the model that has been post-trained (PSD), and the SOTA model (UtMP) to perform retrieval on the subset and count the retrieval results.

The results are shown in Fig. 4(right). Our method achieves 2.1% and 2.9% improvements when the conversation history involves one document and three documents, respectively. This shows that our method is not only more helpful in handling difficult cases where a conversation involves multiple documents, but also performs better for queries on different passages in a single document.

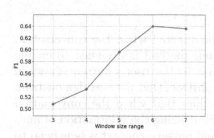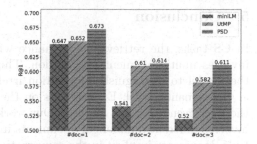

Fig. 4. Effectiveness of Multi-Document Conversation Segmentation (left) and improvements for CS with different numbers of documents (right).

4.4 Ablation Experiment

To further explore the effects of each training task in the PSD post-training method, as shown in Table 2, we use the model that has not been post-trained and directly fine-tuned as the baseline (miniLM), and add our training tasks before fine-tuning one by one. According to the results, it can be seen that both the passage-segment task (+pseg) and the passage-document task (+pdoc) bring certain improvements, indicating the effectiveness of the task. Among them, the passage-segment task brings a more significant improvement, indicating that making the model better model the correlation between conversation and passages of multiple documents is more conducive to retrieval in CS tasks. Moreover, the joint training of the two tasks using the multi-task learning method has a greater improvement, indicating that the two tasks have a potential correlation to promote each other, allowing the model to learn features that are more conducive to retrieval.

Table 2. Multi-task ablation experiment.

Models	MRR@10	R@1	R@5	R@10
miniLM	0.691	0.593	0.818	0.879
+pseg	0.725(+3.4%)	0.630(+3.7%)	0.850(+3.2%)	0.903(+2.4%)
+pdoc	0.704(+1.3%)	0.606(+1.3%)	0.834(+1.6%)	0.896(+1.7%)
+PSD(Ours)	0.738(+4.7%)	0.644(+5.1%)	0.857(+3.9%)	0.911(+3.2%)

5   Conclusion

In CS tasks, the retrieval performance will be affected when the conversation involves multi-document information. Thus, our method is designed to enable the model to distinguish information irrelevant to the user's current question at the semantic-level. Experiments on the MultiDoc2Dial dataset demonstrate the effectiveness of our method. Our work verifies that when the conversation involves different numbers of documents in CS, utilizing the information of multiple documents implicit in the conversation and passage context is beneficial to improve retrieval accuracy. Furthermore, we explore the utilization of LLMs to facilitate CS retrieval from a new perspective, which can bring some inspiration to CS, IR, and other tasks. In future work, we will further investigate more efficient multi-document information extraction methods and post-training tasks and explore more complex cases where there is more than one correct passage.

References

1. Bansal, S., et al.: R3: refined retriever-reader pipeline for multidoc2dial. In: Proceedings of the Second DialDoc Workshop on Document-grounded Dialogue and Conversational Question Answering, DialDoc@ACL 2022, Dublin, Ireland, May 26, 2022, pp. 148–154 (2022)
2. Bonifacio, L., Abonizio, H., Fadaee, M., Nogueira, R.: Inpars: data augmentation for information retrieval using large language models (2022)
3. Feng, S., Patel, S.S., Wan, H., Joshi, S.: Multidoc2dial: modeling dialogues grounded in multiple documents, pp. 6162–6176 (2021)
4. Han, J., Hong, T., Kim, B., Ko, Y., Seo, J.: Fine-grained post-training for improving retrieval-based dialogue systems. In: Proceedings of the 2021 Conference of the North American Chapter of the Association for Computational Linguistics: Human Language Technologies, pp. 1549–1558 (2021)
5. He, S., Zhang, S., Zhang, X., Feng, Z.: Conversational search based on utterance-mask-passage post-training. In: China Conference on Knowledge Graph and Semantic Computing (2023)
6. Ishii, E., Xu, Y., Cahyawijaya, S., Wilie, B.: Can question rewriting help conversational question answering? pp. 94–99 (2022)
7. Jang, Y., et al.: Improving multiple documents grounded goal-oriented dialog systems via diverse knowledge enhanced pretrained language model. In: Proceedings of the Second DialDoc Workshop on Document-grounded Dialogue and Conversational Question Answering, pp. 136–141 (2022)
8. Jeronymo, V., et al.: Inpars-v2: large language models as efficient dataset generators for information retrieval. arXiv preprint arXiv:2301.01820 (2023)
9. Jiang, Y., Hadifar, A., Deleu, J., Demeester, T., Develder, C.: Ugent-t2k at the 2nd dialdoc shared task: a retrieval-focused dialog system grounded in multiple documents. In: Proceedings of the Second DialDoc Workshop on Document-grounded Dialogue and Conversational Question Answering, pp. 115–122 (2022)
10. Karpukhin, V., et al.: Dense passage retrieval for open-domain question answering, pp. 6769–6781 (2020)
11. Kim, S., Kim, G.: Saving dense retriever from shortcut dependency in conversational search, pp. 10278–10287 (2022)

12. Li, K., et al.: Grounded dialogue generation with cross-encoding re-ranker, grounding span prediction, and passage dropout. In: Proceedings of the Second DialDoc Workshop on Document-Grounded Dialogue and Conversational Question Answering, pp. 123–129 (2022)
13. Liu, S., Johns, E., Davison, A.J.: End-to-end multi-task learning with attention. In: Proceedings of the IEEE/CVF Conference on Computer Vision and Pattern Recognition, pp. 1871–1880 (2019)
14. Mao, K., et al.: Learning denoised and interpretable session representation for conversational search. In: Proceedings of the ACM Web Conference, pp. 3193–3202 (2023)
15. Shen, T., Long, G., Geng, X., Tao, C., Zhou, T., Jiang, D.: Large language models are strong zero-shot retriever. arXiv preprint arXiv:2304.14233 (2023)
16. Tran, N., Litman, D.: Getting better dialogue context for knowledge identification by leveraging document-level topic shift. In: Proceedings of the 23rd Annual Meeting of the Special Interest Group on Discourse and Dialogue, pp. 368–375 (2022)
17. Wang, Z., Tu, Y., Rosset, C., Craswell, N., Wu, M., Ai, Q.: Zero-shot clarifying question generation for conversational search. In: Proceedings of the ACM Web Conference 2023, pp. 3288–3298 (2023)
18. Wu, Z., et al.: CONQRR: conversational query rewriting for retrieval with reinforcement learning. arXiv preprint arXiv:2112.08558 (2021)
19. Xu, Y., Zhao, H., Zhang, Z.: Topic-aware multi-turn dialogue modeling. In: Proceedings of the AAAI Conference on Artificial Intelligence, pp. 14176–14184 (2021)
20. Yu, W., et al.: Generate rather than retrieve: large language models are strong context generators. arXiv preprint arXiv:2209.10063 (2022)
21. Zhang, S., Du, Y., Liu, G., Yan, Z., Cao, Y.: G4: grounding-guided goal-oriented dialogues generation with multiple documents. In: Proceedings of the Second DialDoc Workshop on Document-grounded Dialogue and Conversational Question Answering, pp. 108–114 (2022)
22. Zhang, T., Ladhak, F., Durmus, E., Liang, P., McKeown, K., Hashimoto, T.B.: Benchmarking large language models for news summarization. arXiv preprint arXiv:2301.13848 (2023)

Recurrent Update Representation Based on Multi-head Attention Mechanism for Joint Entity and Relation Extraction

Shengjie Jia[1], Jikun Dong[1], Kaifang Long[1], Jiran Zhu[1], Hongyun Du[1], Guijuan Zhang[1], Hui Yu[2], and Weizhi Xu[1(✉)]

[1] School of Information Science and Engineering, Shandong Normal University, Jinan 250014, China
xuweizhi@sdnu.edu.cn
[2] Business School, Shandong Normal University, Jinan 250014, China

Abstract. Joint extraction of entities and relations from unstructured text is an important task in information extraction and knowledge graph construction. However, most of the existing work only considers the information of the context in the sentence and the information of the entities, with little attention to the information of the possible relations between the entities, which may lead to the failure to extract valid triplets. In this paper, we propose a recurrent update representational method based on multi-head attention mechanism for relation extraction. We use a multi-head attention mechanism to interact the information between the relational representation and the sentence context representation, and make the feature information of both fully integrated by cyclically updating the representation. The model performs relation extraction after the representation is updated. Using this approach we are able to leverage the relationship information between entities for relational triple extraction. Our experimental results on four public datasets show that our approach is effective and the model outperforms all baseline models.

Keywords: Relation extraction · Multi-head attention mechanism · Relational representation · Recurrent update

1 Introduction

Relational Triple Extraction(RTE) is a key task in Natural Language Processing (NLP), involving the identification and extraction of entity and relation between the entities in unstructured text, and it has a wide range of applications in various fields, such as information extraction, question and answer, and social network analysis. Relational triple extraction is more challenging than separate entity or relation extraction because it requires to consider the dependency relations between entities and entities, especially the problem is more challenging when entities have multiple relations or relations involve multiple entities.

The currently main approach for RTE is the joint extraction method, which uses an end-to-end approach for joint entity relation extraction. Some recent joint

B. Luo et al. (Eds.): ICONIP 2023, CCIS 1967, pp. 16–27, 2024.
https://doi.org/10.1007/978-981-99-8178-6_2

extraction methods [10,14,15] have shown strong extraction ability in dealing with complex sentences containing overlapping or multiple triples. Among these joint extraction methods, an annotation-based approach is increasing research interest. These methods usually use a binary annotation sequence to determine the start position of entities, or to determine the relation between entities. For example, [17] used a span-based labeling scheme to further decompose the two subtask into multiple sequential labeling problems, and employed hierarchical boundary labeling and multiple span decoding algorithms to extract relational triples. [15] map relations to objects by mapping functions, extracting subjects followed by relations and objects. Despite the great success of these methods, they are far from reaching their full potential.

In this paper, we propose a recurrent update representational method based on multi-head attention mechanism for joint entity and relation extraction. We use the multi-head attention to interact with information about the relation representation and the sentence context representation, and update the feature information of both by cyclically updating the representation. The model performs the relation extraction after the representation is updated. First, we use a subject tagger to identify all possible subjects in the sentence, then we combine each sentence representation with candidate subject and relation representations and use an object tagger to identify possible relations and corresponding objects. Using multi-head attention mechanism, we are able to interact with rich contextual and relational information. This way the sentence context is able to integrate specific relation information and each specific relation also incorporates the sentence context information, facilitating the extraction of overlapping triples, reducing the extraction of redundant triples and improving the efficiency of relation triples.

In summary, the main contributions of this paper are as follows.

(1) We propose to use multi-head attention mechanism to facilitate the interaction and integration of sentence context and relational semantic information.
(2) We propose a recurrent update representation method to enhance the representation of different types of information and improve the performance of RTE tasks.
(3) Our method effectively extracts relation triples, and experimental results demonstrate its success on all four public datasets.

2   Related Work

Currently, several methods for joint entity and relation extraction have been proposed. Based on the extraction approaches, we can categorize them into the following three general types.

Tagging Based Methods. Binary marker sequences are commonly used in these methods to identify the start and end positions of entities. Occasionally, relationships between two entities are established. The paper [20] first proposed a new annotation model that directly transforms the information extraction task into a sequence annotation task to extract entities and relations. Based on this, in recent years, researchers have started to explore an annotation approach based on a unidirectional extraction framework. [15] proposed a tagging-based approach that models relations as functions mapping subjects to objects. This provides a new perspective on the RTE task. Experiments show that [15] not only achieves competitive results, but also robustly extracts triples from sentences with overlapping or multiple triples.

Seq2Seq Based Methods. This type of methods usually treats the triad as a sequence of tokens and converts the RTE task into a generation task that generates the triad in some order, such as generating relations first, then entities, etc. For example, [19] is an end-to-end joint extraction model for sequence-to-sequence learning based on the Copy Mechanism, which is able to solve the relations overlap problem. In [8], the joint extraction of entity relations is viewed as an ensemble prediction problem without considering the order among multiple triples. Using a non-autoregressive parallel decoding approach, it is able to directly output the final predicted set of triples at once. [16]viewed relations to triple extraction as a sequence generation problem and, inspired by current research on transformer-based natural language generation, it proposed a comparative triple extraction model based on the generative transformer.

Table Filling Based Methods. Such methods [3,14] maintain a table for each kind of relations that exist, and each item in the table is usually the start position of two specific entities. Thus, the RTE task in this approach translates into the task of filling out these tables accurately and efficiently.

In addition to the above, researchers have explored many other approaches for joint extraction. For example, [1] proposed the use of a graph convolution neural network-based approach to perform joint extraction.

3   Task Formulation

In the joint task of entity and relation extraction, the objective is to identify all possible triples (subject, relation, object) from unstructured text. To accomplish this, we build upon previous work [15] by directly modeling triples and designing training objectives focused on the triple level.

Our objective is to maximize the maximum likelihood of all sentences in the training set N, which consists of annotated sentences x_i and their corresponding possible triples $T_i = \{(s, r, o)\}$.

$$
\prod_{i=1}^{|N|} \left[\prod_{(s,r,o) \in T_i} p((s,r,o) \mid x_i) \right]
$$

$$
= \prod_{i=1}^{|N|} \left[\prod_{s \in T_i} p(s \mid x_i) \prod_{(r,o) \in T_i | s} p((r,o) \mid s, x_i) \right] \tag{1}
$$

$$
= \prod_{i=1}^{|N|} \left[\prod_{s \in T_i} p(s \mid x_i) \prod_{r \in T_i | s} p_r(o \mid s, x_i) \prod_{r \in R \setminus T_i | s} p_r(o_\emptyset \mid s, x_i) \right]
$$

where $s \in T_i$ denotes the subject in the triple T_i. $T_i|s$ is the set of triples in T_i with s as the subject. $(r, o) \in T_i|s$ is an (r, o) pair in the triple with s as the subject. r is the set of all relations in the training set. $R \backslash T_i|s$ denotes all relations in the triple T_i except those guided by s as the subject. ϕ is the "empty" object, denoting all relations in the triple except those guided by s as the subject.

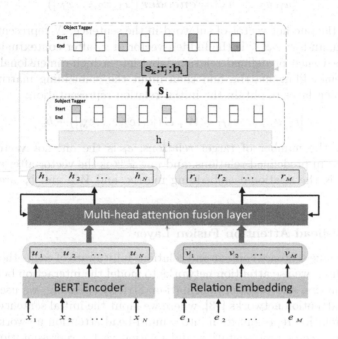

Fig. 1. The overall architecture of the model RURE.

4   Methodology

In this section, we will detail the general structure of the model, which is shown in Fig. 1. It consists of three main parts:

Representation Construction. Given a sentence and a target set of relations, we encode the sentence as a vector and embed each relation as a vector to obtain an initial representation of the sentence and the relations, which is inputted to the multi-head attention fusion layer.

Multi-head Attention Fusion Layer. We propose a multi-head attention fusion layer to cyclically update the sentence representation and the relational representation.

Relation Extraction. After the representation is updated, we perform relational triple extraction.

4.1   Representation Construction

First, we encode each sentence in the training set using the pre-training model of Bert [5], and then we use all word embeddings output by the last hidden layer of Bert as the initial representation of the whole sentence:

$$[u_1, u_2, \ldots u_N] = \boldsymbol{encoder}\left([x_1, x_2, \ldots x_N]\right) \tag{2}$$

where x_i is the one-hot vector of subword in the sentence, N represents the sentence length, and $u_i \in \mathbb{R}^{d_h}$ is the hidden vector of x_i after contextual encoding.

We embed each predefined relation label into a high-dimensional vector to obtain an embedding matrix, and then input the embedding matrix into the linear mapping layer to obtain the initial relation representation:

$$[v_1, v_2, \ldots, v_M] = \boldsymbol{W_r} \boldsymbol{E}\left([e_1, e_2, \ldots, e_M]\right) + \boldsymbol{b_r} \tag{3}$$

where M is the number of target relations, e_i is the one-hot vector of relational indices in predefined relations, and $v_i \in \mathbb{R}^{d_h}$ is the vector after embedding mapping. e is the relational embedding matrix, and $\boldsymbol{W_r}$ and $\boldsymbol{b_r}$ are trainable parameters.

4.2   Multi-head Attention Fusion Layer

In order to make sentence context and relations fully aware of each other's semantic information, we use attention networks to model the interaction between sentence information and relation information. Strictly speaking, we use canonical multi-head attention networks [13], where we input the initial sentence representation and relation representation into a multi-head attention network to obtain new sentence vector representation and relation vector representation. Specifically, we first use the relational representation v_i as a query vector to obtain the relation-to-sentence context-aware representation:

$$v_i' = \mathbf{MHA}\left(\boldsymbol{W}^q v_i, \boldsymbol{W}^k u_i, \boldsymbol{W}^v u_i\right) \tag{4}$$

where v_i and u_i are the initial relation vector representation and sentence vector representation, respectively. $\mathbf{MHA}$ denotes the processing of the multi-head attention. $\boldsymbol{W}^q$, $\boldsymbol{W}^k$, $\boldsymbol{W}^v$ are the different trainable parameters and $v_i' \in \mathbb{R}^{d_h}$ is the output of the multi-head attention. We use a gate mechanism instead of the activation function to maintain the nonlinear capability, as follows:

$$\begin{aligned} g_i &= \text{sigmoid}\left(\boldsymbol{W_g}\left[v_i; v_i'\right]\right) \\ s_i &= (1 - g_i) \odot v_i' + g_i \odot v_i \end{aligned} \tag{5}$$

where $\boldsymbol{W_g}$ is the trainable parameter, g_i is a scalar, $\odot$ is the Hadamard product, and [;] denotes the splicing operation. $s_i \in \mathbb{R}^{d_h}$ is the final output. Afterwards, we add a residual join to avoid the gradient vanishing problem, and the updated final representation is as follows:

$$\boldsymbol{r}_i = s_i + v_i \tag{6}$$

For convenience we will define the above process as follows:

$$r_i = \textbf{MHAF}\,(v_i, u_i, u_i) \tag{7}$$

v_i and u_i are the initial relation vector representation and sentence vector representation, respectively, and **MHAF** represents the above procedure. $r_i \in \mathbb{R}^{d_h}$ is the new relation representation of the output.

Recurring Update Representation. To realize the integration of both semantic information, we adopt the strategy of cyclic update. First, we can get the updated sentence context representation h_i according to Eq. 7 as follows:

$$h_i = \textbf{MHAF}\,(u_i, v_i, v_i) \tag{8}$$

After getting the new representation, we take a circular update, and the update process for layer l is as follows:

$$\begin{aligned} h_i^l &= \textbf{MHAF}\left(h_i^{l-1}, r_i^{l-1}, r_i^{l-1}\right) \\ r_i^l &= \textbf{MHAF}\left(r_i^{l-1}, h_i^{l-1}, h_i^{l-1}\right) \end{aligned} \tag{9}$$

4.3 Relation Extraction

Similar to previous work [15], we use a subject tagger that first extracts all possible entities, which are two identical binary classifiers, to identify the start and end positions of the subject, as follows:

$$\begin{aligned} p_i^{start\_s} &= \sigma\left(W_{start\_s}\tanh\left(h_i\right) + b_{start\_s}\right) \\ p_i^{end\_s} &= \sigma\left(W_{end\_s}\tanh\left(h_i\right) + b_{end\_s}\right) \end{aligned} \tag{10}$$

h_i denotes the last layer output of the multi-head attention fusion layer, W_{start_s}, b_{start_s}, W_{end_s}, b_{end_s} are trainable parameters. $p_i^{start\_s}$ and $p_i^{end\_s}$ denote the probability of recognizing the start position and end position of the subject, and if the probability exceeds the set threshold, it is marked as 1 at the position of the corresponding word. σ is the sigmoid activation function. The subject tagger optimizes the following likelihood function to determine the span of subject s in sentence x:

$$p_{\eta_s}(s \mid x) = \prod_{t \in \{start_s, end_s\}} \prod_{i=1}^{N} \left(p_i^t\right)^{I\{y_i^t = 1\}}\left(1 - p_i^t\right)^{I\{y_i^t = 0\}} \tag{11}$$

where η_s is the set of all arguments in the subject tagger, N is the sentence length, and $I\{z\} = 1$ if z is true and 0 otherwise.

Unlike the subject tagger, on the object tagger, we take into account the obtained relation information and the candidate subject information except for using the contextual information of the sentence, as follows:

$$\begin{aligned} p_i^{start\_o} &= \sigma\left(W_{start\_o}\tanh\left[s_k; r_i; h_i\right] + b_{start\_o}\right) \\ p_i^{end\_o} &= \sigma\left(W_{end\_o}\tanh\left[s_k; r_i; h_i\right] + b_{end\_o}\right) \end{aligned} \tag{12}$$

where $p_i^{start\_o}$ and $p_i^{end\_o}$ denote the probability of sentence sequence word recognition as object start position and end position. σ is the sigmoid activation function, s_k is the detected subject encoding vector, r_i and h_i denote the updated relation representation vector and context representation vector. $[; ;]$ denotes the splicing operation, and $W_{start\_o}, b_{start\_o}, W_{end\_o}$, and b_{end_o} are all trainable parameters. Given a sentence x, a subject s, and a relation r, the object tagger optimizes the following likelihood function to determine the span o of the object:

$$p_{\eta_o}(s \mid x, s, r) = \prod_{t \in \{start_o, end_o\}} \prod_{i=1}^{N} \left(p_i^t\right)^{I\{y_i^t=1\}} \left(1 - p_i^t\right)^{I\{y_i^t=0\}} \tag{13}$$

η_o denotes the set of all parameters in the object tagger. For an empty object, the position markers are all 0. According to Eq 1, we are able to obtain the objective function:

$$\mathcal{L} = \log \prod_{(s,r,o) \in T_i} p((s,r,o) \mid x_i)$$

$$= \sum_{s \in T_i} \log p_{\eta_s}(s \mid x) + \sum_{r \in T_i \mid s} \log p_{\eta_o}(o \mid x, s, r) +$$

$$\sum_{r \in R \backslash T_i \mid s} \log p_{\eta_o}(o_\varnothing \mid x, s, r) \tag{14}$$

When training the model, we use the gradient descent algorithm [4] to maximize this loss function $\mathcal{L}$.

5   Experiments

5.1   Experiment Settings

Table 1. Statistics of datasets. EPO and SEO refer to the entity pair overlapping and single entity overlapping respectively. Note a sentence can belong to both EPO and SEO.

Category	WebNLG(*)		NYT10		NYT*	
	Train	Test	Train	Test	Train	Test
Normal	1596	246	59396	2963	37013	3266
EPO	227	26	5376	715	9782	978
SEO	3406	457	8772	742	14735	1297
ALL	5019	703	70339	4006	56195	5000

Datasets. We evaluat our model on four publicly available datasets NYT* [7], WebNLG [2], WebNLG* [2], and NYT10 [7]. The WebNLG* dataset which is another version of the WebNLG dataset. The NYT* and WebNLG* datasets annotated with only the last word of the entity. Table 1 shows the statistics for these datasets.

Evaluation Metrics. We evaluate the results using standard micro-precision, recall, and F1 scores. The RTE task has two criteria for matching: partial match and exact match. A triple is considered partially matched if its extracted relations and the subject and object entities are correct. An exact match is only achieved when the entities and relations of the triple are exactly the same as those of a correct triple. We utilize partial match on the WebNLG* and NYT* datasets, exact match on the WebNLG dataset, and both types of matching on the NYT10 dataset.

Table 2. Results of WebNLG* and WebNLG datasets

| Model | Partial Match | | | Exact Match | | |
| | WebNLG* | | | WebNLG | | |
	Prec	Rec	F1	Prec	Rec	F1
NovelTagging [20]	–	–	–	52.5	19.3	28.3
CopyRE [19]	37.7	36.4	37.1	–	–	–
GraphRel [1]	44.7	41.1	42.9	–	–	–
OrderCopyRE [18]	63.3	59.9	61.6	–	–	–
CGT [16]	92.9	75.6	83.4	–	–	–
RIN [9]	87.6	87.0	87.3	77.3	76.8	77.0
ETL-span [17]	84.0	91.5	87.6	84.3	82.0	83.1
CasRel [15]	**93.4**	90.1	91.8	88.3*	84.6*	86.4*
TPLinker [14]	91.8	92.0	91.9	88.9	84.5	86.7
PMEI [10]	91.0	**92.9**	92.0	80.8	82.8	81.8
StereoRel [12]	91.6	92.6	92.1	–	–	–
Our Model	93.3	91.9	**92.6**	**89.9**	**87.9**	**88.9**

Implementation Details. We use SGD to optimize the parameters of our model, with the learning rate set to 0.1 and batch size set to 6. We set the maximum length of the input sentences to 150. During the training phase, we set the threshold for determining the start and end labels of words to 0.5. Additionally, we use Bert (base-cased) as the pre-trained model. We trained the dataset for 50 epochs. To avoid overfitting, we stopped the training process when the results on the validation set remained unchanged for at least 10 consecutive epochs.

Baselines. For comparison, we used 13 strong baseline models, including: NovelTagging [20], CopyRE [19], GraphRel [1], OrderCopyRE [18], HRL [11], CGT [16], RIN [9], ETL-span [17], CasRel [15], Tplinker [14], PMEI [10], Stereorel [12], BiRTE [6]. Most of the results of these baselines were directly replicated from their original papers, and some baselines in some datasets did not report results. * indicates the result we get by running the source code.

Table 3. Results of NYT* dataset

Model	Partial Match NYT*		
	Prec	Rec	F1
NovelTagging [20]	62.4	31.7	42.0
CopyRE [19]	61.0	56.6	58.7
GraphRel [1]	63.9	60.0	61.9
OrderCopyRE [18]	77.9	67.2	72.1
ETL-span [17]	84.9	72.3	78.1
RIN [9]	87.2	87.3	87.3
CGT [16]	**94.7**	84.2	89.1
CasRel [15]	89.7	89.5	89.6
PMEI [10]	90.5	89.8	90.1
TPLinker [14]	91.3	**92.5**	91.9
Our Model	93.1	90.9	**92.0**

Table 4. Results of NYT10 dataset

Model	NYT10 Partial Match			Exact Match		
	Prec	Rec	F1	Prec	Rec	F1
CopyRE [19]	56.9	45.2	50.4	–	–	–
ETL-Span [17]	–	–	–	74.5*	57.9*	65.2*
HRL [11]	71.4	58.6	64.4	–	–	–
$PMEI_{LSTM}$ [10]	79.1	67.2	72.6	75.4	65.8	70.2
CasRel [15]	77.7	68.8	73.0	76.8*	68.0*	72.1*
StereoRel [12]	**80.0**	67.4	73.2	–	–	–
$BiRTE_{LSTM}$ [6]	79.0	68.8	73.5	76.1	67.4	71.5
Our Model	77.6	**70.5**	**73.9**	80.3	**69.1**	**74.3**

5.2   Experimental Results

Table 2 shows the results of our model for relation extraction on the WebNLG dataset and the WebNLG* dataset, and the results show that our model improves significantly on the F1 score, by 0.5% and 2.2%, respectively, over the best baseline model. Table 3 shows the results of our model for relation extraction on the NYT* dataset, and the results show that our model outperforms all the baseline models, improving by 0.1% over the best baseline model in terms of F1 score. Table 4 shows the results of our model for relation extraction on the NYT10 dataset, and the results show that our approach improves by 0.4% and 2.8% for the two matching methods, respectively. These results fully demonstrate the superiority of our model. In this section, we conduct ablation experiments on the WebNLG and WebNLG* datasets to explore the impact of different components on the model. First, we explore the impact of different numbers of multiple attention fusion layers on the model. The results show that the best performance of the model is achieved when the number of layers of multiple attention fusion layers is 2. Secondly, we investigate the effect of the number of heads in the multi-head attention on the model, and the experimental results show that the model achieves better performance when the number of attention heads is 24.

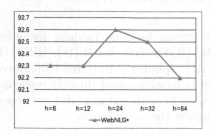 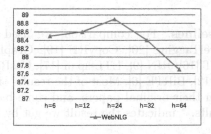

Fig. 2. Effect of different number of heads in multi-head attention on results on WebNLG* and WebNLG datasets

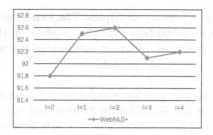 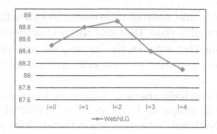

Fig. 3. Effect of the number of layers of the multi-head attention fusion layer on the results on the WebNLG* and WebNLG datasets

5.3   Number of Heads in Multi-head Attention

To explore the effect of the number of heads in the multi-head attention on the model performance, we keep increasing the number of heads of multi-head attention, and the experimental results are shown in Fig. 2. The F1 score is highest when the number of heads of attention is 24. We found that the experimental performance did not always improve as the number of heads increased, and the F1 score decreased significantly when the number of heads was increased from 24 to 32. Therefore, we set the number of heads of attention to 24 for all tasks.

5.4   Layers of Multi-head Attention Fusion Layers

To determine the number of layers of multi-head attention fusion layers, we first fix the number of attention heads to 24, and the experimental results are shown in Fig. 3. With the increase of the number of layers, the F1 score of the dataset gradually improves, and the model performance reaches the best when the number of fusion layers is 2. The performance gradually decreases when the number of fusion layers is greater than 2. We also observe that the model performance is significantly improved when the number of fusion layers is from 0 to 1. This fully illustrates that our proposed multi-head attention fusion layer greatly improves the ability of the model to extract triples.

6   Conclusion

In this paper, we propose a recurrent update representation based on a multi-head attention for the joint entity and relation extraction task, and verify the effectiveness of the method through extensive experiments. We introduce the relational embedding representation and feature aggregation between the relational representation and the sentence representation through a multi-head attention fusion layer, and we also use the cyclic update method to update the feature information of both. This method reduces the extraction of redundant triples and greatly improves the extraction efficiency of triples. The experimental results show that our method achieves an advantage over the baseline on the public datasets. In the future, we will explore more different methods of feature fusion and extend the cyclic update method to other models.

Acknowledgements. This work was supported in part by Natural Science Foundation of Shandong Province (No. ZR2022MF328 and No. ZR2019LZH014), and in part by National Natural Science Foundation of China (No. 62172265, No. 61602284 and No. 61602285).

References

1. Fu, T.J., Li, P.H., Ma, W.Y.: Graphrel: modeling text as relational graphs for joint entity and relation extraction. In: Proceedings of the 57th Annual Meeting of the Association for Computational Linguistics, pp. 1409–1418 (2019)
2. Gardent, C., Shimorina, A., Narayan, S., Perez-Beltrachini, L.: Creating training corpora for NLG micro-planning. In: 55th Annual Meeting of the Association for Computational Linguistics (ACL) (2017)
3. Gupta, P., Schütze, H., Andrassy, B.: Table filling multi-task recurrent neural network for joint entity and relation extraction. In: Proceedings of COLING 2016, the 26th International Conference on Computational Linguistics: Technical Papers, pp. 2537–2547 (2016)
4. Kingma, D.P.: A method for stochastic optimization. ArXiv Prepr (2014)
5. Lee, J., Toutanova, K.: Pre-training of deep bidirectional transformers for language understanding. arXiv preprint arXiv:1810.04805 (2018)
6. Ren, F., Zhang, L., Zhao, X., Yin, S., Liu, S., Li, B.: A simple but effective bidirectional framework for relational triple extraction. In: Proceedings of the Fifteenth ACM International Conference on Web Search and Data Mining, pp. 824–832 (2022)
7. Riedel, S., Yao, L., McCallum, A.: Modeling relations and their mentions without labeled text. In: Balcázar, J.L., Bonchi, F., Gionis, A., Sebag, M. (eds.) ECML PKDD 2010, Part III. LNCS (LNAI), vol. 6323, pp. 148–163. Springer, Heidelberg (2010). https://doi.org/10.1007/978-3-642-15939-8_10
8. Sui, D., Zeng, X., Chen, Y., Liu, K., Zhao, J.: Joint entity and relation extraction with set prediction networks. IEEE Trans. Neural Networks Learn. Syst. (2023)
9. Sun, K., Zhang, R., Mensah, S., Mao, Y., Liu, X.: Recurrent interaction network for jointly extracting entities and classifying relations. In: Proceedings of the 2020 Conference on Empirical Methods in Natural Language Processing (EMNLP), pp. 3722–3732 (2020)
10. Sun, K., Zhang, R., Mensah, S., Mao, Y., Liu, X.: Progressive multi-task learning with controlled information flow for joint entity and relation extraction. In: Proceedings of the AAAI Conference on Artificial Intelligence, vol. 35, pp. 13851–13859 (2021)
11. Takanobu, R., Zhang, T., Liu, J., Huang, M.: A hierarchical framework for relation extraction with reinforcement learning. In: Proceedings of the AAAI Conference on Artificial Intelligence, vol. 33, pp. 7072–7079 (2019)
12. Tian, X., Jing, L., He, L., Liu, F.: Stereorel: relational triple extraction from a stereoscopic perspective. In: Proceedings of the 59th Annual Meeting of the Association for Computational Linguistics and the 11th International Joint Conference on Natural Language Processing (Volume 1: Long Papers), pp. 4851–4861 (2021)

13. Vaswani, A., et al.: Attention is all you need. In: Advances in Neural Information Processing Systems, vol. 30 (2017)
14. Wang, Y., Yu, B., Zhang, Y., Liu, T., Zhu, H., Sun, L.: Tplinker: single-stage joint extraction of entities and relations through token pair linking. In: Proceedings of the 28th International Conference on Computational Linguistics, pp. 1572–1582 (2020)
15. Wei, Z., Su, J., Wang, Y., Tian, Y., Chang, Y.: A novel cascade binary tagging framework for relational triple extraction. In: Proceedings of the 58th Annual Meeting of the Association for Computational Linguistics, pp. 1476–1488 (2020)
16. Ye, H., et al.: Contrastive triple extraction with generative transformer. In: Proceedings of the AAAI Conference on Artificial Intelligence, vol. 35, pp. 14257–14265 (2021)
17. Yu, B., et al.: Joint extraction of entities and relations based on a novel decomposition strategy. In: ECAI 2020, pp. 2282–2289. IOS Press (2020)
18. Zeng, X., He, S., Zeng, D., Liu, K., Liu, S., Zhao, J.: Learning the extraction order of multiple relational facts in a sentence with reinforcement learning. In: Proceedings of the 2019 Conference on Empirical Methods in Natural Language Processing and the 9th International Joint Conference on Natural Language Processing (EMNLP-IJCNLP), pp. 367–377 (2019)
19. Zeng, X., Zeng, D., He, S., Liu, K., Zhao, J.: Extracting relational facts by an end-to-end neural model with copy mechanism. In: Proceedings of the 56th Annual Meeting of the Association for Computational Linguistics (Volume 1: Long Papers), pp. 506–514 (2018)
20. Zheng, S., Wang, F., Bao, H., Hao, Y., Zhou, P., Xu, B.: Joint extraction of entities and relations based on a novel tagging scheme. In: Proceedings of the 55th Annual Meeting of the Association for Computational Linguistics (Volume 1: Long Papers), pp. 1227–1236 (2017)

Deep Hashing for Multi-label Image Retrieval with Similarity Matrix Optimization of Hash Centers and Anchor Constraint of Center Pairs

Ye Liu[1,3,4,5], Yan Pan[1,3], and Jian Yin[2,3(✉)]

[1] School of Computer Science and Engineering, Sun Yat-sen University, Guangzhou, China
liuye7@mail2.sysu.edu.cn, panyan5@mail.sysu.edu.cn
[2] School of Artificial Intelligence, Sun Yat-sen University, Zhuhai, China
issjyin@mail.sysu.edu.cn
[3] Guangdong Key Laboratory of Big Data Analysis and Processing, Guangzhou, China
[4] Artificial Intelligence Department, Lizhi Inc., Beijing, China
[5] Big Data Department, Lizhi Inc., Guangzhou, China

Abstract. Deep hashing can improve computational efficiency and save storage space, which is the most significant part of image retrieval task and has received extensive research attention. Existing deep hashing frameworks mainly fall into two categories: single-stage and two-stage. For multi-label image retrieval, most single-stage and two-stage deep hashing methods usually consider two images to be similar if one pair of the corresponding category labels is the same, and do not make full use of the multi-label information. Meanwhile, some novel two-stage deep hashing methods proposed in recent years construct hash centers firstly and then train through deep neural networks. For multi-label processing, these two-stage methods usually convert multi-label into single-label objective, which also leads to insufficient use of label information. In this paper, a novel multi-label deep hashing method is proposed by constructing the similarity matrix and designing the optimization algorithm to construct the hash centers, and the proposed method constructs the training loss function through the multi-label hash centers constraint and anchor constraint of center pairs. Experiments on several multi-label image benchmark datasets show that the proposed method can achieve the state-of-the-art results.

Keywords: Deep hashing · Multi-label image · Hash centers · Matrix optimization · Anchor constraint

1 Introduction

With the rapid development of Internet area, a large amount of image data has been generated, so image retrieval has become a research hotspot in the field of

B. Luo et al. (Eds.): ICONIP 2023, CCIS 1967, pp. 28–47, 2024.
https://doi.org/10.1007/978-981-99-8178-6_3

computer vision. In recent years, the models based on deep neural network have become the mainstream, so the key of image retrieval task includes the selection of deep neural network backbone model and the framework of hashing method based on deep model [3, 19, 22].

In the traditional image classification task, the original images and corresponding labels are used as input to the deep model through supervised learning [18]. After the training process, the classification model based on the deep network can be obtained, and the feature representations of the corresponding images can be extracted through the model. Based on the training framework of deep learning, the deep model learned for supervised image classification tasks can be extended to multiple downstream image processing tasks through the pre-training and fine-tuning paradigm with the saved deep model parameters. Therefore, the image feature representations required in image retrieval tasks can be quickly and effectively obtained from the pre-trained deep models.

The methods by obtaining features with pre-training deep models have some shortcomings, mainly reflect in that high-dimensional image features consume more computing resources and also occupy a large amount of storage. Therefore, converting high-dimensional image features into low-dimensional hash codes can save computing and storage resources. Compared with traditional hashing methods, deep hashing based on deep neural network model has achieved absolute advantages in effect. From a framework point of view, deep hashing includes two main categories, one is single-stage and the other is two-stage.

Single-stage deep hashing mainly constructs the loss function constraint of the deep neural network model through the similarity relationship between local images, and the construction of loss function includes pointwise, pairwise, triplewise and some other forms. NINH [10] is a single-stage end-to-end deep hashing model originally proposed, which greatly improves the structure of the deep hashing framework by designing an integrated structure of deep model and loss function.

In the two-stage deep hashing framework, the hash codes corresponding to the training images are usually generated by the construction method, and then the images and hash codes are used as the input of the deep model training. In the work of CNNH [24], hash codes for supervised learning are obtained based on the construction of similarity matrix and optimization solution. Then, the generated hash codes and image classification are used as labels for the training of the deep neural network. The disadvantages of this method include that the convergence speed of similarity optimization algorithm is slow, and it is not an end-to-end framework. This kind of the deep hashing methods sample from training images to construct similarity constraint, so it will lead to slow training process, affect the convergence speed, and fail to solve the imbalance of data distribution. A recent two-stage deep hashing method is CSQ [25], which proposes the concept of hash center through the idea of agent, generates multiple hash center codes according to the category number of the image data, and uses these hash center codes for training. For multi-label image retrieval task, two-stage hashing methods such as CSQ adopt a compromise treatment, which actually simplifies the multi-label scenario into the single label scenario before training deep model.

Inspired by previous work, we propose a novel deep hashing framework directly for multi-label image hashing tasks. Our proposed framework belongs to the two-stage deep hashing paradigm. Firstly, we apply the similarity matrix construction and optimization solution to the hash center encoding learning. Secondly, we directly match the hash center codes to each image, so that each multi-label image can directly correspond to multiple hash centers. And a novel anchor center pairs loss function constraint form of two anchors of hash centers to one image is designed for multiple hash centers of each image. Based on the deep hashing framework proposed in this paper, we carry out several types of experimental analysis. Experimental results on several datasets (NUS-WIDE [2], MS-COCO [13], FLICKR-25K [8]) show that the proposed deep hashing method can achieve better results in the image retrieval task than the state-of-the-art methods. Figure 1 shows the overview of our proposed deep hashing training framework for multi-label image retrieval.

2   Related Work

Since the sustainable development of Internet technology and multimedia, the amount of image data on the Internet is growing rapidly. The image retrieval task is to search for images that are visually or semantically similar to the query image in the image dataset. Traditional image retrieval methods are facing great challenges because of the demand for large-scale image retrieval tasks based on a million scale image database. Hashing based image retrieval methods map high-dimensional features of an image to the binary space to generate low-dimensional binary coded features, which greatly reduces the feature dimension and avoids the dimension disaster [19]. Through Hamming distance measurement [16], the fast search of image coding features can greatly improve the retrieval efficiency.

Before the emergence of deep hashing algorithm, the traditional hashing algorithms KSH [14], ITQ [5] and SDH [21] have achieved certain results in image retrieval, which are mainly realized by directly constructing the optimization objective and solution algorithm of hash coding. Single-stage deep hashing is often referred to as end-to-end deep hashing method, and NINH [10] is the first single-stage hashing framework proposed to achieve effective results, while DHN [26], HashNet [1] and so on are subsequent variants and improvements. Two-stage deep hashing was first proposed in the research of CNNH [24] and MBH [15]. At the beginning, this kind of method solves the hash code of each image in the training dataset by optimizing the similarity matrix, and the computational efficiency is limited under large-scale data. In the latest research, CSQ [25] introduces the concept of agent, which is the hash center. By constructing a single hash center for each image and training deep model with hash center codes, these methods such as CSQ [25], DPN [4], OrthoHash [7], MDSH [23] can achieve considerable results on single-label image data, but do not make good use of the information of multi-label images.

3   The Proposed Deep Hashing Framework

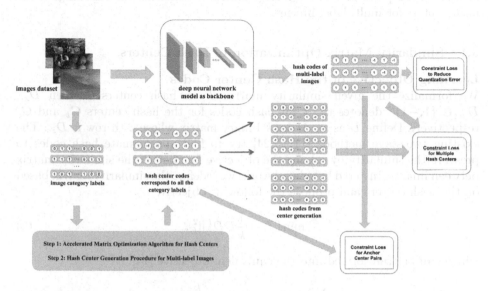

Fig. 1. Overview of the proposed deep hashing training framework for multi-label image retrieval with similarity matrix optimization of hash centers and anchor constraint of center pairs.

In the following construction and analysis process about the deep hashing framework, we refer to the definition of symbols in [10,15,23,24]. Given the training dataset consisting of N images $\mathcal{G} = \{G_1, G_2, ..., G_N\}$ and the number of category labels for the image dataset is defined as M. Then we formally define the notion of hash centers as $\mathcal{C} = \{C_1, C_2, ..., C_M\}$, so the similarity matrix T constructed with the corresponding encodings of the hash centers can be defined by:

$$T_{xy} = \begin{cases} +1, & x = y \\ -1, & x \neq y \end{cases} \qquad (1)$$

where $T_{xy} \in [-1, 1]$ denotes the similarity matrix between the x-th and the y-th hash centers. The main goal of deep hashing task is to obtain the hash mapping function by learning based on $\mathcal{G}$, $\mathcal{C}$ and T, which is a way of mapping the input query image to get k bit number codes in $\{1, -1\}$.

Deep hashing frameworks usually construct a single objective task for learning, which leads to complex and difficult optimization problems. In order to simplify the complexity of solving the deep hashing problem, the two-stage hashing methods [11,15,24] divide the training procedure and learn the training hash codes by optimizing the similarity matrix, then associate the corresponding hash codes with the deep network loss functions. The proposed deep hashing framework belongs to the two-stage methods, and the main difference is that the previous two-stage deep hashing mainly constructs the similarity matrix based on the training images, while our proposed framework is inspired by the agent idea [4,7,23,25] and uses the similarity matrix to construct the encodings of the hash centers for multi-label images.

3.1   Similarity Matrix Optimization of Hash Centers

Learning Objective of the Hash Center Codes

We formalize the given similarity matrix T on hash centers $\mathcal{C}$, then $D_{x\cdot}$, $D_{y\cdot} \in \{1, -1\}^k$ denotes the k-bit hash codes for the hash centers C_x and C_y respectively. Define D as an M by k binary matrix whose z-th row is $D_{z\cdot}$. The existing two-stage methods [11,15,24] try to find approximate hash codes to preserve the similarity by solving the objective function of the similarity matrix between images. Inspired by related work, we redefine the similarity matrix based on the hash centers and propose the following objective:

$$\min_D ||T - \frac{1}{k}DD^T||_F^2, \tag{2}$$

which can be transformed into the equivalent objective [14]:

$$\min_{D_{x\cdot}, D_{y\cdot}} \sum_{x,y}(||D_{x\cdot} - D_{y\cdot}||_{\mathcal{H}} - \frac{k}{2}(1 - T_{xy}))^2, \tag{3}$$

where Hamming distance is expressed by $|| \cdot ||_{\mathcal{H}}$. In the optimization objective of formula (3), the hash center similarity or dissimilarity of C_x and C_y is determined by category label information, and then the Hamming distance of the corresponding x-th and y-th hash codes is minimized or maximized. Thus we can get the following expressions:

$$||D_{x\cdot} - D_{y\cdot}||_{\mathcal{H}} \to 0 \; if \; T_{xy} = 1$$
$$||D_{x\cdot} - D_{y\cdot}||_{\mathcal{H}} \to k \; if \; T_{xy} = -1. \tag{4}$$

However, in the above optimization formula, this constraint is actually restrictive for the purpose of solving the maximum Hamming distance for the dissimilar hash center pairs. A more feasible method is to transform the optimization constraint. For the dissimilar hash centers C_x and C_y, the Hamming distance calculated by the corresponding hash codes only needs to be large enough within a certain range. Then the updated target formula can be obtained:

$$||D_{x\cdot} - D_{y\cdot}||_{\mathcal{H}} = 0 \; if \; T_{xy} = 1$$
$$||D_{x\cdot} - D_{y\cdot}||_{\mathcal{H}} \geq dist \; if \; T_{xy} = -1, \tag{5}$$

where $dist$ is the distance range parameter that meets the $0 \leq dist \leq k$ condition. In practice, when $dist = k$, the optimization objective (4) is a special form of the formula (5).

Based on setting the distance parameter in the objective formula (5), we can learn the approximate estimated hash codes corresponding to the set $\mathcal{C}$ of hash centers by minimizing the following optimization formula:

$$
\begin{aligned}
& \min_{D \in \{-1,1\}^{M \times k}} [\sum_{x \neq y} \max(0, dist - 2||D_x - D_y.||_{\mathcal{H}})^2 \\
& + \sum_{x=y} \max(0, 2||D_x - D_y.||_{\mathcal{H}})^2] \\
= & \min_{D \in \{-1,1\}^{M \times k}} [\sum_{x \neq y} \max(0, dist - \frac{1}{2}||D_x - D_y.||_{\mathcal{F}}^2)^2 \\
& + \sum_{x=y} \max(0, \frac{1}{2}||D_x - D_y.||_{\mathcal{F}}^2)^2] \\
= & \min_{D \in \{-1,1\}^{M \times k}} [\sum_{x \neq y} \max(0, D_x.D_y^T - (k - dist))^2 \\
& + \sum_{x=y} \max(0, k - D_x.D_y^T)^2] \\
= & \min_{D \in \{-1,1\}^{M \times k}} || \max(0, P \odot (DD^T - Q))||_{\mathcal{F}}^2,
\end{aligned}
\tag{6}
$$

where we define the Frobenius norm using the symbol $|| \cdot ||_{\mathcal{F}}$, and $\mathbf{0}$ is denoted as an all-zero matrix of M times M. The element-wise multiplication operator is represented by the $\odot$ symbol. In the process of deducing the optimization formula, we use the transformation of the definition formula $||i - j||_{\mathcal{H}} = \frac{1}{4}||i-j||_{\mathcal{F}}^2$ with $i, j \in \{-1, 1\}^k$. In the formula details, $P_{xy} = -1$ when the hash centers C_x and C_y are similar, and $P_{xy} = 1$ when the hash centers C_x and C_y are dissimilar, so P is defined as the indicator matrix. The matrix Q is defined synchronously such that $Q_{xy} = k$ when the hash centers C_x and C_y are similar, and $Q_{xy} = dist - k$ when C_x and C_y are dissimilar.

Similarity Matrix Optimization Procedure

Although the above optimization objective (6) is clearly defined, it is tricky to optimize the objective directly because of the integer constraint form on D. In order to better solve this problem, a general and better way is to make the solution target relaxed by replacing the integer constraint $D \in \{-1, 1\}^{M \times k}$ with the range constraint $D \in [-1, 1]^{M \times k}$, then we can get the relaxation form below:

$$
\min_D || \max(0, P \odot (DD^T - Q))||_{\mathcal{F}}^2
$$
$$
s.t.\ D \in [-1, 1]^{M \times k}.
\tag{7}
$$

With in-depth analysis, it can be seen that due to the non-convex form DD^T and the element-wise operator P, the relaxation form (7) is still difficult to optimize the calculation. In order to solve the optimization form (7), a novel random coordinate descent algorithm with Newton directions is proposed.

When the proposed algorithm is calculated, one of the entries in D is selected randomly or sequentially for updating, and all the other entries are kept fixed meanwhile. In each iteration, we select only one element D_{xy} in D and leave the other elements fixed and unchanged. So, we can define $D = [D_{\cdot 1}, D_{\cdot 2}, \ldots, D_{\cdot k}]$, where the y-th column in D is represented by $D_{\cdot y}$. Therefore, the original optimization goal (7) can be redefined as:

$$\| \max(\mathbf{0}, P \odot (D_{\cdot y} D_{\cdot y}^T - (Q - \sum_{c \neq y} D_{\cdot c} D_{\cdot c}^T))) \|_{\mathcal{F}}^2$$

$$= \| \max(\mathbf{0}, P \odot (D_{\cdot y} D_{\cdot y}^T - L)) \|_{\mathcal{F}}^2 \tag{8}$$

$$= \sum_{l=1}^{M} \sum_{k=1}^{M} (\max(0, P_{lk}(D_{ly} D_{ky} - L_{lk})))^2$$

where $L = Q - \sum_{c \neq y} H_{\cdot c} H_{\cdot c}^T$ is used to simplify the expression in the formula. It is obvious that L is symmetric, since the matrix Q is symmetric. Using our proposed method to fix the other elements in D, the optimization formula of D_{xy} can be redesigned as:

$$g(D_{xy}) = \sum_{l=1}^{M} \sum_{k=1}^{M} (\max(0, P_{lk}(D_{ly} D_{ky} - L_{lk})))^2$$

$$= \max(0, P_{xy}(D_{xy}^2 - L_{xx}))^2 \tag{9}$$

$$+ 2 \sum_{k \neq x} \max(0, P_{xk}(D_{xy} D_{ky} - L_{xk}))^2 + constant,$$

where we use the property that L is symmetric in the above description.

According to the principle of optimization algorithm, we generally update D_{xy} to $D_{xy} + d$, then the optimal solution problem can be defined as follows:

$$\min_{d} g(D_{xy} + d) \quad s.t. \ -1 \leq D_{xy} + d \leq 1. \tag{10}$$

In order to obtain d faster, we use a quadratic function via Taylor expansion to approximate $g(D_{xy} + d)$:

$$\hat{g}(D_{xy} + d) = g(D_{xy}) + g'(D_{xy})d + \frac{1}{2} g''(D_{xy})d^2, \tag{11}$$

where the first order derivative of g at D_{xy} is $g'(D_{xy})$ and the second order derivative of g at D_{xy} is $g''(D_{xy})$:

$$g'(D_{xy}) = 4 \sum_{k=1}^{M} \max(0, P_{xk}(D_{xy} D_{ky} - L_{xk})) P_{xk} D_{ky} \tag{12}$$

$$g''(D_{xy}) = P_{xx}^2(12D_{xy}^2 - 4L_{xx})\mathbf{I}_{P_{xx}(D_{xy}D_{xy}-L_{xx})>0}$$
$$+ 4\sum_{k\neq x} P_{xx}^2 D_{ky}^2 \mathbf{I}_{P_{xk}(D_{xy}D_{ky}-L_{xk})>0}$$
$$= (12D_{xy}^2 - 4L_{xx})\mathbf{I}_{P_{xx}(D_{xy}D_{xy}-L_{xx})>0}$$
$$+ 4\sum_{k\neq x} D_{ky}^2 \mathbf{I}_{P_{xk}(D_{xy}D_{ky}-L_{xk})>0} \tag{13}$$
$$= (8D_{xy}^2 - 4L_{xx})\mathbf{I}_{P_{xx}(D_{xy}D_{xy}-L_{xx})>0}$$
$$+ 4\sum_{k=1}^{M} D_{ky}^2 \mathbf{I}_{P_{xk}(D_{xy}D_{ky}-L_{xk})>0}$$

where an indicator function $\mathbf{I}_x$ is defined in the formula that $\mathbf{I}_x = 1$ if x is true and otherwise $\mathbf{I}_x = 0$, and we simplify $g''(D_{xy})$ in the form of $P_{xy} \in \{1, -1\}$.

Next, we set the derivative of the formula (11) to 0 with respect to d, then $d = -\frac{g'(D_{xy})}{g''(D_{xy})}$ is derived. Thus, the solution of the optimization objective

$$\min_{d} \hat{g}(D_{ij} + d) \quad s.t. \ -1 \le D_{xy} + d \le 1. \tag{14}$$

can be obtained by

$$d = \max(-1 - D_{xy}, \min(-\frac{g'(D_{xy})}{g''(D_{xy})}, 1 - D_{xy})), \tag{15}$$

which can be solved by updating rule $D_{xy} \leftarrow D_{xy} + d$.

Referring to the recent research [9,11,14,15], the optimization objective can be solved by multiple iterations of the coordinate descent algorithm.

Accelerated Optimization Algorithm

Based on the optimization objective derived above, calculating the optimization matrix $L = kQ - \sum_{c\neq x} D_{\cdot c}D_{\cdot c}^T \in \mathcal{R}^{M \times M}$ directly needs $O(M^2)$ time, which is time-consuming with large number M of hash centers. In order to better solve this problem, we design a new calculation method, which does not need to explicitly calculate L in the optimization process. In detail, we define the matrix $R = DD^T - Q$ so that we can update R with low cost in each optimization iteration.

Then $g'(D_{xy})$ and $g''(D_{xy})$ can be calculated with the defined matrix R:

$$g'(D_{xy}) = 4\max(\mathbf{0}, P_{x\cdot} \odot R_{x\cdot})(P_{\cdot x} \odot D_{\cdot y}),$$
$$g''(D_{xy}) = 4\sum_{k=1}^{M} D_{ky}^2 \mathbf{I}_{P_{xk}R_{xk}>0} + 4(D_{xy}^2 + R_{xx})\mathbf{I}_{P_{xx}R_{xx}>0}, \tag{16}$$

where $R_{x\cdot}$ is denoted as the x-th row of R. Consequently, $g'(D_{xy})$ and $g''(D_{xy})$ can be calculated in $O(M)$ time.

According to the analysis of the optimization process, when D_{xy} is updated, only the values of the x-th row and x-th column of the matrix R will be affected. Likewise, the time complexity of updating the matrix R is also $O(M)$. For instance, given $D_{xy} \leftarrow D_{xy} + d$, the matrix R can be updated by:

$$R_{x\cdot} \leftarrow R_{x\cdot} + dD_{\cdot y}^T, \ R_{\cdot x} \leftarrow R_{\cdot x} + dD_{\cdot y}, \ R_{xx} \leftarrow R_{xx} + d^2. \tag{17}$$

The procedure of the efficient matrix optimization algorithm for hash centers is shown in Algorithm 1.

Algorithm 1. Efficient Matrix Optimization Algorithm for Hash Centers

---

Input: T, $P \in \{-1,1\}^{M \times M}$, the hash code number of bits k, the tolerance error ϵ, maximum iterations $ITER$.

Initialize: randomly initialize $D \in [-1,1]^{M \times k}$; $R \leftarrow DD^T - T$, $t \leftarrow 0$, $S_{(0)} \leftarrow \| \max(\mathbf{0}, P \odot R)\|_F^2$, $D_{(0)} \leftarrow D$, $R_{(0)} \leftarrow R$.

for $t=1$ to $ITER$ **do**

  Decide the order of $M \times k$ indices (x,y) by random permutation ($x = 1, 2, ..., M$ and $y = 1, 2, ..., k$).

  for each of the $M \times k$ indices (x,y) **do**

    Select the entry D_{xy} to update.

    Calculate $g'(D_{xy})$ and $g''(D_{xy})$ by (16).

    Calculate d by (15) and update D_{xy} by $D_{xy} \leftarrow D_{xy} + d$.

    Update R by (17).

  end for

  $S_{(t)} \leftarrow \| \max(\mathbf{0}, P \odot R)\|_F^2$.

  if $S_{(t)} \leq S_{(t-1)}$ **then** $D_{(t)} \leftarrow D$, $R_{(t)} \leftarrow R$,

  else $D_{(t)} \leftarrow D_{(t-1)}$, $R_{(t)} \leftarrow R_{(t-1)}$, **continue**.

  if the relatively change $\frac{S_{(t-1)} - S_{(t)}}{S_{(t-1)}} \leq \epsilon$, **then break**.

end for

Output: the sign matrix D with either 1 or -1 elements.

3.2   Loss Function of Multi-label Deep Hashing Model

We choose a pre-trained deep neural network as backbone for training multi-label image hash codes, while carefully designing a novel multi-objective loss function for the deep hashing training procedure. The proposed multi-objective loss function mainly includes three loss constraints: (1) Image data often has multiple category labels attached to each image, so each image may have multiple hashing centers, then multiple hash center loss function is constructed according to the relationship between the image and its multiple hash centers. (2) For multiple hash centers of an image, we design a hash center anchor constraint to construct an image corresponding to two hash centers, one hash center is similar to the image, and the other is not similar. (3) Furthermore, quantization loss function constraint is added to reduce quantization error.

Constraint of Multi-label Hash Centers

In order to design the constraint on the hash centers, we firstly need to formally express the encodings of the image hash centers in the dataset, so we define $C = \{C_{x,y}\}_{x=1,2,\ldots,N \ and \ y=1,2,\ldots,NUM(x)}$ as the hash codes of hash centers. In the definition of C as the set of hash center codes, $NUM(x)$ is used to represent the number of category labels for image x, and $C_{x,y}$ is denoted as the k bits hash codes of the y-th hash center with image x. Analogously, we define the expression $H = \{H_x\}_{x=1,2,\ldots,N}$ to represent the set of hash codes corresponding to the image dataset, where H_x is denoted as the hash codes containing k bits. Complementally, We choose the symbol c to represent the c-th bit in the hash codes, then we further combine this expression with the definition of hash codes to get $C_{x,y,c}$ and $H_{x,c}$ in the following expressions.

Based on the formal definition above, the loss function constraint L_{ml} of multiple hash centers is defined as follows:

$$L_{ml} = \frac{1}{k} \sum_{x=1}^{N} \sum_{y=1}^{NUM(x)} \sum_{z=1}^{k} [\frac{\frac{C_{x,CLS(x,y),z}+1}{2} \log \frac{H_{x,z}+1}{2}}{NUM(x)} + \frac{(1 - \frac{C_{x,CLS(x,y),z}+1}{2}) \log (1 - \frac{H_{x,z}+1}{2})}{NUM(x)}]$$

(18)

where the denoted symbol $NUM(x)$ represents the amount of category labels with image x, and the function $CLS(x,y)$ is defined as the position of the y-th category label corresponding to image x in all class labels.

With efficient optimization algorithm for encoding of hash centers, each image in the multi-label image dataset can correspond to multiple hash centers with different category labels. Based on this approach, the proposed multi-label image hash center generation procedure is shown in Algorithm 2.

Constraint of the Image and Anchor Center Pairs

For images with multiple hash centers, we construct a loss constraint such that the hash code of each image is close to the hash center of each category label where the image resides, while it is far away from other hash centers. The anchor center pairs loss function constraint based on multi-label image hash center can be denoted as:

$$L_{ac} = \sum_{x=1}^{N} \sum_{y=1}^{M} \sum_{z=1}^{M} \frac{TRM(x,y,z)}{M^2}$$

(19)

where the notation $TRM(x,y,z)$ is the triplet margin loss function first proposed in FaceNet [20].

The margin distance constraint is introduced for the hash codes H_x of each image and any pair of anchor hash centers, then we can get the equation of $TRM(x,y,z)$:

$$TRM(x,y,z) = \max(\|H_x - C_{x,y}\|_2^2 - \|H_x - C_{x,z}\|_2^2 + margin, 0)$$

(20)

Algorithm 2. Hash Center Generation Procedure for Multi-label Images

---

Input: the number of all images N in the dataset, the number of categories M in the image dataset, the number of bits k by hash coding, maximum iterations $ITER$, the threshold ϵ_1 and the threshold ϵ_2, the category labels $B = \{B_{x,y}\}_{x=1,2,...,N \ and \ y=1,2,...,M} \in \{0,1\}^{N \times M}$ of the corresponding multi-label images.

Initialize: create matrix variables for all hash centers $D \in \{-1,1\}^{M \times k}$, $CNT = 1$.

while $CNT <= M$ **do**

    Assign matrix $D' \in \{-1,1\}^{M \times k}$ with the Algorithm 1.

    for $t=1$ to M **do**

        Calculate the minimal distance $DIST_{min}$ for $D'_{c=t}$ and $D'_{c \neq t}$, and calculate the mean distance $DIST_{mean}$ for $D'_{c=t}$ and $D'_{c \neq t}$.

        if $DIST_{min} > \epsilon_1$ **and** $DIST_{mean} > \epsilon_2$ **then** $D_{CNT} = D'_t$, $CNT+ = 1$.

    end for

end while

for $t=1$ to N **do**

    for $c=1$ to M **do**

        Assign the c-th row of matrix D to $C_{t,c}$,

    end for

end for

Output: the hash centers of all the multi-label images in the dataset $C = \{C_{x,y}\}_{x=1,2,...,N \ and \ y=1,2,...,M} \in \{-1,1\}^{N \times M \times k}$.

Constraint of Quantization Loss

For the training procedure of deep hashing models, the hash codes in the process of model training are generally real numbers, and the actual output form to be obtained should be Hamming codes. Therefore, the quantization constraint needs to be added as the loss function for reducing the impact of quantization error.

In order to reduce the quantization error of image hash codes, we define the constraint of quantization loss function:

$$L_{qu} = \frac{1}{k} \sum_{x=1}^{N} \sum_{y=1}^{k} (\|H_{x,y}\|_1 - 1)^2 \tag{21}$$

Multi-objective Loss Function

To sum up, we can get the multi-objective loss function of the deep model training process, as shown below:

$$L = \lambda_1 L_{ml} + \lambda_2 L_{ac} + \lambda_3 L_{qu} \tag{22}$$

where λ_1, λ_2 and λ_3 are defined as the parameters of the proposed multi-objective loss function.

4 Experiments

In the experiments that follow in this section, our proposed method is evaluated against traditional hashing methods and state-of-the-art deep hashing methods on several multi-label image datasets. By analyzing experimental results on multiple datasets, the effectiveness of the hash codes obtained by the proposed deep hash coding method for image retrieval is verified.

4.1 Baselines and Datasets

In the comparison experiment, eight hashing algorithms are selected for comparison, which are described as follows.

- **KSH** [14] is a kernel based supervised hashing method, which uses the equivalence relation between the inner product of binary codes and Hamming distance to optimize the model and build a compact hash codes.
- **ITQ** [5] is an iterative quantization hashing algorithm. When the dataset contains the information of image labels, ITQ-CCA hashing algorithm is obtained by combining with canonical correlation analysis (CCA).
- **SDH** [21] is the abbreviation of supervised discrete hashing, which constructs the objective function by generating binary codes through discrete optimization, then the function can be efficiently solved by regularization method.
- **CNNH** [24] firstly obtains the training hash codes by constructing the similarity matrix and optimizing the similarity preserving function, and then the training hash codes are used as the supervised labels for the training of deep neural network, which is a two-stage deep hashing algorithm.
- **NINH** [10] constructs a one-stage algorithm framework by combining the Network-in-Network (NIN) [12] structure and triple loss function, and implements the end-to-end deep hash coding method.
- **DHN** [26] is a pair-wise deep hashing method based on the cross entropy loss function. It proposes the pair quantization loss and the pair cross entropy loss is optimized so as to generate a compact and concentrated hash codes.
- **HashNet** [1] retains similarity by designing a new weighted pair cross entropy loss, which can accurately learn binary codes by quantization of continuous coding to generate binary codes and minimize quantization error.
- **CSQ** [25] proposes a deep hashing framework with the hash centers. The algorithm can construct the hash centers efficiently, and uses the hash center as the supervised training tags of deep neural network, so as to obtain effective image hash codes.

In the experimental verification, three multi-label image benchmark datasets (NUS-WIDE [2], MS-COCO [13], and FLICKR-25K [8]) are selected for detailed evaluation of our proposed method. The description and statistical data of the these datasets are summarized as follows.

- **NUS-WIDE** [2] is a multi-label image dataset containing images from websites with multiple labels. We use 81 categories as multi-label data in the experiment, randomly selected 10,000 images as the training set, 5,000 images as the query set, and then selected 168,692 images from the remaining part as the database retrieval set.

- **MS-COCO** [13] is a large-scale dataset for image recognition and image segmentation, which contains 80 different image categories. In the comparative
  experiment, 10,000 images are randomly choosen as the dataset for training,
  and 5,000 images are selected as the dataset for query test. In addition, a
  total of 117,218 images are used for image retrieval as the retrieval database.
- **FLICKR-25K** [8] is an image subset selected from the Flickr dataset that
  contains 38 categories and a total of 25,000 images. In a similar manner, 4,000
  images are selected as the training set in the experiment, while 1,000 images
  are used as the query dataset. Furthermore, the remaining 20,000 images are
  selected to build the database for the retrieval task in the experiment.

4.2   Experimental Details and Hyperparameter Settings

All the comparative experiments are chosen to run on server clusters using
GeForce RTX series GPUs for training and retrieval of hashing algorithms, and
the concrete implementation of these experiments is through the framework of
Pytorch [17].

For experimental details, we choose ResNet [6] as the backbone of the deep
neural network for deep hashing, and run 20 epochs on ResNet during each
training process. For each round of deep hashing comparison experiments, the
initial learning rate of the deep networks is set to 1×10^{-5}, and we set the amount
of images to 64 in each mini-batch. Referring to the usual setting method, in
order to match ResNet as the backbone of the deep network, the input training
images are adjusted to 256×256 and cropped to 224×224.

4.3   Evaluation Criteria

In the hashing retrieval experiments, we choose the mean average precision
(mAP) evaluation commonly used by hash coding methods [10,24,25] as the
comparison standard for the experimental results. Formally, the metric of mAP
is defined by:

$$mAP = \frac{1}{Q} \sum_{i=1}^{Q} AP_i \tag{23}$$

where Q represents the number of images in the query set and AP_i is expressed
as the average precision of the i-th image.

Then we define the $SIM(i,j)$ function for the formal expression that image
i and image j are similar or dissimilar, and we can get the calculation formula
of AP_i:

$$AP_i = \frac{1}{R_i} \sum_{j=1}^{N_{db}} \frac{R_{i,j}}{j} SIM(i,j) \tag{24}$$

where R_i indicates the number of images relevant to the i-th image in the retrieval database, $R_{i,j}$ indicates the number of images that are similar to the i-th image among the top-j images in the retrieval returned results, and N_{db} represents the amount of images in the database for retrieval.

In the experimental session, we adopt mAP@5000 for the multi-label image datasets of NUS-WIDE and MS-COCO following the setting in [25], and we use mAP@ALL on the FLICKR-25K image dataset.

4.4   Experimental Comparison Results

The comparative experiment is conducted on NUS-WIDE dataset to verify the effectiveness of our proposed deep hashing framework. According to the experimental results in Table 1, it can be seen that the proposed hash coding algorithm achieves the best results with mAP measurement among all the hashing baseline algorithms on 16/32/64 bits hash codes, and the proposed method also achieves 3.35%/4.93%/6.93% improvement over the best results of the baseline algorithms.

Table 1. mAP in % of Hamming ranking with respect to (w.r.t) different number of bits on NUS-WIDE.

Method	NUS-WIDE (mAP)		
	16 bits	32 bits	64 bits
KSH	35.62	33.26	33.63
ITQ	45.97	40.49	34.62
SDH	47.53	55.41	58.11
CNNH	56.87	58.22	59.87
NINH	59.72	61.53	63.83
DHN	63.72	66.29	67.13
HashNet	66.25	69.82	71.61
CSQ	72.51	74.51	74.78
Ours	**74.94**	**78.18**	**79.96**

We also compare the proposed method with eight baseline hashing methods on MS-COCO dataset, as shown in Table 2. The experimental results can be concluded that the optimal baseline experiment results of 16/32/64 bits hash codes can improve by 4.41%/1.09%/2.94% compared with the baseline methods, that is, our proposed deep hashing method achieves the best results on the dataset with mAP metric.

Table 2. mAP in % of Hamming ranking with respect to (w.r.t) different number of bits on MS-COCO.

Method	MS-COCO (mAP)		
	16 bits	32 bits	64 bits
KSH	52.32	53.56	53.62
ITQ	56.71	56.32	51.25
SDH	55.49	56.41	58.12
CNNH	60.01	61.83	62.21
NINH	64.47	65.21	64.77
DHN	72.08	73.22	74.52
HashNet	74.56	77.29	78.79
CSQ	76.12	85.21	86.53
Ours	**79.48**	**86.14**	**89.07**

Table 3. mAP in % of Hamming ranking with respect to (w.r.t) different number of bits on FLICKR-25K.

Method	FLICKR-25K (mAP)		
	16 bits	32 bits	64 bits
CSQ	68.52	69.35	69.43
Ours	**74.43**	**77.61**	**78.93**

Then the research experiment is continued on FLICKR-25K dataset, and our proposed method gains the best results. Due to the larger increase in experimental results on the dataset, we simplify the data results presentation of the baselines in Table 3, showing only the best results of the baseline methods. It can be summarized from the experimental results that our proposed method can achieve a further improvement of 8.63%/11.91%/13.68% with 16/32/64 bits hash codes compared with the best results of the contrastive baselines.

In addition, the proposed method is compared with the benchmark algorithms in terms of precision and recall metrics. Figure 2, Fig. 3 and Fig. 4 show the experimental results on Precesion-recall curves, Recall@top-N curves and Precision@top-N with hash codes @16/32/64bits of our proposed method and CSQ method on three image datasets.

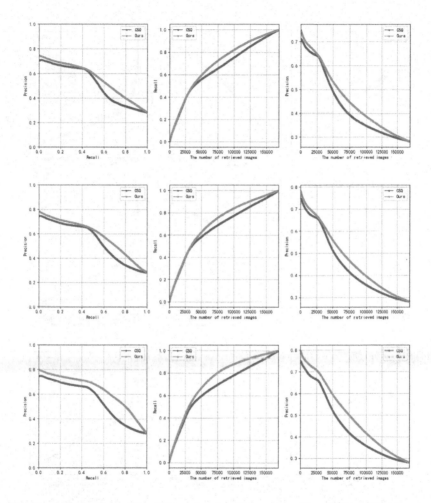

Fig. 2. Experimental results (Precesion-recall curves, Recall@top-N curves and Precision@top-N) with hash codes @16/32/64bits of our proposed method and CSQ method on NUS-WIDE datasets.

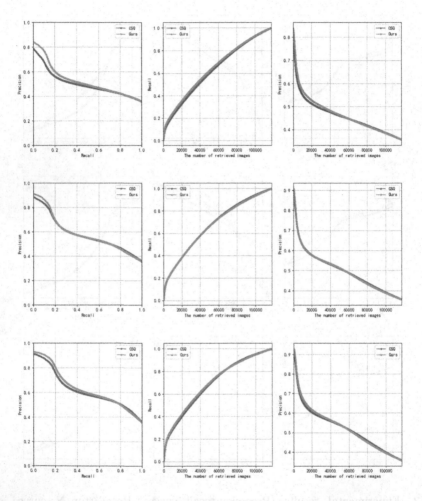

Fig. 3. Experimental results (Precesion-recall curves, Recall@top-N curves and Precision@top-N) with hash codes @16/32/64bits of our proposed method and CSQ method on MS-COCO datasets.

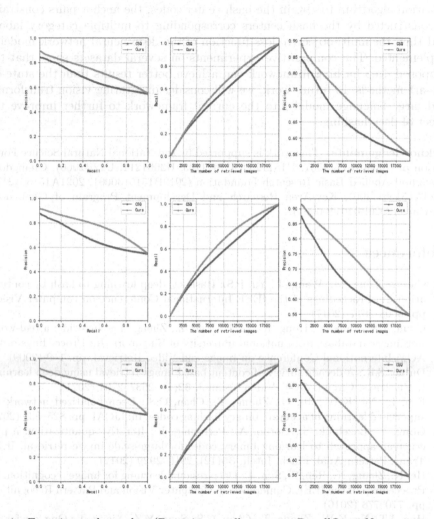

Fig. 4. Experimental results (Precesion-recall curves, Recall@top-N curves and Precision@top-N) with hash codes @16/32/64bits of our proposed method and CSQ method on FLICKR-25K datasets.

5 Conclusion

In this paper, we propose a novel deep hashing framework for multi-label image retrieval, which makes full use of the information from multi-class category labels. By constructing the hash center similarity matrix and designing an optimization algorithm to obtain the hash center codes, the anchor pairs constraint is constructed by the hash centers corresponding to multiple category labels, and then the multi-objective loss function of the deep neural network model is implemented. The comparative experiments on several datasets show that the proposed deep hashing framework can achieve better results than the state-of-the-art methods. In future work, we will consider integrating vision transformer and large language model into the current framework to further improve the effect of deep hashing.

Acknowledgements. This work is supported by the National Natural Science Foundation of China (61772567, U1811262, U1911203, U2001211, U22B2060), Guangdong Basic and Applied Basic Research Foundation (2019B1515130001, 2021A1515012172, 2023A1515011400), Key-Area Research and Development Program of Guangdong Province (2020B0101100001).

References

1. Cao, Z., Long, M., Wang, J., Yu, P.S.: Hashnet: deep learning to hash by continuation. In: Proceedings of the IEEE International Conference on computer Vision, pp. 5608–5617 (2017)
2. Chua, T.S., Tang, J., Hong, R., Li, H., Luo, Z., Zheng, Y.: Nus-wide: a real-world web image database from national university of Singapore. In: Proceedings of the ACM International Conference on Image and Video Retrieval, pp. 1–9 (2009)
3. Dubey, S.R.: A decade survey of content based image retrieval using deep learning. IEEE Trans. Circuits Syst. Video Technol. **32**(5), 2687–2704 (2021)
4. Fan, L., Ng, K.W., Ju, C., Zhang, T., Chan, C.S.: Deep polarized network for supervised learning of accurate binary hashing codes. In: IJCAI, pp. 825–831 (2020)
5. Gong, Y., Lazebnik, S., Gordo, A., Perronnin, F.: Iterative quantization: a procrustean approach to learning binary codes for large-scale image retrieval. IEEE Trans. Pattern Anal. Mach. Intell. **35**(12), 2916–2929 (2012)
6. He, K., Zhang, X., Ren, S., Sun, J.: Deep residual learning for image recognition. In: Proceedings of the IEEE Conference on Computer Vision and Pattern Recognition, pp. 770–778 (2016)
7. Hoe, J.T., Ng, K.W., Zhang, T., Chan, C.S., Song, Y.Z., Xiang, T.: One loss for all: Deep hashing with a single cosine similarity based learning objective. Adv. Neural. Inf. Process. Syst. **34**, 24286–24298 (2021)
8. Huiskes, M.J., Lew, M.S.: The MIR Flickr retrieval evaluation. In: Proceedings of the 1st ACM International Conference on Multimedia Information Retrieval, pp. 39–43 (2008)
9. Kulis, B., Darrell, T.: Learning to hash with binary reconstructive embeddings. In: Proceedings of the Advances in Neural Information Processing Systems, pp. 1042–1050 (2009)

10. Lai, H., Pan, Y., Liu, Y., Yan, S.: Simultaneous feature learning and hash coding with deep neural networks. In: Proceedings of the IEEE Conference on Computer Vision and Pattern Recognition, pp. 3270–3278 (2015)
11. Lin, G., Shen, C., Suter, D., van den Hengel, A.: A general two-step approach to learning-based hashing. In: Proceedings of the IEEE Conference on Computer Vision, Sydney, Australia (2013)
12. Lin, M., Chen, Q., Yan, S.: Network in network. arXiv preprint arXiv:1312.4400 (2013)
13. Lin, T.-Y.: Microsoft COCO: common objects in context. In: Fleet, D., Pajdla, T., Schiele, B., Tuytelaars, T. (eds.) ECCV 2014. LNCS, vol. 8693, pp. 740–755. Springer, Cham (2014). https://doi.org/10.1007/978-3-319-10602-1_48
14. Liu, W., Wang, J., Ji, R., Jiang, Y.G., Chang, S.F.: Supervised hashing with kernels. In: 2012 IEEE Conference on Computer Vision and Pattern Recognition, pp. 2074–2081. IEEE (2012)
15. Liu, Y., Pan, Y., Lai, H., Liu, C., Yin, J.: Margin-based two-stage supervised hashing for image retrieval. Neurocomputing 214, 894–901 (2016)
16. Norouzi, M., Fleet, D.J., Salakhutdinov, R.R.: Hamming distance metric learning. In: Advances in Neural Information Processing Systems, vol. 25 (2012)
17. Paszke, A., et al.: Pytorch: an imperative style, high-performance deep learning library. In: Advances in Neural Information Processing Systems, vol. 32 (2019)
18. Rawat, W., Wang, Z.: Deep convolutional neural networks for image classification: a comprehensive review. Neural Comput. 29(9), 2352–2449 (2017)
19. Rodrigues, J., Cristo, M., Colonna, J.G.: Deep hashing for multi-label image retrieval: a survey. Artif. Intell. Rev. 53(7), 5261–5307 (2020)
20. Schroff, F., Kalenichenko, D., Philbin, J.: FaceNet: a unified embedding for face recognition and clustering. In: Proceedings of the IEEE Conference on Computer Vision and Pattern Recognition, pp. 815–823 (2015)
21. Shen, F., Shen, C., Liu, W., Tao Shen, H.: Supervised discrete hashing. In: Proceedings of the IEEE Conference on Computer Vision and Pattern Recognition, pp. 37–45 (2015)
22. Wang, J., Zhang, T., Sebe, N., Shen, H.T., et al.: A survey on learning to hash. IEEE Trans. Pattern Anal. Mach. Intell. 40(4), 769–790 (2017)
23. Wang, L., Pan, Y., Liu, C., Lai, H., Yin, J., Liu, Y.: Deep hashing with minimal-distance-separated hash centers. In: Proceedings of the IEEE/CVF Conference on Computer Vision and Pattern Recognition, pp. 23455–23464 (2023)
24. Xia, R., Pan, Y., Lai, H., Liu, C., Yan, S.: Supervised hashing for image retrieval via image representation learning. In: Twenty-Eighth AAAI Conference on Artificial Intelligence (2014)
25. Yuan, L., et al.: Central similarity quantization for efficient image and video retrieval. In: Proceedings of the IEEE/CVF Conference on Computer Vision and Pattern Recognition, pp. 3083–3092 (2020)
26. Zhu, H., Long, M., Wang, J., Cao, Y.: Deep hashing network for efficient similarity retrieval. In: Proceedings of the AAAI conference on Artificial Intelligence, vol. 30 (2016)

MDAM: Multi-Dimensional Attention Module for Anomalous Sound Detection

Shengbing Chen, Junjie Wang(✉), Jiajun Wang, and Zhiqi Xu

School of Artificial Intelligence and Big Data, Hefei University, Hefei 230000, China
156567930810163.com

Abstract. Unsupervised anomaly sound detection (ASD) is a challenging task that involves training a model to differentiate between normal and abnormal sounds in an unsupervised manner. The difficulty of the task increases when there are acoustic differences (domain shift) between the training and testing datasets. To address these issues, this paper proposes a state-of-the-art ASD model based on self-supervised learning. Firstly, we designed an effective attention module called the Multi-Dimensional Attention Module (MDAM). Given a shallow feature map of sound, this module infers attention along three independent dimensions: time, frequency, and channel. It focuses on specific frequency bands that contain discriminative information and time frames relevant to semantics, thereby enhancing the representation learning capability of the network model. MDAM is a lightweight and versatile module that can be seamlessly integrated into any CNN-based ASD model. Secondly, we propose a simple domain generalization method that increases domain diversity by blending the feature representations of different domain data, thereby mitigating domain shift. Finally, we validate the effectiveness of the proposed methods on DCASE 2022 Task 2 and DCASE 2023 Task 2.

Keywords: Anomalous sound detection · Self-supervised learning · Attention mechanism · Domain shift

1 Introduction

Anomalous sound detection (ASD) has attracted significant interest from researchers due to its wide application in machine condition monitoring [4,6, 9,11,22,23,26]. The goal of ASD is to distinguish between normal and abnormal sounds emitted by machines. In practical scenarios, abnormal sounds are

This work was supported by the following grants:
- Universities Natural Science Research Project of Anhui Province (KJ2021ZD0118)
- Program for Scientific Research Innovation Team in Colleges and Universities of Anhui Province (2022AH010095)
- The second key orientation of open bidding for selecting the best candidates in innovation and development project supported by speech valley of china : the collaborative research of the multi-spectrum acoustic technology of monitoring and product testing system for key equipment in metallurgical industry (2202-340161-04-04-664544).

B. Luo et al. (Eds.): ICONIP 2023, CCIS 1967, pp. 48–60, 2024.
https://doi.org/10.1007/978-981-99-8178-6_4

often infrequent and diverse, making it challenging to collect a sufficient number of anomalous sound samples. In contrast, it is relatively easier to gather training datasets that consist only of normal samples. Therefore, unsupervised or self-supervised methods are particularly crucial for ASD.

Unsupervised methods [4, 6] utilize autoencoders (AE) to reconstruct spectrograms, using the reconstruction error as an anomaly indicator. The conventional interpretation is that since the model is trained only on normal samples, it is expected to have larger reconstruction errors for abnormal cases. However, Jiang [11] found that most abnormal spectrograms can still be well reconstructed by AE, indicating that reconstruction error alone may not be an effective anomaly indicator.

In order to improve the effectiveness of anomaly sound detection, many researchers have devoted themselves to self-supervised learning-based approaches [9, 22, 23, 26]. Self-supervised learning typically involves constructing auxiliary tasks to extract representations that contain discriminative information, which are then inputted into an outlier detection algorithm for anomaly scoring. The assumption is that in order to train the auxiliary task, the representations should capture sufficient information from the raw data, and this information should also be sufficient for detecting anomalies.For example, Inoue et al. [9] used data enhancement techniques to generate pseudo-classes from normal sounds by training a classifier to predict which data enhancement technique was used for each sound sample to learn the representation, however, choosing a reasonable data enhancement technique is challenging.

A more popular approach [22, 23, 26] is to use metadata such as machine type, machine identity (ID), and attribute labels to build an auxiliary classification task. This involves training a classifier to predict the machine type and ID for each type of normal sound. Although these methods perform much better than the reconstruction-based models, their network architecture has relatively weak learning ability, and the extracted representations do not focus on the critical frequency bands in the audio signals and the semantically relevant time segments. In particular, due to differences in machine operating conditions or environmental noise, there is a domain shift between the source domain data and target domain data, resulting in poor generalization performance of the representations in unknown domains.

The main contributions of this paper are as follows: 1)Proposing a Multi-Dimensional Attention Module (MDAM) specifically designed for ASD, which adaptively refines contextual information of time, frequency, and channel features to capture discriminative representations. 2)Designing a simple domain generalization method to alleviate domain shift by blending representations from different data domains, thereby increasing domain diversity and improving ASD performance. 3)Comparing and selecting a suitable outlier detection algorithm to generalize across multiple machine types.

2   Related Work

2.1   Attention Mechanism

Attention mechanisms have been proven to be a potent means of enhancing deep convolutional neural networks (CNNs). In the field of image recognition, SENet [8] first introduced channel attention mechanism, which reweights the importance of each channel feature based on global information. This allows the model to focus on useful information and ignore irrelevant information, thereby improving model performance. Building upon this, ECANet [20] proposed an efficient method for channel attention calculation, reducing computational complexity while maintaining performance. CBAM [25] introduced the importance of spatial attention, simultaneously learning weights for both channel and spatial dimensions to enhance feature representation. In the domain of environmental sound classification (ESC), the Time-Frequency CNN (TFCNN) [17] was proposed, which considers both the time and frequency dimensions, allowing it to focus on key frequency bands and time frames in the spectrogram. In the field of anomaly sound detection (ASD), Wang [21] inserted CBAM into the network to help the model acquire more precise features. However, these attention mechanisms have shown limited performance in the ASD domain. We propose MDAM, which combines the channel attention mechanism and the time-frequency attention mechanism, a module that simultaneously models explicitly the interdependencies between features within each dimension in three dimensions: time and frequency, and channel, to adaptively recalibrate each feature response. Specifically designed for ASD tasks.

2.2   Analysis of Temporal and Frequency Dimension Characteristics

We analyzed the response of each frequency band and time period to the spectrogram. First, the samples of the categories "ToyCar" and "ToyTrain" on the DCASE 2022 Task 2 dataset [5] were subjected to short-time Fourier transform (STFT) to extract the amplitude spectrogram $S \in \mathbb{R}^{513 \times 313}$. The frequency activation matrix $A_f \in \mathbb{R}^{513 \times 1}$ was obtained by stitching and averaging all S in each category in the time dimension, respectively, as shown in (a)(c) in Fig. 1. Bands with high activation values can be considered as active. It can be observed that the active bands are concentrated in the low frequency region, and the active bands differ between different machines, with "ToyCar" having active bands mainly in the range of 0 to 100, and "ToyTrain" having active bands in the range of 0 to 200. Similarly, the temporal activation matrix $A_t \in \mathbb{R}^{313 \times 1}$ was obtained by stitching and averaging in the frequency dimension, as shown in (b) and (d) in Fig. 1. It can be observed that the active periods of "ToyCar" have longer durations, while the active periods of "ToyTrain" have shorter durations. The above analysis indicates that the activity level of different frequency bands varies over time, and previous studies [17] have shown that time periods or bands with higher activity levels are more important. Therefore, it is necessary to utilize attention mechanisms to focus on time periods and frequency bands that contain discriminative information and disregard time periods and frequency bands that are not relevant to the task.

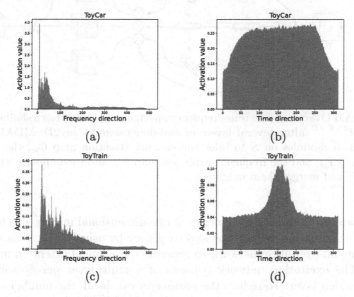

(a) (b)

(c) (d)

Fig. 1. (a) show the frequency activation matrix for "ToyCar" , and (b) show the temporal activation matrix for "ToyCar".(c) show the frequency activation matrix for "ToyTrain" , and (d) show the temporal activation matrix for "ToyTrain".

3 Method

3.1 Multi-Dimensional Attention Module

As shown in Fig. 2, the time-frequency spectrum is obtained after several layers of two-dimensional convolutional networks (Conv2D) to obtain the shallow feature map $S \in \mathbb{R}^{T \times F \times C}$, where C is the number of channels, T is the time, and F is the frequency. After applying MDAM, the feature map S is transformed to the multi-dimensional attention spectral map $\tilde{S}$, and the process can be summarized as follows:

$$\tilde{S} = \text{MDAM}(S) \tag{1}$$

Channel Attention Module. We use the inter-channel relationships of the features to generate the channel attention graph. To efficiently compute channel attention, we use global maximum pooling (GMP) to compress the time-frequency dimension of the input feature graph to aggregate the time-frequency contextual information. Compared with channel attention using global average pooling (GAP) such as SENet and ECANet, we believe that GMP is more suitable for the channel dimension of the sound feature map. We use experiments to confirm that using GMP improves the representational power of the network over GAP (see Sect. 4.3), which demonstrates the validity of our design choices.

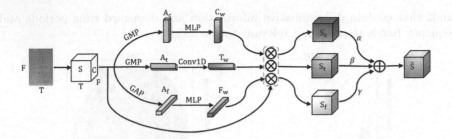

Fig. 2. MDAM Overview. The time-frequency spectrum is obtained as a shallow feature map $S \in \mathbb{R}^{T \times F \times C}$ after several layers of two-dimensional Conv2D. MDAM applies three attention modules on S to infer the channel attention map C_w, the temporal attention map T_w, and the frequency attention map F_w, and then applies attention to S separately and merges them weighted.

As shown in Fig. 2, we first generate a one-dimensional matrix A_c containing contextual information across time and frequency by using GMP. This matrix is then fed into an excitation network to generate our channel attention map $C_w \in \mathbb{R}^{1 \times 1 \times C}$. The excitation network consists of a multi-layer perceptron (MLP) with one hidden layer. To reduce the parameter overhead, the number of hidden neurons is dimensioned down and denoted as $R^{1 \times 1 \times C/r}$, where r is the scaling ratio. Briefly, the channel attention is calculated as follows:

$$C_w = \sigma(MLP_c(MaxPool(S))) = \sigma(W_c^1 \delta(W_c^0(A_c))) \qquad (2)$$

where δ is the ELU [3] activation function, $W_c^0 \in \mathbb{R}^{\frac{C}{r} \times C}$, $W_c^1 \in \mathbb{R}^{C \times \frac{C}{r}}$, and σ is the sigmoid activation function,r is the dimensionality reduction factor.

Time and Frequency Attention Module. We use the temporal and frequency information of the features to generate temporal attention maps and frequency attention maps. The channel and temporal dimensions of the input feature map S are compressed using GMP to obtain the one-dimensional matrix A_f, and the channel and frequency dimensions of the input feature map S are compressed using GAP to obtain the one-dimensional matrix A_t. A_f is then sent to an excitation network consisting of MLP layers to generate a frequency attention map $F_w \in \mathbb{R}^{1 \times F \times 1}$. In contrast, we use Conv1D to generate the temporal attention map $T_w \in \mathbb{R}^{T \times 1 \times 1}$ by interacting locally across time slots on A_t. Temporal information is usually continuous, and a single time slot may only be related to information in adjacent time slots and less relevant to more distant time slots, so we use Conv1D instead of MLP, which can also reduce the computational effort. In short, the temporal-frequency attention is calculated as follows:

$$T_w = \sigma(Conv_t(AvgPool(S))) = \sigma(Conv1D_k(A_t)) \qquad (3)$$

$$F_w = \sigma(MLP_f(MaxPool(S))) = \sigma(W_c^1 \delta(W_c^0(A_f))) \qquad (4)$$

where δ is the ELU activation function, $W_f^0 \in \mathbb{R}^{\frac{F}{r} \times F}$, $W_f^1 \in \mathbb{R}^{F \times \frac{F}{r}}$, k denotes the size of the 1-dimensional convolution kernel, σ is the sigmoid activation function, and r is the dimensionality reduction factor.

Finally, the attentional maps on the 3 dimensions are applied to S to obtain 3 attentional spectral maps, which are fused into the final multidimensional attentional spectral map $\tilde{S}$ in a learnable scale. This is done as follows:

$$\tilde{S} = \alpha S_c + \beta S_t + \gamma S_f \tag{5}$$

where α, β, and γ are the 3 learnable weights and $\alpha + \beta + \gamma = 1$.

3.2 Domain Mixture

A simple domain generalization method, Domain Mixture (DM) is proposed for the domain shift of machine sound. After obtaining the source domain embedding μ_s and the target domain embedding μ_t of the sound signal in the auxiliary classification task, the data imbalance between the source and target domains is first handled by oversampling the target domain embedding using the SMOTE [2] algorithm. Then, the mixed-domain mean μ_{mix} is calculated, which is obtained from Equation (6). The overall embedding x after SMOTE is then centralized by first subtracting the overall mean μ and then adding the mixed domain mean μ_{mix} to indirectly generate the embedding Mix of the new domain, as shown in Equation (7).

$$\mu_{\text{mix}} = \lambda \mu_s + (1 - \lambda)\mu_t \tag{6}$$

$$Mix = x - \mu + \mu_{\text{mix}} \tag{7}$$

where $\lambda \sim \text{Beta}(\alpha, \alpha)$, where $\alpha \in (0, \infty)$ is the hyperparameter.

4 Experiments

4.1 Experimental Setup

Data Sets and Evaluation Metrics. We validate our method on the datasets of DCASE 2022 Task 2 and DCASE 2023 Task 2. The DCASE 2022 Task 2 dataset consists of seven different types of machines, namely "ToyCar" and "Toy-Train" from the MIMII DG dataset [5], and "Bearing", "Fan", "Gearbox", "Slider", and "Valve" from the ToyADMOS2 dataset [7]. Each machine type has six different subsets called "sections," corresponding to different domain shift. For each section, there are 990 normal training samples belonging to the source domain and 10 normal training samples belonging to the target domain. In contrast to the DCASE 2022 Task 2 dataset, the DCASE 2023 Task 2 dataset includes seven additional machines in the development and evaluation sets, namely "Toy-Drone", "ToyNscale", "ToyTank", "Vacuum", "Bandsaw", "Grinder", and "Shaker". Each machine type in the DCASE 2023 Task 2 dataset has only one section. To train the network, the machine types, sections, and different attribute information of all machines belonging to the development and evaluation sets are utilized as classes for the auxiliary task. We use the Area Under Curve (AUC) and the partial Area Under Curve (pAUC) to evaluate the performance of the model, and use the hmean metric to evaluate the average performance of the model on all machines.

Input Features and Base Network. We use the network proposed by Wilkinghoff [7] as the base network, a two-branch network consisting of a modified ResNet [18] and a multilayer one-dimensional convolution.To obtain the linear magnitude spectrogram, we apply the STFT to obtain a 513-dimensional representation. The sampling window size is set to 1024, the hop size is 512, the maximum frequency is set to 8000 Hz, and the minimum frequency is 200 Hz. The entire signal's magnitude spectrum (8000) is used to achieve a high frequency resolution, enabling better capture of stationary sounds. For model training, we utilize the Adam optimizer [13] with a default initial learning rate of 0.001. The SCAdaCos loss function [22] is used, with the number of classes set to the joint categories of machine IDs and attributes.The dimensionality reduction factor r is set to 6 in both the channel and frequency attention modules in MDAM, and the size of the Conv1D convolution kernel in the temporal attention module is set to 3. Initialize $\alpha = 0.4$, $\beta = 0.3$, and $\gamma = 0.3$.

4.2 Comparison of Different Outlier Detection Algorithms

First, to compare the performance of different outlier detection algorithms, we apply seven different outlier detection algorithms to the same embedding, as shown in Table 1. First, the algorithm based on cosine distance [23] is significantly better than LOF [1] and GMM [19] in terms of AUC and pAUC, but still not better than KNN [18]. Second, as k increases from 1 to 7, AUC gradually increases and pAUC gradually decreases, which may be related to the characteristics of the distribution of outliers in the dataset. Considered together, KNN (k=5) is used as the outlier detection algorithm in this paper in all the following experiments.

Table 1. Performance comparison of different outlier detection algorithms.

algorithms	AUC (%)	pAUC (%)
LOF	65.61	58.51
GMM	71.03	58.22
Cosine distance	71.98	61.14
KNN (k=1)	71.17	**61.68**
KNN (k=3)	72.22	61.61
KNN (k=5)	72.31	61.51
KNN (k=7)	**72.36**	61.45

4.3 Pooling Selection

GMP and GAP are commonly used to obtain global information about features, the use of GMP and GAP produces a significant difference in the impact on the performance of the model. As can be observed from Table 2, using GMP performs better in the frequency and channel dimensions. This may be due to the fact that the frequency and channel dimensions usually contain local features, while the global maximum pooling is able to highlight these important features. The temporal dimension is more suitable for using GAP than the frequency and channel dimensions because the temporal dimension usually contains the overall features and using global average pooling can preserve the overall feature information.

Table 2. Performance comparison of different pooling.

		AUC(%)	pAUC(%)
channel	GAP	71.59	60.42
	GMP	**72.11**	**61.57**
frequency	GAP	71.71	60.63
	GMP	**72.08**	**61.23**
time	GAP	**72.02**	**60.88**
	GMP	71.68	60.22

4.4 Comparison with Other Attention Modules

We verified the effectiveness of MDAM and the results are shown in Table 3. We find that compared to the original model, inserting both SENet and ECANet decreases the AUC. MDAM improves the AUC by 0.53% and the pAUC by 0.95%. The poor performance of SENet and ECANet in machine voice detection can be attributed to the use of average pooling for aggregating temporal and frequency information. Additionally, the shallow convolutional feature maps retain some temporal and frequency information. Therefore, applying attention to the temporal-frequency and channel dimensions proves to be more effective than considering the channel dimension alone.To visualize the improvement, we use t-SNE [16] to visualize the representation. In it, we can clearly see that normal and abnormal sounds partially overlap before attention is added (Fig. 3(a)), while they are completely separated after MDAM is added (Fig. 3(b)). This finding shows again that MDAM can help the model to obtain a representation that contains discriminative information.

Table 3. Performance comparison of different attention module.

	AUC (%)	pAUC (%)
No-Attention	71.78	60.56
SENet	71.59	60.42
ECANet	71.62	60.47
MDAM	**72.31**	**61.51**

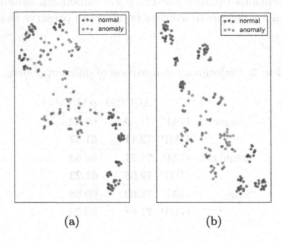

(a)   (b)

Fig. 3. The t-SNE visualization of the vector is represented on the test dataset of Slider ID 00. a: No-Attention; b: MDAM.

4.5 Effectiveness of Domain Mixture

In order to explore the domain generalization effect of Domain Mixture (DM), we conducted an experimental comparison. The results are shown in Table 4. After using DM, the AUC is improved by 0.26% and the pAUC is improved by 0.14%, which indicates that DM can alleviate the domain shift and improve the domain generalization ability of the model to some extent.

Table 4. Performance of Domain Mixture.

	AUC (%)	pAUC (%)
-	72.31	61.51
DM	**72.57**	**61.65**

4.6   Comparison with Other Anomaly Detection Models

Comparing our model with other advanced models, it can be seen from Table 5 that our performance on seven machines in DCASE 2022 Task 2 is significantly better than other advanced models, with average AUC and average pAUC of 72.57% and 61.65% respectively, and an official score of 68.52%. On the "Toy-Train" where other advanced models did not perform well, our AUC and pAUC are 71.13% and 60.43% respectively. Although our performance on "Fan" is not good, it is still higher than the Kevin [23]. From Table 6, it can be observed that our performance in DCASE 2023 Task 2 is equally outstanding with an official score of 66.97%. The average AUC on the seven machines is lower than Lv [15] by 0.3%, but the pAUC is 1.91% higher than it. The AUC of "Vacuum" is as high as 96.18%, significantly better than other advanced models.

Table 5. Comparison of different models on the DCASE 2022 Task 2 dataset.

	Methods	Baseline [4]	Kevin [23]	Guan [26]	Ibuki [14]	**Ours**
ToyCar	AUC(%)	42.79	90.12	90.90	81.67	**93.00**
	pAUC(%)	53.44	72.10	**78.42**	72.16	76.99
ToyTrain	AUC(%)	51.22	68.98	61.66	55.02	**71.13**
	pAUC(%)	50.98	58.07	55.16	51.78	**60.43**
Fan	AUC(%)	50.34	43.86	55.35	**57.19**	44.11
	pAUC(%)	**55.22**	49.75	53.72	53.02	50.09
Gearbox	AUC(%)	51.34	83.57	81.56	84.51	**84.60**
	pAUC(%)	48.49	60.00	58.08	**65.10**	56.41
Bearing	AUC(%)	58.23	**79.32**	67.23	72.30	78.12
	pAUC(%)	52.16	59.37	52.71	57.90	**60.24**
Slider	AUC(%)	62.42	76.31	75.15	**80.62**	76.64
	pAUC(%)	53.07	56.55	**60.63**	58.09	59.79
Valve	AUC(%)	72.77	87.36	89.29	**91.01**	88.39
	pAUC(%)	65.16	75.81	79.23	**83.23**	76.95
All (hmean)	AUC(%)	54.19	71.78	72.22	72.19	**72.57**
	pAUC(%)	53.67	60.56	60.97	61.46	**61.65**
Official score(%)		54.02	67.61	68.04	68.22	**68.52**

Table 6. Comparison of different models on the DCASE 2023 Task 2 dataset.

Methods		Baseline [6]	Kevin [24]	Jiang [10]	Lv [15]	**Ours**
ToyDrone	AUC(%)	**58.39**	53.90	55.83	54.84	58.03
	pAUC(%)	51.42	50.21	49.74	49.37	**51.58**
ToyNscale	AUC(%)	50.73	87.14	73.44	82.71	**89.03**
	pAUC(%)	50.89	76.58	61.63	57.00	**77.74**
ToyTank	AUC(%)	57.89	63.43	63.03	**74.80**	60.33
	pAUC(%)	53.84	62.21	59.74	**63.79**	61.53
Vacuum	AUC(%)	86.84	83.26	81.98	93.66	**96.18**
	pAUC(%)	65.32	74.00	76.42	**87.42**	85.32
Bandsaw	AUC(%)	69.10	66.06	**71.10**	58.48	65.66
	pAUC(%)	**59.55**	52.87	56.64	50.30	53.35
Grinder	AUC(%)	**72.28**	67.10	62.18	66.69	66.63
	pAUC(%)	62.33	62.11	62.41	61.22	**62.45**
Shaker	AUC(%)	72.77	65.91	**75.99**	74.24	68.08
	pAUC(%)	65.16	50.24	64.68	**65.24**	55.97
All(hmean)	AUC(%)	65.01	67.95	68.03	**70.05**	69.75
	pAUC(%)	57.77	59.58	60.71	60.12	**62.03**
Official score(%)		61.05	64.91	65.40	66.39	**66.97**

5 Conclusion

In this paper, we focus on the attention mechanism for anomaly sound detection and propose a multi-dimensional attention module. This module applies attention to sound features in the dimensions of time, frequency, and channels simultaneously. It focuses on specific frequency bands and semantically relevant time frames within the features that carry discriminative information, thereby reducing the impact of domain shift. Additionally, This module is a lightweight plug-and-play module that improves the performance of various deep CNN-based ASD models. Furthermore, we propose a simple domain generalization method that enhances domain diversity by mixing feature representations to alleviate domain shift. The resulting model can be seen as a generalized anomaly detection model, where all hyperparameter settings remain the same for each machine type, eliminating the need for adjusting hyperparameters based on machine type. We applied our proposed method to the DCASE2023 Task 2 challenge and achieved outstanding results [12], confirming the effectiveness of our approach.

References

1. Breunig, M.M., Kriegel, H.P., Ng, R.T., Sander, J.: LOF: identifying density-based local outliers. In: Proceedings of the 2000 ACM SIGMOD International Conference on Management of Data, pp. 93–104 (2000)

2. Chawla, N.V., Bowyer, K.W., Hall, L.O., Kegelmeyer, W.P.: SMOTE: synthetic minority over-sampling technique. J. Artif. Intell. Res. **16**, 321–357 (2002)
3. Clevert, D.A., Unterthiner, T., Hochreiter, S.: Fast and accurate deep network learning by exponential linear units (ELUs). arXiv preprint arXiv:1511.07289 (2015)
4. Dohi, K., et al.: Description and discussion on DCASE 2022 challenge task 2: unsupervised anomalous sound detection for machine condition monitoring applying domain generalization techniques. arXiv preprint arXiv:2206.05876 (2022)
5. Dohi, K., et al.: MIMII DG: sound dataset for malfunctioning industrial machine investigation and inspection for domain generalization task. arXiv preprint arXiv:2205.13879 (2022)
6. Harada, N., Niizumi, D., Ohishi, Y., Takeuchi, D., Yasuda, M.: First-shot anomaly sound detection for machine condition monitoring: a domain generalization baseline. arXiv preprint arXiv:2303.00455 (2023)
7. Harada, N., Niizumi, D., Takeuchi, D., Ohishi, Y., Yasuda, M., Saito, S.: Toyadmos2: another dataset of miniature-machine operating sounds for anomalous sound detection under domain shift conditions. arXiv preprint arXiv:2106.02369 (2021)
8. Hu, J., Shen, L., Sun, G.: Squeeze-and-excitation networks. In: Proceedings of the IEEE Conference on Computer Vision and Pattern Recognition, pp. 7132–7141 (2018)
9. Inoue, T., et al.: Detection of anomalous sounds for machine condition monitoring using classification confidence. In: DCASE, pp. 66–70 (2020)
10. Jiang, A., et al.: Thuee system for first-shot unsupervised anomalous sound detection for machine condition monitoring. Technical report, DCASE2023 Challenge (2023)
11. Jiang, A., Zhang, W.Q., Deng, Y., Fan, P., Liu, J.: Unsupervised anomaly detection and localization of machine audio: a GAN-based approach. In: 2023 IEEE International Conference on Acoustics, Speech and Signal Processing (ICASSP), ICASSP 2023, pp. 1–5. IEEE (2023)
12. Jie, J.: Anomalous sound detection based on self-supervised learning. Technical report, DCASE2023 Challenge (2023)
13. Kingma, D.P., Ba, J.: Adam: a method for stochastic optimization. arXiv preprint arXiv:1412.6980 (2014)
14. Kuroyanagi, I., Hayashi, T., Takeda, K., Toda, T.: Two-stage anomalous sound detection systems using domain generalization and specialization techniques. Technical report, DCASE2022 Challenge, Technical report (2022)
15. Lv, Z., Han, B., Chen, Z., Qian, Y., Ding, J., Liu, J.: Unsupervised anomalous detection based on unsupervised pretrained models. Technical report, DCASE2023 Challenge (2023)
16. Van der Maaten, L., Hinton, G.: Visualizing data using T-SNE. J. Mach. Learn. Res. **9**(11) (2008)
17. Mu, W., Yin, B., Huang, X., Xu, J., Du, Z.: Environmental sound classification using temporal-frequency attention based convolutional neural network. Sci. Rep. **11**(1), 21552 (2021)
18. Pedregosa, F., et al.: Scikit-learn: machine learning in python. J. Mach. Learn. Res. **12**, 2825–2830 (2011)
19. Reynolds, D.A., et al.: Gaussian mixture models. Encycl. Biometrics **741**(659–663) (2009)
20. Wang, Q., Wu, B., Zhu, P., Li, P., Zuo, W., Hu, Q.: ECA-Net: efficient channel attention for deep convolutional neural networks. In: Proceedings of the IEEE/CVF Conference on Computer Vision and Pattern Recognition, pp. 11534–11542 (2020)

21. Wang, Y., et al.: Unsupervised anomalous sound detection for machine condition monitoring using classification-based methods. Appl. Sci. **11**(23), 11128 (2021)
22. Wilkinghoff, K.: Sub-cluster AdaCos: learning representations for anomalous sound detection. In: 2021 International Joint Conference on Neural Networks (IJCNN), pp. 1–8. IEEE (2021)
23. Wilkinghoff, K.: Design choices for learning embeddings from auxiliary tasks for domain generalization in anomalous sound detection. In: 2023 IEEE International Conference on Acoustics, Speech and Signal Processing (ICASSP), ICASSP 2023, pp. 1–5. IEEE (2023)
24. Wilkinghoff, K.: Fraunhofer FKIE submission for task 2: first-shot unsupervised anomalous sound detection for machine condition monitoring. Technical report, DCASE2023 Challenge (2023)
25. Woo, S., Park, J., Lee, J.Y., Kweon, I.S.: CBAM: convolutional block attention module. In: Proceedings of the European Conference on Computer Vision (ECCV), pp. 3–19 (2018)
26. Xiao, F., Liu, Y., Wei, Y., Guan, J., Zhu, Q., Zheng, T., Han, J.: The dcase2022 challenge task 2 system: Anomalous sound detection with self-supervised attribute classification and GMM-based clustering. Challenge Technical report (2022)

A Corpus of Quotation Element Annotation for Chinese Novels: Construction, Extraction and Application

Jinge Xie, Yuchen Yan, Chen Liu, Yuxiang Jia(✉), and Hongying Zan

School of Computer and Artificial Intelligence, Zhengzhou University,
Zhengzhou, China
{ieyxjia,iehyzan}@zzu.edu.cn

Abstract. Quotations or dialogues are important for literary works, like novels. In the famous Jin Yong's novels, about a half of all sentences contain quotations. Quotation elements like speaker, speech mode, speech cue and the quotation itself are very useful to the analysis of fictional characters. To build models for automatic quotation element extraction, we construct the first quotation corpus with annotation of all the four quotation elements, and the corpus size of 31,922 quotations is one of the largest to our knowledge. Based on the corpus, we compare different models for quotation element extraction and conduct extensive experiments. For the application of extracted quotation elements, we explore character recognition and gender classification, and find out that quotation and speech mode are effective for the two tasks. We will extend our work from Jin Yong's novels to other novels to analyze various characters from different angles based on quotation structures.

Keywords: Quotation Element Extraction · Fictional Character Analysis · Gender Classification · Chinese Novels

1 Introduction

Quotations or dialogues play a vital role in literary works, carrying information encompassing emotions, thoughts, culture, and more about fictional characters [13]. Quotations are pervasive in novels. As shown in Table 1, on average, about a half of all sentences in Jin Yong's novels contain quotations, which is similar to many other literary works [5]. The three Jin Yong's novels we begin with are 射雕英雄传 (The Legend of the Condor Heroes, LCH), 神雕侠侣 (The Return of the Condor Heroes, RCH) and 倚天屠龙记 (Heaven Sword and Dragon Sabre, HSDS). Hence, the extraction of quotations and other elements like the speakers is important for fictional character analysis, sentiment analysis, dialogue corpus construction, and high quality audiobook generation for literary works.

Different from previous studies, which mainly focus on quotation attribution, i.e., identifying the speaker of the quotation, we extract four quotation elements,

B. Luo et al. (Eds.): ICONIP 2023, CCIS 1967, pp. 61–72, 2024.
https://doi.org/10.1007/978-981-99-8178-6_5

Table 1. Statistics of quotations in Jin Yong's novels. Char is short for Chinese character while sent is short for sentence.

Novel	#Char	Quote Char.%	#Sent	Quote Sent.%
射雕英雄传 (LCH)	949,789	32.13%	23,794	45.54%
神雕侠侣 (RCH)	891,366	34.86%	22,739	47.16%
倚天屠龙记 (HSDS)	954,596	38.83%	22,510	46.93%
Total	2,795,751	35.28%	69,043	46.53%

including quotation, speaker, mode and cue, covering "who said what and how". Figure 1 shows an example from 倚天屠龙记 (HSDS). The quotation is enclosed in the pair of quotation marks, the speaker, mode and cue are in the left clauses of the quotation. 周芷若 (Zhou Zhiruo) is the speaker, 道 (say) is a common cue of quotation and the mode 颤声 (trembling) conveys an emotion of worry.

[周芷若]$_{speaker}$跳起身来，[颤声]$_{mode}$[道]$_{cue}$：["谢大侠
仁侠仗义，对咱们后辈更是慈爱，怎会去杀殷姑
娘。"]$_{quotation}$

Fig. 1. An example of quotation elements

We build a quotation element annotation corpus containing 31,922 quotations from three of Jin Yong's novels. For each quotation, we annotate a quadruple of (quotation, speaker, mode, cue). Based on the corpus, we explore different quotation element extraction models and study applications like fictional character recognition and gender classification. The main contributions of this paper are as follows:

- We introduce a new task named quotation element extraction by expanding the traditional quotation attribution task with new elements like mode and cue, which are proved useful for quotation-based character analysis.
- We build the largest Chinese literary quotation element annotation corpus and conduct preliminary experiments on quotation element extraction. We will release the corpus to the public later.
- We study character recognition and gender classification based on quotation elements, and find out that quotation and mode are effective features for those fictional character profiling tasks.

2   Related Work

Previous studies mainly focus on quotation attribution, i.e., speaker identification. Methods can be categorized into rule-based approach [6] and machine learning-based approaches [7,11,12]. Rule-based methods quantify the impact of different features on candidate speakers and assign scores to the candidate speakers based on the occurrence of these features. Machine learning-based methods typically utilize manually labeled training data to build classifiers for identifying the speakers of quotations. However, all of these studies still rely on handcrafted linguistic features, which are heuristically designed and may not comprehensively capture information of the input text.

With the recent advancement in deep learning, Chen et al. [3] propose a neural network-based approach for speaker identification. They treat the speaker identification task as a scoring problem and develop a Candidate Scoring Network (CSN) to compute scores for each candidate. The candidate with the highest score is selected as the final speaker. Recently, Yu et al. [20] introduce an end-to-end approach for quotation attribution.

Corpus construction is the fundamental work for quotation element extraction. Table 2 provides a summary of some existing corpora for quotations. Most of them are on English with different genres and different elements annotated. Most corpora including Chinese are of literary genre, indicating that quotations are very crucial for literary text analysis. Chinese corpora are based on two famous novels, 平凡的世界 (World of Plainness) [2] and 射雕英雄传 (LCH) [9].

Table 2. A summary of some prominent existing corpora for quotations

Corpus	Language	#Quote	Genre	Element
CQSAC [5]	English	3,176	Literature	speaker
PARC3 [15]	English	19,712	Newspaper	speaker, cue
ACDS [10]	English	1,245	Bible	speaker, cue, listener
RiQuA [14]	English	5,963	Literature	speaker, cue, listener
PDNC [19]	English	35,978	Literature	speaker, listener
平凡的世界 [2]	Chinese	2,548	Literature	speaker
射雕英雄传 [9]	Chinese	9,721	Literature	speaker

3   Corpus Construction

Jin Yong's novels are among the most influential literary works in the Chinese-speaking world. They have shaped numerous iconic characters and hold significant literary value. Therefore, we choose the three novels of 射雕三部曲 (Condor

Trilogy) as the texts for annotation. Based on our observation of Jin Yong's novels, we only consider the current quotation and the sentence before the colon, which cover nearly all the quotation elements.

3.1 Annotation Schema

As shown by the example in Fig. 1, we annotate four elements for each sentence, quotation, speaker, mode and cue. We introduce each type of element in detail as follows.

Quotation. Quotations are defined as linguistic units that provide the content of speech events. For direct quotations, the span of a quotation is indicated by a pair of quotation marks.

Speaker. Speakers are characters that utter quotations. Since we annotate at the mention level, the speaker element can be classified into five distinct categories, person name, pronoun, common noun, multiple persons and implicit, as illustrated in Table 3, where person name is the dominant category and implicit means no character is mentioned as the speaker in the context.

Table 3. Speaker categories

Category	Example	Proportion
Person name	杨过$_{speaker}$叫声：“啊哟。”（Yang Guo$_{speaker}$ shouts: "ah yo."）	84.27%
Pronoun	她$_{speaker}$连运三下劲，始终无法取过短剑，说道：“好啊，你是显功夫来着。”（She$_{speaker}$ exerts three consecutive bursts of strength but can't take away the short sword, says: "alright, you're quite skilled."）	3.50%
Common noun	那僧人$_{speaker}$道：“方丈请首座去商议。”（The monk$_{speaker}$ says: "abbot, please go to discuss it with the head."）	4.71%
Multiple persons	纪晓芙$_{speaker}$和张无忌$_{speaker}$齐声道：“还请保重，多劝劝师母。”（Ji Xiaofu$_{speaker}$ and Zhang Wuji$_{speaker}$ say in unison: "please take care and persuade our master mother."）	0.26%
Implicit	“贫尼法名静空。各位可见到我师傅吗？”（"My monastic name is Jing Kong. Have any of you seen my master?"）	7.26%

Mode. The mode element primarily indicates the manner of speech, i.e., how the speaker expresses the quotation in terms of emotion, tone, or intonation, which is particularly useful for emotion analysis of the quotation. Table 4 shows some examples of mode element of quotation. The mode element may not exist for a quotation.

Table 4. Examples of mode element

Example	Mode
双方渐渐行近，一名尼姑尖声_{mode}叫道："是魔教的恶贼。"(They approach each other, and a nun exclaims sharply_{mode}: "It's the evil thief from the Demonic Sect.")	尖声/sharply
张无忌大吃一惊_{mode}，道："是殷六侠？受伤了么？" (Zhang Wuji is greatly surprised_{mode} and says: "Is it Uncle Yin? Are you injured?")	大吃一惊/greatly surprised
杨过喜_{mode}道："你当真带我去？"(Yang Guo happily_{mode} says: "Are you really taking me there?")	喜/happily

Cue. The cue element primarily expresses or implies the occurrence of a verbal event. In Jin Yong's novels, common cue words include verbs such as 道 (say), 说道 (say), 大叫 (shout), etc. The cue word can be omitted for a quotation.

3.2 Corpus Analysis

We take a multi-round iterative annotation scheme. One annotator performs initial annotation, another verifies it, and the third annotator reviews and solves disagreement through discussions between annotators. The inter-annotator agreement is evaluated using F1 score [8], calculated separately for the four elements as well as for the overall annotations, as shown in Table 5. All F1 scores are high with a slightly lower value for mode element, indicating the reliability of the constructed corpus [1] and the difficulty of mode annotation.

Table 5. Inter-annotator agreement

Quotation	Speaker	Mode	Cue	Overall
100%	99.21%	94.06%	99.07%	98.02%

The overall statistics of our corpus are shown in Table 6. Based on three of Jin Yong's novels, 31,922 quotations are annotated, of which 92.74% have speakers, 50.13% contain mode expressions, and 95.21% contain cue words.

Table 6. Statistics of the corpus

Novel	#Quote	With speaker	With mode	With cue
射雕英雄传 (LCH)	10,752	9,999/92.99%	5,244/48.77%	10,293/95.73%
神雕侠侣 (RCH)	10,701	9,820/91.77%	5,639/52.70%	10,097/94.36%
倚天屠龙记 (HSDS)	10,469	9,784/93.46%	5,120/48.91%	10,002/95.54%
Total	31,922	29,603/92.74%	16,003/50.13%	30,392/95.21%

Table 7. F1 score of intra-novel quotation element extraction(%)

Novel	Model	Speaker	Mode	Cue	Quotation	Overall
射雕英雄传 (LCH)	BiLSTM-CRF	95.21	82.40	96.45	**99.38**	94.19
	BERT	96.62	83.02	96.71	99.15	95.18
	BERT-CRF	96.81	83.31	97.14	98.94	95.31
	BERT-BiLSTM-CRF	**96.85**	**84.10**	**97.31**	99.24	**95.61**
神雕侠侣 (RCH)	BiLSTM-CRF	71.82	74.43	94.57	96.74	86.42
	BERT	95.34	85.03	97.79	99.42	95.43
	BERT-CRF	96.04	85.95	97.42	99.35	95.65
	BERT-BiLSTM-CRF	**96.11**	**87.22**	**97.49**	**99.53**	**95.99**
倚天屠龙记 (HSDS)	BiLSTM-CRF	90.74	80.51	96.52	**99.46**	93.15
	BERT	95.46	84.18	96.67	98.85	94.82
	BERT-CRF	**95.51**	**85.54**	96.90	98.68	94.92
	BERT-BiLSTM-CRF	95.50	84.68	**96.97**	99.30	**95.14**

4 Quotation Element Extraction

We take quotation element extraction as a token-labeling task, labeling each Chinese character as one of the 13 labels, including B-Quotation, I-Quotation, S-Quotation, B-Speaker, I-Speaker, S-Speaker, B-Mode, I-Mode, S-Mode, B-Cue, I-Cue, S-Cue, and O, respectively representing the beginning, internal, single character of each element, and others. Based on our constructed corpus, we investigate four models for quotation element extraction, namely BiLSTM-CRF, BERT [4], BERT-CRF and BERT-BiLSTM-CRF.

The hidden size of the BiLSTM encoder is 384 and the dropout rate is 0.5. To leverage the advantages of pre-trained model, we fine-tune the BERT-base-Chinese model to encode each sentence. The model is trained for 50 epochs and the batch size is set to 8. We employ AdamW as the optimizer, and set the learning rate to 1e-5 and weight decay to 1e-2. We randomly divide the dataset into training, validation, and test sets with ratio of 3:1:1. F1 score is used as the evaluation metric. The intra-novel experimental results are presented in Table 7, with the best results in bold. As can be seen, the introduction of the pre-trained model BERT significantly improves the performance. The extraction of mode element is more difficult than other elements.

To validate the models' performance in cross-novel setting, we take the same training set of 射雕英雄传 (LCH) and the same test set of 倚天屠龙记 (HSDS) as in Table 7. The experimental results are presented in Table 8, with the best results in bold. As can be seen, BERT-based models achieve competitive results nearly on par with those in intra-novel setting. However, the performance declines sharply for BiLSTM-CRF model, especially in speaker and mode extraction, which reflects the good domain adaptability brought by pre-trained models like BERT.

Table 8. F1 score of cross-novel quotation element extraction(%)

Model	Speaker	Mode	Cue	Quotation	Overall
BiLSTM-CRF	23.09	51.80	92.98	97.60	73.79
BERT	94.44	82.31	96.52	99.01	94.24
BERT-CRF	**95.45**	83.15	**96.64**	98.87	94.63
BERT-BiLSTM-CRF	95.09	**83.71**	96.31	**99.23**	**94.67**

Table 9 presents two cases of quotation element extraction. In the first case, the quotation is spoken by 郭襄 (Guo Xiang) to 何足道 (He Zudao) and the actual speaker is 郭襄 (Guo Xiang), not 何足道 (He Zudao). In the second case, 爽朗嘹亮 (bright and clear) is an idiom that describes a sound, but the actual mode for the quotation is 一怔之下 (startled). In both cases, the BiLSTM-CRF model wrongly recognizes both 郭襄 (Guo Xiang) and 何足道 (He Zudao) as the speaker, and both 爽朗嘹亮 (bright and clear) and 一怔之下 (startled) as mode of quotation. On the contrary, the BERT-BiLSTM-CRF model correctly recognizes all quotation elements and discriminates confusing candidates.

Table 9. Case study of quotation element extraction

Input	郭襄向何足道一笑，心道："你这张嘴倒会说话，居然片言折服老和尚。"(Guo Xiang smiles at He Zudao and thinks: "you have a silver tongue. Surprisingly, with just a few words, you manage to impress the old monk.")
Ground Truth	郭襄$_{speaker}$向何足道一笑$_{mode}$，心道$_{cue}$："你这张嘴倒会说话，居然片言折服老和尚。"
BiLSTM-CRF	郭襄$_{speaker}$向何足道$_{speaker}$一笑$_{mode}$，心道$_{cue}$："你这张嘴倒会说话，居然片言折服老和尚。" ✘
BERT-BiLSTM-CRF	郭襄$_{speaker}$向何足道一笑$_{mode}$，心道$_{cue}$："你这张嘴倒会说话，居然片言折服老和尚。" ✔
Input	张翠山听得那爽朗嘹亮的嗓音很耳熟，一怔之下，叫道："是俞莲舟俞师哥么？"(Hearing the bright and clear voice, Zhang Cuishan feels a sense of familiarity. Startled, he calls out: "Is it senior brother Yu Lianzhou?")
Ground Truth	张翠山$_{speaker}$听得那爽朗嘹亮的嗓音很耳熟，一怔之下$_{mode}$，叫道$_{cue}$："是俞莲舟俞师哥么？"
BiLSTM-CRF	张翠山$_{speaker}$听得那爽朗嘹亮$_{mode}$的嗓音很耳熟，一怔之下$_{mode}$，叫道$_{cue}$："是俞莲舟俞师哥么？" ✘
BERT-BiLSTM-CRF	张翠山$_{speaker}$听得那爽朗嘹亮的嗓音很耳熟，一怔之下$_{mode}$，叫道$_{cue}$："是俞莲舟俞师哥么？" ✔

5   Applications of Quotation Elements

Applications of quotation elements include fictional character recognition, gender classification, personality prediction, etc. We explore character recognition and gender classification based on quotation elements in this paper.

For the two tasks, we employ models based on BERT [4] and ERNIE [17]. We design two experimental settings for both tasks: single-quotation and multi-quotation. For the single-quotation setting, we assign a category to each quotation. For the multi-quotation setting, we consider a set of five quotations as a whole belonging to a specific character or gender and input them together for classification.

The experimental parameters are set as follows. For single-quotation setting, the batch size is set to 8 and the padding size is set to 64, while for multi-quotation setting, the batch size is set to 4 and the padding size is set to 256. For both settings, the learning rate is set to 1e-5, the optimizer chosen is Adam, and each model is trained for 30 epochs.

5.1   Character Recognition Based on Quotation Elements

Previous studies [16,18] employ only quotation information for character recognition. We introduce mode information by utilizing both quotation and its corresponding mode to identify the speaker. For each novel, we select the top 10 characters with the most quotations. Taking 神雕侠侣 (RCH) as an example, 杨过(Yang Guo) and 小龙女(Xiaolongnv) have significantly more quotations than other characters. This class imbalance can lead to uneven feature distribution. Therefore, we set a quotation threshold of 500, and when the number of quotation exceeds this threshold, we randomly select 500 quotations to build the training data for that character. Thus, in 神雕侠侣 (RCH), the distribution of quotation counts of top characters is outlined as follows: 杨过 (Yang Guo) 500, 小龙女 (Xiaolongnv) 500, 黄蓉 (Huang Rong) 500, 郭靖 (Guo Jing) 416, 郭芙 (Guo Fu) 350, etc. The dataset is then randomly divided into training, validation, and test sets with a ratio of 8:1:1. The evaluation metric is accuracy.

The character recognition results are shown in Table 10, with the best results in bold. The baseline model predicts the speaker as the dominant character in the training set. We can see that the introduction of mode element improves the performance significantly, especially for multi-quotation setting. It shows that models can achieve relative high accuracy to recognize the speaker based on five quotations. Even only based on a single quotation, the performance is much better than that of the baseline model.

To better illustrate salient features of different characters, we use TF-IDF to generate word clouds. Figure 2 shows word clouds of 杨过 (Yang Guo) and 小龙女 (Xiaolongnnv). It can be observed that for the quotation feature, the key words for 杨过 (Yang Guo) are 姑姑 (aunt), 郭伯伯 (Uncle Guo), etc., while the key words for 小龙女 (Xiaolongnv) are 过儿 (Guoer), 婆婆 (grandma), etc. In terms of the mode feature, the main modes of 杨过 (Yang Guo) include 朗声

Table 10. Accuracy of character recognition (%)

Model	射雕英雄传(LCH)		神雕侠侣(RCH)		倚天屠龙记(HSDS)	
	single	multi	single	multi	single	multi
Baseline	15.53	15.33	13.77	13.63	16.23	16.00
ERNIE	35.00	57.35	38.95	66.23	45.10	61.67
BERT	33.54	62.69	38.12	66.23	51.14	61.54
ERNIE-mode	35.94	63.89	**39.5**	**70.13**	46.73	73.85
BERT-mode	**37.03**	**64.18**	39.06	68.83	**52.13**	**83.08**

(loud), 大声 (loud) and 喝 (shout) while the main modes of 小龙女 (Xiaolongnv) include 微笑 (smile), 微微一笑 (smile) and 柔声 (gentle).

5.2 Gender Classification Based on Quotation Elements

The goal of gender classification is to determine whether the speaker of a given quotation is male or female. The gender classification dataset used in this study consists of the top 20 characters from each of the three novels. After removing quotations without mode element, we get a total of 9,564 quotations, where male is the gender of majority. The dataset is randomly divided into training, validation, and test sets with a ratio of 8:1:1. Accuracy is used as the evaluation metric.

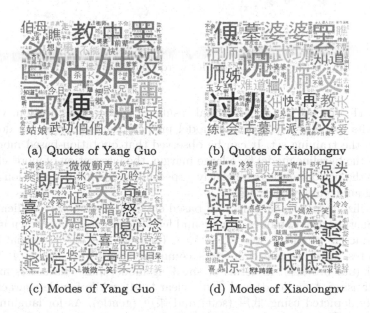

(a) Quotes of Yang Guo   (b) Quotes of Xiaolongnv

(c) Modes of Yang Guo   (d) Modes of Xiaolongnv

Fig. 2. Comparison of character word clouds based on TF-IDF

Table 11. Accuracy of gender classification (%)

Model	single-quotation	multi-quotation
Baseline	61.83	61.81
ERNIE	69.56	82.99
BERT	69.53	87.05
ERNIE-mode	72.38	88.54
BERT-mode	**72.77**	**92.15**

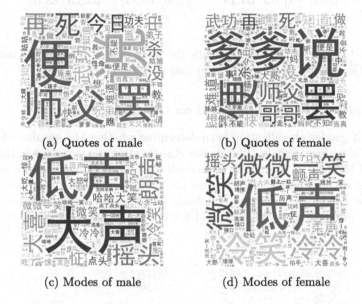

(a) Quotes of male   (b) Quotes of female

(c) Modes of male   (d) Modes of female

Fig. 3. Comparison of gender word clouds based on TF-IDF

Table 11 presents the experimental results of gender classification, with the best results in bold. The baseline model predicts the gender as the dominant gender in the training set. It can be observed that quotation-based model outperforms the baseline model by a large margin, the integration of mode elements improves the performance significantly, especially in the multi-quotation setting, achieving an accuracy of 92.15%.

We utilize word clouds generated based on TF-IDF to observe salient words used in quotations and modes of male and female characters. As shown in Fig. 3, male characters often use words like 师父 (master) and 武功 (martial arts) in their quotes, while female characters commonly use words like 爹爹 (father) and 哥哥 (old brother). In terms of speaking mode, male characters are often described using 大声 (loud) and 朗声 (clear loud), while female characters are commonly depicted using 低声 (soft) and 柔声 (gentle). As for laughing style, words like 大喜 (great joy) and 哈哈大笑 (laugh out loud) are commonly used for male characters, while 微笑 (smile) and 微微一笑 (a faint smile) are often used for female characters.

6 Conclusion

This paper introduces a new task named quotation element extraction by expanding the traditional quotation attribution task with new elements like mode and cue, and builds a rich annotated quotation corpus of Chinese novels. We compare different models for quotation element extraction and conduct cross-novel extraction experiments. Additionally, we apply quotation element extraction to character recognition and gender classification, demonstrating its effectiveness for literary text analysis.

In the future, to investigate the generalization of the quotation element extraction task, we will expand our corpus to other novels, especially online novels. We will further explore fictional character profiling tasks based on extracted quotation elements, e.g., emotion classification and personality classification.

References

1. Artstein, R., Poesio, M.: Inter-coder agreement for computational linguistics. Comput. Linguist. **34**(4), 555–596 (2008)
2. Chen, J.X., Ling, Z.H., Dai, L.R.: A Chinese dataset for identifying speakers in novels. In: INTERSPEECH, Graz, Austria, pp. 1561–1565 (2019)
3. Chen, Y., Ling, Z.H., Liu, Q.F.: A neural-network-based approach to identifying speakers in novels. In: Interspeech, pp. 4114–4118 (2021)
4. Devlin, J., Chang, M.W., Lee, K., Toutanova, K.: Bert: pre-training of deep bidirectional transformers for language understanding. arXiv preprint arXiv:1810.04805 (2018)
5. Elson, D., McKeown, K.: Automatic attribution of quoted speech in literary narrative. In: Proceedings of the AAAI Conference on Artificial Intelligence, vol. 24, pp. 1013–1019 (2010)
6. Glass, K., Bangay, S.: A naive salience-based method for speaker identification in fiction books. In: Proceedings of the 18th Annual Symposium of the Pattern Recognition Association of South Africa (PRASA 2007), pp. 1–6. Citeseer (2007)
7. He, H., Barbosa, D., Kondrak, G.: Identification of speakers in novels. In: Proceedings of the 51st Annual Meeting of the Association for Computational Linguistics (Volume 1: Long Papers), pp. 1312–1320 (2013)
8. Hripcsak, G., Rothschild, A.S.: Agreement, the f-measure, and reliability in information retrieval. J. Am. Med. Inform. Assoc. **12**(3), 296–298 (2005)
9. Jia, Y., Dou, H., Cao, S., Zan, H.: Speaker identification and its application to social network construction for Chinese novels. Inter. J. Asian Lang. Process. **30**(04), 2050018 (2020)
10. Lee, J.S., Yeung, C.Y.: An annotated corpus of direct speech. In: Proceedings of the Tenth International Conference on Language Resources and Evaluation (LREC 2016), pp. 1059–1063 (2016)

11. Muzny, G., Fang, M., Chang, A., Jurafsky, D.: A two-stage sieve approach for quote attribution. In: Proceedings of the 15th Conference of the European Chapter of the Association for Computational Linguistics: Volume 1, Long Papers., pp. 460–470 (2017)

12. O' Keefe, T., Pareti, S., Curran, J.R., Koprinska, I., Honnibal, M.: A sequence labelling approach to quote attribution. In: Proceedings of the 2012 Joint Conference on Empirical Methods in Natural Language Processing and Computational Natural Language Learning, pp. 790–799 (2012)

13. Page, N.: Speech in the English novel. Springer (1988). https://doi.org/10.1007/978-1-349-19047-8

14. Papay, S., Padó, S.: Riqua: a corpus of rich quotation annotation for english literary text. In: Proceedings of the Twelfth Language Resources and Evaluation Conference, pp. 835–841 (2020)

15. Pareti, S.: Parc 3.0: a corpus of attribution relations. In: Proceedings of the Tenth International Conference on Language Resources and Evaluation (LREC 2016), pp. 3914–3920 (2016)

16. Schmerling, E.: Whose line is it?-quote attribution through recurrent neural networks

17. Sun, Y., et al.: Ernie: enhanced representation through knowledge integration. arXiv preprint arXiv:1904.09223 (2019)

18. Vishnubhotla, K., Hammond, A., Hirst, G.: Are fictional voices distinguishable? classifying character voices in modern drama. In: Proceedings of the 3rd Joint SIGHUM Workshop on Computational Linguistics for Cultural Heritage, Social Sciences, Humanities and Literature, pp. 29–34 (2019)

19. Vishnubhotla, K., Hammond, A., Hirst, G.: The project dialogism novel corpus: a dataset for quotation attribution in literary texts. arXiv preprint arXiv:2204.05836 (2022)

20. Yu, D., Zhou, B., Yu, D.: End-to-end chinese speaker identification. In: Proceedings of the 2022 Conference of the North American Chapter of the Association for Computational Linguistics: Human Language Technologies, pp. 2274–2285 (2022)

Decoupling Style from Contents
for Positive Text Reframing

Sheng Xu [ID], Yoshimi Suzuki [ID], Jiyi Li [ID], and Fumiyo Fukumoto[✉] [ID]

Integrated Graduate School of Medicine, Engineering, and Agricultural Sciences,
University of Yamanashi, Kofu, Japan
{g22dts03,ysuzuki,jyli,fukumoto}@yamanashi.ac.jp

Abstract. The positive text reframing (PTR) task, where the goal is
to generate a text that gives a positive perspective to a reader while
preserving the original sense of the input text, has attracted considerable
attention as one of the natural language generation (NLG). In the PTR
task, large annotated pairs of datasets are not available and would be
expensive and time-consuming to create. Therefore, how to interpret a
diversity of contexts and generate a positive perspective from a small size
of the training dataset is still an open problem. In this work, we propose
a simple but effective Framework for Decoupling the sentiment Style
from the Contents of the text (FDSC) for the PTR task. Different from
the previous work on the PTR task that utilizes Pre-trained Language
Models (PLM) to directly fine-tune the task-specific labeled dataset such
as Positive Psychology Frames (PPF), our FDSC fine-tunes the model
for the input sequence with two special symbols to decouple style from
the contents. We apply contrastive learning to enhance the model that
learns a more robust contextual representation. The experimental results
on the PPF dataset, show that our approach outperforms baselines by
fine-turning two popular Seq2Seq PLMs, BART and T5, and can achieve
better text reframing. Our codes are available online (https://github.
com/codesedoc/FDSC).

Keywords: Natural Language Generation · Positive Text Reframing ·
Seq2Seq Model · Contrastive Learning

1 Introduction

Text style transfer (TST) has been first attempted as the frame language-based
systems by McDonald et al. [11] and schema-based Natural Language Generation
by Hovy [3] in the 1980s, TST research has been one of the major research topics
and has been explored for decades. The goal of the TST task is to generate a
new text by changing the linguistic style, such as formality and politeness while
keeping the main semantic contents of the given text. With the success of deep
learning (DL) techniques, many authors have attempted to apply DL to the
task and have successfully gained significant performance [25]. To date, there

© The Author(s), under exclusive license to Springer Nature Singapore Pte Ltd. 2024
B. Luo et al. (Eds.): ICONIP 2023, CCIS 1967, pp. 73–84, 2024.
https://doi.org/10.1007/978-981-99-8178-6_6

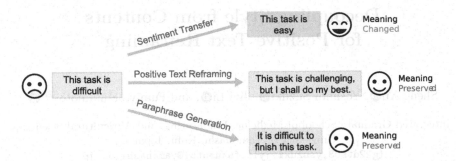

Fig. 1. The task of Sentiment Transfer, Positive Text Reframing, and Paraphrase Generation. The meaning of the input text is preserved in the PTR and PG tasks, while ST generates new text by changing the meaning of the input text.

has been very little work on the positive text reframing (PTR) task. To address the issue, Ziems et al. first created a parallel dataset, positive psychology frames (PPF), and serve as a benchmark dataset that will enable sustained work on the PTR task [26]. They experimented with many of the leading PLMs including BART [6], and T5 [14], i.e., directly applied these models for the PTR task, and showed that these models can learn to shift from a negative to a more positive perspective, while they reported that these models still struggle to generate reasonable positive perspectives.

The PTR task requires both the Sentiment Transfer (ST) and Paraphrase Generation (PG) tasks, i.e., transferring the sentiment of the source text, and more on the linguistic diversity of generation while preserving the meaning of the source text. The PTR is a sub-branch of Sentiment Style Transfer (SST) in terms of sentiment polarity. As mentioned by Ziems et al. [26], the target of PTR is to neutralize the given negative point of view and generate a more positive perspective without contradicting the original meaning, while the SST takes care more of the polarity of the sentence, and changes the meaning of source sentence with reversing its sentiment. For example, in Fig. 1, the target sentence by PTR, "This task is challenging but I shall do my best." shows a positive perspective while preserving the same sense of the input sentence, "This task is difficult" It also shows high linguistic diversity compared with the input sentence as they have large syntactic/lexical differences.

The sentiment is a content-related attribute of text, and the requirement of transferring the sentiment of the given input for the PTR task is a fine-grained transfer. Many of the attempts on the TST task and even less work on the PTR task are either too complex or unable to handle subtle transitions in style. Even if ignoring the cost and leveraging the large language models (LLMs) in text generation tasks, the outputs are abstractive and still limited for generating diverse contexts.

In this paper, we propose a simple but effective **F**ramework to **D**ecouple **S**tyle from **C**ontents (FDSC) for the PTR task. The method decomposes an input sentence into the sentiment attribute and primary content that refers to the factual information and transfers the sentiment feature in the hidden space while preserving the semantic content of the input sentence. We leverage contrastive learning to disentangle sentiment features from the content and learn more fine-grained and robust representations of sequences for preserving semantic meaning. The motivation for such a strategy is based on the assumption that the PTR is relatively more fine-grained than the common SST and other TST tasks. The experimental results by using different PLM backbones by BART [6] and T5 [14] show that our framework improves the baseline results obtained by directly fine-tuning the models. The ablation study also shows that the module on decoupling style from the content improves performance, especially contributing to the fluency of given input. The contributions of our work can be summarized:

1. We propose a simple but effective PTR framework to decompose an input sentence into the sentiment style and primary content, instead of directly fine-tuning the model.
2. We utilize contrastive learning to enhance the model to learn a more fine-grained and robust contextual representation to preserve the original meaning of the given input sentence.
3. The results of experiments show the efficacy of our proposed approach, especially the ablation study indicates that the module of decoupling style from the content contributes to the fluency of output sentences.

2   Related Works

2.1   Text Style Transfer

With the success of deep learning (DL) techniques, TST has been extensively explored recently, as many authors have attempted to apply DL to NLG including the TST tasks. One attempt is the standard seq2seq models based on the supervised learning paradigm which can be directly applied to parallel datasets for TST. Xu et al. [22] utilized a multi-task learning-based approach that combines three loss functions for each requirement for the TST task. The methods by using a data augmentation strategy were also proposed for the scenarios in which the large scale of human-annotated corpora are unavailable. For example, Rao et al. [16] created a pseudo parallel dataset by using the back translation recipe from the Yahoo Answers L6 corpus[1].

For the non-parallel corpora, the majority of approaches for TST are based on an unsupervised learning paradigm. One direction is based on the disentanglement strategy, by removing the source style feature during the encoding phase, and the target style is injected into a decoding phase in order to keep the content

---

[1] https://webscope.sandbox.yahoo.com/catalog.php?datatype=l.

and transfer the style of a given text such as [2], and [9]. Another noteworthy strategy is the prototype-based editing approach by Li et al. [7]. The approach comprises three phases for the TST task: **deleting**, i.e., the markers of source style are deleted, **retrieving**, the candidates with target style are retrieved, and **generating**, the rest part of the text is infilled and completed by using the proper candidates from the second phase to make the generation fluent.

Many of these methods attained significant performance on the TST task while they still fail to handle the fine-grained transfer, i.e., transferring the sentiment of the source text, and more on the linguistic diversity of generation with preserving the meaning of the source text that is required for the PTR task.

2.2   Pre-trained Language Models

The pre-trained language model (PLM) based learning paradigms have a long history in the field of Natural Language Processing (NLP). With the explosive increase of data scale and computational efficiency, large language models (LLMs) such as LLaMA [20], recently gained breakthroughs, and are adopted for a wide range of downstream tasks such as chain-of-thought prompting for enhancing reasoning in language models [21] and the text style transfer [17].

The major attempt is to utilize PLMs to directly fine-tune the task-specific labeled dataset. However, it is challenging to apply the PTR task as it requires fine-grained sentiment transfer with more on the linguistic diversity of generation and preservation of the sense of the source sentence. Inspired by the previous mainstream approaches, we utilize the pre-trained transformers as a backbone and fine-tune the model so that it decouples style from the contents, transfers style, and preserves semantic content for PTR.

2.3   Contrastive Learning

Contrastive learning (CL) which is one of the common learning strategies is originally proposed in Computer Vision (CV) research field and promotes massive progress on the CV to the outstanding attempt for visual representation [1]. Because of its applicability and effectiveness, CL has also been applied in NLP as one of the fundamental techniques. As much of the previous works including [19] have been reported, CL can work well for mining better text representations in different contexts and domains. For the text generation task, CL has also been utilized with different tricks. Lee et al. [5] leverages CL to train a seq2seq model which maximizes the similarity between ground truth input and output sequence and minimizes the similarity between the input and an incorrect output sequence. Su et al. [18], have also injected the CL objective into the encoder, and utilized a contrastive search method to assist the decoder for text generation tasks.

To preserve the primary content that refers to the factual information of the input text, we also use the CL technique to train the model so that it preserves the semantic contexts of the input text, while transferring the sentiment attribute of the input text.

3 Method

Given a sequence/sentence with a negative sentiment, the goal of PTR is to reframe the source input into a target sequence/sentence with a relatively positive perspective while preserving the original sense of the given input. For this task, we propose an approach, FDSC, by leveraging PLM as the backbone model. As illustrated in Fig. 2, FDSC consists of three modules, (i) sequence-to-sequence (Seq2Seq) text generation, (ii) decoupling style from contents, and (iii) preserving primary content. These components are simultaneously trained by adopting loss functions.

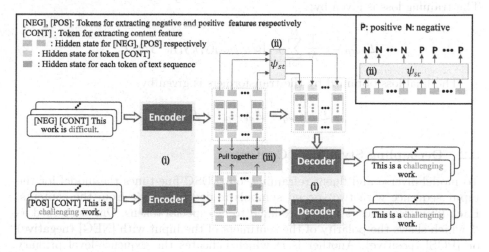

Fig. 2. The architecture and data flow of FDSC which consists of (i) sequence-to-sequence text generation, (ii) decomposing style from contents, and (iii) preserving primary content.

3.1 Seq2Seq Text Generation

Ziems et al. [26] mentioned that the base requirement for PTR is that the model is diverse and fluent in reframing while preserving the original sense of the given sentence with positive perspectives. Following their work, we propose a simple but effective PTR approach, fine-turning the pre-trained transformers for annotated data. As shown in (i) of Fig. 2, we choose transformers consisting of an encoder-decoder structure as the backbone of FDSC and directly control the model to decouple sentiment from the primary content of the input sequence. Let $\mathbf{x}$ be a source sequence and $\mathbf{y}$ be its annotated target, the parallel dataset with N samples can be formulated as $D = \{(\mathbf{x}_i, \mathbf{y}_i)|1 \leqslant i \leqslant N, i \in \mathbb{Z}\}$. The set of all $\mathbf{x}$, and $\mathbf{y}$ in D can be labeled as X and Y respectively, therefore, the population of sequences in the dataset is represented as $S = X \bigcup Y$. Let the prediction of the transformer referring to $\mathcal{G}_{seq2seq}(\mathbf{s})$, $\mathbf{s} \in S$. The cross entropy loss between $\mathbf{y}$ and

$\mathcal{G}_{seq2seq}(\mathbf{x})$ is given by $f_{ce}[\mathcal{G}_{seq2seq}(\mathbf{x}), \mathbf{y}]$. The objective of the backbone during fine-turning is to minimize the following training loss:

$$\mathcal{L}_{neg2pos} = -\frac{1}{N} \sum_{i=1}^{N} f_{ce}[\mathcal{G}_{seq2seq}(\mathbf{x}_i), \mathbf{y}_i)], \quad (\mathbf{x}_i, \mathbf{y}_i) \in D. \tag{1}$$

The transfer of sentiment attribute is unidirectional in PTR, which exclusively changes the polarity of the sentence from negative to relatively positive. To make the model robust, and leverage the manual annotated parallel corpus, for each target reference $\mathbf{y}_i$ of each $\mathbf{x}_i$, $\mathbf{y}_i$ is also pushed into the backbone of FDSC and reconstructs itself which is similar to the auto-encoding learning strategy. The training loss is given by:

$$\mathcal{L}_{auto} = -\frac{1}{N} \sum_{i=1}^{N} f_{ce}[\mathcal{G}_{seq2seq}(\mathbf{y}_i), \mathbf{y}_i)], \quad \mathbf{y}_i \in Y. \tag{2}$$

The final loss obtained by the transformers is given by:

$$\mathcal{L}_{seq2seq} = \mathcal{L}_{neg2pos} + \mathcal{L}_{auto}. \tag{3}$$

3.2   Decoupling Style from Contents

To model diverse and fluent reframing, our FDSC fine-tunes the model for the input sequence with two special symbols to decouple style from the contents. Each input sequence, $s \in S$, is pre-fixed to two special tokens. One is marked as t_s^s which shows the polarity of the sentiment of the input with [NEG] (negative) or [POS] (positive). Another is t_c^s which indicates the sequence-level primary content with [CONT].

Let $(t_1^s, t_2^s, ..., t_n^s)$ be the tokenized sequence of s with the length of n. We create the entire input sequence of the model, $\mathbf{t}^s = (t_s^s, t_c^s, t_1^s, t_2^s, ..., t_n^s)$. For a given $\mathbf{t}^s$, the hidden state output from the encoder is indicated as $\mathbf{H}^s = (\mathbf{h}_s^s, \mathbf{h}_c^s, \mathbf{h}_1^s, \mathbf{h}_2^s, ..., \mathbf{h}_n^s) \in \mathbb{R}^{n \times d}$, where d denotes the dimensions for each token. We utilized $\psi_{sc} : \mathbb{R}^d \to (0, 1)$ which is illustrated in (ii) of the right-hand side of Fig. 2 so that $\mathbf{h}_s^s$ is capable to represent the sentiment feature. Let $\psi_{sc}(s) \in (0, 1)$ be the probability of positive polarity, and the binary cross entropy is chosen as the loss function. Thus, the training objective for decoupling sentiment style is to minimize the following equation:

$$\mathcal{L}_{sc} = -\frac{1}{2N} \sum_{i=1}^{2N} \{\mathbb{1}(s) \cdot \log[\psi_{sc}(s)] + [1 - \mathbb{1}(s)] \cdot \log[(1 - \psi_{sc}(s)]\}, \quad s \in S, \tag{4}$$

$$\mathbb{1}(s) = \begin{cases} 0, & \text{if } s \in X, \\ 1, & \text{else if } s \in Y. \end{cases}$$

Assume that the hidden states of the sentiment which belong to the same polarity group are close to each other in the hidden space. We thus add a sentiment transfer module $\psi_{st} : \mathbb{R}^d \to \mathbb{R}^d$ to FDSC to transfer the negative polarity of the input into the positive one which is illustrated in (ii) of the left-hand side of Fig. 2. The loss for the sentiment style transfer is given by:

$$\mathcal{L}_{st} = -\frac{1}{N} \sum_{i=1}^{N} |\psi_{st}(\mathbf{h}_s^{\mathbf{x}_i}) - \mathbf{h}_s^{\mathbf{y}_i}|^2, \quad (\mathbf{x}_i, \mathbf{y}_i) \in D, \tag{5}$$

where $\mathbf{h}_s^{\mathbf{x}_i}$ refers to the hidden state of the sentiment token in $\mathbf{x}_i$, and $\mathbf{h}_s^{\mathbf{y}_i}$ indicates the hidden state of the counterpart in $\mathbf{y}_i$. The final loss for decoupling style from the content and transferring it to positive polarity can be given by:

$$\mathcal{L}_{dsc} = \mathcal{L}_{sc} + \mathcal{L}_{st}. \tag{6}$$

3.3 Preserving Primary Content

The primary content shares the meaning of the factual information obtained by each pair of $\mathbf{x}$ and $\mathbf{y}$. Therefore, their representations, $\mathbf{h}_c^{\mathbf{x}}$ and $\mathbf{h}_c^{\mathbf{y}}$, should be close to each other in the hidden space. Let $B_x = \{\mathbf{h}_c^{\mathbf{x}_1}, \mathbf{h}_c^{\mathbf{x}_2}, ..., \mathbf{h}_c^{\mathbf{x}_b}\}$ be a mini-batch with the size of b during fine-turning. Every two hidden states from an arbitrary two input sequences, $\mathbf{h}_c^{\mathbf{x}_i}$ and $\mathbf{h}_c^{\mathbf{x}_j}$ ($i \neq j$) should be apart from each other as they obviously represent different content. Likewise, the parallel batch $B_y = \{\mathbf{h}_c^{\mathbf{y}_1}, \mathbf{h}_c^{\mathbf{y}_2}, ..., \mathbf{h}_c^{\mathbf{y}_b}\}$ consisting of the counterpart of each element of $\mathbf{x}$ in B_x has the same manner. To enhance the model to learn a more robust contextual representation and generate the output sequence while preserving the original sense of the given input sequence, we apply the CL strategy which is illustrated in (iii) of Fig. 2.

Following the popular version of contrastive loss proposed by Chen et. al. [1], we also leverage the NT-Xent (the normalized temperature-scaled cross entropy) loss as the training objective. For $B = B_x \bigcup B_y$, the loss is given by:

$$\mathcal{L}_{contr} = -\frac{1}{2b} \sum_{i=1}^{2b} \log \frac{\exp[\phi(\mathbf{h}_c^{\mathbf{x}_i}, h_c^{\mathbf{y}_i})]/\tau}{\sum_{\mathbf{h}_c \in B_i} \exp[\phi(\mathbf{h}_c^{\mathbf{x}_i}, \mathbf{h}_c)]/\tau} \tag{7}$$

$$B_i = B - \{\mathbf{h}_c^{\mathbf{x}_i}, \mathbf{h}_c^{\mathbf{y}_i}\}, \quad \phi(\mathbf{u}, \mathbf{v}) = \mathbf{u} \cdot \mathbf{v}/\|\mathbf{u}\|\|\mathbf{v}\|$$

where $\phi(\mathbf{u}, \mathbf{v})$ refers to the cosine similarity function between two embedding $\mathbf{u}$ and $\mathbf{v}$, and τ is a temperature hyper-parameter. The entire training loss of our FDSC is given by:

$$\mathcal{L} = \mathcal{L}_{seq2seq} + \mathcal{L}_{dsc} + \mathcal{L}_{contr} \tag{8}$$

4 Experiment

4.1 Experimental Settings

We chose BART [6] and T5 [15] models as the PLM in our method since Ziems et al. [26] reported that they provided the best quality of positive reframes among other PLMs including GPT-2 [13]. We utilized the version "facebook/bart-bas", and "t5-bas" on Hugging Face[2] as the backbones. We used the PPF dataset[3] to evaluate our method. It consists of 8,349 sentence pairs with manual annotation. The dataset is divided into three folds: 6,679 for the train set, and 835 for each of the development set and test set. The statistics of the length of tokenized sequences on PPF are summarised in Fig. 3.

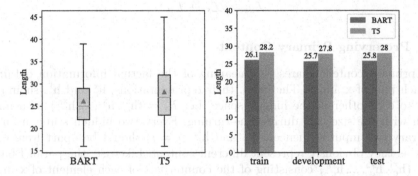

Fig. 3. The statistics of the length of tokenized sequences on the PPF dataset. The right-hand side of Fig. 3 illustrates the average sequence length, and the left-hand-side shows the length distributions obtained by T5 and BART, respectively.

The right-hand side of Fig. 3 shows that the average length obtained by T5 is slightly larger than that of BART, while the left-hand side of Fig. 3 indicates that the length distribution by T5 is larger than that of BART. We tuned the hyper-parameters on the development set as follows: the batch size is 4, 8, 16, 32, the number of epochs is from 2 to 5, the number of layers n is 12, and the value of the learning rate is from 1e-5 to 1e-4. The procedure of tuning hyper-parameters is automatically conducted by the Ray-Tune[4] library. The size of the hidden layer of the sentiment transfer, ψ_{st}, in the FDSC, is set to 512, and other hyper-parameters are consistent with the baseline and the default setting of the transformer package from Hugging Face. We implemented our model and experimented with Pytorch on NVIDIA GeForce RTX 3090ti (24GB memory).

For a fair comparison with the baseline by [26], we used the eight metrics, which are (1) ROUGE-1, -2, -LCS (longest common subsequence) [8], BLEU [12]

---

[2] https://huggingface.co/models.

[3] https://github.com/SALT-NLP/positive-frames.

[4] https://docs.ray.io/en/latest/tune/index.html.

and BERT-Score [24] referring to the gold reference for assessing the performance on content preservation, (2) The ΔTextBlob value [10] for assessing the positivity transfer effectiveness, and (3) The Average Length and Perplexity [23], followed by [4] for measuring the fluency of the output sentences.

Table 1. Main results against the baseline [26] on PPF dataset. R-1, R-2, and R-L refer to ROUGE-1, 2, and LCS. BSocre indicates BERT-Score and Avg.Len shows the Average length. The bold font indicates the best result obtained by each backbone.

Method		R-1	R-2	R-LCS	BLEU	BScore	ΔTB	Avg.Len	PPL
BART	Baseline [26]	27.7	10.8	24.3	10.3	**89.3**	**0.23**	24.4	–
	FDSC	**32.7**	**13.4**	**27.0**	**10.4**	88.5	0.21	**29.8**	**24.0**
T5	Baseline [26]	27.4	9.8	23.8	**8.7**	88.7	0.38	**35.3**	–
	FDSC	**30.4**	**10.9**	**25.1**	8.1	**88.8**	**0.39**	25.0	**13.1**

4.2 Results

Table 1 shows the results of our FDSC against the baseline which directly fine-tuned the PLMs. The performances of FDSC by R-1, R-2, and R-CLS are better than the baseline by both of the PLMs, BART and T5, especially, our FDSC with BART and T5 by R-1 metric attained 18.1%, and 10.9% improvements, respectively. This indicates that FDSC contributes to preserving the meaning of the input. For BLEU, BScore, and ΔTB metrics, both FDSC and the baseline models are similar performance. Although the BScore obtained by FDSC with BART, and BLEU by FDCS with T5 are slightly worse than the baseline, there are no significant differences between them as both of the gaps are less than 0.8%. Ziems et al. reported that the Avg.Len obtained by T5 gains good performance compared with other models such as 21.1 tokens by GPT-2 and 16.1 tokens by CopyNMt as the length by T5 is longer than 35 tokens [26]. Our FDSC has a similar tendency, i.e., the Avg.Len obtained by our model with BART achieves good performance compared with the baseline since the generated sequence by our model is 5.4 tokens longer than the baseline. The improvements obtained by our method based on BART are better than those of the T5 backbone for 5 metrics, i.e., R-1, R-2, R-LCS, BLEU, and Avg.Len. It shows that our method is suitable for BART compared with T5. One of the possible reasons is that the BART used both encoder and decoder in the pre-training phase can learn more latent patterns correlated with those potentially affecting the PTR task.

4.3 Ablation Study

We conducted an ablation study to examine the effects of each module of our FDSC. The result is shown in Table 2. We can see that the model without decoupling style from the content (w/o decoup) can perform almost best on R-1, R-2, R-LCS, and BLEU, while the worst on BScore, ΔTB, Avg.Len and PPL. This

Table 2. Ablation study of our FDSC. "w/o decoup" refers to removing the module on decoupling style from the content (training loss $\mathcal{L}_{dsc}$ in Eq. (8)), and "w/o contr" indicates removing the contrastive learning module employed for preserving invariant contents (training loss $\mathcal{L}_{contr}$ in Eq. (8)). The bold font shows the best result. Note that the smaller value of PPL indicates a better performance.

Method		R-1	R-2	R-LCS	BLEU	BScore	ΔTB	Avg.Len	PPL
BART	FDSC	32.7	13.4	27.0	10.4	**88.5**	**0.21**	**29.8**	**24.0**
	w/o decoup	32.7	**13.9**	**27.3**	**11.0**	88.1	0.10	26.2	28.6
	w/o contr	**32.9**	13.4	27.0	10.5	88.4	0.19	27.5	24.2
T5	FDSC	30.4	10.9	25.1	8.1	**88.8**	**0.39**	**25.0**	**13.1**
	w/o decoup	**30.6**	**11.1**	**25.4**	**8.7**	**88.8**	0.37	24.3	14.0
	w/o contr	29.9	10.6	24.7	8.3	88.7	**0.39**	24.7	13.4

indicates that the component for contrastive learning can contribute to preserving the meaning of source sentences. In contrast, the model without contrastive learning (w/o contr) obtains generally better results on ΔTB, Avg.Len and PPL but the worse result on R-1, R-2, and R-LCS compared with the variant "w/o decoup".

Avg.Len obtained by FDSC is better than those obtained by the other two models on both BART and T5, especially the improvement compared with the model without decoupling style from the content (w/o decoup) by BART and T5 are 13.7% and 2.9%, respectively. Likewise, the improvement of PPL obtained by FDSC works well, especially the improvement compared with the model without decoupling style from the content (w/o decoup) with BART and T5 are 19.2% and 7.6% respectively. Our FDSC works well on BScore, ΔTB, Avg.Len. PPL is relatively middle results by the other matrices. From these observations, we can conclude that the module of decoupling style from the content contributes to the fluency of output sentences and strikes a balance between style transfer and meaning preservation.

5   Conclusion

We proposed a simple but effective text reframing framework, FDSC by leveraging the Pre-trained Language Models as a backbone to decouple the sentiment style from the text content for the PTR task. The experimental results on PPF dataset, showed that our approach outperforms baselines by fine-turning two popular seq2seq PLMs, BART and T5, and achieved better text reframing. Throughout the ablation study, we found that the module of decoupling sentiment from the contents contributes to improving the performance of AVG.Len. and PPL, i.e., the fluency of the output sentences. Future work will include: (1) exploring effective augmentation strategies by leveraging a huge number of unlabeled datasets, (2) exploring the method based on LLMs including LLaMa

[20] with in-context-learning, and (3) applying our method to other TST tasks such as formality and politeness.

Acknowledgments. We would like to thank anonymous reviewers for their comments and suggestions. This work is supported by SCAT, JKA, Kajima Foundation's Support Program, and JSPS KAKENHI (No. 21K12026, 22K12146, and 23H03402). The first author is supported by JST, the establishment of university fellowships towards the creation of science technology innovation, Grant Number JPMJFS2117.

References

1. Chen, T., Kornblith, S., Norouzi, M., Hinton, G.: A simple framework for contrastive learning of visual representations. In: Proceedings of the 37th International Conference on Machine Learning. Proceedings of Machine Learning Research, vol. 119, pp. 1597–1607 (July 2020)
2. Fu, Z., Tan, X., Peng, N., Zhao, D., Yan, R.: Style transfer in text: exploration and evaluation. Proc. AAAI Conf. Artif. Intell. **32**(1), 663–670 (2018)
3. Hovy, E.: Generating natural language under pragmatic constraints. J. Pragmat. **11**(6), 689–719 (1987)
4. Jin, D., Jin, Z., Hu, Z., Vechtomova, O., Mihalcea, R.: Deep learning for text style transfer: a survey. Comput. Linguist. **48**(1), 155–205 (2022)
5. Lee, S., Lee, D.B., Hwang, S.J.: Contrastive learning with adversarial perturbations for conditional text generation. In: International Conference on Learning Representations (2021)
6. Lewis, M., et al.: BART: denoising sequence-to-sequence pre-training for natural language generation, translation, and comprehension. In: Proceedings of the 58th Annual Meeting of the Association for Computational Linguistics, pp. 7871–7880 (2020)
7. Li, J., Jia, R., He, H., Liang, P.: Delete, retrieve, generate: a simple approach to sentiment and style transfer. In: Proceedings of the 2018 Conference of the North American Chapter of the Association for Computational Linguistics: Human Language Technologies, Volume 1 (Long Papers), pp. 1865–1874. Association for Computational Linguistics (2018)
8. Lin, C.Y.: ROUGE: a package for automatic evaluation of summaries. In: Text Summarization Branches Out, pp. 74–81 (2004)
9. Liu, D., Fu, J., Zhang, Y., Pal, C., Lv, J.: Revision in continuous space: unsupervised text style transfer without adversarial learning. In: Proceedings of the AAAI Conference on Artificial Intelligence, vol. 34, pp. 8376–8383 (2020)
10. Loria, S.: textblob documentation. Release 0.16 2 (2018)
11. McDonald, D.D., Pustejovsky, J.D.: A computational theory of prose style for natural language generation. In: Second Conference of the European Chapter of the Association for Computational Linguistics, pp. 187–193 (1985)
12. Papineni, K., Roukos, S., Ward, T., Zhu, W.J.: Bleu: a method for automatic evaluation of machine translation. In: Proceedings of the 40th Annual Meeting of the Association for Computational Linguistics, pp. 311–318 (2002)
13. Radford, A., Wu, J., Child, R., Luan, D., Amodei, D., Sutskever, I.: Language models are unsupervised multitask learners. OpenAI Blog **1**(8), 9 (2019)
14. Raffel, C., et al.: Exploring the limits of transfer learning with a unified text-to-text transformer. J. Mach. Learn. Res. **21**(140), 1–67 (2020)

15. Raffel, C., et al.: Exploring the limits of transfer learning with a unified text-to-text transformer. J. Mach. Learn. Res. **21**(140), 1–67 (2020)

16. Rao, S., Tetreault, J.: Dear sir or madam, may I introduce the GYAFC dataset: corpus, benchmarks and metrics for formality style transfer. In: Proceedings of the 2018 Conference of the North American Chapter of the Association for Computational Linguistics: Human Language Technologies, Volume 1 (Long Papers), pp. 129–140 (2018)

17. Reif, E., Ippolito, D., Yuan, A., Coenen, A., Callison-Burch, C., Wei, J.: A recipe for arbitrary text style transfer with large language models. In: Proceedings of the 60th Annual Meeting of the Association for Computational Linguistics (Volume 2: Short Papers), pp. 837–848. Association for Computational Linguistics (2022)

18. Su, Y., Lan, T., Wang, Y., Yogatama, D., Kong, L., Collier, N.: A contrastive framework for neural text generation. In: Oh, A.H., Agarwal, A., Belgrave, D., Cho, K. (eds.) Advances in Neural Information Processing Systems (2022). https://openreview.net/forum?id=V88BafmH9Pj

19. Tan, W., Heffernan, K., Schwenk, H., Koehn, P.: Multilingual representation distillation with contrastive learning. In: Proceedings of the 17th Conference of the European Chapter of the Association for Computational Linguistics, pp. 1477–1490 (May 2023)

20. Touvron, H., et al.: Llama: open and efficient foundation language models. arXiv preprint arXiv:2302.13971 (2023)

21. Wei, J., et al.: Chain of thought prompting elicits reasoning in large language models. arXiv preprint arXiv:2201.11903 (2022)

22. Xu, R., Ge, T., Wei, F.: Formality style transfer with hybrid textual annotations. arXiv preprint arXiv:1903.06353 (2019)

23. Yang, Z., Hu, Z., Dyer, C., Xing, E.P., Berg-Kirkpatrick, T.: Unsupervised text style transfer using language models as discriminators. In: Advances in Neural Information Processing Systems, vol. 31 (2018)

24. Zhang, T., Kishore, V., Wu, F., Weinberger, K.Q., Artzi, Y.: Bertscore: evaluating text generation with bert. In: International Conference on Learning Representations, pp. 74–81 (2020)

25. Zhang, Y., Ge, T., Sun, X.: Parallel data augmentation for formality style transfer. In: Proceedings of the 58th Annual Meeting of the Association for Computational Linguistics, pp. 3221–3228 (2020)

26. Ziems, C., Li, M., Zhang, A., Yang, D.: Inducing positive perspectives with text reframing. In: Proceedings of the 60th Annual Meeting of the Association for Computational Linguistics (Volume 1: Long Papers), pp. 3682–3700 (2022)

Multi-level Feature Enhancement Method for Medical Text Detection

Tianyang Li[1,2]($\boxtimes$), Jinxu Bai[1], Qingzhu Wang[1], and Hanwen Xu[1]

[1] Computer Science, Northeast Electric Power University, Jilin 132012, Jilin, China
tianyangli@neepu.edu.cn
[2] Jiangxi New Energy Technology Institute, Xinyu 338000, Jiangxi, China

Abstract. In recent years, Segmentation-based text detection methods have been widely applied in the field of text detection. However, when it comes to tasks involving text-dense, extreme aspect ratios, and multi-oriented text, the current detection methods still fail to achieve satisfactory performance. In this paper, we propose an efficient and accurate text detection system that incorporates an efficient segmentation module and a learnable post-processing method. More specifically, the segmentation head consists of an Efficient Feature Enhancement Module (EFEM) and a Multi-Scale Feature Fusion Module(MSFM). The EFEM is a cascaded U-shaped module that incorporates spatial attention mechanism to introduce multi-level information for improved segmentation performance. MSFM integrates features from the EFEM at different depths and scales to generate the final features for segmentation. Moreover, a post-processing module employing a differentiable binarization approach is utilized, allowing the segmentation network to adaptively set the binarization threshold. The proposed model demonstrates excellent performance on medical text image datasets. Multiple benchmark experiments further validate the superior performance of the proposed method. Code is available at: https://github.com/csworkcode/EFDBNet.

Keywords: Text detection · Medical text · Feature Enhancement · Differentiable Binarization

1 Introduction

The rapid development of computer technology has led to an explosive growth of information and fundamentally changed the way people access information. Textual information in images contains rich and accurate high-level semantic information, serving as a crucial element for understanding images. Text detection in images is a fundamental and critical task in computer vision, as it plays a key role in many text-related applications such as text recognition, license plate recognition, autonomous driving, electronic archives, and intelligent healthcare. Medical images can reveal important information related to diagnostic outcomes. Therefore, the rapid and accurate extraction of such information from text holds practical significance.

B. Luo et al. (Eds.): ICONIP 2023, CCIS 1967, pp. 85–96, 2024.
https://doi.org/10.1007/978-981-99-8178-6_7

Due to the recent advancements in object detection [2,4,21] and segmentation [3,9,20] based on Convolutional Neural Networks (CNN) [7], significant progress has been achieved in the field of scene text detection [12,14]. However, there are still numerous challenges when it comes to certain specific text environments. Typically, in general object and scene text detection, where the number of target objects is small, the spacing between objects is significant, and the segmentation is clear, accurate detection can be achieved. However, in certain special cases, such as images with a large volume of text, dense distribution of text, and diverse content, such algorithms fail to accurately locate all the text.

Compared to general object and scene text detection, medical image text detection needs to address the following issues: (1) When scanning or capturing original medical images, text content may become skewed due to non-flat surfaces. (2) Medical images often contain a large amount of text, with dense text distribution being prevalent. (3) Medical text can include Chinese characters, alphabets, identifiers, as well as images and tables, leading to incomplete text boundary detection.

In summary, our contributions are three-fold:

- We propose an Efficient Feature Enhancement Module (EFEM) and Multi-Scale Feature Fusion Module (MSFM) which are two high-efficiency modules that can improve the feature representation of the network.
- Our detection model can simultaneously handle the challenges of rotation, extreme aspect ratios, dense text instances in multi-oriented medical text.
- Using the proposed method, we achieve competitive results in terms of efficiency, accuracy, F-score and robustness on two public natural scene text detection datasets.

The rest paper is organized as follows: Sect. 2 reviews the relevant text detection methods. Our proposed approach is described in Sect. 3. The experiments are discussed and analyzed in Sect. 4. The conclusions are summarized in Sect. 5.

2   Related Works

Recently, deep learning has emerged as the predominant approach in the field of scene text detection. Deep learning methods for scene text detection can be broadly categorized into three groups based on the granularity of the predicted target: regression-based methods and segmentation-based methods.

Regression-based methods encompass a series of models that directly estimate the bounding boxes of text instances. Methods like EAST [22] and Deep-Reg [6] are anchor-free approaches that employ pixel-level regression for detecting multi-oriented text instances. DeRPN [16] introduced a dimension-decomposition region proposal network to address the scale variation challenge in scene text detection. These regression-based methods often benefit from straightforward post-processing algorithms, such as non-maximum suppression. However, they face limitations in accurately representing irregular shapes, such as curved shapes, with precise bounding boxes.

Segmentation-based methods typically employ a combination of pixel-level prediction and post-processing algorithms to obtain the bounding boxes. For instance, PSENet [15] introduced progressive scale expansion by segmenting text instances using different scale kernels. Tian et al [13] proposed pixel embedding to group pixels based on the segmentation results. Additionally, PSENet [15] and SAE [13] presented novel post-processing algorithms to refine the segmentation results, albeit at the expense of reduced inference speed. In contrast, DBNet [6] focuses on enhancing the segmentation results by incorporating the binarization process during training, without sacrificing inference speed.

In comparison to previous rapid scene text detectors, our approach exhibits enhanced performance in terms of both speed and the ability to detect text instances of medical text and arbitrary shapes. Notably, we introduced a strategy that incorporates a segmentation head with low computational requirements and employs learnable post-processing techniques for text detection. This methodology resulted in an efficient and accurate detector for medical text and arbitrary-shaped text.

3 Methodology

3.1 Overall Network Architecture

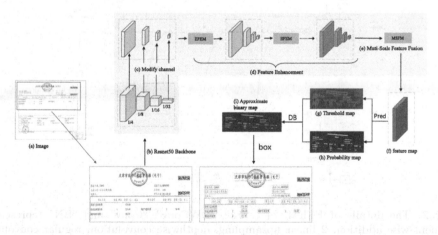

Fig. 1. Architecture of our proposed method. (a) Input image. (b) Feature extraction backbone network. (c) reduce the channel number of each feature map to 256. The 1/4, 1/8, 1/16, 1/32 indicate the scale ratio compared to the input image.

The architecture of our proposed model is shown in Fig. 1. The pipeline can be divided into four parts: a feature extraction backbone, a feature enhancement backbone, a Multi-scale feature fusion model and a post-processing procedure. Firstly, similar to the traditional CNN backbone, the input image is fed to an

FPN structure in order to obtain multi-level features (see Fig. 1(a), (b)). Secondly, the pyramid features are up-sampled to the same scale and then input into the Efficient Feature Enhancement Module(EFEM). The EFEM module is designed to be cascaded, offering the advantage of low computational cost. It can be seamlessly integrated behind the backbone network to enrich and enhance the expressive power of features at different scales (see Fig. 1(c), (d)). Subsequently, we incorporate the Multi-scale Feature Fusion Module (MSFM) to effectively merge the features generated by the EFEMs at various depths, producing a comprehensive final feature representation for segmentation (see Fig. 1(e)). Then, the extracted feature F is utilized to predict both the probability map (P) and the threshold map (T). Subsequently, the approximate binary map ($\hat{B}$) is computed using the probability map and feature F (see Fig. 1(f), (g), (h), (i)). During the training phase, supervision is applied to the probability map, threshold map, and approximate binary map. Notably, the probability map and approximate binary map share the same supervision. During the inference phase, the bounding boxes can be effortlessly obtained from either the approximate binary map or the probability map using a dedicated box formulation module.

3.2   Efficient Feature Enhancement Module

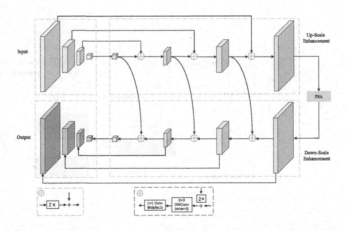

Fig. 2. The details of EFEM. "+", "2×", "DWConv", "Conv" and "BN" represent element-wise addition, 2 linear upsampling, depthwise convolution, regular convolution and Batch Normalization respectively.

The EFEM is depicted as a U-shaped module (see Fig. 2). It comprises three phases: up-scale enhancement, Pyramid Squeeze Attention(PSA) [19] module and down-scale enhancement. The up-scale enhancement operates on the input feature pyramid, where iterative enhancement is applied to the feature maps with strides of 32, 16, 8, and 4 pixels. PSA is an efficient attention block that enables the effective extraction of multi-scale spatial information at a finer granularity, while also establishing long-range channel dependencies. In the down-scale phase, the input is the feature pyramid generated by the up-scale enhancement, and the enhancement is performed from 4-stride to 32-stride.

EFEM enhances features of different scales by integrating low-level and high-level information. However, the EFEM module offers two additional advantages compared to FPN. Firstly, EFEM is a cascadable module, meaning that it can be cascaded multiple times. As the cascade number increases, the fusion of feature maps from different scales becomes more comprehensive, and the receptive fields of the features expand. This allows for better integration of information across scales. Secondly, EFEM is computationally efficient, resulting in a lower computational cost. This efficiency makes it suitable for practical applications where computational resources may be limited.

3.3   Multi-scale Feature Fusion Module

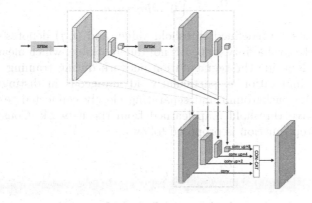

Fig. 3. Illustration of the MSFM module. "+" is element-wise addition.

Context plays a crucial role in semantic segmentation techniques. Both context and scale are interdependent concepts, as context aids in perceiving objects and scenes on a larger scale. Incorporating multi-scale fusion strategies is often employed to enhance contextual information. Consequently, the integration of multi-scale features has become a common practice in semantic segmentation methodologies. Multi-scale Feature Fusion Module is applied to fuse the feature of different depths. Given the significance of both low-level and high-level semantic information in semantic segmentation, a straightforward and efficient approach to merge these feature pyramids is through upsampling and concatenation. Therefore, we propose an alternative fusion method (see Fig. 3). Initially, we merge the feature maps of corresponding scales through element-wise addition. Subsequently, the resulting feature maps are upsampled and concatenated to create a final feature map.

3.4   Post-processing

To enhance the expressiveness of the text regions obtained, it is essential to employ suitable post-processing strategies for the acquired features. During the parsing of these features into the visualized text regions, it is necessary to execute binarization and label generation operations on the features.

Differentiable Binarization. The technique of differentiable binarization for the probability map was initially introduced by Liao et al [6], offering a trainable method to generate a segmentation mask. The output of the segmentation network is a probability map $P \in R^{H \times W}$, where H and W represent the height and width of the map, respectively. For the task of determining whether an area contains text or not, a binary map $B \in R^{H \times W}$ is required, where pixels with a value of 1 indicate valid text areas. Typically, the standard binarization process can be formulated as follows:

$$B_{i,j} = \begin{cases} 1 \ if P_{i,j} >= t, \\ 0 \ otherwise. \end{cases} \tag{1}$$

Here t represents a predefined threshold value, and (i, j) denotes the value of pixel (x, y) in the probability map. 1 is not differentlable, which means it can not be optimized alongside the segmentation network during training. In contrast, differentiable binarization is particularly advantageous in distinguishing text regions from the background and separating closely connected text instances. T is the adaptive threshold map learned from the network. Consequently, an approximate step function is proposed follows:

$$\hat{B}_{i,j} = \frac{1}{1 + e^{-k(P_{i,j} - T_{i,j})}}. \tag{2}$$

3.5   Deformable Convolution and Label Generation

Deformable convolution [23] can offer a versatile receptive field for the model, which is particularly advantageous for text instances with extreme aspect ratios. In line with [24], modulated deformable convolutions are employed in all the 3×3 convolutional layers within the conv3, conv4, and conv5 stages of the ResNet-50 backbone.

The label generation for the probability map is influenced by PSENet [15]. Typically, post-processing algorithms display the segmentation results using a collection of vertices that define a polygon:

$$G = \{S_k\}_{k=1}^n. \tag{3}$$

n represents the number of vertices, which typically varies depending on the labeling rules in different datasets, and S denotes the segmentation results for each image. Mathematically, the offset D can be calculated as follows:

$$D = \frac{\text{Area}(P) \times (1 - r^2)}{\text{Perimeter}}. \tag{4}$$

In this context, Area($\cdot$) refers to the calculation of the polygon's area, and Perimeter($\cdot$) denotes the calculation of the polygon's perimeter. The shrink ratio, denoted as "r" is empirically set to 0.4. By employing graphics-related operations, the shrunken polygons can be derived from the original ground truth, serving as the fundamental building block for each text region.

4 Experimental Results

Our proposed model achieves excellent results in medical text detection, and to further validate the robustness and generalisation performance of the model, we add two sets of comparison experiments on publicly available scene text datasets.

4.1 Datasets

The text datasets used in the experiments are described as follows.

MEBI-2000 Dataset is derived from the public dataset of Ali Tianchi competition [1]. A dataset created to solve the task of detection and recognition of medical ticket text. MEBI-2000 consists of a total of 2000 scanned images and photos taken with mobile phones from real-life environments. Among them, 1500 images are used for training data, and 500 images are used for testing data.

ICDAR 2015 Dataset [8] is a commonly used dataset for text detection. It contains a total of 1500 images, 1000 of which are used for training and the remaining are for testing.

MSRA-TD500 Dataset [18] is a dataset with multi-lingual, arbitrary-oriented and long text lines. Because the training set is rather small, we follow the previous works [10,11] to include the 400 images from HUST-TR400 [17] as training data.

4.2 Implementation Details

We pre-train our network on SynthText [5] and then finetune it on the real datasets (MEBI, ICDAR2015, MSRA-TD500). Following the pre-training phase, we proceed with fine-tuning the models for 1200 epochs on the corresponding real-world datasets. During training, our primary data augmentation techniques encompass random rotation, random cropping, random horizontal and vertical flipping. Additionally, we resize all images to 640 × 640 to enhance training efficiency. For all datasets, the training batch size is set to 16, and we adhere to a "poly" learning rate policy to facilitate gradual decay of the learning rate. Initially, the learning rate is set to 0.007, accompanied by an attenuation coefficient of 0.9. Our framework employs stochastic gradient descent (SGD) as the optimization algorithm, with weight decay and momentum values set to 0.0001 and 0.9, respectively.

4.3   Ablation Study

We conduct an ablation study on the MSRA-TD500 dataset and the MEBI dataset to show the effectiveness of the modules including deformable convolution, Efficient Feature Enhancement Module and Multi-Scale Feature Fusion Module. The detailed experimental results are shown in Table 1.

Table 1. Detection results with different settings of Deformable Convolution, EFEM and MSFM.

Backbone	DConv	EFEM	MSFM	MSRA-TD500				MEBI			
				P	R	F	FPS	P	R	F	FPS
ResNet-50	×	×	×	84.5	73.2	78.4	**55**	81.5	72.8	76.9	**39.3**
ResNet-50	✓	×	×	88.6	78.2	83.1	49.5	85.8	75	80	35.6
ResNet-50	×	×	✓	86.3	75.5	80.5	47	86.1	78.3	82.0	34.1
ResNet-50	✓	×	✓	90.6	78.1	83.9	40.7	86.9	83.1	85	28.7
ResNet-50	×	✓	×	88.4	**81.7**	84.9	48.1	85	80.9	82.9	35
ResNet-50	✓	✓	×	92.1	79.2	85.2	39	88.1	85.3	86.7	27.2
ResNet-50	✓	✓	✓	**92.2**	81.4	**86.5**	36	**88.5**	**86.3**	**87.4**	25

Deformable Convolution. As shown in Table 1, for the MSRA-TD500 dataset, the deformable convolution increase the F-measure by 4.7%. For the MEBI dataset, 3.1% improvements are achieved by the deformable convolution.

The Effectiveness of EFEM. As shown in Table 1, for the MSRA-TD500 dataset, The effectiveness of EFEM increase the F-measure by 6.5%, 6.8% improvements are achieved by the EFEM+DConv. For the MEBI dataset, The effectiveness of MSFM increase the F-measure by 6%, 9.8% improvements are achieved by the EFEM+DConv.

Table 2. Detection results on the MEBI dataset.

Method	P	R	F	FPS
EAST (Zhou et al. 2017)	78.7	70.4	74.3	15
PSE-Net (Wang et al. 2019)	84.9	85.7	85.3	4
SAST (Wang et al. 2019)	85.1	80.4	82.7	–
PAN (Wang et al. 2019)	87.6	82.1	84.8	–
DBNet (Liao et al. 2020)	82.3	74.2	78.1	20
DBNet++ (Liao et al. 2022)	83.3	80.1	81.7	15
Ours	**88.5**	**86.3**	**87.4**	**25**

The Effectiveness of MSFM. As shown in Table 1, for the MSRA-TD500 dataset, The effectiveness of MSFM increase the F-measure by 2.1%, 5.5% improvements are achieved by the MSFM+DConv. For the MEBI dataset, The effectiveness of MSFM increase the F-measure by 5.1%, 8.1% improvements are achieved by the MSFM+DConv.

4.4   Comparisons with Previous Methods

We compare our proposed method with previous method on three standard benchmarks, including one benchmark for medical text, and one benchmark for multi-oriented text, and one multi-language benchmarks for long text lines. Some qualitative results are visualized in Fig. 5.

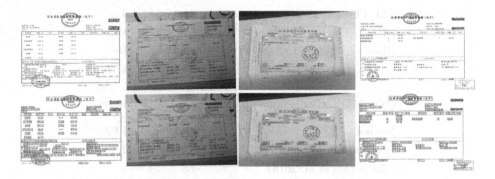

Fig. 4. Some visualization results on text instances of MEBI.

Fig. 5. Some visualization results on text instances of various shapes, including curved text, multi-oriented text, vertical text.

Medical Text Detection. The MEBI dataset is a medical text dataset that contains lots of diverse scales, irregular shapes, and extreme aspect ratios of textual instances. In Table 2 Comparisons with previous methods, the model of ours attains state-of-the-art performance in terms of accuracy, f-measure, recall and fps. Some visualization results on medical text instances of MEBI (see Fig. 4). To further validate the robustness of the model, we added two sets of comparison trials on the publicly available scene text dataset.

Table 3. Detection results on the ICDAR 2015 dataset.

Method	P	R	F	FPS
CTPN (Tian et al. 2016)	74.2	51.6	60.9	7.1
EAST (Zhou et al. 2017)	83.6	73.5	78.2	13.2
SSTD (He et al. 2017)	80.2	73.9	76.9	7.7
WordSup (Hu et al. 2017)	79.3	77	78.2	–
TB (Liao, Shi, and Bai 2018)	87.2	76.7	81.7	11.6
PSE-Net (Wang et al. 2019)	86.9	84.5	85.7	1.6
SPCNet (Xie et al. 2019)	88.7	**85.8**	87.2	–
LOMO (Zhang et al. 2018)	91.3	83.5	87.2	–
PAN (Wang et al. 2019b)	84.0	81.9	82.9	**26.1**
SAE (Tian et al. 2019)	85.1	84.5	84.8	3
DBNet (Liao et al. 2020)	91.8	83.2	87.3	12
DBNet++ (Liao et al. 2022)	90.9	83.9	87.3	10
Ours	**92.5**	83.1	**87.5**	15.4

Table 4. Detection results on the MSRA-TD500 dataset.

Method	P	R	F	FPS
DeepReg (He et al. 2017)	77	70	74	1.1
RRPN (Ma et al. 2018)	82	68	74	–
RRD (Liao et al. 2018)	87	73	79	10
MCN (Liu et al. 2018)	88	79	83	–
PixelLink (Deng et al. 2018)	83	73.2	77.8	3
CRAFT (Baek et al. 2019)	88.2	78.2	82.9	8.6
SAE (Tian et al. 2019)	84.2	81.7	82.9	–
PAN (Wang et al. 2019)	84.4	83.8	84.1	30.2
DBNet (Liao et al. 2020)	91.5	79.2	84.9	32
DRRG (Zhang et al. 2020)	88.1	82.3	85.1	–
MOST (He et al. 2021)	90.4	82.7	86.4	–
TextBPN (Zhang et al. 2021)	86.6	**84.5**	85.6	–
DBNet++ (Liao et al. 2022)	91.5	83.3	**87.2**	29
Ours	**92.2**	81.4	86.5	**36**

Multi-oriented and Multi-language Text Detection. The ICDAR 2015 dataset is a multi-oriented text dataset that contains lots of small and low-resolution text instances. In Table 3 the model of ours attains state-of-the-art performance in terms of accuracy and f-measure. Compared to the DBNet++(Liao et al. 2022), our method outperforms it by 1.6% on accuracy. Our method is robust on multi-language text detection. As shown in Table 4.

Our method is superior to previous methods on accuracy and speed. For the accuracy, our method surpasses the previous state-of-the-art method by 0.7% on the MSRA-TD500. For the speed, our method runs at 36 FPS.

5   Conclusion

In this paper, we present an efficient framework for detect medical text and arbitrary-shaped text in real-time. We begin by introducing Efficient Feature Enhancement Module and Multi-scale Feature Fusion Module. This design enhances feature extraction while introducing minimal additional computational overhead. Through extensive experimentation on MEBI, ICDAR 2015, and MSRA-TD500 datasets, our proposed method showcases notable advantages in terms of both speed and accuracy when compared to previous state-of-the-art text detectors.

Acknowledgement. This work was supported by the Scientific Research Funds of Northeast Electric Power University (No. BSZT07202107).

References

1. Medical Inventory Invoice OCR Element Extraction Task (CMedOCR). https://tianchi.aliyun.com/dataset/131815
2. Fan, D.P., Cheng, M.M., Liu, J.J., Gao, S.H., Hou, Q., Borji, A.: Salient objects in clutter: bringing salient object detection to the foreground. In: Proceedings of the European Conference on Computer Vision (ECCV), pp. 186–202 (2018)
3. Fan, D.P., Gong, C., Cao, Y., Ren, B., Cheng, M.M., Borji, A.: Enhanced-alignment measure for binary foreground map evaluation. arXiv preprint arXiv:1805.10421 (2018)
4. Fan, D.P., Wang, W., Cheng, M.M., Shen, J.: Shifting more attention to video salient object detection. In: Proceedings of the IEEE/CVF Conference on Computer Vision and Pattern Recognition, pp. 8554–8564 (2019)
5. Gupta, A., Vedaldi, A., Zisserman, A.: Synthetic data for text localisation in natural images. In: Proceedings of the IEEEConference on Computer Vision and Pattern Recognition, pp. 2315–2324 (2016)
6. He, W., Zhang, X.Y., Yin, F., Liu, C.L.: Deep direct regression for multi-oriented scene text detection. In: Proceedings of the IEEE International Conference on Computer Vision, pp. 745–753 (2017)
7. Huang, G., Liu, Z., Pleiss, G., Van Der Maaten, L., Weinberger, K.Q.: Convolutional networks with dense connectivity. IEEE Trans. Pattern Anal. Mach. Intell. **44**(12), 8704–8716 (2019)
8. Karatzas, D., et al.: ICDAR 2015 competition on robust reading. In: 2015 13th International Conference on Document Analysis and Recognition (ICDAR), pp. 1156–1160. IEEE (2015)
9. Liao, M., Wan, Z., Yao, C., Chen, K., Bai, X.: Real-time scene text detection with differentiable binarization. In: Proceedings of the AAAI Conference on Artificial Intelligence, vol. 34, pp. 11474–11481 (2020)

10. Long, S., Ruan, J., Zhang, W., He, X., Wu, W., Yao, C.: Textsnake: a flexible representation for detecting text of arbitrary shapes. In: Proceedings of the European Conference on Computer Vision (ECCV), pp. 20–36 (2018)
11. Lyu, P., Yao, C., Wu, W., Yan, S., Bai, X.: Multi-oriented scene text detection via corner localization and region segmentation. In: Proceedings of the IEEE Conference on Computer Vision and Pattern Recognition, pp. 7553–7563 (2018)
12. Tian, Z., Huang, W., He, T., He, P., Qiao, Yu.: Detecting text in natural image with connectionist text proposal network. In: Leibe, B., Matas, J., Sebe, N., Welling, M. (eds.) ECCV 2016. LNCS, vol. 9912, pp. 56–72. Springer, Cham (2016). https://doi.org/10.1007/978-3-319-46484-8_4
13. Tian, Z., et al.: Learning shape-aware embedding for scene text detection. In: Proceedings of the IEEE/CVF Conference on Computer Vision and Pattern Recognition, pp. 4234–4243 (2019)
14. Wang, W., et al.: Shape robust text detection with progressive scale expansion network. In: Proceedings of the IEEE/CVF Conference on Computer Vision and Pattern Recognition, pp. 9336–9345 (2019)
15. Wang, W., et al.: Shape robust text detection with progressive scale expansion network. In: Proceedings of the IEEE/CVF Conference on Computer Vision and Pattern Recognition, pp. 9336–9345 (2019)
16. Xie, L., Liu, Y., Jin, L., Xie, Z.: DERPN: taking a further step toward more general object detection. In: Proceedings of the AAAI Conference on Artificial Intelligence, vol. 33, pp. 9046–9053 (2019)
17. Yao, C., Bai, X., Liu, W.: A unified framework for multioriented text detection and recognition. IEEE Trans. Image Process. **23**(11), 4737–4749 (2014)
18. Yao, C., Bai, X., Liu, W., Ma, Y., Tu, Z.: Detecting texts of arbitrary orientations in natural images. In: 2012 IEEE Conference on Computer Vision and Pattern Recognition, pp. 1083–1090. IEEE (2012)
19. Zhang, H., Zu, K., Lu, J., Zou, Y., Meng, D.: Epsanet: an efficient pyramid squeeze attention block on convolutional neural network. In: Proceedings of the Asian Conference on Computer Vision, pp. 1161–1177 (2022)
20. Zhao, H., Shi, J., Qi, X., Wang, X., Jia, J.: Pyramid scene parsing network. In: Proceedings of the IEEE Conference on Computer Vision and Pattern Recognition, pp. 2881–2890 (2017)
21. Zhao, J.X., Cao, Y., Fan, D.P., Cheng, M.M., Li, X.Y., Zhang, L.: Contrast prior and fluid pyramid integration for RGBD salient object detection. In: Proceedings of the IEEE/CVF Conference on Computer Vision and Pattern Recognition, pp. 3927–3936 (2019)
22. Zhou, X., et al.: East: an efficient and accurate scene text detector. In: Proceedings of the IEEE Conference on Computer Vision and Pattern Recognition, pp. 5551–5560 (2017)
23. Zhu, X., Hu, H., Lin, S., Dai, J.: Deformable convnets v2: more deformable, better results. In: Proceedings of the IEEE/CVF Conference on Computer Vision and Pattern Recognition, pp. 9308–9316 (2019)
24. Zhu, X., Hu, H., Lin, S., Dai, J.: Deformable convnets v2: More deformable, better results. In: Proceedings of the IEEE/CVF Conference on Computer Vision and Pattern Recognition (CVPR) (June 2019)

Neuron Attribution-Based Attacks Fooling Object Detectors

Guoqiang Shi[1], Anjie Peng[1,2(✉)], Hui Zeng[1], and Wenxin Yu[1]

[1] Southwest University of Science and Technology, Mianyang 621010, Sichuan, China
penganjie200012@163.com
[2] Engineering Research Center of Digital Forensics, Ministry of Education, Nanjing University of Information Science and Technology, Nanjing, China

Abstract. In this work, we propose a neural attribution-based attack (NAA) to improve the transferability of adversarial examples, aiming at deceiving object detectors with different backbones or architectures. To measure the neuron attribution (importance) for a CNN layer of detector, we sum the classification scores of all positive proposal boxes to calculate the integrated attention (IA), then get the neuron attribution matrix via element-wise multiplying IA with the feature difference between the clean image be attacked and a black image. Considering that the summation may bias importance values of some neurons, a mask is designed to drop out some neurons. The proposed loss calculated from the rest of neurons is minimized to generated adversarial examples. Since our attack disturbs the upstream feature outputs, it effectively disorders the outputs of downstream tasks, such as box regression and classification, and finally fool the detector. Extensive experiments on PASCAL VOC and COCO dataset demonstrate that our method achieves better transferability compared to the state-of-the-arts.

Keywords: object detectors · CNN · adversarial examples · transferability

1 Introduction

Convolutional Neural Networks (CNNs) have been widely used in computer vision tasks [1–3]. However, CNNs have been shown to be vulnerable to adversarial attacks [4], which perturb the original images with adversarial noise and mislead the CNNs in the object detections [5, 6]. A transferable attack can generate adversarial examples from a known CNN to fool other CNN models with different architectures, which has received extensive attentions [7–12].

DFool [13] is the first work to attack the two-stage detector Faster R-CNN [24]. Since then, a series of adversarial attacks [5, 10, 14–17] targeting object detection have been proposed. Some attacks focus on to attack the region proposal network (RPN), which provides fundamental proposal boxes for subsequent object recognition tasks. Attacking the RPN causes Faster R-CNN to misclassify foreground as background or output incorrect bounding box positions. DAG [5] is a typical method that disturbs the outputs of RPN and has shown excellent performance under white-box settings. In

B. Luo et al. (Eds.): ICONIP 2023, CCIS 1967, pp. 97–108, 2024.
https://doi.org/10.1007/978-981-99-8178-6_8

addition to attacking the RPN, some methods try to destroy some specific components, such as the attack against non-maximum suppression (NMS) [19] scheme or category-specific attacks against anchor-free detectors. These methods also have demonstrated superior attack performance in white-box settings. However, when attacking different backbone networks or object detection models with different architectures under black-box settings, the transferable performance of the above-mentioned attacks still needs to be improved.

DAG   Ours

Fig. 1. The transferable performance of DAG [5] and our attack for attacking an un-seen Faster-RCNN detector with MobileNetv3 as backbone (FR-M3). FR-R50 as a white box.

To improve the transferability of adversarial examples under black-box settings, we try to disrupt the crucial features related to the object in the outputs of CNN backbone network. As indicated by the adversarial attacks against image classifications [20–22], disrupting the shared crucial features benefits for transferability. This motivates us to propose a feature-level attack to disorder the object-aware feature of detector. Inspired by the neuron attribution-based attack (NAA) [23], we perform neuron attribution [22] on the output of specific layers of the backbone to measure the critical information of object. Considering that the positive proposal boxes control the output of object detectors, the proposed neuron attributions are calculated by the summation of classification outputs of all positive proposal boxes. We design a new feature-based loss function based on some important neurons and minimize the loss to generate adversarial examples. Due to disturb the shared crucial feature of "sheep", Fig. 1 shows that our method generates a transferable adversarial example to fool a Faster R-CNN with MobileNetv3 as backbone under black-box settings.

2   Background

Object Detections. CNN-based object detection is a fundamental task in computer vision that locates instances of visual objects in digital images or video frames. Faster R-CNN [24], SSD [25], YOLOv5 [28] and their derivation versions [29, 30] have been widely applied in the object detection. Faster R-CNN is a two-stage object detector, which first generates proposal boxes by a region proposal network (RPN) and then utilizes a detection head to locates the objects from regions of interest (ROI) selected from all proposal boxes. Figure 2. Shows the detection process of a Faster-RCNN with

ResNet50 as the backbone network (FR-R50). FR-R50 first uses the ResNet50 to extract feature, then inputs feature maps into the RPN to search positive proposals which contain the objects, finally outputs classification and location results from ROIs. To filter out redundant proposal boxes and produce the concise detection results, NMS scheme is applied in the detection head. For the one-stage detectors, such as YOLOv5, they run detection directly over a dense sampling of locations after extracting features from backbone. Obviously, extracting the feature from backbone is important for the downstream tasks, such as RPN and the detection head. In literature, researchers design backbones to pursue better feature representations to improve the detection performance. For example, YOLOv1 uses ResNet-50 as the backbone, while YOLOv5 uses the combination of CSPNets as backbones. Therefore, attacking the backbone of the detector will disturb the inputs for the downstream tasks and finally defeat the detectors.

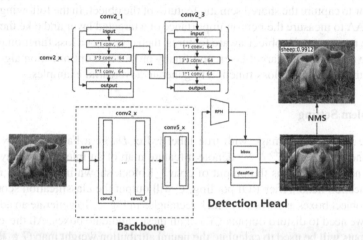

Fig. 2. The detection process of a Faster-RCNN detector with ResNet50 as backbone.

Adversarial Attack for Object Detection. Dense Adversary Generation (DAG) [5] employs an iterative optimization-based attack to make proposal boxes to be incorrectly predicted. To ensure fool the detector successfully, DAG increases the threshold of intersection over union (IOU) of NMS in RPN to attack proposal boxes as many as possible. Enhanced DAG (DAG+) [18] ensembles a feature-based loss with the loss of DAG to enhance the transferability of adversarial examples. ShapeShifter [17] leverages C&W [26] and expectation over transformation [27], both of which are attacks on image classification, to attack object detector. RAP [14] introduces a joint loss function based on classification and regression boxes to mislead the classification results and cause imprecise localization of objects. Contextual Adversarial Perturbation (CAP) [15] incorporates the contextual background loss with the loss of RAP to improve the attack performance. The targeted adversarial object gradients (TOG) [10] attack constructs three types of attacks: object-disappearance, object-fabrication, and incorrect object labeling. Recently, some works design adversarial patches on local regions, physical attack to fool object detectors [31, 32]. In this work, we will launch an imperceptible transfer-based attack, aim at improving transferable performance among different detectors under black-box cases.

3   Methodology

Inspired by the feature-level attack NAA [23] fooling image classification, we disturb the feature output of backbone to fool the current detector and other unknown object detectors. The backbone networks of detectors, whether they have similar architectures, pursuit to learn similar sematic features which are object-aware discriminative for object detection. This means that, if the adversarial example destroys the sematic feature output of the current detector, it will probably fool other detectors, i.e., transferability. The key issue is how to capture the shared sematic features of the object. In the following, we first employ NAA to measure the neuron importance of a targeted layer and take the features of positive neurons as the object-aware features, then construct a loss function to disturb the feature outputs of the target layer, and finally propose an optimization algorithm to minimize the NAA-based loss function to generate adversarial examples.

3.1   Problem Setting

We denote x as a clean image with true label y. Let D_θ be a Faster R-CNN detector with parameter θ, $F \in \mathbb{R}^{H \times W \times N}$ denote a feature map of an intermediate layer of the backbone network (such as the output of conv2_1 block shown in Fig. 2), C represent the classification head. After ROI pooling, C will output the classification scores of all candidate object boxes as shown by black rectangles in Fig. 2. To generate an adversarial example, we need to disturb outputs $C(.)$ of all these candidate boxes. All these classification outputs will be used to calculate the neural attribution weight map $G \in \mathbb{R}^{H \times W \times N}$ of the feature map F. Our objective is to design a loss function $L(x, G, F)$ to generate an adversarial perturbation r, such that $x^{adv} = x + r$ cannot be correctly recognized by the object detection model. Additionally, to ensure visual invisibility of x^{adv}, r is constrained by a L_∞-norm distance $||r||_\infty \leq \varepsilon$.

3.2   Our Approach

We propose a feature-level attack to perturb the feature output of a specific layer F in the backbone network. Since subsequent tasks, such as classification and box regression, rely on the feature output of the backbone network, disturbing the feature output will result in error recognition results. Inspired by NAA [23], we first measure the neuron attribution G of the specific layer F, then craft adversarial examples to suppress the useful features and enlarge the harmful feature.

$$G = \sum_{i=1}^{N} (x_i - x_{i\prime}) \int_0^1 \frac{\partial C(x\prime + \alpha(x - x\prime))}{\partial x_i} d\alpha$$

$$where \quad C(x) = \sum_{j=1}^{L} C^j(x)$$

(1)

How to calculate the neuron attribution G is the key point to generate feature-level adversarial attack. According to the pipeline of detector, the final locating box and classification results directly depend on the outputs of the classification head. Therefore, we focus on all outputs of the classification head for all candidate proposal boxes to calculate G. According the multiple outputs of the classification head, the proposed G is calculated by Eq. (1) using a path integration of the partial derivative of C. In Eq. (1), N is the number of pixels of the image x, $x\prime$ is a baseline image, $C^j(x)$ is the classification score at the index of true label for the jth candidate object box. The fast approximation of Eq. (1) for the layer F named G^F is given in Eq. (2), where F and F' is the feature output at the layer F for the image x to be attacked and a black image $x\prime = 0$ respectively. The expression $\frac{\partial C(x\prime + m/n(x-x\prime))}{\partial F}$ is the gradients of the layer F with a virtual image $x\prime + m/n(x - x\prime)$. To calculated the integrated attention (IA), there are $n = 30$ virtual images are used.

$$G^F = (F - F')IA(F)$$

$$where \ \ IA(F) = \frac{1}{n}\sum_{m=1}^{n}\frac{\partial C(x' + m/n(x - x'))}{\partial F} \tag{2}$$

The neuron attribution matrix G^F is calculated by the summation classification outputs of L candidate object boxes. Due to without NMS scheme, the number of candidate boxes L is usually very large ($L > 100$). The summation of L classification outputs may bias the importance of some neurons. So, we drop out some neurons to leave behind the important neurons for decision-making. To do this, we sort all neuron attribution values in a descending order and only keep the top $k\%$. We select the value at the $k\%$ percentile as the threshold η. A mask M with the same dimension of G^F is calculated as in Eq. (3). In the following, the mask will be used to design the loss function.

$$M_i = \begin{cases} 1 & G_i^F > \eta \\ 0 & G_i^F \leq \eta \end{cases} \tag{3}$$

We design a loss function $L(x, G, F)$ for the specific layer F in Eq. (4). Here, G^F is a weighted feature map whose weights are determined by IA, $\odot$ represents element-wise multiplication, $||.||_2$ is L_2 norm. After executing $G^F \odot M$, we select the feature points that are crucial for the final detection results and build a loss function from them. By minimizing the Eq. (4), we decrease the positive feature points which are important for detections, leading to errors in the final detection results.

$$L(x, G, F) = ||G^F \odot M||_2 \tag{4}$$

$$\arg\min_{x^{adv}} L(x, G, F)$$

$$s.t. ||x^{adv} - x||_\infty \leq \varepsilon \tag{5}$$

The optimization for generating adversarial example is summarized in Eq. (5). Algorithm 1 outlines the details of generating adversarial perturbations using a gradient descent method like DAG [5]. The intermediate adversarial perturbation $r*$ is normalized as $\frac{\gamma \times r*}{\|r*\|_\infty}$ with the learning rate $\gamma = 0.5$. The final adversarial perturbation r is obtained by accumulating each intermediate adversarial perturbation $r*$. To ensure imperceptibility to the human eye, the normalized $r*$ is clipped within an L_∞ norm ε-ball. The final adversarial image $x^{adv} = x + r$ is obtained by truncating pixel values within the range of [0, 255].

4 Experiments

4.1 Experimental Setting

We conduct experiments using the validation set of 20-class PASCAL VOC 2012 which consists of 5,823 images, as well as the validation set of 80-class COCO 2017 consisting of 5,000 images. AP and AP^{50} are selected as evaluation metrics. AP^{50} is a mean average precision (mAP) with an IoU threshold of 0.5, which is the primary evaluation metric in PASCAL VOC 2012. AP used in the COCO 2017 dataset is the average mAP with IoU thresholds ranging from 0.5 to 0.95 with a step size of 0.05. After attacking, a lower value of AP or AP^{50} indicates a better attack method.

To test the effectiveness of adversarial attack against the two-stage detector Faster R-CNN, three different backbone networks are used, namely ResNet50 (FR-R50), ResNet50v2 (FR-R50v2) and MobileNetv3 (FR-M3). For the one-stage object detection models, SSD with VGG16 (SSD_VGG16) and YOLOv5s are employed to evaluate the transferability of adversarial examples across different detection frameworks. Specifically, adversarial examples generated by the two-stage object detectors are used to attack the one-stage object detection models.

Table 1. AP50(%) and AP (%) of attacked detectors(column) on the validation set of COCO 2017. FR-R50* as a white box, others are black-box attack.

	FR-R50*		FR-R50V2		FR-M3		SSD_VGG16		YOLOv5s	
	AP^{50}	AP	AP^{50}	AP	AP^{50}	AP	AP^{50}	AP	AP^{50}	AP
Clean	58.5	36.9	50.1	37.1	52.5	32.8	41.5	25.1	48.1	32.6
DAG	**0.0**	**0.0**	14.1	9.8	29.0	17.0	34.5	20.5	21.5	14.1
DAG +	**0.0**	**0.0**	12.5	8.6	23.3	12.6	30.2	17.3	13.8	8.6
RAP	0.1	0.0	20.0	13.7	31.2	18.3	36.1	21.4	27.8	18.1
CAP	0.3	0.1	20.3	14.0	34.8	20.3	36.2	21.4	25.8	16.8
TOGV	0.0	0.0	28.1	20.4	40.6	24.4	37.7	22.4	33.3	22.2
Ours	1.5	0.9	**7.8**	**4.8**	**10.7**	**6.0**	**25.1**	**14.1**	**7.5**	**4.8**

We compare our method with typical iterative optimization-based attacks, namely DAG [5], DAG+ [18], RAP [14], CAP [15], and TOGV [10], and set the maximum iteration number $T = 150, 20, 210, 200$ and 10 respectively. To ensure image quality, the perturbations constraint ϵ is set to be 0.08 for all methods. Other attack parameters of compared methods are set according to the original papers. For our method, when attacking FR-R50 under white-box cases, we will keep top 40% features of conv4_3 of ResNet50 to generate adversarial examples via Algorithm 1. The maximum iteration number T of our method is set to be 20.

Algorithm 1. Generating adversarial perturbation r

    Input: a clean image x, a two-stage object detector D_θ, the feature layer F, the reserving ratio k of the feature at layer F, the maximal iteration T, the learning rate $\gamma=0.5$, the perturbation constraint ϵ.

    Output: the adversarial perturbation r

(1) $t=0$, $r=0$

(2)   Calculate the neuronal attribution matrix G^F using $n=30$ virtual images as in Eq. (2)

(3)   Sort matrix elements of G^F and set the value at the $k\%$ percentile as the threshold η

(4)   Calculate the mask matrix M as in Eq. (3)

(5)   while $t \le T$ do:

(6) $x^{adv} = x + r$

(7)   Calculate the loss $L(x,G,F)$ as Eq. (4)

(8) $r^* = \nabla_x L(x,G,F)$

(9) $r = r + \dfrac{\gamma \times r^*}{\| r^* \|_\infty}$

(10) $t = t + 1$

(11)   end while

(12)   return r

4.2 Comparison with Other Methods

In this section, we compare our method with relevant state-of-the-arts. Table 1 and Table 2 demonstrate the results on the validation sets of COCO 2017 and PASCAL VOC 2012, respectively. As seen, our method exhibits much better attack performance under black-box settings, i.e., better transferability. As shown in Table 2 for attacking YOLOv5s, our method reduces AP^{50} to 17.6%, while the second-best DAG+ reduces AP^{50} to 31.2%. These results means that our method achieves at least 13.6% attack revenue. The results also show that the feature-level attacks, such as our method and DAG+, exhibits better transferability than other attacks which mainly destroy the downstream outputs of detectors. These results indicate that the feature-level attack benefits for transferability of adversarial examples.

To further validate the transferability of our method, we attack the 9th layer of MobileNetv3 of FR-M3 under the white-box setting, and test the transferability of adversarial examples to fool other models. For brevity, we compared our approach with DAG and DAG+ which performed well in Table 1 and Table 2. As shown in Table 3 for attacking FR-R50, FR-R50v2, and YOLOv5, our method achieves 12.7% ~ 34.6% higher of AP and 8.6% ~ 24.7% higher of AP^{50} respectively. These results again verify the superior transferability of our attack method.

Table 2. AP^{50}(%) and AP (%) of attacked detectors(column) on the validation set of **PASCAL VOC 2012**, FR-R50* as a white box, others are black-box attack.

	FR-R50*		FR-R50V2		FR-M3		SSD_VGG16		YOLOv5s	
	AP^{50}	AP	AP^{50}	AP	AP^{50}	AP	AP^{50}	AP	AP^{50}	AP
Clean	89.4	64.8	94.2	75.5	84.6	62.1	66.2	41.8	72.9	50.2
DAG	**0.0**	**0.0**	42.2	25.5	51.0	30.9	57.5	34.9	44.2	27.4
DAG +	**0.0**	**0.0**	34.9	21.2	36.7	20.7	**47.6**	**26.9**	31.2	18.0
RAP	3.0	1.3	26.9	16.9	43.8	25.2	54.5	32.8	39.4	23.0
CAP	0.3	1.0	48.7	29.9	57.1	35.0	58.6	35.5	46.9	28.9
TOGV	0.8	0.4	64.0	44.6	63.8	41.5	59.9	36.8	56.4	36.3
Ours	2.6	1.9	**7.8**	**5.0**	**23.8**	**13.0**	48.9	28.3	**17.6**	**10.2**

4.3 Ablation Study

We conduct ablation experiments to demonstrate the impact of different parameters of our attack. The considered parameters contain the selected layer, keeping rate of neurons and perturbations constraint ϵ. They are fixed as conv4_3, $k = 40\%$ and $\epsilon = 0.08$, when varying the other parameter for discussion. The dataset used in this section is the validation set of COCO 2017.

Influence of Attacking Different Layers. In this subsection, we evaluate attack revenues for attacking different layers of ResNet50 of FR-R50. The block with shortcut connection is the basic unit of ResNet50. So, we try to disturb feature outputs of the intermediate block, including conv3_1, conv3_2, conv3_3, conv4_1, conv4_2, conv4_3, conv5_1, conv5_2 and conv5_3. Please refer to Fig. 2 to see the details of blocks. For white-box cases shown in Fig. 3(a), attacking conv4_2 and conv4_3 reduces mAPs the most. For black-box cases shown in Fig. 3(b)-(d), attacking conv4_3 achieves the nearly best transferable performance. The middle-level feature of conv4_3 may contain some shared semantic information among different CNN-based backbones, thus disturbing feature map of conv4_3 will defeat different detectors. Figure 3 (b)-(d) also indicates that the proposed feature-level attack can transfer their adversarial examples to fool detectors with different network architectures or different framework.

Influence of Different Keeping Rates of Neurons. After calculating the neuron attribution matrix G of conv4_3, we select the top $k\%$ features to construct loss function, where k is defined as a keeping rate. By focusing on these $k\%$ top-ranked neurons and their corresponding features, we can capture crucial features for the proposed feature-level attack. For the white-box attack against the conv4_3 layer of FR-R50 in Fig. 4(a), keeping 20%-100% of neurons in the conv4_3 layer obtains nearly perfect attack performance which making AP be less than 1%. The higher keeping rate achieves a better attack revenue under white-box case. However, the results of black-box attacks shown in Fig. 4(b)-(d) are different from that of white-box attacks. As seen, keeping 20%-50% of neurons obtains the better attack performance than keeping other rates. For example, the attack with $k\% = 40\%$ achieves the best transferability performance against

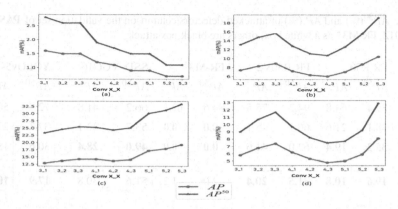

Fig. 3. The mAPs(%) on COCO 2017 of (a) FR-R50 undergone the proposed attacks against different layers of white-box case. (b) FR-M3, (c) SSD_VGG16, (d) YOLOv5s under black-box cases.

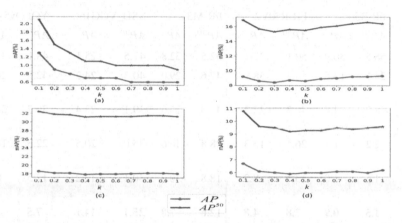

Fig. 4. The mAPs(%) on COCO 2017 of (a) FR-R50 undergone the white-box attacks with different keeping rates of features of conv4_3. (b) FR-M3, (c) SSD_VGG16, (d) YOLOv5s under black-box cases.

SSD_VGG16 and YOLOv5, reducing the clean AP from 25.1%, 32.6% to 17.8%, 5.9% respectively. The results in Fig. 4 (b)-(d) verify that the proposed selecting scheme of important neurons are helpful for improving transferability of adversarial examples.

Influence of Different Perturbation Constraints. The perturbation constraint ϵ refers to the attack strength, which is important for the transferable performance. We apply our method with normalized perturbation constraints of 0.01, 0.02, 0.04, 0.06, and 0.08. As shown in Table 4, with increasing the perturbation magnitude, AP and AP^{50} of all detectors reduces more, indicating that the adversarial examples generated by our method demonstrates a more significant adversarial performance. Considering that the attack with $\epsilon = 0.08$ has the best attack performance and will not decrease image quality sharply, so we set $\epsilon = 0.08$ for our method.

Table 3. $AP^{50}(\%)$ and AP (%) of attacked detectors(column)on the validation set of **PASCAL VOC 2012**, FR-M3* as a white box, others are black-box attack.

	FR-R50		FR-R50V2		FR-M3*		SSD_VGG16		YOLOv5s	
	AP^{50}	AP	AP^{50}	AP	AP^{50}	AP	AP^{50}	AP	AP^{50}	AP
Clean	89.4	64.8	94.2	75.5	84.6	62.1	66.2	41.8	72.9	50.2
DAG	38.1	21.6	66.1	45.1	**0.0**	**0.0**	53.7	32.1	37.3	22.7
DAG +	35.2	19.4	62.0	41.6	**0.0**	**0.0**	**49.0**	**28.4**	30.6	18.0
Ours	**19.6**	**10.8**	**31.5**	**20.4**	2.0	1.2	51.6	30.8	**17.9**	**10.4**

Table 4. $AP^{50}(\%)$ and AP (%) of attacked detectors(column) on the validation set of **COCO 2017**, FR-R50* as a white box under different perturbation constraint ϵ.

	FR-R50*		FR-R50V2		FR-M3		SSD_VGG16		YOLOv5s	
	AP^{50}	AP	AP^{50}	AP	AP^{50}	AP	AP^{50}	AP	AP^{50}	AP
Clean	58.5	36.9	50.1	37.1	52.5	32.8	41.5	25.1	48.1	32.6
$\epsilon = 0.01$	24.1	13.6	58.3	38.9	47.6	29.0	40.3	24.2	42.3	28.4
$\epsilon = 0.02$	8.7	4.8	49.3	32.2	43.0	25.6	39.1	23.4	36.7	24.5
$\epsilon = 0.04$	2.2	1.3	26.1	16.3	28.8	16.6	35.0	20.5	22.3	14.5
$\epsilon = 0.06$	1.6	1.0	13.1	8.2	18.8	10.0	30.1	17.2	12.5	8.1
$\epsilon = 0.08$	**1.5**	**0.9**	**7.8**	**4.8**	**12.6**	**7.0**	25.1	14.1	**7.5**	**3.8**

5   Conclusion

In this study, we propose a neuron attribution-based attack to improve the transferability of adversarial examples for fooling object detectors with different backbones or architectures. Through analyzing the detection process, we propose a new method to calculate the neuron attribution tailored for the detection mechanism of object detector. Furthermore, when constructing loss function, we propose a filter scheme to filter out some unimportant neurons. The main contribution of this paper lies that we design a neuron attribution-based transferable attack adapted for object detectors. Extensive experiments on PASCAL VOC and COCO dataset demonstrates that our method achieves better transferable performance than state-of-the-arts when fooling detectors with different backbones or different detection frameworks.

Acknowledgement. This work was partially supported by NFSC No.62072484, Sichuan Science and Technology Program (No. 2022YFG0321, No. 2022NSFSC0916), the Opening Project of Engineering Research Center of Digital Forensics, Ministry of Education.

References

1. He, K., Zhang, X., Ren, S., Sun, J.: Deep residual learning for image recognition. In: Proceedings of the IEEE Conference on Computer Vision and Pattern Recognition, pp.770–778 (2016)
2. Redmon, J., Farhadi, A.: YOLOv3: An incremental improvement. arXiv preprint arXiv:1804. 02767 (2018)
3. Ronneberger, O., Fischer, P., Brox, T.: U-Net: Convolutional networks for biomedical image segmentation. In: Medical Image Computing and Computer-Assisted Intervention–MICCAI 2015, pp. 234–241 (2015)
4. Goodfellow, I.J., Shlens, J., Szegedy, C.: Explaining and harnessing adversarial examples. arXiv preprint arXiv:1412.6572 (2014)
5. Xie, C., Wang, J., Zhang, Z., Zhou, Y., Xie, L., Yuille, A.: Adversarial examples for semantic segmentation and object detection. In: Proceedings of the IEEE International Conference on Computer Vision, pp. 1369–1378 (2017)
6. Liu, X., Yang, H., Liu, Z., Song, L., Li, H., Chen, Y.: DPATCH: An adversarial patch attack on object detectors. arXiv preprint arXiv:1806.02299 (2018)
7. Waseda, F., Nishikawa, S., Le, T.N., Nguyen, H.H., Echizen, I.: Closer look at the transferability of adversarial examples: how they fool different models differently. In: Proceedings of the IEEE/CVF Winter Conference on Applications of Computer Vision, pp. 1360–1368 (2023)
8. Zhang, C., Benz, P., Karjauv, A., Cho, J.W., Zhang, K., Kweon, I.S.: Investigating top-k white-box and transferable black-box attack. In: Proceedings of the IEEE/CVF Conference on Computer Vision and Pattern Recognition, pp. 15085–15094 (2022)
9. Tramèr, F., Kurakin, A., Papernot, N., Goodfellow, I., Boneh, D., McDaniel, P.: Ensemble adversarial training: Attacks and defenses. arXiv preprint arXiv:1705.07204 (2017)
10. Chow, K.H., Liu, L., Gursoy, M.E., Truex, S., Wei, W., Wu, Y.: TOG: targeted adversarial objectness gradient attacks on real-time object detection systems. arXiv preprint arXiv:2004. 04320 (2020)
11. Liao, Q., Wang, X., Kong, B., Lyu, S., Zhu, B., Yin, Y., Wu, X.: Transferable adversarial examples for anchor free object detection. In: 2021 IEEE International Conference on Multimedia and Expo (ICME), pp. 1–6. (2021)
12. Huang, H., Chen, Z., Chen, H., Wang, Y., Zhang, K.: T-SEA: transfer-based self-ensemble attack on object detection. In: Proceedings of the IEEE/CVF Conference on Computer Vision and Pattern Recognition, pp. 20514–20523 (2023)
13. Lu, J., Sibai, H., Fabry, E.: Adversarial examples that fool detectors. arXiv preprint arXiv: 1712.02494 (2017)
14. Li, Y., Tian, D., Chang, M.C., Bian, X., Lyu, S.: Robust adversarial perturbation on deep proposal-based models. arXiv preprint arXiv:1809.05962 (2018)
15. Zhang, H., Zhou, W., Li, H.: Contextual adversarial attacks for object detection. In: 2020 IEEE International Conference on Multimedia and Expo (ICME), pp. 1–6 (2020)
16. Wu, X., Huang, L., Gao, C., Lee, W. S., Suzuki, T.: G-UAP: generic universal adversarial perturbation that fools RPN-based detectors. In: ACML, pp. 1204–1217 (2019)

17. Chen, S.T., Cornelius, C., Martin, J., Chau, D.H.: Shapeshifter: Robust physical adversarial attack on faster R-CNN object detector. In: Machine learning and knowledge discovery in databases: European Conference, ECML PKDD, pp. 52–68 (2019)
18. Shi, G., Peng, A., Zeng, H.: An enhanced transferable adversarial attack against object detection. In: International Joint Conference on Neural Networks. in press (2023)
19. Wang, D., Li, C., Wen, S., Han, Q.L., Nepal, S., Zhang, X., Xiang, Y.: Daedalus: breaking non-maximum suppression in object detection via adversarial examples. In: IEEE Trans. Cybern. **52**(8), pp.7427–7440 (2021)
20. Ganeshan, A., BS, V., Babu, R. V.: FDA: feature disruptive attack. In: Proceedings of the IEEE/CVF International Conference on Computer Vision, pp. 8069–8079 (2019)
21. Naseer, M., Khan, S. H., Rahman, S., Porikli, F.: Task-generalizable adversarial attack based on perceptual metric. arXiv preprint arXiv:1811.09020 (2018)
22. Dhamdhere, K., Sundararajan, M., Yan, Q.: How important is a neuron? arXiv preprint arXiv: 1805.12233 (2018)
23. Zhang, J., Wu, W., Huang, J.T., Huang, Y., Wang, W., Su, Y., Lyu, M.R.: Improving adversarial transferability via neuron attribution-based attacks. In: Proceedings of the IEEE/CVF Conference on Computer Vision and Pattern Recognition, pp.14993–15002 (2022)
24. Ren, S., He, K., Girshick, R., Sun, J.: Faster R-CNN: towards real-time object detection with region proposal networks. In: Advances in Neural Information Processing Systems, vol. 28 (2015)
25. Liu, W., Anguelov, D., Erhan, D., Szegedy, C., Reed, S., Fu, C.Y., Berg, A.C.: SSD: single shot multibox detector. In: ECCV 2016, pp. 21–37 (2016)
26. Carlini, N., Wagner, D.: Towards evaluating the robustness of neural networks. In: 2017 IEEE Symposium on Security and Privacy, pp. 39–57 (2017)
27. Athalye, A., Engstrom, L., Ilyas, A., Kwok, K.: Synthesizing robust adversarial examples. In: International Conference on Machine Learning PMLR, pp. 284–293 (2018)
28. Zhu, X., Lyu, S., Wang, X., Zhao, Q.: TPH-YOLOv5: Improved YOLOv5 based on transformer prediction head for object detection on drone-captured scenarios. In: Proceedings of the IEEE/CVF International Conference on Computer Vision, pp. 2778–2788 (2021)
29. Lin, T., Dollár, P., Girshick, R., He, K., Hariharan, B., Belongie, S.: Feature pyramid networks for object detection. In: IEEE Conference on Computer Vision and Pattern Recognition, pp. 936–944 (2016)
30. He, K., Gkioxari, G., Dollár, P., Grishick, R.: Mask R-CNN. IEEE Trans. Pattern Anal. Mach. Intell. **42**, 386–397 (2017)
31. Liu, X., Yang, H., Liu, Z., Song, L., Chen, Y., Li, H.H.: DPATCH: an adversarial patch attack on object detectors. In: Proceedings of the IEEE/CVF Conference on Computer Vision and Pattern Recognition (2018)
32. Hu, Z., Huang, S., Zhu, X., Sun, F., Zhang, B., Hu, X.: Adversarial texture for fooling person detectors in the physical world. In: Proceedings of the IEEE/CVF Conference on Computer Vision and Pattern Recognition, pp. 307–316 (2022)

DKCS: A Dual Knowledge-Enhanced Abstractive Cross-Lingual Summarization Method Based on Graph Attention Networks

Shuyu Jiang[1], Dengbiao Tu[2], Xingshu Chen[1,3]($\boxtimes$), Rui Tang[1], Wenxian Wang[3], and Haizhou Wang[1]($\boxtimes$)

[1] School of Cyber Science and Engineering, Sichuan University, Chengdu, China
{chenxsh,whzh.nc}@scu.edu.cn
[2] National Computer Network Emergency Response Technical Team Coordination Center of China, Beijing, China
[3] Cyber Science Research Institute, Sichuan University, Chengdu, China

Abstract. Cross-Lingual Summarization (CLS) is the task of generating summaries in a target language for source articles in a different language. Previous studies on CLS mainly take pipeline methods or train an attention-based end-to-end model on translated parallel datasets. However, challenges arising from lengthy sources and non-parallel mappings hamper the accurate summarization and translation of pivotal information. To address this, this paper proposes a novel **Dual Kknowledge-enhanced abstractive CLS** model (DKCS) via a graph-encoder-decoder architecture. DKCS implements a clue-focused graph encoder that utilizes a graph attention network to simultaneously capture inter-sentence structures and significant information guided by extracted salient internal knowledge. Additionally, a bilingual lexicon is introduced in the decoder with an attention layer for enhanced translation. We construct the first hand-written CLS dataset for evaluation as well. Experimental results demonstrate the model's robustness and significant performance gains over the existing SOTA on both automatic and human evaluations.

Keywords: knowledge-enhanced summarization · cross-lingual summarization · graph attention · internal knowledge · bilingual semantics

1 Introduction

With the constant acceleration of information globalization, an ever-increasing mass of information in various languages has flooded into people's views. Whether it's scientific research or daily reading, it will always encounter the need to quickly grasp the gists of such massive articles. However, it is hard for people to accomplish this for an article written in an unfamiliar language. At this time, the need for cross-lingual summarization (CLS) has become extremely urgent. Cross-lingual summa-

© The Author(s), under exclusive license to Springer Nature Singapore Pte Ltd. 2024
B. Luo et al. (Eds.): ICONIP 2023, CCIS 1967, pp. 109–121, 2024.
https://doi.org/10.1007/978-981-99-8178-6_9

rization is the task to compress an article in a source language (e.g. English) into a summary in a different target language (e.g. Chinese).

Compared with monolingual summarization (MS), CLS requires models to have translation capabilities except for summarization capabilities. Since source and target words are not in a one-to-one correspondence coupled with instances where some target words are absent in source texts, the challenge of mapping bilingual semantics in CLS becomes notably more intricate than conventional machine translation (MT) tasks. Recently, most end-to-end abstractive models [15,23,24] mainly use the attention mechanism [16] to implicitly model important information and build the mapping of bilingual semantics. However, as the length of input increases, the attention of models faces the problem of long-distance dependency truncation and is easily distracted or misled by more redundant information, leading to a decline in the ability to summarize and translate.

To tackle these problems, we propose a **D**ual **K**nowledge-enhanced **C**ross-lingual **S**ummarizer, namely **DKCS**, to enhance summarization and translation capabilities of CLS by simultaneously incorporating internal and external knowledge into DKCS of the proposed graph-encoder-decoder architecture.

Specifically, DKCS first extracts salient internal knowledge such as keywords, topic words and named entities from the source article as key clues to explicitly guide CLS since these important words can provide significant prompts for the summary [21]. Then it introduces a Clue-Focused Graph (CFG) encoder, which transforms the source article into a clue-focused graph based on the co-occurrence of key clues and leverages graph attention networks (GAT) [17] to synchronously learn inter-sentence structures and aggregate important information. The original encoder of Transformer is used as Clue encoder to encode key clues to mitigate the oblivion of important information. Subsequently, the hidden states of both encoders are sent into the decoder to jointly generate summaries. Meanwhile, DKCS exploits the attention distribution of extracted internal knowledge to directly capture salient target words from the external bilingual knowledge, to enhance the ability of translation. In summary, our contributions are as follows:

1. To the best of our knowledge, we are the first to construct and publish the hand-written cross-lingual dataset (namely **CyEn2ZhSum**[1]) to avoid the "translationese" [8] phenomenon for the evaluation of CLS. It contains 3,600 English news about cybersecurity, 3,600 Chinese hand-written summaries with publication dates, bilingual titles, etc.
2. To solve the performance degradation caused by over-length source information and non-one-to-one bilingual mappings, we propose a DKCS model of a graph-encoder-decoder architecture to simultaneously incorporate internal and external knowledge by adding GAT into Transformer and introducing a bilingual lexicon with a cross-attention layer.
3. Experimental results on two benchmark translated parallel datasets and our hand-written dataset show that this method has stronger robustness for longer inputs and substantially outperforms the existing SOTA on both automatic and human evaluations.

---

[1] CyEn2ZhSum dataset is available in https://github.com/MollyShuu/CyEn2ZhSum.

2   Related Work

Early research on CLS usually takes a pipeline-based approach, either summarize-then-translate [22], or translate-then-summarize [14], with various strategies to integrate bilingual features. Although pipeline-based approaches are intuitive, they still suffer from recurring latency and error propagation [20].

To alleviate these problems, many advanced end-to-end CLS methods have emerged in recent years. At first, most studies focus on training end-to-end models using zero-shot learning [5] or transfer learning [3], due to lacking training corpus. As large-scale CLS corpora became available, increasing studies have focused on designing various strategies to improve the summarization and translation ability of CLS [1,10,15,19,24]. Liang et al. [10], Okazaki et al. [15] and Zhu et al. [23] unitized extra MS and MT data based on the multi-task framework to respectively reinforce the summarization and translation ability of CLS. For translation improvement, Zhu et al. [24] incorporated external bilingual alignment information to aid models in language mapping. Wang et al. [19] explored extending multi-lingual pre-trained language models to CLS. To enhance summarization capabilities, Bai et al. [1] made the MS task a CLS prerequisite, introducing the MCLAS framework for the unified generation of monolingual and cross-lingual summaries.

Different from these studies, we have explored simultaneously integrating internal knowledge and external knowledge to enhance the capability of summarization and translation of CLS, instead of only utilizing the external knowledge from extra MS or MT data.

3   The Framework

The architecture of the proposed DKCS method is depicted in Fig. 1. Here, let V_s and V_t be the vocabulary of the source language and target language respectively. Given an input article as a set of sentences $D = \{s_1, s_2, ..., s_k\}$ where s_i represents a sentence, we first extract and model the internal knowledge including keywords, topic words, and named entities as key clues $C = \{c_1, c_2..., c_n\}$ for explicit guidance as described in Sect. 3.1 Then, we transform the article into a clue-focused graph $G = (V, E)$ based on the co-occurrence of C, and implement the graph attention network to synchronously learn article structures and aggregate important information as described in Sect. 3.2. At last, the embeddings of C and G are sent into one decoder to jointly decide on the generation of target summary $S = \{y_1, y_2, ..., y_m\}$ where y_i represents a word and $(y_i, S \subseteq V_t)$. In particular, an additional translation layer is added to the decoder to use the external translation knowledge as described in Sect. 3.4.

3.1   Key Clues Extraction

Keywords and topics, considered as a condensed form of essential content, are proven to provide crucial clues for maintaining the general idea or gist of the target text [21]. In addition, named entities usually appear as the subjects or objects

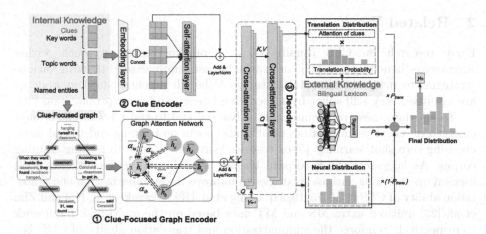

Fig. 1. The framework of DKCS. $\vec{h_i}$ represents the i-th vertex embedding. Q, K, and V represent the query, key, and value of the attention mechanism respectively. Y_{m-1} is the previous m-1 target words and p_{trans} is the probability of translating clue words from an external bilingual lexicon. Words in the blue, green, and purple boxes in the clue-focused graph represent extracted internal knowledge.

in the summary, but are often omitted or incorrectly translated due to their low frequency. Therefore, we take the sets of these three kinds of internal knowledge as key clues: (1) keywords; (2) topic words; (3) named entities. Specifically, we first respectively apply TextRank [13], TopicRank [2], and a named entity recognition model[2] to obtain these three kind of internal knowledge. These extracted named entities, words, and phrases consist of a coarse set of key clues. denoted as $C = \{c_1, c_2..., c_n\}$. Then, repeated items in this coarse set are de-duplicated and merged to get the fine-grained one. Lastly, we reorder the sentences in the article with TextRank and re-rank the de-duplicated key clues in the order where they appear in the reordered article.

3.2   Clue-Focused Graph Encoder

This section introduces how to construct and encode the clue-focused graph. Here, every vertex is initialized by the self-attention score of a corresponding sentence and dynamically updated with the weight of graph attention.

(1) **Transform an Article Into a Clue-Focused Graph.** Firstly, every sentence in the source article is set as a vertex. Then the vertex v_i and v_j are connected by edge e_{ij} if they share at least one key clue and the weight of e_{ij} (denoted as ω_{ij})equals the number of key clues they both contain. We also add a self-loop edge for every vertex to enhance the attention of self-information. The weight ω_{ii} of i-th self-loop edge equals the number of key clues its vertex contains. The weight of every vertex, $\omega(v_i)$, equals the sum of all the connected

---

[2] https://spacy.io/models/.

edges' weights, representing the degree of clue aggregation for the corresponding sentence. The importance of a vertex is represented by its weight and the vertex with a weight of 0 is removed to reduce the impact of noisy or useless sentences.

(2) Encode CFG with Graph Attention Network. As stated above, a vertex is associated with a sentence that consists of a word sequence $s = \{x_1, x_2, ..., x_t\}$. To capture the positional information of every word in the sentence, we use positional encoding [16] to get its positional embedding. The final embedding of i-th word is the sum of its original word embedding $\mathbf{x}_i$ and its positional embedding $\mathbf{p}_i$. Then we concatenate that as the sentence embedding and feed it into a multi-head attention (MHA) layer to obtain the embedding h_i for each vertex v_i as follows:

$$h_i = \text{MHA}(\|_{i-1}^t (\mathbf{x}_i + \mathbf{p}_i), \|_{i-1}^t (\mathbf{x}_i + \mathbf{p}_i), \|_{i=1}^t (\mathbf{x}_i + \mathbf{p}_i)), \tag{1}$$

where $\|$ represents the concatenation operation.

After getting the embedding of each vertex, we implement multi-head graph attention to stabilize the learning process and capture the article graph structure as well as sentence semantics. The graph attention of every head is independently calculated and then concatenated together. In this process, every vertex v_i updates its embedding using the attention over itself and 1-hop neighbors:

$$\overrightarrow{h_i} = \|_{h=1}^H f \left(\sum_{j \in \mathcal{N}(v_i)} \alpha_{i,j}^h \mathbf{W}^h h_j \right),$$
$$\alpha_{i,j} = \frac{\exp \left(LeakyReLU \left(a^T \cdot [\mathbf{W}h_i \oplus \mathbf{W}h_j] \right) \right)}{\sum_{j' \in \mathcal{N}} \exp \left(LeakyReLU \left(a^T \cdot [\mathbf{W}h_i \oplus \mathbf{W}h_{j'}] \right) \right)}, \tag{2}$$

where $f(\cdot)$ denotes the activation function; H denotes the number of heads; $\alpha_{i,j}$ denotes the normalized attention between vertex v_i and v_j; and $\cdot^T$ represents transposition. $a \in \mathbb{R}^{2d}$ is a learnable weight vector used to calculate the attention score and $\mathbf{W} \in \mathbb{R}^{d \times d}$ is the weight matrix for feature transformation.

In the last layer of the graph attention network, averaging is employed instead of concatenation to combine multi-head graph attention outputs:

$$\overrightarrow{v}_i = f \left(\frac{1}{H} \sum_{f=1}^H \sum_{j \in \mathcal{N}(v_i)_i} \alpha_{ij}^h \mathbf{W}^h \overrightarrow{h_j} \right). \tag{3}$$

To avoid the over-smoothing problem, we employ a residual connection and layer normalization around an MLP layer to get the final outputs $Z^g = [z_1^g; z_2^g; ...; z_t^g]$ of the Graph encoder as follows:

$$z_i^g = \text{LayerNorm} \left(\overrightarrow{v}_i + \text{MLP} \left(\overrightarrow{v}_i \right) \right). \tag{4}$$

3.3   Clue Encoder

Although the Graph encoder has captured the global article structure and sentence semantics, some important information is still omitted because a vertex can not represent different word semantics in one sentence. Therefore, we use an additional encoder to capture the key clues C. The word embedding of clues c_i are directly concatenated as the input since there is no sentence structure within clue words. Then the encoder of Transformer [16] is utilized as the Clue Encoder to model clues into deep context embeddings Z^c.

3.4 Decoder

The decoder shares the same architecture as Clue encoder, except that two extra cross-attention blocks are added to model the dependencies between the input sequences and the target summary. In the decoding process, the final probability distribution $P(w)$ of target words consists of the neural distribution P_N and the translation distribution of key clues P_T.

Neural Distribution. As shown in Fig. 1, the graph representation Z^g and the attention distribution of clues Z^c are passed through two cross-attention blocks to get the decoder hidden state Z^{dec}. Then the final neural distribution $P_N(w)$ on V_t is calculated as follows:

$$P_N(w) = \text{softmax}\left(\mathbf{W}_{dec} Z^{dec}\right), \tag{5}$$

$\mathbf{W}_{dec} \in \mathbb{R}^{|V_t| \times d}$ is a learnable matric, used to project Z^{dec} to the vector of the same size of V_t. $|V_t|$ represents the size of the target vocabulary V_t.

Translation Distribution. We apply the fast-align tool [7] to extract the word alignments on LDC translation corpora and only save the alignments exiting in both source-to-target and target-to-source direction as the bilingual lexicon. The average of the maximum likelihood estimation on these word alignments L of both directions is used as the translation probability $P^L(w_1 \Rightarrow w_2)$. Then the translation distribution P_T is calculated as follows [24]:

$$P_T(w_1 \Rightarrow w_2) = \frac{P^L(w_1 \Rightarrow w_2)}{\sum_{w_j} P^L(w_1 \Rightarrow w_j)}. \tag{6}$$

Final Probability Distribution. To determine whether generating a word w from the neural distribution sampling or the translation distribution of clue words, P_N and P_T are weighted by the translating probability $p_{\text{trans}} \in [0, 1]$. p_{trans} is computed from the cross-attention of clues $\overrightarrow{Z^c}$ via a dynamic gate:

$$p_{\text{trans}} = \text{sigmoid}\left(\mathbf{W}_2\left(\mathbf{W}_1 \overrightarrow{Z^c} + b_1\right) + b_2\right), \tag{7}$$

where $b_1 \in \mathbb{R}^d$ and $b_2 \in \mathbb{R}^1$ are learnable bias vectors, $\mathbf{W}_1 \in \mathbb{R}^{d \times d}$ and $\mathbf{W}_2 \in \mathbb{R}^{1 \times d}$ are learnable matrices. Finally, the final probability distribution $P(w)$ is calculated from $P_N(w)$, P_T and p_{trans} as follows:

$$P(w) = p_{\text{trans}} \sum_{i:w_i = w_{\text{clue}}} z_{t,i}^c P_T(w_{\text{clue}} \Rightarrow w) + (1 - p_{\text{trans}}) P_N(w). \tag{8}$$

where $P_T(w_{\text{clue}} \Rightarrow w)$ is the translation probability of source clue word w_{clue} to target word w.

4   Experimental Setup

4.1   Datasets

We first evaluate our method on the benchmark English-to-Chinese summarization dataset **En2ZhSum** and the Chinese-to-English summarization dataset **Zh2EnSum** created by Zhu et al. [23]. The summaries of these two datasets are both translated from original monolingual summaries with a round-trip translation strategy. To avoid the "translationese" in translated texts, we constructed a small but precise human-written dataset, namely **CyEn2ZhSum**, to further evaluate the performance of models on human-written CLS. The statistics of training, validation, and test sets for these datasets are presented in Table 1.

CyEn2ZhSum contains 3,600 English news about cybersecurity and 3,600 Chinese summaries which are first written by a postgraduate, and then reviewed and revised by researchers. The statistics of CyEn2ZhSum dataset are presented in Table 2.

Table 1. The statistics of training, validation and test sets.

Datasets	Source	Train	Valid	Test
En2ZhSum	News	364,687	3,000	3,000
Zh2EnSum	Microblog	1,693,713	3,000	3,000
CyEn2ZhSum	News	3,000	300	300

Table 2. The statistics of CyEn2ZhSum. Ctitle, Etitle, Doc and Sum represent the Chinese title, the English title, the source article and the target summary of CyEn2ZhSum dataset. AvgChars and AvgSents respectively represent the average number of characters and sentences. AvgSentsChar refers to the average number of characters in a sentence. TotalNum is the total number of CLS pairs.

Statistic	Ctitile	Etitle	Doc	Sum
AvgChars	20.27	11.37	593.60	135.81
AvgSents	1.01	1.06	19.27	3.69
AvgSentsChar	20.17	10.84	35.45	36.89
TotalNum	3600			

4.2   Baselines

We compare our method with and two relevant pipeline-based methods, **Pipe-TS** [3] (a translate-then-summarize method) and **Pipe-ST** [3] (a summarize-then-translate method), and following strong end-to-end methods: **VHM** [10], **TNCLS** [23], **ATS-NE** [24], **ATS-A** [24], models based on pre-trained language models including **mBART** [12] and **GPT3.5** (text-davinci-003 backed)[3] [18];

---

[3] https://platform.openai.com/docs/models/gpt-3-5.

4.3   Experimental Settings

Following configuration *transformer_base* [16], the embedding dimensions d and the inner dimension d_{ff} are respectively set to 512 and 2048, and Clue encoder and decoder both have 6 layers of 8 heads for attention. For Graph encoder, it is set to 1 layer of 3 heads for attention. We use the batch size of 1024 for all datasets. The vocabulary setting follows [23]. During training, we apply Adam with $\beta_1 = 0.9$, $\beta_2 = 0.998$ and $\epsilon = 10^{-9}$ as the optimizer to train parameters. During the evaluation stage, we decode the sequence using beam search with a width of 4 and length penalty of 0.6. For CyEn2ZhSum dataset, we carry out the experiments of several end-to-end methods on it through fine-tuning the model trained from En2ZhSum dataset.

5   Results and Analysis

Similar to previous studies [3,6,23,24], we assessed all models with the F1 scores of standard **ROUGE-1/2/L** [11] and **METEOR** metrics [4].

5.1   Automatic Evaluation

Results on En2ZhSum and Zh2EnSum. We have replicated baselines on the En2ZhSum and Zh2EnSum datasets. As shown in Table 3, our model DKCS can substantially outperform all baselines in both English-to-Chinese and Chinese-to-English summarization, respectively achieving an improvement of +3.39 ROUGE-1 and +1.71 METEOR scores.

There is an interesting phenomenon that mBART and GPT3.5 perform much better on the En2ZhSum dataset than on the Zh2EnSum dataset. We think it is reasonable because most target words in En2ZhSum dataset can be directly translated from the source text while most target words in Zh2EnSum dataset are not in the source text. This requires higher summarizing ability while these two pre-trained language models have better translation ability than summarization ability, since they have learned bilingual mappings from a large number of multilingual datasets in advance.

We find the performance of both pipeline-based methods on the Zh2EnSun dataset is much better than that on the En2ZhSum dataset. This is mostly because the source article in the En2ZhSum dataset is much longer than that in the Zh2EnSum dataset. As longer texts contain richer semantics, the longer the text is, the more semantic deviations are generated during the translation process, resulting in poorer performance for cross-lingual summarization.

Table 3. Automatic evaluation experimental results. The best results of baselines are underlined with wavy lines and those of all models are shown in bold. R1, R2, RL and ME represent ROUGE-1, ROUGE-2, ROUGE-L and METEOR respectively. * (†) indicates that the improvements over baselines are statistically significant with $p < 0.001(0.05)$ for t-test.

Metrics		Pipe-TS	Pipe-ST	TNCLS	ATS-NE	ATS-A	mBART	VHM	GPT3.5	CGS
En2Zh Sum	R1	25.21	27.23	34.93	36.94	38.01	43.02	41.51	41.85	**45.56**†
	R2	7.41	8.91	15.04	17.23	18.37	22.58	22.98	18.70	**23.97**†
	RL	20.01	19.50	27.13	28.25	29.37	31.85	31.23	28.48	**32.86**
	ME	10.81	11.25	14.94	15.48	16.51	18.97	20.01	20.48	**20.72***
Zh2En Sum	R1	26.08	31.39	35.21	36.31	37.41	28.31	40.62	24.15	**40.80**†
	R2	10.86	14.28	18.59	20.12	21.08	11.89	21.97	6.70	**22.05**†
	RL	24.34	29.15	33.01	34.35	35.60	25.86	37.26	19.43	**37.88**†
	ME	12.02	16.00	17.82	18.04	18.54	15.98	18.98	16.66	**20.69***
CyEn2 ZhSum	R1	–	–	37.49	39.01	39.18	42.73	40.09	42.75	**42.91**†
	R2	–	–	18.69	20.86	21.43	22.26	21.91	22.32	**22.67**
	RL	–	–	32.65	34.91	36.72	35.38	34.07	**36.99**	35.36
	ME	–	–	16.12	17.16	17.49	17.83	19.30	20.91	**21.33**†

Results on CyEn2ZhSum. We implement end-to-end baselines and our method on the human-written CyEn2ZhSum dataset. As shown in Table 3, DKCS outperforms baselines in all metrics except ROUGE-L metric. As the CyEn2ZhSum dataset belongs to the cybersecurity domain, it contains lots of proper nouns in the field of cybersecurity. The translations of these proper nouns are much different from daily words. Consequently, we speculate that ATS-A, mBART, and GPT3.5 achieve higher ROUGE-L scores due to dynamic translation probability adjustment and pre-training on multilingual data.

The higher METEOR scores of DKCS indicate the overall semantics of summaries generated by DKCS are closer to reference summaries although more the longest common sequences appear in the summaries of GPT3.5, mBART and ATS-A. This is most likely because some reference words in the summary generated by DKCS are represented by synonyms or the same morpheme with different forms. In summary, our DKCS model significantly outperforms baselines in automated evaluation.

5.2 Ablation Study

In this section, we carry out an ablation study to analyze the influence of each key component of DKCS. DKCS-C and DKCS-G respectively refer to the variants of DKCS model that remove CFG encoder and Clue encoder. Experimental results are shown in Table 4. It reveals that both ablation operations will cause a decrease in the model performance. We find DKCS-C performs much better than DKCS-G. This is mostly because DKCS-C directly learns salient information from key clues, and is affected by less noisy information.

Table 4. Ablation study on three datasets. * (†) means the statistically significant improvements over two variants with DKCS from t-test at $p < 0.01(0.05)$

Method	En2ZhSum				Zh2Ensum				CyEn2ZhSum			
	R1	R2	RL	ME	R1	R2	RL	ME	R1	R2	RL	ME
DKCS-G	21.10	3.66	14.71	8.22	18.30	2.93	15.76	7.87	31.33	14.43	27.77	15.38
DKCS-C	44.48	22.42	30.44	18.96	38.22	19.20	36.25	19.18	37.13	19.20	32.09	15.96
DKCS	46.56†	23.97†	32.86†	20.72*	40.80*	22.05*	37.88†	20.69*	42.91	22.67	35.36*	21.33†

5.3 Human Judgement

Since ROUGE and METEOR metrics can't assess the conciseness, informativeness and fluency, we randomly select 60 samples form test sets and ask three graduate researchers who are proficient in Chinese and English to rank generated summaries based on these aspects using the Best-Worst Scaling method [9].

Figure 2 shows the rank distribution for each method. Human evaluation yielded Fleiss' Kappa of 0.26, 0.31, and 0.20 for datasets, indicating strong annotator consistency. Results imply annotators think the summaries generated by DKCS can achieve the best performance on all datasets to the greatest extent and the fewest summaries generated by DKCS are chosen as the worst ones. Similar to results of automatic evaluation, for summaries generated by GPT3.5 and mBART, the annotators selected more samples on the Zh2EnSum dataset as the worst ones as well.

5.4 The Impact of Article Length

To explore the impact of article length on each model, we divide the articles in the En2ZhSum test set into 10 collections according to the article length. These 10 collections are shown in the legend in Fig. 3, where $[a, b)$ represents a set of articles whose number of words is bigger than or equal to a and smaller than b. We compare the performance of DKCS with end-to-end baselines on these 10 collections. The ROUGE-L score differences of $[0, 100)$ set and other sets are shown in Fig. 3.

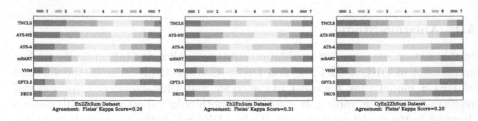

Fig. 2. Human evaluation results. Rank distribution for generated summaries on all datasets, with numbers in the legend indicating annotator-assigned ranks. Higher numbers correspond to better quality summaries.

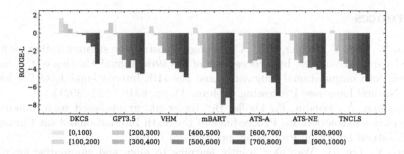

Fig. 3. ROUGE-L score difference of summaries generated from articles of different lengths. The legend indicates the range of the number of source words. Vertical coordinates represent the difference of ROUGE-L scores between each set and the [0,100) set.

Results show that the performance of all models decreases with the increase in article length. However, our DKCS decreases much slower than other models, indicating DKCS is more robust to the length of input texts. The overall downward trend most likely results from the increasing redundant information distracting the attention of models. The reason why our DKCS can greatly relieve the impact of the change of article length is that it extracts and encodes the salient internal knowledge as explicit guidance and the noisy sentences and important sentences are treated differently when constructing CFG. So the attention of the DKCS model is largely undistracted by useless information.

6   Discussion and Conclusion

This paper proposed a novel dual knowledge-enhanced end-to-end CLS summarization method. Apart from implicitly modeling with the attention mechanism, it extracts and encodes salient internal knowledge as key clues to explicitly guide the encoding stage and utilizes the attention distribution of clues to directly get salient target words from external bilingual knowledge. Experimental results show that the model performance decreases with the increase of article length while our proposed method can take advantage of internal knowledge to achieve better robustness than baselines. In future work, we would like to supplement and reorganize these clues into meta-event graphs to summarize the content about specific types of events.

Acknowledgements. This work is partially supported by the National Natural Science Foundation of China (No. U19A2081), the National Key Research and Development Program of China (No. 2022YFC3303101), the Key Research and Development Program of Science and Technology Department of Sichuan Province(No. 2023YFG0145), Science and Engineering Connotation Development Project of Sichuan University (No. 2020SCUNG129) and the Fundamental Research Funds for the Central Universities (No. 2023SCU12126).

References

1. Bai, Y., Gao, Y., Huang, H.Y.: Cross-lingual abstractive summarization with limited parallel resources. In: Proceedings of the 59th Annual Meeting of the Association for Computational Linguistics and the 11th International Joint Conference on Natural Language Processing (Volume 1), pp. 6910–6924 (2021)
2. Bougouin, A., Boudin, F., Daille, B.: Topicrank: graph-based topic ranking for keyphrase extraction. In: Proceedings of the Sixth International Joint Conference on Natural Language Processing, pp. 543–551 (2013)
3. Cao, Y., Liu, H., Wan, X.: Jointly learning to align and summarize for neural cross-lingual summarization. In: Proceedings of the 58th Annual Meeting of the Association for Computational Linguistics, pp. 6220–6231 (2020)
4. Denkowski, M., Lavie, A.: Meteor universal: language specific translation evaluation for any target language. In: Proceedings of the Ninth Workshop on Statistical Machine Translation, pp. 376–380 (2014)
5. Dou, Z.Y., Kumar, S., Tsvetkov, Y.: A deep reinforced model for zero-shot cross-lingual summarization with bilingual semantic similarity rewards. In: Proceedings of the Fourth Workshop on Neural Generation and Translation, pp. 60–68 (2020)
6. Duan, X., Yin, M., Zhang, M., Chen, B., Luo, W.: Zero-shot cross-lingual abstractive sentence summarization through teaching generation and attention. In: Proceedings of the 57th Annual Meeting of the Association for Computational Linguistics, pp. 3162–3172 (2019)
7. Dyer, C., Chahuneau, V., Smith, N.A.: A simple, fast, and effective reparameterization of IBM model 2. In: Proceedings of the 2013 Conference of the North American Chapter of the Association for Computational Linguistics: Human Language Technologies, pp. 644–648 (2013)
8. Graham, Y., Haddow, B., Koehn, P.: Translationese in machine translation evaluation. In: Proceedings of the 2020 Conference on Empirical Methods in Natural Language Processing (EMNLP), pp. 72–81 (2020)
9. Kiritchenko, S., Mohammad, S.: Best-worst scaling more reliable than rating scales: a case study on sentiment intensity annotation. In: Proceedings of the 55th Annual Meeting of the Association for Computational Linguistics, pp. 465–470 (2017)
10. Liang, Y., et al.: A variational hierarchical model for neural cross-lingual summarization. In: Proceedings of the 60th Annual Meeting of the Association for Computational Linguistics (Volume 1: Long Papers), pp. 2088–2099 (2022)
11. Lin, C.Y.: Rouge: a package for automatic evaluation of summaries. In: Proceedings of the ACL-04 Workshop, pp. 74–81 (2004)
12. Liu, Y., et al.: Multilingual denoising pre-training for neural machine translation. Trans. Assoc. Comput. Linguist. 8, 726–742 (2020)
13. Mihalcea, R., Tarau, P.: Textrank: bringing order into text. In: Proceedings of the 2004 Conference on Empirical Methods in Natural Language Processing, pp. 404–411 (2004)
14. Ouyang, J., Song, B., McKeown, K.: A robust abstractive system for cross-lingual summarization. In: Proceedings of the 2019 Conference of the North American Chapter of the Association for Computational Linguistics: Human Language Technologies, Volume 1 (Long and Short Papers), pp. 2025–2031 (2019)
15. Takase, S., Okazaki, N.: Multi-task learning for cross-lingual abstractive summarization. In: Proceedings of the Thirteenth Language Resources and Evaluation Conference, pp. 3008–3016 (2022)

16. Vaswani, A., et al.: Attention is all you need. In: Proceedings of the 31st International Conference on Neural Information Processing Systems, pp. 6000–6010 (2017)

17. Veličković, P., Cucurull, G., Casanova, A., Romero, A., Liò, P., Bengio, Y.: Graph attention networks. International Conference on Learning Representations (2018)

18. Wang, J., Liang, Y., Meng, F., Li, Z., Qu, J., Zhou, J.: Zero-shot cross-lingual summarization via large language models. arXiv preprint arXiv:2302.14229 (2023)

19. Wang, J., et al.: Clidsum: a benchmark dataset for cross-lingual dialogue summarization. arXiv preprint arXiv:2202.05599 (2022)

20. Wang, J., et al.: A survey on cross-lingual summarization. Trans. Assoc. Comput. Linguist. **10**, 1304–1323 (2022)

21. Yu, W., et al.: A survey of knowledge-enhanced text generation. ACM Comput. Surv. (2022)

22. Zhang, J., Zhou, Y., Zong, C.: Abstractive cross-language summarization via translation model enhanced predicate argument structure fusing. IEEE/ACM Trans. Audio Speech Lang. Process. **24**(10), 1842–1853 (2016)

23. Zhu, J., et al.: NCLS: neural cross-lingual summarization. In: Proceedings of the 2019 Conference on Empirical Methods in Natural Language Processing and the 9th International Joint Conference on Natural Language Processing (EMNLP-IJCNLP), pp. 3045–3055. Association for Computational Linguistics, Hong Kong (2019)

24. Zhu, J., Zhou, Y., Zhang, J., Zong, C.: Attend, translate and summarize: an efficient method for neural cross-lingual summarization. In: Proceedings of the 58th Annual Meeting of the Association for Computational Linguistics, pp. 1309–1321 (2020)

A Joint Identification Network for Legal Event Detection

Shutao Gong and Xudong Luo[✉]

Guangxi Key Lab of Multi-Source Information Mining & Security, School of
Computer Science and Engineering, Guangxi Normal University, Guilin, China
luoxd@mailbox.gxnu.edu.cn

Abstract. Detecting legal events is a crucial task in legal intelligence
that involves identifying event types related to trigger word candidates
in legal cases. While many studies have focused on using syntactic and
relational dependencies to improve event detection, they often overlook
the unique connections between trigger word candidates and their surrounding features. This paper proposes a new event detection model that
addresses this issue by incorporating an initial scoring module to capture global information and a feature extraction module to learn local
information. Using these two structures, the model can better identify
event types of trigger word candidates. Additionally, adversarial training enhances the model's performance, and a sentence-length mask is
used to modify the loss function during training, which helps mitigate
the impact of missing trigger words. Our model has shown significant
improvements over state-of-the-art baselines, and it won third prize in
the event detection task at the Challenge of AI in Law 2022 (CAIL 2022).

Keywords: Legal Event Detection · Global Pointer · GCN ·
DeBERTa · FGM

1 Introduction

Event detection [1] in information extraction refers to the process of identifying
and extracting specific events, occurrences, or actions from unstructured text
data. This task is a crucial component of Natural Language Processing (NLP)
and text mining applications, as it enables the extraction of meaningful and
structured information from raw text. Event detection has numerous applications, including (but not limited to) news analysis, social media analysis, text
summarisation, question-answering systems, and legal case analysis. In particular, legal event detection aims to automatically identify the types of event
trigger word candidates in a legal case to quickly reconstruct case facts and help
machines and humans better understand the legal issues concerned.

The state-of-the-art event detection techniques may employ machine learning
algorithms, such as deep learning, to improve the accuracy and efficiency of the
process. For example, Yan et al. [43] used Convolutional Neural Network (CNN)

B. Luo et al. (Eds.): ICONIP 2023, CCIS 1967, pp. 122–139, 2024.
https://doi.org/10.1007/978-981-99-8178-6_10

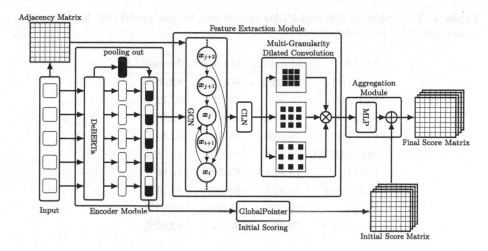

Fig. 1. The overall architecture of our event detection model ($\otimes$ and $\oplus$ are concatenation operations of element-wise addition, (x_i, x_{i+1}) represents a trigger word, (x_j, x_{j+1}, x_{j+2}) represents another trigger word, and the arrows between them represent the direction of node information convergence)

[20] to model spatial relations in fraction matrices. Cui et al. [5] used Graph Convolutional Network (GCN) [18] to exploit syntactic labels and efficiently model inter-word relations. However, the above methods ignore the special connection between trigger word candidates and their neighbouring sequence fragments, which is necessary for event detection.

To address the issues, in this paper, we proposed a novel event detection model. The model consists of four main modules: the encoder module, the initial scoring module, the feature extraction module, and the aggregation module (Fig. 1). The encoder module uses Pre-trained Language Model (PLM) DeBERTa [12], because it works well in NLP. It encodes the input sentences into contextualised word representations, then inputs the encoded representations to the extraction module and uses the Global Pointer [36] to obtain the initial score. Then the two modules score the word pair representation in that sentence, respectively. Finally, the aggregation module reason the relationship between all word pairs according to the outputs of the extraction module and initial scoring module.

The main contributions of this paper are summarised as follows. 1) We propose a new event detection model that can model trigger word candidates and their surrounding features and better capture global and local features. 2) Based on the legal event (LEVENT) dataset [44], we use the ZLPR (zero-bounded logsum-exp & pairwise rank-based) loss function [37] to capture better label dependencies and the ranking relation between positive and negative categories. 3) In training, we apply a sentence-length mask to revise the loss function to reduce the influence of the trigger words in the dataset with missed markers on the model and use Fast Gradient Method (FGM) [29] adversarial training to greatly

Table 1. The results of different PLMs as encoders for our model (the highest scores among all PLMs marked in bold)

Model	Macro-F1	Micro-F1	Averaged-F1
DeBERTa	**80.22%**	**86.18%**	**83.20%**
RoBERTa	79.80%	85.64%	82.72%
NEZHA	79.90%	85.90%	82.90%
UBERT	79.98%	85.66%	82.82%
StructBERT	79.33%	84.83%	82.08%
ERNIE	79.91%	85.69%	82.80%
RoFormer	76.40%	84.01%	80.21%
BERT	79.91%	85.90%	82.91%
MacBERT	79.91%	85.33%	82.62%

enhance the robustness of the model and improve its performance. 4) We have done extensive experiments to show that our model outperformed the baseline models. Our model also won the third prize at the Challenge of AI in Law 2022 (CAIL 2022), further confirming the effectiveness of our model.

The rest of the paper is organised as follows. Section 2 details each module of our event detection model. Section 3 discusses our model training. Section 4 presents the experimental evaluation of our model. Section 5 compares our work with related research to demonstrate how ours advances the state-of-the-art in the field. Finally, Sect. 6 concludes the paper with future work.

2 Model Structure

This section will detail the structure of our model as shown in Fig. 1.

2.1 Encoding Module

This subsection will detail how to encode an input sentence.

We use DeBERTa to encode the input text sentence by sentence. We choose this PLM because it surpasses the human performance on the SuperGLUE benchmark for the first time in terms of macro-average score [12]. Moreover, in our task, compared with other Chinese PLMs such as RoBERTa [25], NEZHA [41], UBERT [26], StructBERT [39], ERNIE [38], RoFormer [35], BERT [7], MacBERT [6], DeBERTa perform better in terms of Macro-F1, Micro-F1, and Averaged-F1 (see Table 1).

Suppose $X = \{x_1, \cdots, x_l\}$ is a sentence of length l. We transform each word x_i of the sentence into a token and input them into DeBERTa for encoding. Then the output of DeBERTa contains the parts, i.e.,

$$(S_{hidden}, P) = \text{DeBERTa}(X), \tag{1}$$

where S_{hidden} (an $l \times d_h$ matrix) is the last hidden state of the DeBERTa layer, and P_1 (a vector with d_h element) is the pooler output of DeBERTa, representing the feature information of the whole sentence. Finally, we concatenate these two outputs to get the final sentence feature representation (a matrix of $l \times 2d_h$), i.e.,

$$H = \text{concat}(S_{hidden}, P). \tag{2}$$

It is a real number matrix of $2d \times l$ and can be denoted as

$$H = (h_1, \cdots, h_l). \tag{3}$$

2.2 Initial Scoring Module

This section will discuss how to extract the global information of a segment of a sentence. Specifically, we will present our method to get the score matrix of a sentence from its hidden states obtained in the previous section. The score reflects the degree to which each segment of a sentence is every type of event from a global view of point.

We use GlobalPointer to do this. First, the module passes the feature vectors h_i and h_j (see formula (3)) at positions i and j, respectively, through two different fully connected layers and obtains query vector q_i and key vector k_i, i.e.,

$$q_i = W_{q,t} h_i + b_{q,t}, \tag{4}$$
$$k_i = W_{k,t} h_j + b_{k,t}, \tag{5}$$

where $W_{q,t}$, W_{kit}, $b_{q,t}$, and $b_{k,t}$ are trainable parameters for event type t.

Then the module uses Rotational Positional Encoding (RoPE) [35] to merge position information into the query vector $q = (q_1, \cdots, q_l)$ and the key vector $k = (k_1, \cdots, k_l)$ for the whole sentence, i.e.,

$$q' = \text{RoPE}(q), \tag{6}$$
$$k' = \text{RoPE}(k). \tag{7}$$

That is, q' and k' are q and k after rotational positional encoding, respectively.

Finally, we calculate the score that the segment from the i-th token to the j-th token in the sentence $X = \{x_1, ..., x_l\}$ is event type t as follows:

$$S_t(i, j) = {q'}_i^\top k'_j. \tag{8}$$

We denote the scoring matrix of a sentence being event type t as $S_t \in R^{l \times l}$. For all the event type considered, a sentence score is $S = (S_1, \cdots, S_{n_t})$, where n_t is the total number of the event type considered.

2.3   Feature Extraction Module

This section will discuss how to extract the local information of a segment of a sentence. After using GlobalPointer to get the score that each segment of a sentence is each type of event, it seems we have recognised all the events in the sentence. However, we notice that trigger word candidates and surrounding features also impact the cognition of events (in the experiment section, we will show putting these features into consideration does improve the accuracy of event detection). So, in this section, we will discuss how to extract these features.

1) **GCN for connection extraction**: We first use GCN to extract the connections between trigger word candidates. Specifically, we treat each word (token) of a sentence as a node in the graph structure of the sentence, and the first and last tokens of the trigger word candidates are connected as edges of the graph. Then, from this graph, GCN can extract the complex connections between trigger word candidates of the sentence. Formally, we have

$$Z = \text{ReLU}(\hat{A}\text{ReLU}(\hat{A}HW^{(0)})W^{(1)}), \tag{9}$$

where $\hat{A}$ denotes the graph's adjacency matrix, H is the encoder's output (see formula (3)), and $W^{(0)}$ and $W^{(1)}$ are both learnable parameters.

We set the value at the corresponding position in adjacency matrix $\hat{A}$ to 1 if the sequence of fragments between two nodes is a trigger word candidate. We also set the value on the diagonal of each node to 1, enabling the GCN to aggregate messages from trigger word candidates and the nodes themselves.

2) **CLN layer**: Now from complex connections between trigger word candidates of the sentence, we use Conditional Layer Normalisation (CLN) [34] to get the feature representation of the segment from the i-th to the j-th tokens as follows:

$$V_{i,j} = \gamma_i \odot \frac{(z_j - \mu_j)}{\sigma_j} + \beta_i, \tag{10}$$

where

- z_i and z_j are the feature representations of the i-th and j-th token of the GCN output Z (see formula (9)), respectively;
- the generating conditions' normalised gain $\gamma_i = W_\alpha z_i + b_\alpha$ and bias $\beta_i = W_\varepsilon z_i + b_\varepsilon$, where W_α, W_ε, b_α, and b_ε are learnable parameters; and
- μ_j and σ_j are the mean and standard deviation of each element z_j of Z, respectively, i.e.,

$$\mu_j = \frac{1}{d_h} \sum_{k=1}^{d_h} z_{j,k},$$

$$\sigma_j = \sqrt{\frac{1}{d_h} \sum_{k=1}^{d_h} (z_{j,k} - \mu)^2},$$

where d_h is the dimensionality of $V_{i,j}$ and $z_{j,k}$ is the k-th dimension of z_j.

The feature representation of the segment from the i-th token to the j-th toke
Then the feature representation of the whole sentence is

$$V = \{V_{i,j} \mid 1 \le i \le j \le l\}. \tag{11}$$

3) **CNN layer**: Since it is easier for CNN to capture the association between adjacent positions [17], we use it to associate the potential features between the trigger word candidates and their adjacent segments and thus obtain richer information when classifying the trigger word candidates. Specifically, we use two types of CNN blocks: one is a typical Conv2d block, and the other is a Conv2d block with multi-granularity dilated, i.e.,

$$Q = \mathrm{GeLU}(\mathrm{Conv2ds}(\mathrm{GeLU}(\mathrm{Conv2d}(V)))), \tag{12}$$

where Conv2d is a 2d CNN layer, Conv2ds is a 2d CNN layer with multi-granularity dilated, GeLU [13] is the activation function, and V is input, the feature representation of sentence (see formula (11)).

It is important to note that all padding in V is padded with 0 to ensure that the shape of the different granularity outputs in Conv2ds is the same because we need to continue with the second layer of the CNN set after the first layer of CNN processing. The finished shape is still a three-dimensional matrix representation $Q \in R^{3 \times d_c \times N \times N}$, where d_c is the number of channels of the convolutional layer output, and 3 represents the length of the granularity set, which in our model is multi-grained values are 1, 3, and 5.

2.4 Aggregation Module

This section will discuss how to aggregate the global information and the local information obtained in the previous two sections and determine the event type of a segment of a sentence.

The Global Pointer module focuses on capturing the global information of a sentence, while the extraction module focuses on the local information between trigger word candidates and their surrounding fragments. Both of these pieces of information are helpful in determining whether or not a segment of the sentence represents an event of a certain type. Therefore, we aggregate both to determine whether a segment is an event or not. Specifically, the final event type of fragment (x_i, x_j) is the index of the highest score category of the fragment, i.e.,

$$T_{i,j}^* = \arg\max\{S_t(i,j) + O_t(i,j) \mid 1 \le t \le n_t\}, \tag{13}$$

where n_t is the total number of all the event types, $S_t(i,j)$ is the global score that segment (x_i, x_j) is an event of type t (see formula (8)), and $O_t(i,j)$ is the local score of fragment (x_i, x_j) being an event of type t, calculated as:

$$O_t(i,j) = \mathrm{MLP}(Q_{c,i,j}), \tag{14}$$

where $Q_{c,i,j}$ is the feature representation of the c channels of fragment (x_i, x_j) in the convolutional layer output Q (see formula (12)), and MLP is a multilayer perceptron [33]. Thus, the scores that all the segments of a sentence are all the event types are

$$O = (O_1, \cdots, O_{n_t}). \tag{15}$$

3 Model Training

This section will discuss how to train our model with the LEVENT dataset.

3.1 Dataset

Our training dataset is the LEVENT dataset provided by the CAIL 2022 competition. It contains 5,301 legal documents containing 41,238 sentences, 297,252 non-event segments, and 98.410 event segments, and the validation dataset contains 1,230 legal documents containing 9,788 sentences, 69,645 non-event segments, and 22,885 event segments. All the events have 108 types, including 64 types of accusation events and 44 types of public affairs.

3.2 Revising the Loss Function by Mask

Due to the label dependency of the dataset, the ranking relationship between positive and negative categories affects the classification of events. Therefore, We use the ZLPR [37] as follows:

$$L = \log \left(1 + \sum_{(i,j) \in U_\alpha} e^{-P_\alpha(i,j)} \right) + \log \left(1 + \sum_{(i,j) \in W_\alpha} e^{P_\alpha(i,j)} \right),$$

where U_α is the head-to-tail set of all positive trigger words of event type α in this sample, W_α is that set of all negative trigger words in the sample, and $P_\alpha(i,j)$ is the score of segment (x_i, x_j) belonging to event type α. Our training aims to minimise the loss function.

However, our experiments found that trigger words in the LEVENT dataset may have missing labels, which may cause our model to fail to identify the event types of the trigger word candidates. Therefore, we modified the loss function based on the mask of sentence length. The mask is a tensor consisting of only 0 and 1. We construct this mask as follows: 1) in the training and validation sets, we set the positions of positive and negative trigger words in the tensor to 1 and the rest to 0; and 2) in the test set, we set the positions of all trigger word candidates in the tensor to 1 and the rest to 0.

After the final score is calculated jointly, we map the mask to the score matrix so that the scores of segments not related to trigger words in the score matrix are all set to negative infinity. When calculating the loss by the loss function, the positions with negative infinity scores are ignored. Only the loss of a subset of trigger word segments is calculated, so the missing trigger words in the data would not affect the model's training.

3.3   Adversarial Training

To enhance the performance and generalisation of our model, we incorporate adversarial training using Fast Gradient Method (FGM). This method is extended based on specific gradients to generate improved adversarial samples. We apply perturbations to the embedding layer of DeBERTa to obtain the adversarial samples. The model is further trained by adding the gradients of the original sample and the counter sample and performing back-propagation. Our ablation experiment in Sect. 4.6 shows that the FGM adversarial training significantly improves the model's performance and its generalisation capability.

4   Experiment

This section will present the experimental evaluation of our model.

4.1   Dataset

Our experiments in this section, we also the LEVENT dataset provided by the CAIL 2022 competition (see Sect. 3.1 for more details of it). CAIL 2022 provides the extra test dataset. It contains 11,052 legal documents containing 79,635 sentences and without annotating 753,897 non-event and event segments. Our test results need to be submitted to the organiser of CAIL 2022 for checking.

4.2   Baseline

Our experiments use the following baselines which are chosen by CAIL 2022:

1) *Two models of token classification:* One uses BiLSTM [14] as the encoder, while another uses BERT as the encoder. Then both uses the same full connection layer to classify the feature representations of trigger word candidates into their corresponding event types.
2) *Two models based on max-pooling:* One, called DMCNN [3], employs CNN as the encoder, while another, called DMBERT [40] uses or BERT as the encoder. Then, according to the extracted features, both use a dynamic pooling layer to predict the event types of trigger word candidates.
3) *Two models based on Condition Radom Field (CRF):* One, called BiLSTM+CRF [16], uses BiLSTM as the encoder, while another, called BERT+CRF [2], uses BERT as the encoder. Then both use CRF [19] to predict the event type of the trigger word candidates.

4.3   Evaluation Criteria

The evaluation criteria we use are Macro-F1, Micro-F1, and Averaged-F1.

Let TP_i be the True Positive of category i (i.e., the true category is i and the predicted category is i); FP_i be the False Positive of category i (i.e., the true category is not i and the predicted category is i); FN_i be the False Negative of category i (i.e., the true category is i and the predicted category is not i). Then Micro-F1 is calculated as follows:

$$\text{Micro-F1} = 2 \times \frac{\text{Micro-Recall} \times \text{Micro-Precision}}{\text{Micro-Recall} + \text{Micro-Precision}}, \tag{16}$$

where

$$\text{Micro-Recall} = \frac{\sum_{i=1}^{t} \text{TP}_i}{\sum_{i=1}^{t} \text{TP}_i + \sum_{i=1}^{t} \text{FN}_i}, \tag{17}$$

$$\text{Micro-Precision} = \frac{\sum_{i=1}^{t} \text{TP}_i}{\sum_{i=1}^{t} \text{TP}_i + \sum_{i=1}^{t} \text{FP}_i}. \tag{18}$$

Table 2. Results on the LEVENT dataset (bold marks the highest score)

Model	Macro-F1	Micro-F1	Averaged-F1
BiLSTM	75.38%	84.14%	79.76%
BERT	78.79%	84.84%	81.82%
DMCNN	75.21%	82.85%	79.03%
DMBERT	77.86%	85.96%	81.91%
BiLSTM+CRF	75.03%	84.17%	79.60%
BERT+CRF	78.99%	85.05%	82.02%
Our Model	**80.22%**	**86.18%**	**83.20%**

Macro-F1 is calculated as follows:

$$\text{Macro-F1} = \frac{\sum_{i=1}^{t} \text{F1-score}_i}{n_t}, \tag{19}$$

where n_t is the total number of event-types, and

$$\text{F1-score}_i = \frac{2\text{TP}_i}{2\text{TP}_i + \text{FP}_i + \text{FN}_i}. \tag{20}$$

Averaged-F1 is obtained by averaging the Macro-F1 and Micro-F1, i.e.,

$$\text{Averaged-F1} = \frac{\text{Micro-F1} + \text{Macro-F1}}{2}. \tag{21}$$

4.4 Setting

We use the Pytorch framework to construct our model, and the DeBERTa model in our experiments comes from the official model library of hugging face.

Since there are 108 event types in the LEVENT dataset, if we constructed a label grid for each category and set the corresponding position of its trigger word corresponding to the class to 1, the amount of data we collect would be vast and challenging to train. Therefore, we set only one label grid for each sample, fill the trigger word's corresponding position with the trigger word's event type, and then convert the label grid when the model is trained.

Due to the difference in sentence lengths in the dataset, we use dynamic padding techniques to improve the model training speed during the training process. We use AdamW as our optimiser and set the learning rate to 10^{-5}, weight decay to 10^{-5}, warmup steps to 1,000, batch size to 32, and dropout rate to 0.2. The head size of the Global Pointer is set to 64, the output dimension of the hidden middle state of GCN is set to 1,024, and the output dimension of the hidden state of the CLN layer is set to 128. We use the Averaged-F1 value on the validation set to achieve an early stop of the model training.

4.5 Benchmark Experiment

Table 2 displays the experimental results of benchmarking our model with the baseline models. As can be seen from the table, our model outperforms all competing baseline models for legal event detection on the LEVENT dataset, achieving the highest Micro-F1 score, Macro-F1 score, and Averaged-F1 score. We attribute the performance improvement to the following factors:

Table 3. The ablation study of our model.

Model	Macro-F1	Micro-F1	Averaged-F1
Best Our Model	**80.22%**	**86.18%**	**83.20%**
-concat	79.53%	85.73%	82.63%
-Initial Scoring Module	78.04%	85.15%	81.60%
-Feature Extraction Module	79.41%	85.31%	82.36%
-GCN	79.25%	85.13%	82.19%
-CNN	79.24%	85.00%	82.12%
-mask	79.30%	85.46%	82.38%
-FGM	78.78%	84.81%	81.80%

1) We combine two outputs of the DeBERTa encoder so that the hidden state of each token reflects sentence-level information (see formula (2)). Therefore, despite using an approach that differs from the max-pooling-based models like DMCNN and DMBERT, our model still achieves excellent results. In fact, our Averaged-F1 score has been improved by 4.17% compared to DMCNN and 1.29% compared to DMBERT.

2) Our extraction method outperforms the baselines in capturing the local information of sentence segments. Additionally, our extraction module effectively captures the unique connections between trigger word candidates and their

neighbouring sequence segments, surpassing the performance of the baseline models BiLSTM and BERT by 3.44% and 1.38%, respectively. These results demonstrate the superior effectiveness of our model in modelling trigger word candidates and their surrounding features.

3) We use a mask to revise the loss function and apply FGM during training. Our model significantly outperforms the baseline models without adjusting the loss function. Compared with the BiLSTM+CRF and BERT+CRF models with loss calculated by CRF, our model improves by 3.60% and 1.18%, respectively, demonstrating the existence of data miss-labelling in the LEVENT dataset and the effectiveness of our method in reducing its impact on the model. Moreover, our model's generalisation ability is significantly higher than the baseline model without FGM.

4.6   Ablation Experiment

To show the effectiveness of each component, we conduct an ablation experiment on our model using the official test set provided by CAIL 2022. Table 3 shows the experimental result, where concat represents the pooling output of DeBERTa and the last layer of hidden states stitched together, GCN and CNN are the structures inside the extraction module, and the mask represents the use of mask revision in the loss function.

1) *Validation of pooling out and merging the two outputs of the DeBERTa encoder*: When we do not pool out one output of DeBERTa and merge it

Table 4. Enter ChatGPT's question template

Your task is to classify each word in the word list according to the list of categories given later, or None if it does not belong to the list of categories. You are given three auxiliary categories of data, a list of categories containing 108 classes, the sentences in which the words are found, and a list of words to be classified. Your answer will output only the results of the categories, separated by commas, and only {count} categories in total. Inside [] are 108 categories; please separate sentences with "' and please separate the list of words with $.
    List of categories:[event type].
    Sentence:"' {sentence}"'
    List of words:List of words:${trigger words}$

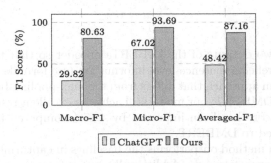

Fig. 2. ChatGPT (GPT 3.5) and Ours three F1 scores

with another output of DeBERTa, the model's averaged-F1 score decreases by 0.57%, indicating that the component of pooling out and merging captures the entire sentence information effectively.

2) *Validation of initial scoring module*: The removal of the initial scoring module decreased the Averaged-F1 score by 1.60%, which is a significant decrease. So using Global Pointer to initial scoring also plays a vital role in our model. And the initial scoring module and extraction modules can complement each other to achieve the best joint prediction.

3) *Validation of feature extraction module*: After removing the whole extraction module, the score of Averaged-F1 is only 82.36%, which is a decrease of 0.84%, which indicates that the feature extraction module has a better local information-capturing ability.

4) *Validation of GCN layer*: When we remove the GCN layer from the extraction module, the Averaged-F1 score decreases by 1.01%, indicating that the GCN effectively captures the associations between trigger word candidates.

5) *Validation of CNN layer*: After we remove the CNN module in extraction, the result shows that the Averaged-F1 score decreases by 1.08%. That indicates that the CNN layer is significant in capturing the information between the trigger word candidates and their adjacent sequence segments.

6) *Validation of revising the loss function*: If we do not use the mask to adjust the loss function, the Averaged-F1 score drops by 0.82%. That shows that the mask revising of the loss function mitigates the impact caused by data miss-mark and can improve our model's performance and stability.

7) *Validation of FGM against training*: When we do not use FGM adversarial training during training, the Averaged-F1 score decreases by 1.40%. So, FGM plays a significant role in our model, which can make our model more robust and significantly improve its performance and generalisation ability.

Table 5. CAIL2022 event detection track competition results (blood is our team)

Ranking	Team Affiliation	Macro-F1	Micro-F1	Averaged-F1
1	Peking University	87.08%	83.72%	85.40%
2	Alibaba Damo Academy	86.98%	83.80%	85.39%
3	Guizhou University	86.87%	82.69%	84.78%
4	Dalian University of Technology	86.53%	82.01%	84.27%
5	**Guangxi Normal University**	85.35%	81.15%	83.25%
6	Beihang University	83.09%	74.95%	79.02%

4.7 Benchmark with ChatGPT

We use ChatGPT, a large language model based on the GPT-3.5 architecture developed by OpenAI, as a benchmark for our model. It is part of the GPT family and is designed to perform conversationally and chat tasks with excellent conversational and generative capabilities, providing powerful support for various conversational applications and tasks. We have experimented with the "GPT-3.5 Turbo" API. Considering the cost and time issues, we used a comparison experiment with a validation set (9,788 sentences, 92,530 trigger word candidates). Our model classifies all trigger word candidates in this sentence by one sentence. For a better comparison, the question templates fed into ChatGPT via the API interface in our experiments are shown in Table 4. Where {count} represents the number of trigger word candidates in the sentence, [event type] represents all temporal categories, {sentence} means the sentence in which the trigger word candidates are located, and {trigger word candidates} represents the set of trigger word candidates in the sentence.

Since the answers it generates are unstable, and even when told to output only as many categories, the number of types it outputs is occasionally more significant than the number of trigger word candidates, we manually truncated the extra-long classification results. Figure 2 shows the experimental results. Our model's results are a big step ahead of ChatGPT in all three F1 scores, and ChatGPT has yet to achieve good results in legal event detection.

4.8   Competition Results

Despite not filtering or analysing the dataset extensively, our model still achieved good results in the CAIL2022 event detection track, though we still needed to catch up on the top spot. Unfortunately, resource and cost limitations prevent us from pursuing further optimisation. Nonetheless, we are proud of our third-place finish in the CAIL2022 event detection track, as shown in Table 5.

5   Related Works

In this section, we will compare our work with related work to demonstrate how our research advances the state-of-the-art in the field of legal intelligence.

5.1   Generative-Based Approach

Researchers in the field of event detection are increasingly exploring generative approaches. For example, Lu et al. [27] used a Text-to-Text Transfer Transformer (T5) pre-training model to transform event extraction into structure generation. Du et al. [9] proposed an end-to-end generative transformer model based on BERT. Li et al. [21] treated event detection as a conditional generation task by incorporating an external knowledge base for knowledge injection. Wei et al. [42] employed simple dialogue generation methods and a hybrid attention mechanism to capture complementary event semantics. However, these generative methods often require assistance in generating standard event structures, resulting in confusing content and posing challenges during model training. In contrast, our approach utilises a conditional generation structure to generate intermediate features, reducing the likelihood of producing confusing content. Moreover, our model does not rely on a specific format and effectively organises the feature representation, enhancing training efficiency.

5.2 Classification-Based Approach

Event detection models use a linear layer to classify captured information and identify events. Different models have varying approaches to capturing sentence information. For example, Feng et al. [11] used CNN to extract chunk information encoded by LSTM, while Hong et al. [15] combined Bi-LSTM with Generative Adversarial Network (GAN) to extract pure sentence features. Nguyen and Grishman [30] learned effective feature representations using pre-trained word embeddings, positional embeddings, and entity-type embeddings with CNN. Ding et al. [8] improved traditional LSTM by integrating semantic information with an additional LSTMCell. Li et al. [23] used a combination of contextual semantic features and statistical features for decision-making. However, these models do not account for missed trigger words, which can significantly impact the accuracy of classification and event detection. On the other hand, our approach uses a sentence-length mask to modify the loss function and reduce the negative impact of missed trigger words on event detection.

5.3 Graph Neural Network-Based Approach

Nguyen and Grishman [31] introduced the integration of syntactic information into GCNs, followed by Cui et al. [4], who combined syntactic structure and dependency labels to update relational representations in GCN. Cui et al. [5] extended this approach by incorporating syntactic dependency structure and relational tags in GCN. Liu, Xu, and Liu [24] utilised attention mechanisms to fuse syntactic structure and potential dependency relations. Mi, Hu, and Li [28] leveraged the complementary nature of syntactic and semantic links to enhance event detection. To capture more diverse information, Peng et al. [32] proposed MarGNN, a multi-intelligent augmented weighted multi-relational graph neural network framework. However, these methods overlook the connections between trigger word candidates. In contrast, our approach utilises GCN to model these connections, enabling richer information extraction.

5.4 Span-Based Approach

Span-based entity recognition and event extraction approaches have been applied to event detection tasks. Eberts and Ulges [10] proposed a span-based method for extracting joint entities and relations, highlighting the benefits of local contextual representation. In the NER domain, Li et al. [22] introduced a formal word-word relationship classification that leverages adjacency relations between entity words. For event detection, Yu, Ji, and Natarajan [45] addressed the issue of severe type imbalance by transferring knowledge between different event types using a span-based approach. However, the span-based method has limitations, and there is room for improving the extraction of local features. Our proposed method focuses on regional features while significantly enhancing the effectiveness compared to the span-based approach.

6   Conclusion

Event detection in legal documents is a crucial aspect of legal intelligence. This paper presents a novel legal event detection model that considers global and local information within a sentence. The model incorporates an initial scoring and extraction module to determine the event type of trigger words jointly. Local information is extracted using a hybrid neural network, while global information is obtained through Global Pointer. To address the issue of missing trigger word tags, a mask is incorporated to revise the ZLPR loss function. FGM adversarial training is included during the training process to enhance the model's generalisation ability. Experimental results on the LEVENT dataset demonstrate that the proposed model outperforms baseline models, and ablation experiments confirm the effectiveness of each model component. Additionally, the model achieved the third prize in the CAIL 2022 event detection track. Future research should consider incorporating syntactic and relational dependency structures and extending the model's scope to encompass the entire context of legal documents.

Acknowledgement. This work was partially supported by a Research Fund of Guangxi Key Lab of Multi-source Information Mining Security (22-A-01-02) and a Graduate Student Innovation Project of School of Computer Science, Engineering, Guangxi Normal University (JXXYYJSCXXM-2021-001) and the Middle-aged and Young Teachers' Basic Ability Promotion Project of Guangxi (No. 2021KY0067).

References

1. Allan, J., Papka, R., Lavrenko, V.: On-line new event detection and tracking. In: Proceedings of the 21st Annual International ACM SIGIR Conference on Research and Development in Information Retrieval, pp. 37–45 (1998)
2. Boros, E., Moreno, J.G., Doucet, A.: Event detection with entity markers. In: Hiemstra, D., Moens, M.-F., Mothe, J., Perego, R., Potthast, M., Sebastiani, F. (eds.) ECIR 2021. LNCS, vol. 12657, pp. 233–240. Springer, Cham (2021). https://doi.org/10.1007/978-3-030-72240-1_20
3. Chen, Y., Xu, L., Liu, K., Zeng, D., Zhao, J.: Event extraction via dynamic multi-pooling convolutional neural networks. In: Proceedings of the 53rd Annual Meeting of the Association for Computational Linguistics and the 7th International Joint Conference on Natural Language Processing. vol. 1, pp. 167–176 (2015)
4. Cui, S., Yu, B., Liu, T., Zhang, Z., Wang, X., Shi, J.: Edge-enhanced graph convolution networks for event detection with syntactic relation. In: Findings of the Association for Computational Linguistics: EMNLP 2020, pp. 2329–2339 (2020)

5. Cui, S., Yu, B., Liu, T., Zhang, Z., Wang, X., Shi, J.: Event detection with relation-aware graph convolutional neural networks. arXiv preprint arXiv:2002.10757 (2020)
6. Cui, Y., Che, W., Liu, T., Qin, B., Wang, S., Hu, G.: Revisiting pre-trained models for Chinese natural language processing. In: Findings of the Association for Computational Linguistics: EMNLP 2020, pp. 657–668 (2020)
7. Devlin, J., Chang, M.W., Lee, K., Toutanova, K.: BERT: pre-training of deep bidirectional transformers for language understanding. In: Proceedings of the 17th Annual Conference of the North American Chapter of the Association for Computational Linguistics: Human Language Technologies, pp. 4171–4186 (2019)
8. Ding, N., Li, Z., Liu, Z., Zheng, H., Lin, Z.: Event detection with trigger-aware lattice neural network. In: Proceedings of the 2019 Conference on Empirical Methods in Natural Language Processing and the 9th International Joint Conference on Natural Language Processing, pp. 347–356 (2019)
9. Du, X., Rush, A.M., Cardie, C.: GRIT: generative role-filler transformers for document-level event entity extraction. In: Proceedings of the 16th Conference of the European Chapter of the Association for Computational Linguistics, pp. 634–644 (2021)
10. Eberts, M., Ulges, A.: Span-based joint entity and relation extraction with transformer pre-training. In: ECAI 2020: 24th European Conference on Artificial Intelligence, Frontiers in Artificial Intelligence and Applications, vol. 325, pp. 2006–2013. IOS Press (2020)
11. Feng, X., Huang, L., Tang, D., Ji, H., Qin, B., Liu, T.: A language-independent neural network for event detection. In: Proceedings of the 54th Annual Meeting of the Association for Computational Linguistics, vol. 2, pp. 66–71 (2016)
12. He, P., Liu, X., Gao, J., Chen, W.: DeBERTa: decoding-enhanced BERT with disentangled attention. In: Proceedings of the 9th International Conference on Learning Representations, pp. 1–23 (2021)
13. Hendrycks, D., Gimpel, K.: Gaussian error linear units (GELUs). arXiv preprint arXiv:1606.08415 (2016)
14. Hochreiter, S., Schmidhuber, J.: Long short-term memory. Neural Comput. 9(8), 1735–1780 (1997)
15. Hong, Y., Zhou, W., Zhang, J., Zhou, G., Zhu, Q.: Self-regulation: employing a generative adversarial network to improve event detection. In: Proceedings of the 56th Annual Meeting of the Association for Computational Linguistics, vol. 1, pp. 515–526 (2018)
16. Huang, Z., Xu, W., Yu, K.: Bidirectional LSTM-CRF models for sequence tagging. arXiv preprint arXiv:1508.01991 (2015)
17. Kim, Y.: Convolutional neural networks for sentence classification. In: Proceedings of the 2014 Conference on Empirical Methods in Natural Language Processing (EMNLP), pp. 1746–1751 (2014)
18. Kipf, T.N., Welling, M.: Semi-supervised classification with graph convolutional networks. In: Proceedings of the 5th International Conference on Learning Representations (2017)
19. Lafferty, J.D., McCallum, A., Pereira, F.C.N.: Conditional random fields: probabilistic models for segmenting and labeling sequence data. In: Proceedings of the Eighteenth International Conference on Machine Learning, pp. 282–289 (2001)
20. LeCun, Y., et al.: Backpropagation applied to handwritten zip code recognition. Neural Comput. 1(4), 541–551 (1989)

21. Li, H., et al.: KiPT: knowledge-injected prompt tuning for event detection. In: Proceedings of the 29th International Conference on Computational Linguistics, pp. 1943–1952 (2022)
22. Li, J., et al.: Unified named entity recognition as word-word relation classification. In: Proceedings of the AAAI Conference on Artificial Intelligence, vol. 36, pp. 10965–10973 (2022)
23. Li, R., Zhao, W., Yang, C., Su, S.: Treasures outside contexts: improving event detection via global statistics. In: Proceedings of the 2021 Conference on Empirical Methods in Natural Language Processing, pp. 2625–2635 (2021)
24. Liu, A., Xu, N., Liu, H.: Self-attention graph residual convolutional networks for event detection with dependency relations. In: Findings of the Association for Computational Linguistics: EMNLP 2021, pp. 302–311 (2021)
25. Liu, Y., et al.: RoBERTa: a robustly optimized BERT pretraining approach. arXiv preprint arXiv:1907.11692 (2019)
26. Lu, J., Yang, P., Zhang, J., Gan, R., Yang, J.: Unified BERT for few-shot natural language understanding. arXiv preprint arXiv:2206.12094 (2022)
27. Lu, Y., et al.: Text2Event: controllable sequence-to-structure generation for end-to-end event extraction. In: Proceedings of the 59th Annual Meeting of the Association for Computational Linguistics and the 11th International Joint Conference on Natural Language Processing, vol. 1, pp. 2795–2806 (2021)
28. Mi, J., Hu, P., Li, P.: Event detection with dual relational graph attention networks. In: Proceedings of the 29th International Conference on Computational Linguistics, pp. 1979–1989 (2022)
29. Miyato, T., Dai, A.M., Goodfellow, I.: Adversarial training methods for semi-supervised text classification. In: Proceedings of the 5th International Conference on Learning Representations (2017)
30. Nguyen, T.H., Grishman, R.: Event detection and domain adaptation with convolutional neural networks. In: Proceedings of the 53rd Annual Meeting of the Association for Computational Linguistics and the 7th International Joint Conference on Natural Language Processing, vol. 2, pp. 365–371 (2015)
31. Nguyen, T.H., Grishman, R.: Graph convolutional networks with argument-aware pooling for event detection. In: Proceedings of the AAAI Conference on Artificial Intelligence, vol. 32, pp. 5900–5907 (2018)
32. Peng, H., Zhang, R., Li, S., Cao, Y., Pan, S., Philip, S.Y.: Reinforced, incremental and cross-lingual event detection from social messages. IEEE Trans. Pattern Anal. Mach. Intell. 45(1), 980–998 (2022)
33. Ramchoun, H., Idrissi, M.J., Ghanou, Y., Ettaouil, M.: Multilayer perceptron: architecture optimization and training with mixed activation functions. In: Proceedings of the 2nd International Conference on Big Data, Cloud and Applications, pp. 1–6 (2017)
34. Su, J.: Conditional layer normalization-based conditional text generation (in Chinese) (2019). https://spaces.ac.cn/archives/7124
35. Su, J., Lu, Y., Pan, S., Murtadha, A., Wen, B., Liu, Y.: RoFormer: enhanced transformer with rotary position embedding. arXiv preprint arXiv:2104.09864 (2021)
36. Su, J., et al.: Global pointer: Novel efficient span-based approach for named entity recognition. arXiv preprint arXiv:2208.03054 (2022)
37. Su, J., Zhu, M., Murtadha, A., Pan, S., Wen, B., Liu, Y.: ZLPR: a novel loss for multi-label classification. arXiv preprint arXiv:2208.02955 (2022)
38. Sun, Y., et al.: ERNIE: enhanced representation through knowledge integration. arXiv preprint arXiv:1904.09223 (2019)

39. Wang, W., et al.: StructBERT: incorporating language structures into pre-training for deep language understanding. In: Proceedings of the 8th International Conference on Learning Representations (2020). https://openreview.net/pdf?id=BJgQ4lSFPH

40. Wang, X., Han, X., Liu, Z., Sun, M., Li, P.: Adversarial training for weakly supervised event detection. In: Proceedings of the 2019 Conference of the North American Chapter of the Association for Computational Linguistics: Human Language Technologies, vol. 1, pp. 998–1008 (2019)

41. Wei, J., et al.: NEZHA: neural contextualized representation for Chinese language understanding. arXiv preprint arXiv:1909.00204 (2019)

42. Wei, Y., et al.: DESED: dialogue-based explanation for sentence-level event detection. In: Proceedings of the 29th International Conference on Computational Linguistics, pp. 2483–2493 (2022)

43. Yan, H., Sun, Y., Li, X., Qiu, X.: An embarrassingly easy but strong baseline for nested named entity recognition. arXiv preprint arXiv:2208.04534 (2022)

44. Yao, F., et al.: LEVEN: a large-scale Chinese legal event detection dataset. In: Findings of the Association for Computational Linguistics: ACL 2022, pp. 183–201 (2022)

45. Yu, P., Ji, H., Natarajan, P.: Lifelong event detection with knowledge transfer. In: Proceedings of the 2021 Conference on Empirical Methods in Natural Language Processing, pp. 5278–5290 (2021)

YOLO-D: Dual-Branch Infrared Distant Target Detection Based on Multi-level Weighted Feature Fusion

Jianqiang Jing, Bing Jia[✉], Baoqi Huang, Lei Liu, and Xiao Yang

College of Computer Science Inner Mongolia University, Hohhot, China
jiabing@imu.edu.cn

Abstract. Infrared distant target detection is crucial in border patrol, traffic management, and maritime search and rescue operations due to its adaptability to environmental factors. However, in order to implement infrared distant target detection for aerial patrols using Unmanned Aerial Vehicles (UAVs), the challenges such as low signal-to-clutter ratio (SCR), limited contrast, and small imaging area have to be addressed. To this end, the paper presents a dual-branch infrared distant target detection model. To be specific, the model incorporates a contour feature extraction branch to improve the network's ability in recognizing distant targets and a multi-level weighted feature fusion method that combines contour features with their original counterparts to enhance target representation. The proposed model is evaluated using the High-altitude Infrared Thermal Dataset for Unmanned Aerial Vehicles (HIT-UAV), which includes persons, cars, and bicycles as detection targets at altitudes ranging from 60 to 130 m. Experimental results show that, in comparison with the state-of-the-art models, our model improves the Average Precision (AP) of persons, bicycles, and cars by 2%, 2.21%, and 0.39% on average, respectively, and improves the mean Average Precision (mAP) of all categories by 1.53%.

Keywords: Dual-branch model · Contour feature extraction · Multi-level weighted feature fusion

1 Introduction

Using Unmanned Aerial Vehicles (UAVs) for infrared distant target detection plays a crucial role in border patrols, traffic management, and maritime search and rescue operations. In these scenarios, UAVs typically operate at altitudes of over 100 m to cover larger areas. However, this leads to smaller target imaging areas, resulting in low detection accuracy with existing methods. With the development of YOLO (You Only Look Once) [1], its detection speed and accuracy have surpassed human performance. However, its sensitivity to external factors such as lighting and visibility limits its applicability. In contrast, infrared imaging technology operates independently of external light sources, exhibiting

B. Luo et al. (Eds.): ICONIP 2023, CCIS 1967, pp. 140–151, 2024.
https://doi.org/10.1007/978-981-99-8178-6_11

robustness to environmental conditions. Therefore, the migration of the YOLO target detection models from the visible light domain to the infrared domain holds significant theoretical and practical significance. Indeed, compared to visible light, infrared target detection has some noticeable disadvantages, including:1)Low signal-to-clutter ratio (SCR) and contrast: This makes it challenging to distinguish targets from the background in infrared images. 2) Lack of texture features: Due to the principles of infrared imaging, infrared images often struggle to capture detailed texture features of the targets. 3)Difficulty in target classification: The limited imaging area hinders the effective extraction of target features, making it challenging to determine the target's class.

In the field of infrared target detection, researchers have made notable advancements in both traditional and deep learning methods to improve detection rates. However, traditional methods have certain limitations that hinder their effectiveness. For instance, filter-based methods [2,3] excel at suppressing uniform background clutter but struggle with complex clutter. Human Visual System (HVS) based methods [4] face limitations in effectively suppressing background clutter. Local contrast-based methods [5–7] are only suitable for high-contrast objects. Methods based on low-rank representations [8] still struggle with high false positive rates when dealing with regularly shaped objects in complex backgrounds. Moreover, traditional methods heavily rely on hand-crafted features, leading to increased workload and poor robustness.

Deep learning methods, particularly those utilizing convolutional neural networks (CNNs), have demonstrated robustness in extracting target features. However, they also possess certain limitations. Two-stage object detection methods [9–12] automatically learn features and reduce noise interference but suffer from computational and training speed limitations, hindering real-time object detection. On the other hand, CNN-based single-stage target detection methods [1,13,14] perform detection directly on the entire image, enabling faster detection speed and simpler network structures, but at the expense of lower detection accuracy. To address the limitations, visible light-infrared image fusion methods have been proposed. Hybrid feature fusion [15] combines features from visible light and infrared images for target classification and localization. Branch fusion [16,17] employs separate branch networks for visible light and infrared images, merging their features to obtain the final detection result. Attention mechanism fusion [18] utilizes weighting through attention mechanisms to fuse features from visible light and infrared images. While visible light-infrared image fusion methods overcome some limitations of individual images, they introduce challenges such as increased computational complexity, longer training and inference times, and difficulties in feature fusion.

This paper proposes a dual-branch model for infrared distant target detection, which combines the original target features and contour features through multi-level weighted feature fusion. One branch of the model is dedicated to extracting the original target features, while the other branch utilizes a combination of bilateral filtering and Sobel filtering operators to extract the target contours. The contour image is then input into a feature extraction network to extract features. The obtained original target features and contour features are fused using the multi-level weighted feature fusion method, improving the utilization of target contour features to ensure accurate localization and recognition of targets while reducing false detections in target detection. The main contributions of this research can be summarized as follows:

(1) The introduced contour extraction method combines bilateral filtering and the Sobel filtering operator, resulting in enhanced intensity and accuracy of detected target edges.

(2) The proposed multi-level weighted feature fusion method ensures the effective integration of target contour features and overall target features, thereby enhancing the representation capability of the targets and improving the robustness of the features.

(3) The dual-branch model improves the accuracy of target detection by integrating multi-level weighted feature fusion. Experimental results show that compared to the baseline models YOLOX [19], YOLOv7 [20], and YOLOv8, the average detection accuracy has been improved by 2.68%, 1.11%, and 1.63%, respectively.

2   Method

2.1   Model Architecture

The proposed infrared distant target detection dual-branch model consists of an original target feature extraction branch and a target contour feature extraction branch. These two branches operate independently and in parallel, avoiding parameter sharing to better adapt to diverse application scenarios. To obtain richer, more accurate, and robust feature representations, multi-level feature fusion is performed using $FP1$ and $FP2$, which serve as inputs for the subsequent feature extraction. Prior to contour feature extraction, the image undergoes contour extraction through bilateral filtering and Sobel filtering. The final three layers of the multi-level weighted feature fusion are fed into a Feature Pyramid Network (FPN) [21] after feature enhancement, serving as the primary features for target detection. The model structure is illustrated in Fig. 1.

Among them, the YOLO Block-i($1 \leq i \leq 5$) in the original target feature and target contour feature extraction model represents the feature extraction module for each YOLO model. In this paper, YOLOX, YOLOv7, and YOLOv8 are selected as reference models for further improvement. The $Origin\_Fi$, Out_Fi, and $Fusion\_Fi$ denote the original target feature, target contour feature, and the fusion feature computed by $FP1$, respectively. $FP2$ is employed to calculate the fusion feature of $Origin\_Fi$ and $Fusion\_Fi$, the process of feature fusion is referred to as the Fusion-Phase.

2.2   Contour Feature Extraction

The quality and effective utilization of contours play a crucial role in infrared distant target detection. To improve the accuracy of contour extraction, we propose a method that combines bilateral filtering and Sobel filtering. By applying

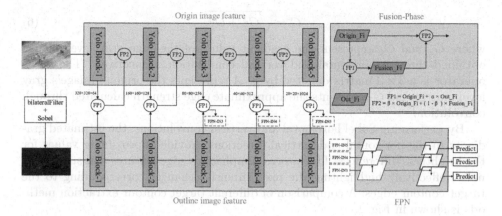

Fig. 1. Model Architecture

bilateral filtering, we reduce noise and enhance target contours. Subsequently, the Sobel operator filter is employed to emphasize significant pixel variations along target edges, thereby achieving higher-quality contour extraction.

The proposed method begins by using bilateral filtering to calculate the gray value of each position, which combines spatial proximity and the similarity of pixel values. The calculation method is as follows:

$$G(i,j) = \omega_s \gamma_p. \tag{1}$$

The calculation of spatial proximity is based on the Euclidean distance between two pixels, which is defined as:

$$\omega_s(i,j,k,l) = e^{\left(-\frac{(i-k)^2 + (j-l)^2}{2\sigma_d^2}\right)}. \tag{2}$$

The method to calculate the similarity of pixel values is:

$$\gamma_p(i,j,k,l) = e^{\left(-\frac{\|f(i,j)-f(k,l)\|^2}{2\sigma_r^2}\right)}, \tag{3}$$

where (k,l) is the coordinates of the center point of the template window, (i,j) is the coordinates of other points, σ is the standard deviation of the Gaussian function, $f(i,j)$ and $f(k,l)$ is the pixel value of the corresponding point.

To enhance the quality of the target contour image, we apply the Sobel operator filter to further process the results of bilateral filtering. This involves using a 3×3 convolutional kernel to perform horizontal and vertical convolutions on the image, generating approximate values for the brightness differences in the horizontal and vertical directions. The calculation method for the Sobel operator filtering is as follows:

$$G_x = \rho \times A, \tag{4}$$

$$G_y = \rho^T \times A, \tag{5}$$

$$G = \sqrt{G_x^2 + G_y^2}, \tag{6}$$

where G_x and G_y represent the gradient values of the edge gray level change obtained through horizontal and vertical convolution, respectively. The Sobel operator is a 3×3 matrix denoted by ρ. A represents the original image's gray level matrix, and G represents the approximate edge strength calculated through convolution.

By performing Sobel horizontal and vertical convolutions, the estimated gradients in the horizontal and vertical directions provide approximate values for the brightness differences. This process emphasizes edge features and suppresses noise, effectively highlighting the pixel intensity changes corresponding to the target contour edges. A comparison of different target contour extraction methods is shown in Fig. 2.

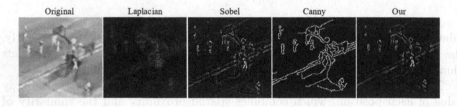

Fig. 2. Contour extraction effect comparison

The comparative analysis reveals that the Canny method produces clear target contours but sacrifices important target-specific feature information. In contrast, our proposed method accurately extracts target contours while preserving their intrinsic features, thereby enhancing the reliability and effectiveness of subsequent target detection.

2.3   Multilevel Feature Fusion

Contour features are important information for infrared distant target detection, and efficiently utilizing contour features is crucial for improving detection accuracy. To address this, this paper proposes a multi-level weighted feature fusion method. Firstly, the five layers of obtained original features and five layers of contour features are fused using weighted feature fusion. Secondly, the resulting fusion features are further fused with the original target features using weighted feature fusion. Finally, the obtained fusion features are input to the next-level feature network for feature extraction. The last three layers of the fusion features $(Fusion - Fi, 3 \leq i \leq 5)$ are used as inputs to the FPN for feature enhancement. The specific process of multi-level weighted feature fusion is illustrated in Fig. 3.

In this feature fusion process, $FP1$ is a weighted feature fusion for contour features and original target features, and its calculation method is as follows.

$$FP1 = Origin\_Fi + \alpha \times Out\_Fi (i \in N^*, 1 \leq i \leq 5). \tag{7}$$

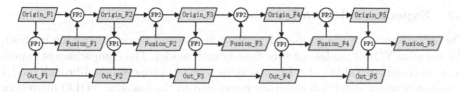

Fig. 3. Multi-level feature fusion process. *Origin_Fi*, *Out_Fi*, and *Fusion_Fi* represents the original features, contour features, and fused features, respectively.

The obtained fusion features and the original features are weighted and fused again and input to the next layer of feature extraction network, and the calculation method is as follows.

$$FP2 = \beta \times Origin\_Fi + (1 - \beta) \times Fusion\_Fi(i \in N^*, 1 \leq i \leq 4), \qquad (8)$$

where α and β both represent the weight of the feature.

Since the contour features are considered low-level features and their representation decreases with the increase in convolutional layers, the weight parameter α in Eq. (7) decreases as the number of layers increases. To ensure that the subsequent feature extraction can better utilize the contour features, the fused features obtained from Eq. (8) are further combined with the target features as the input for the next level of feature extraction. In this case, the weight parameter β increases as the number of layers increases.

3   Experiments

3.1   Datasets and Evaluation Metrics

In the experiment, the paper utilizes 2,898 infrared images sourced from the High-altitude Infrared Thermal Dataset for Unmanned Aerial Vehicles (HIT-UAV) [22]. These images are captured in various scenes including schools, parking lots, roads, and playgrounds. The targets of interest for detection in this study are persons, bicycles, and cars.

The primary evaluation metrics used are mAP for all categories, AP for each category, and F1 score, Recall, and Precision for each model. These metrics are compared to assess the performance of the models and demonstrate the effectiveness of the proposed method.

3.2   Experiment Setup

We train our model using Stochastic Gradient Descent (SGD) optimizer with a learning rate of 0.02 for 300 epochs, applying a weight decay of 5^{-4} and a batchsize of 16, while dividing the dataset into training, validation, and test sets in an 8:1:1 ratio. We assess the performance of the YOLOX, YOLOv7, and YOLOv8 models and their respective dual-branch counterparts by substituting the YOLO Block-i($1 \leq i \leq 5$) in the network architecture with the feature extraction module from the YOLO backbone network.

3.3 Experimental Data Results

The effectiveness of the proposed method is assessed through training on both the baseline YOLO models and its dual-branch model. The comparison of experimental results in Table 1 and Fig. 4 reveals that our proposed dual-branch model achieves superior detection accuracy compared to the baseline YOLO models in the context of infrared distant target detection.

Table 1. Comparison of detection results between the baseline model and the dual-branch model.

models	AP person(%)	AP bicycle(%)	AP car(%)	mAP 0.5(%)
YOLOX-Tiny	71.85	65.21	83.25	73.44
YOLOX-D-Tiny	**73.93**(+2.08)	**67.10**(+1.89)	**84.83**(+1.58)	**75.29**(+1.85)
YOLOX-S	74.31	70.42	88.99	77.91
YOLOX-D-S	**77.79**(+3.48)	**75.02**(+4.60)	88.11(-0.88)	**80.31**(+2.40)
YOLOX-M	78.88	74.90	91.11	81.63
YOLOX-D-M	**79.67**(+0.79)	**78.11**(+3.21)	**91.37**(+0.27)	**83.05**(+1.42)
YOLOX-L	80.99	78.13	90.03	83.05
YOLOX-D-L	**82.16**(+1.17)	**79.57**(+1.44)	**91.29**(+1.26)	**84.34**(+1.29)
YOLOX-X	82.22	81.74	94.11	86.03
YOLOX-D-X	**82.72**(+0.5)	**85.15**(+3.41)	93.44(-0.67)	**87.11**(+1.08)
YOLOv7-L	86.45	92.25	95.64	91.45
YOLOv7-D-L	**89.57**(+3.12)	**94.02**(+1.77)	**95.87**(+0.23)	**93.15**(+1.70)
YOLOv7-X	90.16	92.52	96.39	93.02
YOLOv7-D-X	**91.11**(+0.95)	**93.33**(+0.81)	96.18(-0.21)	**93.54**(+0.52)
YOLOv8-N	85.84	90.23	95.10	90.39
YOLOv8-D-N	**90.49**(+4.65)	**93.47**(+3.24)	**97.33**(+2.23)	**93.76**(+3.37)
YOLOv8-S	89.11	91.11	96.73	92.32
YOLOv8-D-S	**92.48**(+3.37)	**94.30**(+3.19)	**97.93**(+1.2)	**94.90**(+2.58)
YOLOv8-M	88.78	92.31	97.96	93.02
YOLOv8-D-M	**90.44**(+1.66)	**93.26**(+0.95)	97.87(-0.09)	**93.86**(+0.84)
YOLOv8-X	90.31	93.12	97.51	93.65
YOLOv8-D-X	**91.23**(+0.92)	**94.57**(+1.45)	97.10(-0.41)	**94.30**(+0.65)
YOLOv8-L	91.22	93.18	97.25	93.88
YOLOv8-D-L	**92.45**(+1.23)	**94.37**(+1.19)	**97.00**(−0.25)	**94.61**(+0.73)

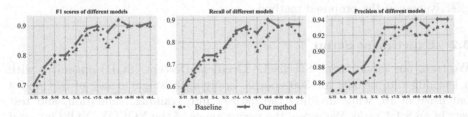

Fig. 4. The performance comparison results of the YOLO model and its dual-branch model. The x-axis represents different YOLO models, such as "YOLOX-Tiny" is represented as "X-Ti."

The experimental results in Table 1 demonstrate the superior detection accuracy of the proposed dual-branch model compared to the baseline YOLO models across all categories, with notable improvements in person and bicycle detection. Figure 4 further highlights the consistent outperformance of the dual-branch model in classification performance and prediction accuracy compared to various versions of the baseline models (YOLOX, YOLOv7, and YOLOv8). These findings solidify the effectiveness of the dual-branch model in enhancing target detection capabilities, particularly for persons and bicycles.

3.4 Experimental Visual Results

To compare the performance visually, we assessed the X versions of YOLOX, YOLOv7, and YOLOv8, as well as their corresponding dual-branch models, on the HIT-UAV dataset. The detection results for targets within the range of 60 m to 130 m are depicted in Fig. 5.

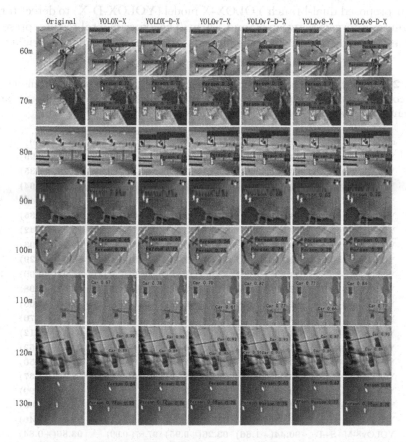

Fig. 5. Comparison of visual results of X version detection of YOLOX, YOLOv7, and YOLOv8.

The visual result graph clearly illustrates the effectiveness of our proposed dual-branch model in accurately detecting targets, even under conditions of low contrast and SCR. The model demonstrates higher accuracy in detecting targets, reducing the error rate, as well as minimizing the false detection rate and missed detection rate. These findings highlight the superior performance of our approach in challenging scenarios with limited visibility and challenging target characteristics.

3.5   Ablation and Validation Experiments

To assess the efficacy of our method, we conducted experiments using two models: one incorporating only Sobel contour extraction, and the other combining Sobel filtering and bilateral filtering for contour extraction. The experimental results, presented in Table 2, demonstrate the superior performance of the combined approach in terms of target detection. Additionally, to evaluate the model's performance across varying heights, we utilized both the benchmark YOLOX-X model and our proposed dual-branch YOLOX-X model (YOLOX-D-X) to detect targets at 10-meter intervals between 60 and 130 m. The experimental results, presented in Table 3, provide insights into the model's effectiveness at different heights.

Table 2. Comparison of the detection accuracy of the contour extraction model with Sobel contour extraction and the combination of Sobel and Bilateral filter. S and B represent the Sobel filter and Bilateral filter, respectively.

models	AP person(%)	AP bicycle(%)	AP car(%)	mAP 0.5(%)
YOLOX-Tiny+S	74.13(+2.28)	64.62(-0.57)	84.54(+1.29)	74.43(+0.99)
YOLOX-Tiny+S+B	73.93(+2.13)	67.10(+1.89)	84.83(+1.58)	75.29(+1.85)
YOLOX-S+S	75.76(+1.45)	74.15(+3.73)	88.74(-0.25)	79.55(+1.64)
YOLOX-S+S+B	77.79(+3.48)	75.02(+4.60)	88.11(-0.88)	80.31(+2.40)
YOLOX-M+S	79.88(+1.00)	77.26(+2.36)	91.80(+0.69)	82.98(+1.35)
YOLOX-M+S+B	79.67(+0.79)	78.11(+3.21)	91.37(+0.27)	83.05(+1.42)
YOLOX-L+S	81.53(+0.54)	78.93(+0.80)	91.48(+1.45)	83.98(+0.93)
YOLOX-L+S+B	82.16(+1.17)	79.57(+1.44)	91.29(+1.26)	84.34(+1.29)
YOLOX-X+S	83.55(+1.33)	83.62(+1.88)	93.62(-0.49)	86.93(+0.90)
YOLOX-X+S+B	82.70(+0.48)	84.54(+2.80)	93.79(-0.32)	87.01(+0.98)
YOLOv7-L+S	89.07(+2.62)	92.58(+0.33)	95.85(+0.21)	92.50(+1.05)
YOLOv7-L+S+B	89.57(+3.12)	94.02(+1.77)	95.87(+0.23)	93.15(+1.70)
YOLOv7-X+S	90.70(+0.54)	92.05(-0.47)	96.66(+0.27)	93.14(+0.12)
YOLOv7-X+S+B	91.11(+0.95)	93.33(+0.81)	96.18(-0.21)	93.54(+0.52)
YOLOv8-N+S	90.06(+4.22)	91.82(+1.52)	96.78(+1.68)	92.89(+2.50)
YOLOv8-N+S+B	90.49(+4.65)	93.47(+3.24)	97.33(+2.23)	93.76(+3.37)
YOLOv8-S+S	90.93(+1.82)	93.82(+2.71)	98.18(+1.45)	94.31(+1.99)
YOLOv8-S+S+B	92.48(+3.37)	94.30(+3.19)	97.93(+1.2)	94.90(+2.58)
YOLOv8-M+S	89.98(+1.20)	92.12(-0.19)	98.33(+0.37)	93.48(+0.46)
YOLOv8-M+S+B	90.44(+1.66)	93.26(+0.95)	97.87(-0.09)	93.86(+0.84)
YOLOv8-X+S	90.82(+0.51)	93.98(+0.86)	97.33(-0.18)	94.04(+0.39)
YOLOv8-X+S+B	91.23(+0.92)	94.57(+1.45)	97.10(-0.41)	94.30(+0.65)
YOLOv8-L+S	91.73(+0.51)	93.48(+0.30)	97.35(+0.10)	94.19(+0.31)
YOLOv8-L+S+B	92.45(+1.23)	94.37(+1.19)	97.00(-0.25)	94.61(+0.73)

Based on the comparison of the experimental results presented above, it can be concluded that both contour extraction methods have a positive impact on the detection of infrared distant targets. Particularly notable is the substantial improvement observed in the detection accuracy of person and bicycles when compared to the benchmark model. These findings highlight the effectiveness of incorporating contour extraction techniques for enhancing the detection performance of infrared distant targets detection systems.

Table 3. Comparison of target detection performance of different models at 60 m–130 m.

Method	Baseline				Our			
Metrics	F1	Recall	Precision	mAP(%)	F1	Recall	Precision	Map(%)
60 m	0.84	0.82	0.87	86.31	0.87	0.85	0.90	92.89
70 m	0.80	0.73	0.91	89.22	0.82	0.74	0.93	90.03
80 m	0.87	0.81	0.95	91.07	0.88	0.82	0.94	91.41
90 m	0.83	0.79	0.89	85.19	0.84	0.81	0.89	86.10
100 m	0.80	0.73	0.91	83.47	0.81	0.73	0.92	84.71
110 m	0.80	0.74	0.87	82.74	0.78	0.69	0.91	83.41
120 m	0.55	0.51	0.61	58.61	0.59	0.53	0.66	63.16
130 m	0.76	0.77	0.80	79.79	0.78	0.77	0.80	80.64

Table 3 demonstrates that our proposed dual-branch model outperforms the baseline model in terms of detection performance at different heights, indicating that the dual-branch model exhibits superior robustness across various detection heights compared to the baseline model. To further demonstrate the robustness of the dual-branch model under different lighting conditions, we compared the detection metrics of the model for nighttime and daytime detection, as shown in Fig. 6.

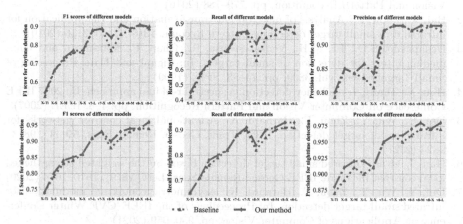

Fig. 6. Nighttime and daytime detection parameters comparison.

By comparing the results, it is evident that our proposed dual-branch model outperforms the baseline model in both nighttime and daytime detection, indicating that our dual-branch model exhibits strong robustness under different lighting conditions. Due to the larger amount of nighttime data compared to daytime data, the F1 score, recall rate, and accuracy are higher in nighttime detection than in daytime detection.

4   Conclusion

We propose a method called YOLO-D to address the challenges of low contrast and SCR in infrared distant target detection. Specifically, we introduce a contour feature extraction branch and a multi-level weighted feature fusion method based on the target contour features and original target features in the model to tackle these issues. In the contour feature extraction branch, we employ Sobel and bilateral filtering techniques to preprocess the images, reducing noise and highlighting the target contours to obtain high-quality contour features. Through multi-level weighted feature fusion, we enhance the target contour features, thereby improving the detection rate and reducing false alarms. Experimental results demonstrate that our proposed dual-branch model has a positive impact on improving the detection performance of YOLO in infrared distant target detection.

Acknowledgements. This work was funded by National Natural Science Foundation of China (Grant No. 42161070), Natural Science Foundation of Inner Mongolia Autonomous Region of China (Grant No. 2023MS06004, 2021ZD13), Key Science Technology Project of Inner Mongolia Autonomous Region (Grant No. 2021GG0163).

References

1. Redmon, J., Divvala, S., Girshick, R., Farhadi, A.: You only look once: unified, real-time object detection. In: Proceedings of the IEEE Conference on Computer Vision and Pattern Recognition, pp. 779–788 (2016)
2. Bai, X., Zhou, F.: Analysis of new top-hat transformation and the application for infrared dim small target detection. Pattern Recogn. **43**(6), 2145–2156 (2010)
3. Deshpande, S.D., Er, M.H., Venkateswarlu, R., Chan, P.: Max-mean and max-median filters for detection of small targets. In: Signal and Data Processing of Small Targets 1999, vol. 3809, pp. 74–83. SPIE (1999)
4. Hou, X., Zhang, L.: Saliency detection: a spectral residual approach. In: 2007 IEEE Conference on Computer Vision and Pattern Recognition, pp. 1–8. IEEE (2007)
5. Dai, Y., Wu, Y., Zhou, F., Barnard, K.: Attentional local contrast networks for infrared small target detection. IEEE Trans. Geosci. Remote Sens. **59**(11), 9813–9824 (2021)
6. Han, J., et al.: Infrared small target detection based on the weighted strengthened local contrast measure. IEEE Geosci. Remote Sens. Lett. **18**(9), 1670–1674 (2020)
7. Dai, Y., Wu, Y., Zhou, F., Barnard, K.: Asymmetric contextual modulation for infrared small target detection. In: Proceedings of the IEEE/CVF Winter Conference on Applications of Computer Vision, pp. 950–959 (2021)

8. Zhang, L., Peng, L., Zhang, T., Cao, S., Peng, Z.: Infrared small target detection via non-convex rank approximation minimization joint l 2, 1 norm. Remote Sensing **10**(11), 1821 (2018)

9. Girshick, R., Donahue, J., Darrell, T., Malik, J.: Rich feature hierarchies for accurate object detection and semantic segmentation. In: Proceedings of the IEEE Conference on Computer Vision and Pattern Recognition, pp. 580–587 (2014)

10. Girshick, R.: Fast r-cnn. In: Proceedings of the IEEE International Conference on Computer Vision, pp. 1440–1448 (2015)

11. Faster, R.: Towards real-time object detection with region proposal networks. In: Advances in Neural Information Processing Systems. vol. 9199(10.5555), pp. 2969239–2969250 (2015)

12. He, K., Gkioxari, G., Dollár, P., Girshick, R.: Mask r-cnn. In: Proceedings of the IEEE International Conference on Computer Vision, pp. 2961–2969 (2017)

13. Liu, W., et al.: SSD: single shot multibox detector. In: Leibe, B., Matas, J., Sebe, N., Welling, M. (eds.) ECCV 2016. LNCS, vol. 9905, pp. 21–37. Springer, Cham (2016). https://doi.org/10.1007/978-3-319-46448-0_2

14. Lin, T.Y., Goyal, P., Girshick, R., He, K., Dollár, P.: Focal loss for dense object detection. In: Proceedings of the IEEE International Conference on Computer Vision, pp. 2980–2988 (2017)

15. Bell, S., Zitnick, C.L., Bala, K., Girshick, R.: Inside-outside net: Detecting objects in context with skip pooling and recurrent neural networks. In: Proceedings of the IEEE Conference on Computer Vision and Pattern Recognition, pp. 2874–2883 (2016)

16. Wang, H., Zhou, L., Wang, L.: Miss detection vs. false alarm: adversarial learning for small object segmentation in infrared images. In: Proceedings of the IEEE/CVF International Conference on Computer Vision, pp. 8509–8518 (2019)

17. Qu, Z., Shang, X., Xia, S.F., Yi, T.M., Zhou, D.Y.: A method of single-shot target detection with multi-scale feature fusion and feature enhancement. IET Image Proc. **16**(6), 1752–1763 (2022)

18. Zhou, K., Chen, L., Cao, X.: Improving multispectral pedestrian detection by addressing modality imbalance problems. In: Vedaldi, A., Bischof, H., Brox, T., Frahm, J.-M. (eds.) ECCV 2020. LNCS, vol. 12363, pp. 787–803. Springer, Cham (2020). https://doi.org/10.1007/978-3-030-58523-5_46

19. Ge, Z., Liu, S., Wang, F., Li, Z., Sun, J.: Yolox: exceeding yolo series in 2021. arXiv preprint arXiv:2107.08430 (2021)

20. Wang, C.Y., Bochkovskiy, A., Liao, H.Y.M.: Yolov7: trainable bag-of-freebies sets new state-of-the-art for real-time object detectors. In: Proceedings of the IEEE/CVF Conference on Computer Vision and Pattern Recognition, pp. 7464–7475 (2023)

21. Lin, T.Y., Dollár, P., Girshick, R., He, K., Hariharan, B., Belongie, S.: Feature pyramid networks for object detection. In: Proceedings of the IEEE Conference on Computer Vision and Pattern Recognition, pp. 2117–2125 (2017)

22. Suo, J., Wang, T., Zhang, X., Chen, H., Zhou, W., Shi, W.: Hit-uav: a high-altitude infrared thermal dataset for unmanned aerial vehicle-based object detection. Scientific Data **10**(1), 227 (2023)

Graph Convolutional Network Based Feature Constraints Learning for Cross-Domain Adaptive Recommendation

Yibo Gao[1], Zhen Liu[1(✉)], Xinxin Yang[1], Yilin Ding[1], Sibo Lu[1], and Ying Ma[2]

[1] School of Computer and Information Technology, Beijing Jiaotong University, Beijing 100044, China
{gaoyibo,zhliu,yangxinxin,ylding,lusibo}@bjtu.edu.cn
[2] State Information Center, Beijing 100045, China
maying@cegn.gov.cn

Abstract. The problem of data sparsity is a key challenge for recommendation systems. It motivates the research of cross-domain recommendation (CDR), which aims to use more user-item interaction information from source domains to improve the recommendation performance in the target domain. However, finding useful features to transfer is a challenge for CDR. Avoiding negative transfer while achieving domain adaptation further adds to this challenge. Based on the superiority of graph structural feature learning, this paper proposes a graph convolutional network based Cross-Domain Adaptive Recommendation model using Feature Constraints Learning (CDAR-FCL). To begin with, we construct a multi-graph network consisting of single-domain graphs and one cross-domain graph based on overlapping users. Next, we employ specific and common graph convolution on the graphs to learn domain-specific and domain-invariant features, respectively. Additionally, we design feature constraints on the features obtained in different graphs and mine the potential correlation for domain adaptation. To address the issue of shared parameter conflicts within the constraints, we develop a binary mask learning approach based on contrastive learning. Experiments on three pairs of real cross-domain datasets demonstrate the effectiveness.

Keywords: Cross-Domain Recommendation · Feature Constraint · Domain Adaptation · Graph Convolutional Network

1 Introduction

The existing recommendation systems are limited by the problem of sparse data. Data sparsity [18] is a common problem in recommendation systems, particularly in cold-start situations. In the real world, users' preferences in one domain are often related to their preferences in other domains, which has motivated the development of Cross-Domain Recommendation(CDR). The basic idea behind

B. Luo et al. (Eds.): ICONIP 2023, CCIS 1967, pp. 152–164, 2024.
https://doi.org/10.1007/978-981-99-8178-6_12

CDR is to use information from multiple domains to make recommendations, making it a helpful tool for addressing data sparsity.

Transferring useful features and avoiding negative transfer [12] of information is a hot topic in CDR research. In a cross-domain scenario involving books and movies as shown in Fig. 1, the user has had interactions in movie domain with science fiction theme, but there is no interaction record in book domain. Therefore, we can recommend science fiction books to the user in the book domain. However, the user may have very different preferences in the book domain. In such a case, this kind of information may not be suitable for transfer between domains.

Fig. 1. A cross-domain scenario.

In this paper, we aim to improve recommendation performance in both domains. To achieve a well-performing dual-target CDR, there are three significant challenges in the literature, which are as follows.

CH1: How to distinguish and obtain domain-specific features and domain-invariant features of two domains?

CH2: After obtaining domain-specific features and domain-invariant features, how to enhance these features to facilitate more effective cross-domain feature transfer?

CH3: The goal of dual-target CDR is to improve recommendation performance in both domains simultaneously. However, a challenge is to handle the conflict problem of shared parameters [2].

In response to the above challenges, this paper proposes a graph convolutional network based Cross-Domain Adaptive Recommendation model using Feature Constraints Learning(CDAR-FCL).

The contributions of this paper are as follows:

– Aiming at the cross-domain recommendation problem with completely overlapping users, we construct single-domain graphs and a cross-domain graph to capture domain-specific and domain-invariant features, respectively. The

cross-domain graph can help to learn domain-invariant features under the same distribution.
- To learn more comprehensive feature representations and improve the effect of cross-domain feature combination, we introduce feature constraints, an approach for domain adaptation by identifying implicit correlations between the same users or items in different graphs. It utilizes linear mapping to mine hidden correlations while preserving feature distinctions in different spaces.
- To solve the sharing parameter conflict problem, we devise a mask matrix learning strategy based on contrastive learning. This strategy involves learning a mask matrix in an unsupervised manner to enable flexible updates of shared parameters.
- We validate the effectiveness of the proposed model by conducting experiments on three pairs of datasets, and report promising results.

2   Related Work

2.1   Cross-Domain Recommendation

To address the problem of data sparsity in recommendation systems, the cross-domain recommendation has emerged [18] to utilize more affluent information. EMCDR [10] uses a multi-layer perceptron(MLP) to capture the nonlinear mapping function between domains, while Conet [4] uses cross-connections between networks. DDTCDR [6] learns latent orthogonal mappings. PTUPCDR [19] utilizes a meta-learner to learn personalized propagation features.

With the recent application of Graph Neural Networks(GNN) in recommendation systems, many GNN-based CDR methods have also emerged. PPGN [17] builds a cross-domain preference matrix to model different interaction domains as a whole. BiTGCF [7] exploits higher-order connectivity in single-domain graphs for feature propagation. MGCDR [16] construct the residual network with the overlapping users to achieve multi-graph convolutional feature transfer.

2.2   Domain Adaptation

Existing works on domain adaptation can be classified into three types: discrepancy-based, adversarial-based, and reconstruction-based methods. Correlation Alignment (CORAL) [11] is quite common discrepancy-based method. [8] achieve domain adaptation by minimizing the Stein Path distance between domain distributions. Adversarial-based methods integrate a domain discriminator for adversarial training, e.g., Adversarial Discriminative Domain Adaptation (ADDA) [14]. Reconstruction-based methods use autoencoders to learn the distribution characteristics of their respective domains, e.g., [5].

3   Methodology

In this section, we first present the problem formulation addressed in this paper. Next, we introduce our proposed model, called CDAR-FCL.

3.1 Problem Formulation

We consider a source domain $\mathcal{S}$ and a target domain $\mathcal{T}$, each with a user set $\mathcal{U}$ and an item set $\mathcal{I}$. In this paper, we focus on the fully-overlapping (common) users between $\mathcal{S}$ and $\mathcal{T}$. To capture domain-specific features, we define specific graphs in the source domain and the target domain, denoted as $\mathcal{G}^S$ and $\mathcal{G}^T$, respectively. Take the specific graph $\mathcal{G}^S$ for example, $\mathcal{G}^S = \{\mathcal{V}^S, \mathcal{E}^S\}$. $\mathcal{V}^S$ is the node set, which has common users $\mathcal{U}$ and specific items $\mathcal{I}^S$ in domain $\mathcal{S}$, and $\mathcal{E}^S$ is the edge set, which includes the interactions between overlapping users $\mathcal{U}$ and specific items $\mathcal{I}^S$ in domain $\mathcal{S}$. $\mathcal{G}^T$ has the similar definition as $\mathcal{G}^S$. Moreover, $\mathcal{G}^C = \{\mathcal{V}^C, \mathcal{E}^C\}$ is defined as the cross-domain graph, where $\mathcal{V}^C = \{\mathcal{U}, \mathcal{I}^S, \mathcal{I}^T\}$, $\mathcal{E}^C$ represents the user-item interactions in domain $\mathcal{S}$ and domain $\mathcal{T}$.

The problem can be defined as learning enhanced feature representations of users $\mathcal{U}$ and items $\mathcal{I}$ by modeling graphs $\mathcal{G}^S$, $\mathcal{G}^T$ and $\mathcal{G}^C$, aiming to improve recommendation in both domains.

3.2 CDAR-FCL Overview

The proposed CDAR-FCL model is presented in this section, as illustrated in Fig 2. In graph embedding layer, we apply the graph convolutional network to propagate node information on graphs $\mathcal{G}^S$, $\mathcal{G}^T$ and $\mathcal{G}^C$ to generate different user and item feature representations. In feature constraint part, three constraints are designed to align the features in different graphs to enhance the feature representation, and a contrastive mask is used to prune shared parameters. Finally, in feature combination part, the embeddings are combined to generate predictions for user-item interactions.

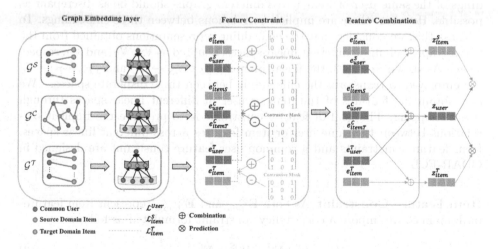

Fig. 2. The framework of CDAR-FCL.

3.3 Graph Embedding Layer

To capture the latent features and relationships between users and items in graph $\mathcal{G}^{\mathcal{S}}$, $\mathcal{G}^{\mathcal{T}}$ and $\mathcal{G}^{\mathcal{C}}$, we use LightGCN [3] as our feature propagation approach. Let the 0-th layer embedding matrix be denoted as $\mathbf{E}^{(0)} \in \mathbb{R}^{(|\mathcal{U}|+|\mathcal{I}|) \times \dim}$, where dim is the embedding size. For a network with l layers, the embedding representations at layer k $(1 \leqslant k \leqslant l)$ can be summarized as follows:

$$\mathbf{E}^{(k)} = (D^{-\frac{1}{2}}AD^{-\frac{1}{2}})\mathbf{E}^{(k-1)} \tag{1}$$

where A is the adjacency matrix of the user-item graph. D is a $(|\mathcal{U}| + |\mathcal{I}|) \times (|\mathcal{U}| + |\mathcal{I}|)$ diagonal matrix, in which each entry D_{ii} denotes the number of nonzero entries in the i-th row vector of the adjacency matrix A. Finally, we can get the final embedding matrix as:

$$\mathbf{E} = \alpha_0 \mathbf{E}^{(0)} + \alpha_1 \mathbf{E}^{(1)} + \alpha_2 \mathbf{E}^{(2)} + \cdots + \alpha_l \mathbf{E}^{(l)} \tag{2}$$

where α_k is the weight coefficient of the k-th convolutional layer. In this way, we can obtain the embeddings of common users $\mathbf{E}^{\mathcal{S}}_{user}$ and items $\mathbf{E}^{\mathcal{S}}_{item}$ from $\mathcal{G}^{\mathcal{S}}$, embeddings of common users $\mathbf{E}^{\mathcal{T}}_{user}$ and items $\mathbf{E}^{\mathcal{T}}_{item}$ from $\mathcal{G}^{\mathcal{T}}$, and embeddings of common users $\mathbf{E}^{\mathcal{C}}_{user}$, embeddings of items $\mathbf{E}^{\mathcal{C}}_{item_{\mathcal{S}}}$ in source domain $\mathcal{S}$ and embeddings of items $\mathbf{E}^{\mathcal{C}}_{item_{\mathcal{T}}}$ in target domain $\mathcal{T}$ from $\mathcal{G}^{\mathcal{C}}$.

3.4 Feature Constraint

The purpose of designing $\mathcal{G}^{\mathcal{S}}$, $\mathcal{G}^{\mathcal{T}}$, and $\mathcal{G}^{\mathcal{C}}$ is to better capture domain-invariant and domain-specific features effectively. This can be considered as the embeddings of the same user or item across different graphs should be as discrepant as possible. However, there are implicit correlations between these embeddings. To address this, we consider that the embedding representations obtained from the three graphs belong to different linear spaces, denoted as $\mathbb{V}_{\mathcal{S}}$, $\mathbb{V}_{\mathcal{T}}$ and $\mathbb{V}_{\mathcal{C}}$, respectively. As shown in Fig. 3, the linear mapping $f_{\mathcal{S}} : \mathbb{V}_{\mathcal{S}} \to \mathbb{V}_{\mathcal{C}}$ and $f_{\mathcal{T}} : \mathbb{V}_{\mathcal{T}} \to \mathbb{V}_{\mathcal{C}}$ are employed to transform the embedding between these disparate spaces. We can obtain the features of the item or user in different linear spaces through linear mapping. The goal of feature constraint is to preserve the implicit correlations between the same user or item features across different linear spaces. Item feature constraints and a common user feature constraint are designed in CDAR-FCL.

Item Feature Constraint As both $\mathbf{E}^{\mathcal{S}}_{item}$ and $\mathbf{E}^{\mathcal{C}}_{item_{\mathcal{S}}}$ represent item features in domain $\mathcal{S}$, we impose a consistency constraint formulated as follows:

$$L^{\mathcal{S}}_{item} = \left\| \mathbf{E}^{\mathcal{S}}_{item} W^{\mathcal{S}} - \mathbf{E}^{\mathcal{C}}_{item_{\mathcal{S}}} \right\|^{2}_{F} \tag{3}$$

where $W^{\mathcal{S}} \in \mathbb{R}^{\dim \times \dim}$ is a learnable parameter. We apply $W^{\mathcal{S}}$ to $\mathbf{E}^{\mathcal{S}}_{item}$ to map it to the same linear space as $\mathbf{E}^{\mathcal{C}}_{item_{\mathcal{S}}}$, and then we minimize the Frobenius

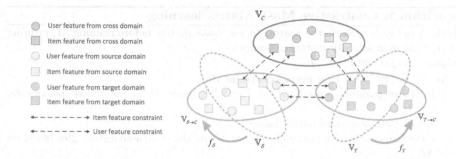

Fig. 3. By applying linear mappings f_S and f_T to $\mathbb{V}_S$ and $\mathbb{V}_T$, respectively, we can obtain two linear spaces, $\mathbb{V}_{S \to C}$ and $\mathbb{V}_{T \to C}$, that represent the same linear space as $\mathbb{V}_C$. Subsequently, we impose constraints on the features of the same item or user in $\mathbb{V}_C$, $\mathbb{V}_{S \to C}$, and $\mathbb{V}_{T \to C}$ to preserve their potential consistency.

norm of the discrepancy between the two embedding matrices to enhance the implicit correlation. In the same way, for the embeddings $\mathbf{E}_{item}^{T}$ and $\mathbf{E}_{itemT}^{C}$, the constraint can be defined as:

$$L_{item}^{T} = \left\| \mathbf{E}_{item}^{T} W^{T} - \mathbf{E}_{itemT}^{C} \right\|_{F}^{2} \tag{4}$$

where $W^{T} \in \mathbb{R}^{\dim \times \dim}$ is a learnable parameter used to map the embedding matrix $\mathbf{E}_{item}^{T}$ to the same linear space as $\mathbf{E}_{itemT}^{C}$. In this way, we do not directly damage the discrepancy between domain-specific features and domain-invariant features but consider the consistency between them.

Common User Feature Constraint We also incorporate a consistency constraint for user embeddings. The constraint is defined as:

$$L^{user} = \left\| \mathbf{E}_{user}^{S} W^{S} - \mathbf{E}_{user}^{T} W^{T} \right\|_{F}^{2} \tag{5}$$

To ensure consistency between the different user embeddings, we use the matrix W^{S} to map $\mathbf{E}_{user}^{S}$ into the same linear space as $\mathbf{E}_{user}^{C}$, and W^{T} to map $\mathbf{E}_{user}^{T}$ into the same linear space as $\mathbf{E}_{user}^{C}$. By doing this, we consider $\mathbf{E}_{user}^{S} W^{S}$ and $\mathbf{E}_{user}^{T} W^{T}$ as different representations of the embedding $\mathbf{E}_{user}^{C}$ in the same linear space, and the constraint is designed to improve their consistency.

Contrastive Mask. In (3), (4), and (5), W^{S} and W^{T} are shared parameters, which create a parameter conflict problem during training. To overcome this issue, we propose a method that uses contrastive learning to train a binary mask matrix. Algorithm 1 provides a detailed explanation. The main concept behind this approach is to identify and remove the shared parameters that have little impact on the current constraint, while keeping the parameters that have a significant impact. Take (3) as an example, the constraint is expressed as:

$$L_{item}^{S} = \left\| \mathbf{E}_{item}^{S} W^{S} \odot M_{1}^{S} - \mathbf{E}_{itemS}^{C} \right\|_{F}^{2} \tag{6}$$

Algorithm 1. Contrastive Mask Matrix learning

Input: A network with shared parameters s, mask matrix before pruning M, pruning rate α, Minimum sparsity ε, iteration epochs T.

Output: the pruned mask M

1: **for** W in $\{W^{\mathcal{S}}, W^{\mathcal{T}}\}$ for every constraint **do**
2:     Initialize parameter mask $M^t \in \{0,1\}^{|W|}$, where W is the shared parameter, $t = 1$.
3:     Select a random mini-batch data x
4:     Generate contrastive parameter mask M' by the contrastive strategies based on M.
5:     **for** $t = 1, ..., T$ **do**
6:         Update the shared parameter W by taking a gradient step to optimize contrastive loss in (8).
7:     **end for**
8:     Prune the remaining parameters from W with the smallest numerical change of α percent. Let $W[j] = 0$ if $M[j]$ is pruned.
9:     **if** $\frac{\|M\|_0}{\|W\|} \le \varepsilon$ **then**
10:         Reset W and go to step 3.
11:     **else**
12:         The pruned parameter mask is M.
13:     **end if**
14: **end for**
15: **return** The pruned mask M

where $\odot$ represents the element-wise product used to filter the useful parameters. To learn every parameter mask M, the noise data is denoted as M', which can be generated using any contrastive strategy. In this method, the parameter mask is randomly reversed with different sampling ratios. Take $M_1^{\mathcal{S}}$ as an example, for noise data, the constraint is defined as:

$$L_{item}^{\mathcal{S}}{}' = \left\| \mathbf{E}_{item}^{\mathcal{S}} W^{\mathcal{S}} \odot M_1^{\mathcal{S}'} - \mathbf{E}_{item\mathcal{S}}^{\mathcal{C}} \right\|_F^2 \tag{7}$$

$$Loss_{Contrastive} = softplus(L_{item}^{\mathcal{S}}{}' - L_{item}^{\mathcal{S}}) \tag{8}$$

The shared parameter $W^{\mathcal{S}}$ will be optimized using Algorithm 1.

3.5   Feature Combination

We combine the final features in an additive fashion. $\mathbf{Z}_{user}$ is obtained by summing $\mathbf{E}_{user}^{\mathcal{S}}$, $\mathbf{E}_{user}^{\mathcal{T}}$ and $\mathbf{E}_{user}^{\mathcal{C}}$. Similarly, $\mathbf{Z}_{item}^{\mathcal{S}}$ is obtained by summing $\mathbf{E}_{item}^{\mathcal{S}}$, and $\mathbf{E}_{item\mathcal{S}}^{\mathcal{S}}$, and $\mathbf{Z}_{item}^{\mathcal{T}}$ is obtained by summing $\mathbf{E}_{item}^{\mathcal{T}}$ and $\mathbf{E}_{item\mathcal{T}}^{\mathcal{T}}$. Finally, cosine similarity is used to predict the ratings of user items.

We use the Bayesian Personalized Ranking (BPR) [13] loss in both domains. The loss function is defined as:

$$L_{BPR} = -\sum_{u=1}^{|U|} \sum_{i \in I} \sum_{j \notin I} \ln \sigma(\hat{y}_{ui} - \hat{y}_{uj}) + \lambda \left\| \mathbf{E}^{(0)} \right\|^2 \tag{9}$$

where σ is the softplus activation function, and $\hat{y}$ is the predicted rating of the user for the item. $\mathbf{E}^{(0)}$ represents the embeddings of the 0-th layer in the graph embedding layer. We apply L_2 regularization on the embedding layer, which is sufficient to prevent overfitting, with λ controlling the strength of regularization. Finally, to jointly optimize the recommendation prediction and feature constraints, we define a joint loss function L_{joint} as follows:

$$L_{joint} = L_{BPR}^{S} + L_{BPR}^{T} + \lambda_1 (L_{item}^{S} + L_{item}^{T} + L^{user}) \qquad (10)$$

where λ_1 controls the strength of the feature constraint. We employ the Adam optimizer and train the model in a mini-batch manner.

4   Experiments

We aim to answer the following research questions by doing experiments:

RQ1: How does the value of the feature constraint hyper-parameter affect the results?

RQ2: Does our model outperform existing single-domain and cross-domain recommendation methods?

RQ3: How to prove that the proposed feature constraints capture the consistency of the embeddings obtained from different graphs?

RQ4: How does the feature constraint contribute to the recommendation prediction?

4.1   Experimental Settings

Datasets. We conducted experiments on real world cross-domain datasets from Amazon-5cores[1], where each user or item has at least five ratings. We selected six popular categories: **Books**, Movies and TV (**Movies** for short), Clothing, Shoes and Jewelry (**Cloths** for short), Sports and Outdoors (**Sports** for short), Cell Phones and Accessories (**Cell** for short), and Electronics (**Elec** for short). We used three pairs of datasets: Books & Movies, Cloths& Sports, and Cell & Elec. For each pair of datasets, we initially filtered the datasets to retain users with ratings greater than three, and labeled their interactions with a rating of 1 to indicate positive interactions. Next, we extracted the overlapping users which had more than 5 positive interactions in both domains. Table 1 summarizes the detailed statistics of the three pairs of datasets.

Baselines. We compare the performance of CDAR-FCL with various single-domain models: NGCF [15] and LightGCN [3]. We also compare the cross-domain recommendation models: DDTCDR [6], PPGN [17], BiTGCF [7], and DisenCDR [1].

---

[1] http://jmcauley.ucsd.edu/data/amazon.

Table 1. Datasets

Task	Dataset	#Users	#Items	#Interactions	Density
Task1	Cell	10421	9679	81934	0.081%
	Elec	10421	40121	190522	0.046%
Task2	Books	7378	154387	410391	0.036%
	Movies	7378	42681	326021	0.103%
Task3	Cloths	3290	11860	29147	0.075%
	Sports	3290	12213	37458	0.093%

Metrics. We adopt leave-one-out method for evaluation. For each test user, we randomly split one interaction with a test item as the test interaction. We randomly sample 99 unobserved interactions for the test user and then rank the test item among the 100 items. For the Top-K recommendation task, we adopt Hit Ratio (HR) and Normalized Discounted Cumulative Gain (NDCG) to evaluate the performance of all models, and K is set to 10 in our experiments.

Implementation Details. In our experiments, we ignore the side information of users/items. We implement DDTCDR by ourselves and maintain the optimal configuration in their papers. For PPGN, the embedding size is fixed to 8. As for other methods, we set the shared hyperparameters as follows: the embedding size is fixed to 128 and the embedding parameters are initialized using the Xavier method. The mini-batch size is fixed to 1024, the negative sampling number is fixed to 1, the Adam optimizer is used to update all parameters, the learning rate is set to $1e^{-3}$, and the L2 regularization coefficient is set to $1e^{-4}$. The number of graph convolutional network layers is set to 3. The L_2 regularization coefficient λ is set to $1e^{-4}$. We combine the values of each layer using an average weighting, with the layer combination coefficient set to $\alpha = 1/3$. The minimum sparsity ε is set to 0.3, the pruning rate α to 0.1, and the iteration epochs t in Algorithm 1 set to 5. We implement CDAR-FCL using the PyTorch framework.

4.2 Effect of Hyper-parameter

λ_1 is the hyper-parameter controlling the strength of the feature constraint. If the feature constraints are too powerful, they may affect the distinction between domain-invariant features and domain-specific features. To answer **RQ1**, we vary λ_1 in the range of $\{0, 0.1, 0.3, 0.5, 1, 2\}$ and evaluate CDAR-FCL on the three pairs of datasets. The results are shown in Fig. 4. We observe that CDAR-FCL performs best on Cell&Elec and Cloths & Sports when $\lambda_1 = 0.3$, and on Books & Movies when $\lambda_1 = 0.1$. Additionally, we find that when the value of λ_1 is either too large or too small, it leads to poor performance of CDAR-FCL.

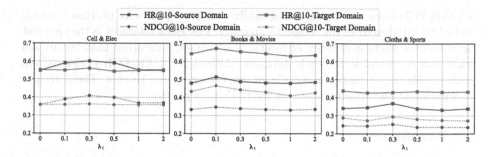

Fig. 4. Impact of λ_1

4.3   Comparison

To answer **RQ2**, we compare CDAR-FCL with the baselines on three pairs of Amazon datasets. The overall comparison is reported in Table 2, and the following observations are made:

Table 2. The experimental results (HR@10 & NDCG@10)

Task	Dataset	Metrics	Single-domain Methods		Cross-domain Methods				
			NGCF	LightGCN	DDTCDR	PPGN	BiTGCF	DisenCDR	CDAR-FCL
Task1	Cell	HR@10	0.5631	0.5874	0.4549	0.5608	0.5428	0.5679	**0.6003**
		NDCG@10	0.3727	0.3920	0.2792	0.3173	0.3461	0.3668	**0.4084**
	Elec	HR@10	0.5093	0.5369	0.5023	0.4964	0.4962	0.5566	**0.5604**
		NDCG@10	0.3352	0.3457	0.3321	0.2958	0.3165	0.3512	**0.3622**
Task2	Books	HR@10	0.5009	0.4983	0.3504	0.5121	0.5211	**0.6083**	0.5135
		NDCG@10	0.3423	0.3558	0.2235	0.3265	0.3425	**0.3967**	0.3470
	Movies	HR@10	0.6474	0.6313	0.4877	0.6345	0.6323	0.6714	**0.6722**
		NDCG@10	0.4232	0.4167	0.2875	0.3706	0.4016	0.4258	**0.4647**
Task3	Cloths	HR@10	0.2836	0.2906	0.3100	0.3256	0.3419	0.3568	**0.3684**
		NDCG@10	0.1627	0.1769	0.1830	0.1853	0.2078	0.1925	**0.2531**
	Sports	HR@10	0.3559	0.3605	0.3231	0.3992	0.4190	0.4021	**0.4292**
		NDCG@10	0.2100	0.2189	0.1838	0.2150	0.2532	0.2365	**0.2949**

- CDAR-FCL has achieved optimal performance in most tasks, thanks to its ability to learn domain-invariant and domain-specific features from different graphs, avoiding negative transfer between domains. Simultaneously, by incorporating feature constraints, it preserves the consistency of embeddings obtained from different graphs.
- PPGN and BiTGCF do not outperform single-domain baselines on Task1, indicating the existence of negative transfer in these algorithms. In contrast, CDAR-FCL outperforms single-domain baselines on every dataset, demonstrating its ability to transfer only useful information to the target domain.

– CDAR-FCL does not perform optimally only on the Books in Task2, which could be attributed to the dataset's sparsity. The Books dataset is the sparsest and has the highest number of interaction entries, indicating that our model may not be suitable for such datasets. Although CDAR-FCL performs less well on Books, it makes a significant improvement on Movies. These results demonstrate the effectiveness of CDAR-FCL in practice.

4.4 Visualization

To answer **RQ3** and more intuitively reflect the effect of our proposed feature constraint, we randomly sampled 200 users and visualize Z_{user} after dimensionality reduction using t-SNE [9]. As shown in Fig. 5, most green points are located next to both a red and a blue point, indicating that CDAR-FCL effectively constrains user features in different domains. The Books&Movies diagram in Fig. 5 reveals several points that are located separately, suggesting that some user features are not adequately constrained. This is also consistent with the poor performance of CDAR-FCL on Books & Movies in Sect. 4.3, and we believe that this may be due to the influence of data sparsity. Applying feature constraints on sparse datasets tends to yield inferior results.

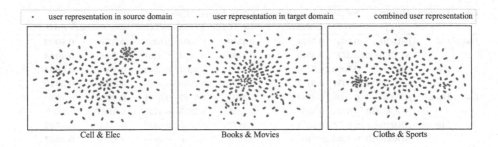

Fig. 5. Visualization of the learned user embedding

4.5 Ablation Study

To answer **RQ4**, we analyze the effectiveness of feature constraints in CDAR-FCL through ablation experiments. We design three variants of CDAR-FCL.

1. CDAR-FCL-NU removes the user feature constraint and only constrains the item embedding. The joint loss is defined as $L_{joint}^{NU} = L_{BPR}^{S} + L_{BPR}^{T} + \lambda_1(L_{item}^{S} + L_{item}^{T})$
2. CDAR-FCL-NI removes the item feature constraint and only constrains the user embedding. The joint loss is defined as $L_{joint}^{NI} = L_{BPR}^{S} + L_{BPR}^{T} + \lambda_2 L^{user}$
3. CDAR-FCL-NA removes all the feature constraints and the joint loss is defined as $L_{joint}^{NA} = L_{BPR}^{S} + L_{BPR}^{T}$.

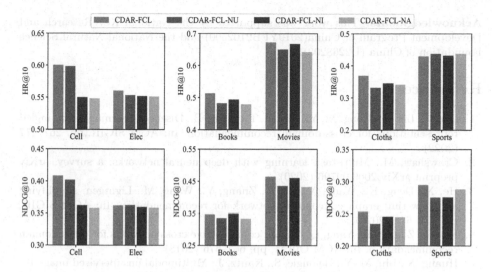

Fig. 6. Performance comparison of model variants

We conduct ablation experiments on three tasks using the three variants above mentioned, and the results are depicted in Fig. 6. Our HR and NDCG analyses reveal the following key findings:

- Except for the sports dataset, CDAR-FCL consistently outperforms all variants in the other datasets, indicating the effectiveness of feature constraints.
- For both CDAR-FCL-NU and CDAR-FCL-NI, we find that they outperform CDAR-FCL-NA in both Cell & Elec and Books& Movies datasets, demonstrating the effectiveness of the single use of the user feature constraint and the item feature constraint. Although single use is not effective on the Cloths & Sports dataset, the best results are obtained when both are used, which illustrates the validity of our shared parameter usage.
- We are unable to discern the performance differences between CDAR-FCL-NU and CDAR-FCL-NI, suggesting that the relative importance of user and item feature constraints may vary across tasks.

5   Conclusion

In this paper, we present CDAR-FCL, a Cross-Domain Adaptive Recommendation model using Feature Constraints Learning. The model uses common and specific graph convolution to learn domain-invariant and domain-specific features. A feature constraint is designed to achieve domain adaptation. CDAR-FCL is a dual-target model capable of mutually transferring useful knowledge between two domains. Our experimental results demonstrate the effectiveness of the proposed model. Moving forward, we plan to extend our approach to partially overlapping cross-domain recommendation scenarios.

Acknowledgments. This work was supported by the National Key Research and Development Program of China(2019YFB2102500) and the National Natural Science Foundation of China (U2268203).

References

1. Cao, J., Lin, X., Cong, X., Ya, J., Liu, T., Wang, B.: Disencdr: learning disentangled representations for cross-domain recommendation. In: ACM SIGIR.,pp. 267–277 (2022)
2. Crawshaw, M.: Multi-task learning with deep neural networks: a survey. arXiv preprint arXiv:2009.09796 (2020)
3. He, X., Deng, K., Wang, X., Li, Y., Zhang, Y., Wang, M.: Lightgcn: simplifying and powering graph convolution network for recommendation. In: ACM SIGIR, pp. 639–648 (2020)
4. Hu, G., Zhang, Y., Yang, Q.: Conet: collaborative cross networks for cross-domain recommendation. In: ACM CIKM, pp. 667–676 (2018)
5. Huang, X., Liu, M.-Y., Belongie, S., Kautz, J.: Multimodal unsupervised image-to-image translation. In: Ferrari, V., Hebert, M., Sminchisescu, C., Weiss, Y. (eds.) ECCV 2018. LNCS, vol. 11207, pp. 179–196. Springer, Cham (2018). https://doi.org/10.1007/978-3-030-01219-9_11
6. Li, P., Tuzhilin, A.: Ddtcdr: deep dual transfer cross domain recommendation. In: ACM WSDM, pp. 331–339 (2020)
7. Liu, M., Li, J., Li, G., Pan, P.: Cross domain recommendation via bi-directional transfer graph collaborative filtering networks. In: ACM CIKM, pp. 885–894 (2020)
8. Liu, W., Su, J., Chen, C., Zheng, X.: Leveraging distribution alignment via stein path for cross-domain cold-start recommendation. Adv. Neural. Inf. Process. Syst. **34**, 19223–19234 (2021)
9. Van der Maaten, L., Hinton, G.: Visualizing data using t-sne. J. Mach. Learn. Res. **9**(11) (2008)
10. Man, T., Shen, H., Jin, X., Cheng, X.: Cross-domain recommendation: an embedding and mapping approach. In: IJCAI, vol. 17, pp. 2464–2470 (2017)
11. Pan, S.J., Tsang, I.W., Kwok, J.T., Yang, Q.: Domain adaptation via transfer component analysis. IEEE T-NN **22**(2), 199–210 (2010)
12. Pan, S.J., Yang, Q.: A survey on transfer learning. IEEE TKDE **22**(10), 1345–1359 (2010)
13. Steffen, R., Christoph, F., Zeno, G., Lars, S.T.: Bayesian personalized ranking from implicit feedback. In: Proceedings of the Twenty-fifth Conference on Uncertainty in Artificial Intelligence. AUAI Press (2009)
14. Tzeng, E., Hoffman, J., Saenko, K., Darrell, T.: Adversarial discriminative domain adaptation. In: IEEE CVPR, pp. 7167–7176 (2017)
15. Wang, X., He, X., Wang, M., Feng, F., Chua, T.S.: Neural graph collaborative filtering. In: ACM SIGIR, pp. 165–174 (2019)
16. Zhang, Y., Liu, Z., Ma, Y., Gao, Y.: Multi-graph convolutional feature transfer for cross-domain recommendation. In: IEEE IJCNN, pp. 1–8 (2022)
17. Zhao, C., Li, C., Fu, C.: Cross-domain recommendation via preference propagation graphnet. In: ACM CIKM, pp. 2165–2168 (2019)
18. Zhu, F., Wang, Y., Chen, C., Zhou, J., Li, L., Liu, G.: Cross-domain recommendation: challenges, progress, and prospects. arXiv preprint arXiv:2103.01696 (2021)
19. Zhu, Y., et al.: Personalized transfer of user preferences for cross-domain recommendation. In: ACM WSDM, pp. 1507–1515 (2022)

A Hybrid Approach Using Convolution and Transformer for Mongolian Ancient Documents Recognition

Shiwen Sun[1,2,3], Hongxi Wei[1,2,3(✉)], and Yiming Wang[1,2,3]

[1] School of Computer Science, Inner Mongolia University, Hohhot 010010, China
cswhx@imu.edu.cn
[2] Provincial Key Laboratory of Mongolian Information Processing Technology,
Hohhot 010010, China
[3] National and Local Joint Engineering Research Center of Mongolian Information
Processing Technology, Hohhot 010010, China

Abstract. Mongolian ancient documents are an indispensable source for studying Mongolian traditional culture. To thoroughly explore and effectively utilize these ancient documents, conducting a comprehensive study on Mongolian ancient document words recognition is essential. In order to better recognize the word images in Mongolian ancient documents, this paper proposes an approach that combines convolutional neural networks with Transformer models. The approach used in this paper takes word images as the input for the model. After passing through a feature extractor composed of convolutional neural networks, the extracted features are fed into a Transformer model for prediction. Finally, the corresponding recognition results of the word images are obtained. Due to the common existence of imbalanced distribution of character classes in recognition tasks, models often tend to excessively focus on common characters while neglecting rare characters. Our proposed approach integrates focal loss to enhance the model's attention towards rare characters, thereby improving the overall recognition performance of the model for all characters. After training, the model is capable of rapidly and efficiently performing end-to-end recognition of words in Mongolian ancient documents. The experimental results indicate that our proposed approach outperforms existing methods for word recognition in Mongolian ancient documents, effectively improving the performance of Mongolian ancient document words recognition.

Keywords: Mongolian ancient documents · Word recognition · Convolutional neural networks · Transformer · Focal loss

1 Introduction

Mongolians have a long and storied history, and their far-reaching culture has left behind numerous valuable ancient documents. These ancient documents serve

B. Luo et al. (Eds.): ICONIP 2023, CCIS 1967, pp. 165–176, 2024.
https://doi.org/10.1007/978-981-99-8178-6_13

as important sources of information for studying the traditional culture of the Mongolian people. There are various kinds of Mongolian ancient documents, including books, scriptures, manuscripts, copies, albums, stone stele rubbing, journals, newspapers, as well as archival materials. In order to further preserve, excavate, and utilize these valuable ancient documents, they are scanned and saved as image formats for relevant research. Mongolian is a phonetic writing system where all the letters of a Mongolian word are written in sequential order from top to bottom along a main line. The letters take different writing forms based on their positions within the word. Adjacent words are separated from each other by a space. Furthermore, the writing system of Mongolian differs from other languages. Its writing order is from top to bottom and column order is from left to right, as shown in Fig. 1.

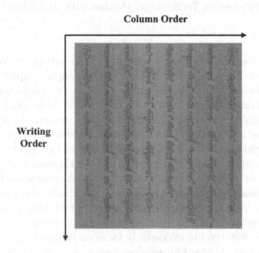

Fig. 1. An example of woodblock-printed Mongolian documents.

The recognition methods of Mongolian ancient documents can be divided into two categories: character-based segmentation methods and whole-word recognition methods. The character-based segmentation method is to divide the Mongolian word into a series of characters along the writing direction, and then recognize each individual character. The segmentation operation requires a higher quality of the word image and is more suitable for modern printed Mongolian documents. However, Mongolian ancient documents often exhibit phenomena such as peeling of text pigments, complex noise, and severe defects, as shown in Fig. 2. The scanning image quality of these documents is poor, making it difficult to achieve satisfactory recognition results using character-based segmentation methods. Therefore, whole-word recognition methods are more suitable for Mongolian ancient documents. Previously, a commonly used whole-word recognition method for Mongolian ancient documents was based on Convolutional Neural Networks (CNN). This method treated the entire recognition task as an image classification task. Although it could achieve whole-word recognition, it was unable to solve the out-of-vocabulary (OOV) problem.

Scanned Image   Binarized Image   Word Image

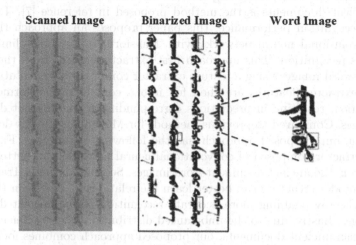

Fig. 2. An example of damage and noise in woodblock-printed Mongolian documents.

In recent years, research on the recognition of Mongolian ancient documents has made some achievements. Gao et al. [5] proposed a Mongolian ancient documents recognition method based on character segmentation principles. This method achieved a recognition accuracy of 71% on a dataset containing 5,500 high-quality images. Subsequently, Su et al. [11,12] improved the accuracy of character recognition by adjusting the segmentation algorithm and achieved a recognition accuracy of 81% on the same dataset. However, due to the long history of Mongolian ancient documents, it is challenging to perform accurate character segmentation on scanned images. Therefore, whole-word recognition methods are adopted. Wei et al. [14] proposed a whole-word recognition method based on CNN for Mongolian ancient documents recognition. They constructed a classifier for the task, but this method failed to address the OOV problem. Later, Kang et al. [7] proposed a sequence-to-sequence (Seq2Seq) whole-word recognition model with an attention mechanism for Mongolian ancient documents recognition. This method can address the OOV problem, and it achieved a recognition accuracy of 88.29%.

With the introduction of the Transformer model [13], many researchers have also applied it to computer vision tasks due to its powerful feature representation capabilities. The main advantage of the Transformer lies in its ability to process longer sequences of data in a shorter amount of time, while effectively capturing the dependencies between elements within the input sequence. When the Transformer is applied to computer vision tasks, it can better handle the global relationships within images. Specifically, it is capable of extracting semantic information from images and learning the dependencies between different features. Additionally, the self-attention mechanism in the Transformer allows the model to focus on the relationships between different parts of the image, thereby avoiding the issue of information loss.

Currently, the best-performing method for whole-word recognition of Mongolian ancient documents is the method proposed in reference [7]. To further improve recognition performance, this paper proposes an approach that combines convolutional neural networks with Transformers for Mongolian ancient documents recognition. This approach first extracts features from the ancient document word images using a feature extractor composed of convolutional layers. The extracted features are then fed into a complete Transformer model for prediction, resulting in predictions corresponding to the ancient document word images. Compared to previous methods for Mongolian ancient documents recognition, our proposed approach has the following advantages: Firstly, the feature extractor composed of convolutional neural networks can better extract features from the ancient document word images. Secondly, using a Transformer encoder-decoder structure can better learn the relationships between the image features, thereby enabling more accurate recognition of the ancient document word images. Lastly, due to the imbalanced distribution of character categories in Mongolian ancient documents, our proposed approach combines focal loss to make the model pay more attention to rare characters, so as to improve the model's character recognition performance. As described above, the proposed approach in this paper possesses stronger representational capabilities, better learning and prediction abilities, and higher recognition performance.

2   Related Work

The Mongolian recognition methods can be mainly divided into two categories: segmentation-based recognition methods and whole-word recognition methods. The segmentation-based recognition method involves splitting the word into a series of graphemes. Subsequently, recognition algorithms are used to individually recognize each grapheme, and the recognition results for each grapheme are combined. Given accurate segmentation, this recognition method only requires recognizing a limited number of glyph types in the subsequent process. However, when it comes to recognizing Mongolian ancient documents, this method faces difficulties in achieving accurate segmentation due to the presence of overlapping and ligature phenomena in the graphemes of these documents. The current state-of-the-art segmentation method is the contour analysis-based grapheme segmentation algorithm [12]. Its main steps include: contour key point detection, main stem localization, and segmentation line generation. For contour key point detection, the algorithm primarily employs the Ramer-Douglas-Peucker algorithm [4,9]. Firstly, an approximate polygon is extracted from the contour of Mongolian words. Then, the extracted approximate polygon is used to assist in locating the main stem of the word, and finally, the segmentation path is generated. The main stem localization is achieved using the vertical projection method, combined with the contour key point information obtained in the previous step to enhance the accuracy of main stem localization. During the generation of segmentation lines, initially, some candidate segmentation lines are generated using the contour key point information and main stem information obtained from the previous two steps. After that, a neural network is employed to filter and retain the valid segmentation lines.

The whole-word recognition method in Mongolian recognition involves directly recognizing Mongolian words without performing any segmentation operation. This method can effectively avoid a series of issues caused by inaccurate word segmentation. The traditional whole-word recognition method initially extracts feature vectors from word images and then matches them against words in a known dictionary using a matching algorithm. Finally, the word with the highest similarity is considered as the recognition result. Deep learning is the current mainstream approach for whole-word recognition, where it utilizes end-to-end models to accomplish the task of whole-word recognition. The aforementioned end-to-end models take word images as input and generate word prediction results as output. There is no need for word segmentation between the input and output.

The commonly used end-to-end optical character recognition (OCR) methods can be primarily divided into two categories: one is based on Convolutional Recurrent Neural Networks (CRNN), and the other is based on attention mechanism. Methods based on CRNN, such as the Connectionist Temporal Classification (CTC) model [10], combine convolutional and Recurrent Neural Networks (RNN). This model first extracts image features using a CNN, then utilizes a RNN to predict the label distribution as a sequence of features. Finally, the CTC network [6] is used to obtain the predicted labels corresponding to the image. Methods based on attention mechanism [1] typically employ CNN and RNN for image feature extraction. Subsequently, the attention mechanism is used to focus on different parts of the feature sequence at different time steps. Finally, the model outputs the recognition results. Both of the aforementioned recognition methods can output variable-length recognition results.

3 The Proposed Approach

The approach proposed in this paper combines CNN and Transformer to address the issues of diverse writing styles and varying character forms in woodblock-printed Mongolian ancient documents. It utilizes whole-word recognition method, which means that Mongolian words are not segmented into individual characters but directly recognized as whole words. Our proposed approach considers Mongolian ancient document words recognition as an image-to-sequence mapping problem, where the input image is a Mongolian ancient document word image, and the output sequence is the text annotation (character sequence) corresponding to the word. It aims to recognize word images by learning the mapping relationship between input images and their corresponding labels. The Mongolian ancient documents recognition model, which combines CNN with Transformer, consists of three components: feature extractor, encoder, and decoder. The structure of the model is shown in Fig. 3, the word image is first fed into the feature extractor, which is composed of CNN, for feature extraction. Then, it is passed through the encoder-decoder structure of Transformer. Finally, the predicted result of the word image is generated as the output.

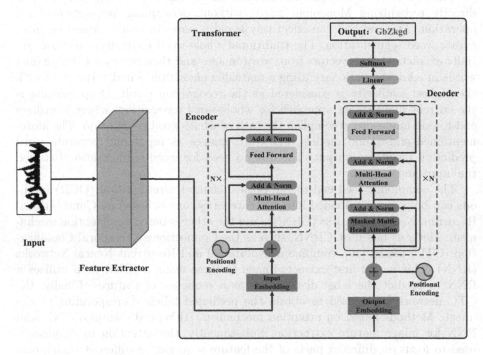

Fig. 3. The structure of the proposed Mongolian ancient documents recognition model.

Our proposed approach utilizes a ResNet-based feature extractor [3] to extract visual features from Mongolian ancient document word images. The structure of the feature extractor is shown in Fig. 4. According to the morphological and top-down writing characteristics of Mongolian words, in order to better preserve the visual information of Mongolian ancient documents, the feature extractor used in this paper first slices the word image into several equally sized image frames along the writing direction from top to bottom. Subsequently, feature extraction is performed on these image frames one by one. The extracted image frame features are concatenated, and then undergo a series of convolution, batch normalization, and pooling operations. Afterwards, they are fed into a ResNet-based feature extraction network. The ResNet-based feature extraction network consists of four stages (Layer1 - Layer4). These structures are responsible for feature extraction at different stages of the input feature maps in order to obtain richer and higher-level abstract features, enabling better capture of semantic information in the images and improving the accuracy of classification or recognition. Finally, after a series of convolution and batch normalization operations, it reaches the encoder-decoder structure of Transformer.

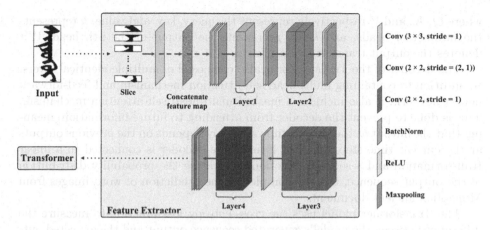

Fig. 4. The structure of the feature extractor.

The approach proposed in this paper employs an encoder-decoder structure composed of Transformer model to learn the mapping relationship between word images and their corresponding labels, thereby achieving the goal of word image recognition. The Transformer model performs exceptionally well in processing sequential data, and its main idea is based entirely on the attention mechanism, abandoning traditional RNN and CNN. It adopts a structure called self-attention mechanism that enables the model to simultaneously consider all positional information in the input sequence. The expression for self-attention mechanism is as follows:

$$Attention(Q, K, V) = softmax(\frac{QK^T}{\sqrt{d_k}})V \tag{1}$$

where Q represents the query matrix, K represents the content to be attended to, and QK^T denotes the dot product operation, which calculates the attention weights for Q over V. The purpose of scaling by $\sqrt{d_k}$ is to avoid large dot products that can result in small gradients during the $softmax$ operation.

The feature maps extracted from word images of Mongolian ancient documents by a feature extractor are first transformed into a fixed-dimensional vector representation through an embedding layer. Then, positional encoding is introduced to preserve the sequential order of elements in the sequence, and the transformed input is fed into the encoder of the Transformer. The encoder of the Transformer consists of multiple identical layers, each layer containing two sub-layers: multi-head self-attention mechanism and feed-forward neural network. The multi-head self-attention mechanism enables the model to attend to different positions in the input sequence simultaneously, while the feed-forward neural network performs representation learning and feature extraction on the input. The expression for multi-head self-attention mechanism is as follows:

$$MultiHead(Q, K, V) = Concat(head_1, \ldots, head_h)W^O$$
$$where \ head_i = Attention(QW_i^Q, KW_i^K, VW_i^V) \tag{2}$$

where Q, K, and V respectively represent the query, key, and value. h represents the number of heads, and $head_i$ represents the output of the i-th head. W^O denotes the output transformation matrix.

The decoder of the Transformer is also composed of multiple identical layers. In addition to containing multi-head self-attention mechanism and feed-forward neural network, it also includes a masked multi-head self-attention mechanism. This is done to prevent the decoder from attending to future information, meaning that the output at the current time step only depends on the previous outputs at the current time step. The last layer of the decoder is connected to a linear transformation and a softmax function, generating the probability distribution of the output sequence, thereby completing the prediction of word images from Mongolian ancient documents.

The Transformer model uses the cross-entropy loss function to measure the difference between the model's generated sequence output and the expected output sequence. By minimizing the cross-entropy loss function, the model adjusts its weights to make the output closer to the true labels. In the task of recognizing word images from Mongolian ancient documents, there is an issue of imbalanced distribution of word image samples and an uneven distribution of character classes. This can cause the model to overly focus on common samples or characters and neglect rare samples or characters during the training process. To address this issue, our proposed approach introduces focal loss as the loss function for the Transformer model. By adjusting the weights of common and rare samples or characters, the model can balance its attention towards them and give more emphasis on rare samples or characters, thereby improving the recognition performance of the model.

Focal loss [8] is a widely used loss function in tasks such as object detection and image segmentation. It was proposed by Lin et al. in 2017. Its design objective is to address the issue of class imbalance in object detection tasks. Focal loss tackles this problem by introducing an adjustment factor that can modify the weights of the loss function based on the difficulty level of the samples. Specifically, focal loss reduces the loss for easy-to-classify samples while amplifying the loss for hard-to-classify samples. This approach directs the model's attention towards the challenging samples, thereby improving its ability to detect positive samples. The expression for focal loss is as follows:

$$FL(p_t) = -\alpha(1 - p_t)^\gamma \log(p_t) \tag{3}$$

where p_t is the predicted probability of the sample by the model, α is a balancing factor used to adjust the weights of positive and negative samples, and γ is the focusing parameter used to adjust the weights of easy and hard samples. When $\gamma = 0$, focal loss degenerates into the regular cross-entropy loss function.

4   Experimental Results

4.1   Dataset and Baseline

To evaluate the performance of the Mongolian ancient documents recognition model proposed in this paper, a set of woodblock-printed Mongolian word images, similar to [7], were collected. This dataset was derived from 100 pages in Mongolian Kanjur, which is a historically significant collection of Mongolian ancient documents. The Mongolian Kanjur was created in 1720 through the woodblock-printed process. The woodblock-printed process is as follows: First, craftsmen engrave Mongolian characters onto wooden blocks. Then, cinnabar ink is applied onto the blocks. Finally, paper is placed onto the blocks, and the characters are printed onto the paper.

The number of word images is 20,948, accompanied by annotations. Through analysis of the annotations, the vocabulary is 1,423. These word images are used to train our proposed model. In this study, 4-fold cross validation is employed to evaluate the recognition performance. Therefore, the dataset is randomly divided into four folds, with each fold containing 5,237 word images. The vocabulary in each fold is shown in Table 1.

Table 1. The division details of the 4-fold cross validation.

Fold	#Vocabulary	#Word images
Fold1	874	5237
Fold2	878	5237
Fold3	883	5237
Fold4	874	5237

In the experiment of this paper, a word recognition method based on the Seq2Seq model with attention mechanism, as described in [7], was used as a baseline for comparison. The baseline method was implemented using the high-quality Python library Keras, with Theano as the backend, and trained on a Tesla K80. The optimizer used was RMSProp with a learning rate of 0.001. The batch size was set to 128, and the number of training epochs was set to 200. With word image frame height and overlap between adjacent frames set to 6 pixels and 3 pixels, respectively, the baseline method achieved the best recognition performance of 88.29%.

4.2 Parameter Settings

The experimental parameter settings are as follows: In the feature extractor, the number of basic blocks in ResNet Layer1 is 1, in Layer2 is 2, in Layer3 is 5, and in Layer4 is 3. In the Transformer structure, both encoder and decoder have 8 heads for the multi-head self-attention mechanism, and the number of sub-layers N is 6. For the focal loss, the parameter α is set to 0.25 and γ is set to 2. The maximum label length is set to 25, the batch size is set to 96, the learning rate is set to 1, and the number of training iterations is set to 500,000.

4.3   Results and Analysis

In the baseline method, the model achieved the highest recognition accuracy when the frame height was set to 6 pixels and the overlap between adjacent frames was set to 3 pixels. Compared with the baseline method, the performance of our proposed approach with cross-entropy loss and with focal loss is shown in Table 2.

Table 2. The comparison of recognition accuracy.

Model	Fold1	Fold2	Fold3	Fold4	Avg.
Seq2Seq [7]	88.10%	87.07%	91.03%	86.94%	88.29%
Ours (with cross-entropy loss)	93.85%	94.41%	92.88%	93.66%	93.7%
Ours (with focal loss)	**94.56%**	**94.55%**	**94.75%**	**94.42%**	**94.57%**

In Table 2, the first row shows the recognition accuracy of the baseline model on each fold and the average recognition accuracy across the four folds. The second row shows the recognition accuracy of the proposed model with cross-entropy loss on each fold and the average recognition accuracy across the four folds. The third row shows the recognition accuracy of the proposed model with focal loss on each fold and the average recognition accuracy across the four folds. By comparing the recognition accuracy, it can be observed that the recognition performance of our proposed approach can significantly outperform the baseline method, and further improve while utilizing focal loss.

To validate the effectiveness of our proposed approach without data augmentation, a comparison with the baseline method utilizing data augmentation was carried out. Two published data augmentation techniques were used: synthetic minority oversampling technique (SMOTE) [2] and Cycle-Consistent Generative Adversarial Network (CycleGAN) [15]. These techniques were applied to generate new samples for each word image in the dataset, effectively doubling the dataset. The comparison is shown in Table 3.

In Table 3, the first row represents the recognition accuracy of the baseline model using the SMOTE data augmentation method on each fold, and the average recognition accuracy across the four folds. The second row represents the

Table 3. The comparison of recognition accuracy with data-augmented baseline model.

Model	Fold1	Fold2	Fold3	Fold4	Avg.
Seq2Seq (SMOTE) [15]	90.80%	92.00%	91.16%	90.38%	91.09%
Seq2Seq (CycleGAN) [15]	90.83%	92.08%	91.50%	92.32%	91.68%
Ours (with cross-entropy loss)	93.85%	94.41%	92.88%	93.66%	93.7%
Ours (with focal loss)	**94.56%**	**94.55%**	**94.75%**	**94.42%**	**94.57%**

recognition accuracy of the baseline model using the CycleGAN data augmentation method on each fold, and the average recognition accuracy across the four folds. The third row represents the recognition accuracy on each fold and the average recognition accuracy across four folds of our proposed approach, which is without data augmentation and adopts the cross-entropy loss. The fourth row represents the recognition accuracy on each fold and the average recognition accuracy across four folds of our proposed approach, which is without data augmentation and adopts focal loss. By comparing the recognition accuracy, it can be observed that the proposed approach can outperform the data-augmented baseline methods. Furthermore, the proposed approach achieves the highest recognition performance when combined with focal loss. Based on the experimental results mentioned above, it can be concluded that the hybrid approach proposed in this paper, which combines convolution with Transformer, performs better in the task of Mongolian ancient document words recognition.

5 Conclusion

In this paper, a hybrid approach based on convolution and Transformer is adopted for Mongolian ancient document words recognition. Specifically, the ResNet-based feature extractor effectively captures the visual features of word images, while the Transformer encoder-decoder structure leverages visual features to achieve words recognition. The use of focal loss, addresses the issue of imbalanced data distribution, enhances the model's robustness and further improves its recognition performance. Compared to other existed methods, the proposed approach achieves higher recognition accuracy. It achieves the state-of-the-art performance in recognizing word images of Mongolian ancient documents, which demonstrates significant advantages in terms of recognition performance. The proposed approach is innovative in its application, enabling efficient end-to-end recognition of Mongolian ancient document word images. This approach carries far-reaching significance for related research on Mongolian ancient documents recognition.

Acknowledgment. This study is supported by the Project for Science and Technology of Inner Mongolia Autonomous Region under Grant 2019GG281, the Natural Science Foundation of Inner Mongolia Autonomous Region under Grant 2019ZD14, and the Program for Young Talents of Science and Technology in Universities of Inner Mongolia Autonomous Region under Grant NJYT-20-A05.

References

1. Bahdanau, D., Cho, K., Bengio, Y.: Neural machine translation by jointly learning to align and translate. arXiv preprint arXiv:1409.0473 (2014)
2. Chawla, N.V., Bowyer, K.W., Hall, L.O., Kegelmeyer, W.P.: Smote: synthetic minority over-sampling technique. J. Artif. Intell. Res. **16**, 321–357 (2002)

3. Cheng, Z., Bai, F., Xu, Y., Zheng, G., Pu, S., Zhou, S.: Focusing attention: towards accurate text recognition in natural images. In: Proceedings of the IEEE International Conference on Computer Vision, pp. 5076–5084 (2017)
4. Douglas, D.H., Peucker, T.K.: Algorithms for the reduction of the number of points required to represent a digitized line or its caricature. Cartographica: Int. J. Geog. Inf. Geovisualization 10(2), 112–122 (1973)
5. Gao, G., Su, X., Wei, H., Gong, Y.: Classical mongolian words recognition in historical document. In: 2011 International Conference on Document Analysis and Recognition, pp. 692–697. IEEE (2011)
6. Graves, A., Fernández, S., Gomez, F., Schmidhuber, J.: Connectionist temporal classification: labelling unsegmented sequence data with recurrent neural networks. In: Proceedings of the 23rd International Conference on Machine Learning, pp. 369–376 (2006)
7. Kang, Y., Wei, H., Zhang, H., Gao, G.: Woodblock-printing mongolian words recognition by bi-lstm with attention mechanism. In: 2019 International Conference on Document Analysis and Recognition, pp. 910–915. IEEE (2019)
8. Lin, T.Y., Goyal, P., Girshick, R., He, K., Dollár, P.: Focal loss for dense object detection. In: Proceedings of the IEEE International Conference on Computer Vision, pp. 2980–2988 (2017)
9. Ramer, U.: An iterative procedure for the polygonal approximation of plane curves. Comput. Graph. Image Process. 1(3), 244–256 (1972)
10. Shi, B., Bai, X., Yao, C.: An end-to-end trainable neural network for image-based sequence recognition and its application to scene text recognition. IEEE Trans. Pattern Anal. Mach. Intell. 39(11), 2298–2304 (2016)
11. Su, X., Gao, G., Wei, H., Bao, F.: Enhancing the Mongolian historical document recognition system with multiple knowledge-based strategies. In: Arik, S., Huang, T., Lai, W.K., Liu, Q. (eds.) ICONIP 2015. LNCS, vol. 9490, pp. 536–544. Springer, Cham (2015). https://doi.org/10.1007/978-3-319-26535-3_61
12. Su, X., Gao, G., Wei, H., Bao, F.: A knowledge-based recognition system for historical Mongolian documents. Int. J. Doc. Anal. Recogn. 19, 221–235 (2016)
13. Vaswani, A., et al.: Attention is all you need. In: Advances in Neural Information Processing Systems, vol. 30 (2017)
14. Wei, H., Gao, G.: A holistic recognition approach for woodblock-print Mongolian words based on convolutional neural network. In: 2019 IEEE International Conference on Image Processing, pp. 2726–2730. IEEE (2019)
15. Wei, H., Liu, K., Zhang, J., Fan, D.: Data augmentation based on CycleGAN for improving woodblock-printing Mongolian words recognition. In: Lladós, J., Lopresti, D., Uchida, S. (eds.) ICDAR 2021. LNCS, vol. 12824, pp. 526–537. Springer, Cham (2021). https://doi.org/10.1007/978-3-030-86337-1_35

Incomplete Multi-view Subspace Clustering Using Non-uniform Hyper-graph for High-Order Information

Jiaqiyu Zhan[✉] and Yuesheng Zhu

Communication and Information Security Lab, School of Electronic and Computer Engineering, Shenzhen Graduate School, Peking University, Beijing, China
{zjqy0429,zhuys}@pku.edu.cn

Abstract. Incomplete multi-view subspace clustering (IMSC) is intended to exploit the information of multiple incomplete views to partition data into their intrinsic subspaces. Existing methods try to exploit the high-order information of data to improve the clustering performance, many tools are used such as tensor factorization and hyper-Laplacian regularization. Compared to using complex mathematical tools to solves problems, why not considering to get considerable improvements through some simple ways? To address this issue, we propose an incomplete multi-view subspace clustering method using non-uniform hyper-graph (NUHG-IMSC) method which makes slightly change to the usual way of constructing uniform hyper-graph. We find a set of data points that have high similarity with the center point of each hyper-edge in high-dimensional space to be its neighbor samples, the cardinality of each hyper-edge is decided based on the distribution of the corresponding center point. This is a simple but effective way to utilize high-order information without bringing computational burden and extra parameters. Besides the advantage that the partial samples can be reconstructed more reasonably, our method also brings benefits to other parts in the whole framework of IMSC, such as learning the view-specific affinity matrices, the weight of each view, and the unified affinity matrix. Experimental results on three multi-view data sets indicate the effectiveness of the proposed method.

Keywords: Incomplete data · Multi-view subspace clustering · High-order information · Non-uniform hyper-graph · Tensor learning

1 Introduction

In practice, objects can be represented by different types of features. Multi-view clustering methods aim to partition these unlabeled multi-view data into respective groups by making full use of complementary multi-view information [1–5]. However, in real application, 'all views are complete' is hard to be achieved because several reasons can lead to randomly missing information from views.

B. Luo et al. (Eds.): ICONIP 2023, CCIS 1967, pp. 177–188, 2024.
https://doi.org/10.1007/978-981-99-8178-6_14

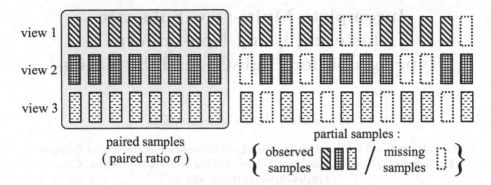

view 1

view 2

view 3

paired samples
(paired ratio σ)

partial samples :

{ observed samples / missing samples }

Fig. 1. An illustration of an incomplete 3-view data set with paired ratio $\sigma = 0.4$. The paired samples are observed in all three views.

Many incomplete multi-view clustering methods have been proposed to deal with these scenarios with incomplete views. Figure 1 shows an incomplete multi-view data with three views, samples contain all views are regarded as the paired samples, the scale of which is controlled by the parameter called paired ratio σ. In this paper, partial samples, which make the data set incomplete, are with only one missing view.

Among all the incomplete multi-view clustering methods, some focus on leaning a low-dimension non-negative consensus representation for clustering based on non-negative matrix factorization [6–8], some anchor-based methods aim to reconstruct the instance-to-instance relationships [9–11], some kernel-based methods are proposed to recover missing entries of the kernel matrix then obtains embedded features for clustering [12–14], and graph-based methods are usually based on the spectral clustering, intended to learn consensus information by using graphs [15–17].

Among multi-view learning problems, multi-view clustering which aims to partition data into their clusters in an unsupervised manner has been widely studied in recent years. Incomplete multi-view subspace clustering (IMSC) is a hot direction in the field of representation learning-based Incomplete multi-view clustering, IMSC often studies subspace representations with the self-expression of paired and partial samples. IMSC assumes that multiple input views are generated from a latent subspace, and targets at recovering the latent subspace from multiple views.

A direct way for IMSC to deal with incomplete data is to replace the missing values with zero or mean values. However, in this way, the correlations between different non-missing data points are failed to take into account, producing deficiently results. An incomplete multi-view clustering method via preserving high-order correlation (HCP-IMSC) [18] was proposed to jointly perform missing view inferring and subspace structure learning, HCP-IMSC introduced a hyper-graph-induced hyper-Laplacian regularization term and a tensor factorization term to capture the high-order correlation.

In this paper, changes are made to the hyper-graph-induced IMSC methods. We proposed incomplete multi-view subspace clustering method using non-uniform hyper-graph (NUHG-IMSC) to get further high-order information. We find a set of data points that have high similarity with the center point of each hyper-edge in high-dimensional space to be its neighbor samples, the cardinality of each hyper-edge is decided based on the distribution of the corresponding center point. NUHG-IMSC reconstructs partial samples more reasonably, and is a simple but effective way to utilize high-order information without bringing computational burden and extra parameters.

2 Background

2.1 Multi-View Subspace Clustering Preliminaries

Multi-view subspace clustering usually learns a unified subspace representation from multiple view-specific subspaces, the objective formulation of multi-view subspace clustering can be written as:

$$\min_{\mathbf{A}^{(v)}} \sum_{v=1}^{V} \left(\mathcal{L}(\mathbf{X}^{(v)}, \mathbf{X}^{(v)}\mathbf{A}^{(v)}) + \lambda \mathcal{R}(\mathbf{A}^{(v)}) \right) \tag{1}$$

where $\mathbf{X}^{(v)} \in \mathbb{R}^{D \times N}$ denotes the v-th view data matrix, $\mathbf{A}^{(v)} \in \mathbb{R}^{N \times N}$ denotes the v-th view affinity matrix, $\mathcal{L}(\mathbf{X}^{(v)}, \mathbf{X}^{(v)}\mathbf{A}^{(v)})$ is instanced with several norms, and $\lambda \mathcal{R}(\mathbf{A}^{(v)})$ stands for the certain regularization term about $\mathbf{A}^{(v)}$.

Equation 1 employs different norms and is under the complete multi-view assumption (paired ratio $\sigma = 1$).

2.2 Tensor Preliminaries

In multi-view subspace clustering, all view-specific data matrices $[\mathbf{X}^{(1)}, \ldots, \mathbf{X}^{(V)}]$ can be regarded as a three-way tensor $\mathcal{X} \in \mathbb{R}^{D \times N \times V}$, so that $\mathbf{X}^{(i)} = \mathcal{X}(:, :, v)$ is the v-th frontal slice of $\mathcal{X}$.

Tensor singular value decomposition (t-SVD) is a generalization of the matrix SVD. t-SVD regards $\mathcal{X}$ as a matrix, each element of which is a tube, and then decomposes $\mathcal{X}$ as:

$$\mathcal{X} = \mathcal{U} * \mathcal{S} * \mathcal{V}^T \tag{2}$$

where $\mathcal{U}$ and $\mathcal{V}$ are orthogonal tensors, $\mathcal{V}^T$ denotes the conjugate transpose of $\mathcal{V}$, $\mathcal{S}$ is a f-diagonal tensor, and $*$ denotes the tensor-to-tensor product. Mathematically, t-SVD is equivalent to a series of matrix SVDs in the Fourier domain:

$$\bar{\mathbf{X}}^{(v)} = \bar{\mathbf{U}}^{(v)} \bar{\mathbf{S}}^{(v)} (\bar{\mathbf{V}}^{(v)})^T \quad v = 1, \ldots, V \tag{3}$$

where $\bar{\mathbf{X}}^{(v)}$ is the v-th frontal slice of $\bar{\mathcal{X}}$, which is the result of performing the discrete Fourier transformation along each tube of $\mathcal{X}$.

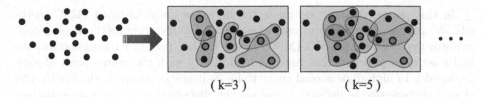

(k=3)   (k=5)

Fig. 2. An illustration of constructing uniform hyper-graphs. Based on the unified affinity matrix, k-nearest neighbors hyper-graphs are constructed where all hyper-edges in a hyper-graph have the same cardinality k.

Further more, tensor tubal rank of $\mathcal{X}$ is defined [19] based on t-SVD as:

$$rank_t(\mathcal{X}) := \{i : \mathcal{S}(i,i,:) \neq 0\} = \max(rank(\bar{\mathbf{X}}^{(1)}), \ldots, rank(\bar{\mathbf{X}}^{(V)})) \quad (4)$$

Eq. 4 represents that the tubal rank of $\mathcal{X}$ is the number of non-zero tubes of $\mathcal{S}$, as mentioned above, $\mathcal{S}$ is a f-diagonal tensor generated from t-SVD applying to $\mathcal{X}$. According to the tensor-to-tensor production, we have $rank_t(\mathcal{X}) \leq \min(rank_t(\mathcal{U} * \mathcal{S}), rank_t(\mathcal{S} * \mathcal{V})) \leq \min(rank_t(\mathcal{U}), rank_t(\mathcal{S}), rank_t(\mathcal{V}))$.

Tensor tubal rank is used to pursue the low rank structure of a tensor, and the tensor nuclear norm (TNN) is its computational convex surrogate, given as:

$$\|\mathcal{X}\|_{TNN} := \sum_{v=1}^{V} \|\bar{\mathbf{X}}^{(v)}\|_* \quad (5)$$

where $\| \cdot \|_*$ denotes the matrix rank, its value is equals to the sum of singular values of the matrix. Equation 5 represents that the nuclear norm of $\mathcal{X}$ is the sum of singular values of all the frontal slices of $\bar{\mathcal{X}}$.

2.3   Hypergraph Preliminaries

Different from an ordinary graph, each edge in which connect two vertices, shows only pair-wise correlation, a hypergraph is its generalization in which an hyper-edge connect an arbitrary number of vertices.

An hypergraph can be denoted as $\mathbf{G} = (\mathbf{V}, \mathbf{E}, \mathbf{W})$, where $\mathbf{V}$ is the set of vertices, defined as: $V = \{v_i \mid i = 1, \ldots, |V|\}$, $\mathbf{E}$ is the set of hyperedges, defined as: $E = \{e_i \mid e_i \subseteq V, e_i \neq \emptyset, i = 1, \ldots, |E|\}$, and $\mathbf{W}$ is the weight matrix with a family of positive numbers associated with the corresponding hyperedges.

In a uniform hypergraph, all hyperedges have the same cardinality, which is the number of vertices in a hyperedge. Figure 2 gives demonstration of uniform hypergraphs. When $|e_1| = \cdots = |e_{|E|}| = k$, the hypergraph is a k-uniform hypergraph, so a 2-uniform hypergraph is a graph.

In the k-uniform hypergraph, the number of vertices in every edge should be exactly equal to k. But in non-uniform hypergraphs, there isn't such restriction, so it can support a more complex data structure.

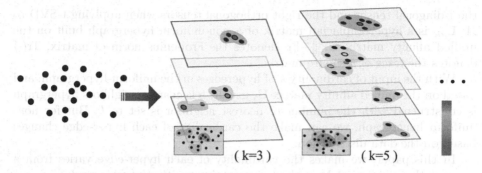

(k=3)   (k=5)

Fig. 3. An illustration of constructing non-uniform hyper-graphs. Based on the unified affinity matrix, k-nearest neighbors hyper-graphs are constructed where hyper-edges in a hyper-graph have different cardinalities based on k.

3   The Proposed NUHG-IMSC Method

In an incomplete multi-view data set, data matrix of the v-th view $\mathbf{X}^{(v)} = [\mathbf{X}_o^{(v)}, \mathbf{X}_m^{(v)}]$ consists of two part, $\mathbf{X}_o^{(v)}$ stands for the observed feature matrix of the v-th view, and $\mathbf{X}_m^{(v)}$ stands for the missing feature matrix of the v-th view. Based on the matrix permutation, $\mathbf{X}^{(v)}$ can be written as:

$$\mathbf{X}^{(v)} = \mathbf{X}_o^{(v)}\mathbf{P}_o^{(v)} + \mathbf{X}_m^{(v)}\mathbf{P}_m^{(v)} \tag{6}$$

where $\mathbf{P}_o^{(v)}$ and $\mathbf{P}_m^{(v)}$ are permutation matrices with elements 0 or 1.

The incomplete multi-view data can be regarded as a three-way tensor $\mathcal{X} = [\mathbf{X}^{(1)}, \ldots, \mathbf{X}^{(V)}] \in \mathbb{R}^{D \times N \times V}$, where D denotes the number of features, N denotes the total number of data points, and V denotes the number of views. For the specific v-th view, $\mathbf{X}^{(v)} \in \mathbb{R}^{d_v \times n_v}$, where $d_v \leq D$ and $n_v \leq N$.

Then based on the knowledge of tensor learning and hypergraphs, the objective function of incomplete multi-view subspace clustering can be formulated as:

$$\min_{\mathbf{X}_m^{(v)}, \mathbf{U}, \mathbf{A}^{(v)}, \mathcal{U}, \mathcal{V}} \sum_{v=1}^{V} (\ \|\mathbf{X}^{(v)} - \mathbf{X}^{(v)}\mathbf{A}^{(v)}\|_F^2 + \omega_v \|\mathbf{A}^{(v)} - \mathbf{U}\|_F^2$$

$$+ \lambda_1 \mathrm{Tr}(\mathbf{X}^{(v)}\mathbf{L}_{nh}\mathbf{X}^{(v)T}))\ + \lambda_2 \|\mathcal{A} - \mathcal{U} * \mathcal{V}\|_F^2 \tag{7}$$

$$\mathrm{s.\,t.} \quad \mathcal{A} = [\mathbf{A}^{(1)}, \ldots, \mathbf{A}^{(V)}], diag(\mathbf{A}^{(v)}) - \mathbf{0}, \mathbf{A}_{ij}^{(v)} \geq 0,$$

$$\mathbf{A}^{(v)} = \mathbf{A}^{(v)T}, \omega_v = \frac{1}{2\|\mathbf{A}^{(v)} - \mathbf{U}\|_F}, \sum_{v=1}^{V} \omega_v = 1.$$

where λ_1 and λ_2 are positive balance parameters, U is the learned unified affinity matrix with block-diagonal structure under ideal condition, $\mathbf{U}$ is the left orthogonal tensors when applying t-SVD to $\mathcal{A}$, $\mathbf{V}$ is the tensor-to-tensor product of the f-diagonal tensor and the right orthogonal tensors when applying t-SVD to $\mathcal{A}$, $\mathbf{L}_{nh}$ is a hyper-Laplacian matrix of a non-uniform hypergraph built on the unified affinity matrix $\mathbf{U}$, $\|\cdot\|_F$ denotes the Frobenius norm of matrix, $\mathrm{Tr}(\cdot)$ denotes the trace of a square matrix.

With the input of cardinality k of hyperedges in the uniform-hypergraph, and based on the unified affinity matrix $\mathbf{U}$, usually a k-nearest neighbors hypergraph is constructed with the number of nearest neighbor is set to k. But for non-uniform hypergraph, we can make the cardinality of each hyper-edge changes based on the data distribution.

In this paper, we makes the cardinality of each hyper-edge varies from a range ψ that consists at least three consecutive positive integers, and:

$$\begin{cases} \mathrm{mean}(\psi) = k \\ \max(\psi) - \min(\psi) = k \end{cases} \tag{8}$$

then we decide the specific cardinality corresponds to each hyper-edge by:

$$\theta(i) = \sum_{j=1}^{|V|} (\mathbf{X}(:,i), \mathbf{X}(:,j)) \tag{9}$$

for vertex v_i, $i = 1, \ldots, |V|$, where $(\cdot, \cdot)$ denotes the vector inner product, and normalize the set θ to ψ. Figure 3 gives an illustration of constructing non-uniform hyper-graphs in our way. This is a simple but effective way to utilize high-order information without bringing computational burden and extra parameters. Besides the advantage that the partial samples can be reconstructed more reasonably, our method also brings benefits to other parts in the whole framework of IMSC, such as learning the view-specific affinity matrices, the weight of each view, and the unified affinity matrix, because L_{nh} participates in the updating process of other variables.

With our method, any IMSC methods that using uniform hypergraphs can make progress and benifit the final clustering results.

4 Experiment and Results

In our experiments, all methods are implemented on a computer with 3.4-GHz CPU and 16-GB RAM. The software environment is Windows 10 system, running MATLAB R2017b. Eight metrics are used to evaluate the clustering performance [20]: clustering accuracy (ACC, the most important evaluation metric for the clustering tasks), normalized mutual information (NMI), precision, purity, recall, F-score, adjusted Rand index (ARI), Entropy. Higher values of these metrics (except for the Entropy) indicate the better performance of clustering.

4.1 Data Sets

We evaluate the proposed method on three complete multi-view data sets, and other details of three complete multi-view data sets are as follows:

Table 1. Data sets used in our experiments.

Data set	#Sample	#Class	#Sample/Class	#View	#Feature/View
MNIST-USPS	5000	10	500	2	784
ORL	400	40	10	3	4096 / 3304 / 6750
Caltech101-7	1474	7	435 / 798 / 52/ 34 / 35 / 64 / 56	6	48 / 40 / 254 / 1984 / 512 / 928

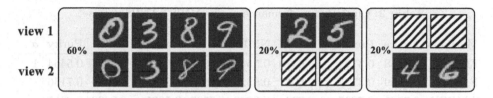

Fig. 4. Demonstration of the synthesis incomplete MNIST-USPS data set with $\sigma = 0.6$. This is a 2-view data set, 60% samples in which are observed both in two views, and 20% are missing in one view and observed in another view 1/2 respectively.

(1)**MNIST+USPS** [21]: Sometimes it is also called Hdigit. It consists of 5000 handwritten digital images (0–9) captured from two sources, i.e., MNIST Handwritten Digits and USPS Handwritten Digits.

(2)**ORL** [22]: It contains 400 face images of 40 individuals under different lighting, times, and facial details. Three feature sets, including 4096 dimension intensity, 3304 dimension LBP, and 6750 dimension Gabor are utilized.

(3)**Caltech101-7** [23]: It consists of 1474 instances in 7 categories, and is a subset of the image data set Caltech101, each image is labelled with a single object, images are of variable sizes, with typical edge lengths of 200–300 pixels.

Table 1 shows various quantitative indicators of the data sets. We made these complete data sets to be incomplete by the way that showed in Fig. 1 and 4, the scale of the paired samples in each incomplete data set is controlled by the parameter paired ratio σ in the set $\{1, 0.8, 0.6, 0.4, 0.2\}$.

4.2   Methods

To validate the effectiveness of our method, we compare it with four methods. These methods are listed as follows:

(1)**EOMSC-CA**, AAAI2022 [24]: In EOMSC-CA, the selection of anchor points and the construction of subspace graphs are optimized jointly to improve the clustering performance in our method. By imposing a connectivity constraint, EOMSC-CA directly generate the clustering result without post-processing. EOMSC-CA can only deal with complete data.

(2)**FPMVS-CAG**, TIP2021 [25]: It firstly jointly conducts anchor selection and latter subspace graph construction into a unified optimization, then the two processes negotiates with each other to promote clustering quality. It also

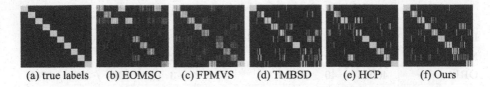

(a) true labels   (b) EOMSC   (c) FPMVS   (d) TMBSD   (e) HCP   (f) Ours

Fig. 5. The visualization of true labels and the clustering results between all compared methods of data set MNIST-USPS. MNIST-USPS is class-balanced.

Table 2. Clustering performance on the MNIST-USPS data set.

Methods	Fscore	Precision	Recall	NMI	ARI	Entropy	ACC	Purity	σ
EOMSC	0.4268	0.3199	0.6410	0.5545	0.3387	1.6584	0.5234	0.5464	1
FPMVS	0.6069	0.5774	0.6967	0.6524	0.5608	3.2502	0.6944	0.7136	
TMBSD	0.7765	0.7697	0.7834	0.8320	0.7527	0.5631	0.8438	0.8455	
TMBSD	0.5282	0.5204	0.5363	0.6166	0.4776	1.2803	0.6715	0.6848	0.8
HCP	0.5707	0.5554	0.5869	0.6638	0.5240	1.1295	0.6765	0.7130	
Ours	0.6337	0.6230	0.6449	0.7218	0.5943	0.9328	0.7475	0.7618	
TMBSD	0.4248	0.4155	0.4345	0.5189	0.3625	1.6065	0.5853	0.6038	0.6
HCP	0.4452	0.4393	0.4512	0.5672	0.3858	1.4432	0.5745	0.6193	
Ours	0.5068	0.4875	0.5279	0.6222	0.4524	1.2724	0.6368	0.6620	
TMBSD	0.3469	0.3415	0.3526	0.4550	0.2768	1.8155	0.5090	0.5248	0.4
HCP	0.3763	0.3701	0.3827	0.5064	0.3092	1.6457	0.4985	0.5193	
Ours	0.3896	0.3693	0.4124	0.5123	0.3212	1.6428	0.5383	0.5508	
TMBSD	0.2340	0.2277	0.2407	0.3153	0.1506	2.2811	0.3848	0.3993	0.2
HCP	0.2676	0.2367	0.3082	0.3874	0.1789	2.0743	0.4243	0.4565	
Ours	0.3318	0.3062	0.3621	0.4729	0.2547	1.7789	0.4628	0.5165	

designs a four-step alternate optimization algorithm with proved convergence, and can automatically learn an optimal low-rank anchor subspace graph without additional hyper-parameters. FPMVS-CAG can only deal with complete data.

(3)**TMBSD**, ICME2021 [26]: It achieved incomplete multi-view clustering by using diffusing multiple block-diagonal structures with a tensor low-rank constraint, and it has an efficient optimization algorithm to solve the resultant model. TMBSD can deal with complete data and incomplete data.

(4)**HCP-IMSC**, TIP2022 [18]: It achieved incomplete multi-view clustering via preserving high-order correlation to jointly perform missing view inferring and subspace structure learning. A hypergraph-induced hyper-Laplacian regularization term and a tensor factorization term were introduced to capture the high-order sample correlation and those view correlation. Also, an efficient optimization algorithm was designed to solve the resultant model. HCP-IMSC can only deal with incomplete data.

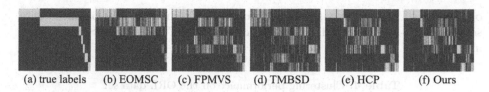

(a) true labels   (b) EOMSC   (c) FPMVS   (d) TMBSD   (e) HCP   (f) Ours

Fig. 6. The visualization of true labels and the clustering results between all compared methods of data set Caltech101-7. Caltech101-7 is class-imbalanced.

Table 3. Clustering performance on the Caltech101-7 data set.

Methods	Fscore	Precision	Recall	NMI	ARI	Entropy	ACC	Purity	σ
EOMSC	0.6113	0.6763	0.5576	0.4685	0.4034	0.9593	0.5929	0.8331	1
FPMVS	0.6104	0.8479	0.4769	0.5700	0.4608	1.1004	0.6147	0.8657	
TMBSD	0.4365	0.7919	0.3013	0.5376	0.2845	0.5663	0.3540	0.8494	
TMBSD	0.4277	0.7759	0.2953	0.4778	0.2733	0.7033	0.3790	0.8316	0.8
HCP	0.5685	0.8651	0.4234	0.6041	0.4221	0.4821	0.5318	0.8859	
Ours	0.5489	0.7977	0.4187	0.5732	0.4159	0.7014	0.5620	0.8028	
TMBSD	0.4226	0.7733	0.2907	0.4815	0.2685	0.6928	0.3446	0.8362	0.6
HCP	0.5335	0.8566	0.3874	0.5820	0.3861	0.4989	0.4934	0.8670	
Ours	0.5227	0.7865	0.3914	0.5997	0.3882	0.6312	0.5335	0.8222	
TMBSD	0.4424	0.7935	0.3067	0.4602	0.2898	0.7460	0.4368	0.8285	0.4
HCP	0.5278	0.8521	0.3823	0.5775	0.3797	0.5074	0.4837	0.8643	
Ours	0.5127	0.7558	0.3879	0.5590	0.3717	0.7330	0.5382	0.7978	
TMBSD	0.4238	0.7687	0.2925	0.4620	0.2683	0.7400	0.3401	0.8289	0.2
HCP	0.5270	0.8517	0.3815	0.5693	0.3789	0.5255	0.4848	0.8629	
Ours	0.5290	0.7851	0.3943	0.5961	0.3900	0.6416	0.5453	0.8175	

4.3 Quantified Experimental Results and Variable Visualizations

The quantified experimental results of three data sets are represented in Table 2, 3 and 4 respectively, and the visualizations of the clustering results between all compared methods of data sets MNIST-USPS and Caltech101-7 are shown in Fig. 5 and 6 respectively.

Because EOMSC-CA and FPMVS-CAG can only deal with the complete data, HCP and our method can only deal with the incomplete data, and TMBSD can deal with both complete data and incomplete data, so when $\sigma = 1$, tables show the clustering results of EOMSC-CA, FPMVS-CAG and TMBSD, when $\sigma = \{0.8, 0.6, 0.4, 0.2\}$, tables show the clustering results of TMBSD, HCP-IMSC and our method NUHG-IMSC.

As shown in Table 1, MNIST-USPS and ORL are class-balanced data sets, i.e. data sets with an approximately equal number of samples from each cluster, meanwhile Caltech101-7 is class-balanced data set. It is noted that some methods, such as TMBSD, may show better clustering results on class-balanced data set than class-imbalanced data set (e.g. compare ACC of the TMBSD in Table 2 and 3).

Table 4. Clustering performance on the ORL data set.

Methods	Fscore	Precision	Recall	NMI	ARI	Entropy	ACC	Purity	σ
EOMSC	0.3573	1.0000	0.2175	0.8637	0.3526	0.0000	0.4450	1.0000	1
FPMVS	0.2762	0.2187	0.3750	0.7201	0.2574	1.9009	0.4950	0.5200	
TMBSD	0.7099	0.6782	0.7448	0.9076	0.7036	0.5178	0.8103	0.8285	
TMBSD	0.6899	0.6600	0.7228	0.8954	0.6832	0.5819	0.7960	0.8128	0.8
HCP	0.7189	0.6853	0.7563	0.9083	0.7129	0.5157	0.8225	0.8390	
Ours	0.7360	0.7023	0.7734	0.9152	0.7304	0.4779	0.8358	0.8485	
TMBSD	0.6265	0.6011	0.6544	0.8708	0.6185	0.7102	0.7565	0.7765	0.6
HCP	0.6821	0.6465	0.7220	0.8960	0.6752	0.5842	0.7923	0.8098	
Ours	0.6853	0.6530	0.7211	0.8972	0.6785	0.5749	0.8000	0.8120	
TMBSD	0.6738	0.6398	0.7119	0.8939	0.6667	0.5945	0.7845	0.8058	0.4
HCP	0.6704	0.6369	0.7078	0.8907	0.6632	0.6108	0.7970	0.8095	
Ours	0.6899	0.6581	0.7251	0.8968	0.6832	0.5755	0.8088	0.8220	
TMBSD	0.6246	0.5892	0.6648	0.8737	0.6164	0.7047	0.7505	0.7723	0.2
HCP	0.6634	0.6310	0.6995	0.8883	0.6562	0.6222	0.7720	0.7935	
Ours	0.6484	0.6084	0.6941	0.8841	0.6407	0.6515	0.7820	0.7968	

The quantified experimental results show that our proposed NUHG-IMSC achieves better results than other compared methods. Under the same paired ratio σ, our method performs better than other IMSC methods in most instances. Sometimes our method even performs better under small paired ratio of data than other methods under large paired ratio. Observed in Table 2 and 4, when σ becomes smaller, the clustering performance of TMBSD becomes worse more quickly than HCP and our method, proves that our method is more stable than TMBSD.

The visualizations show that NUHG-IMSC makes the variables have better block-diagonal structure as the intrinsic nature of data, which proves that our method gives better subspace-preserving (an idealized state for subspace clustering tasks where there are no connection between points from different subspaces or samples which are belonged to different clusters) affinity [27], in this case subspaces are more independent.

5   Conclusions

We propose an incomplete multi-view subspace clustering (IMSC) method using non-uniform hyper-graph (called NUHG-IMSC) which makes slightly change to the traditional methods of IMSC that always build uniform hyper-graphs, and constructs non-uniform hyper-graphs instead. We find a set of data points that have high similarity with the center point of each hyper-edge in high-dimensional space to be its neighbor samples, the cardinality of each hyper-edge is decided based on the distribution of the corresponding center point. This is a simple but effective way to utilize high-order information without bringing computational burden and extra parameters. Besides the advantage that the partial samples can be reconstructed more reasonably, our method also brings benefits to other parts in the whole framework of IMSC.

Acknowledgements. This work was supported in part by the National Innovation 2030 Major S&T Project of China under Grant 2020AAA0104203, and in part by the Nature Science Foundation of China under Grant 62006007.

References

1. Zhan, K., Niu, C., Chen, C., Nie, F., Zhang, C., Yang, Y.: Graph structure fusion for multiview clustering. IEEE Trans. Knowl. Data Eng. **31**(10), 1984–1993 (2018)
2. Tang, C., et al.: Learning a joint affinity graph for multiview subspace clustering. IEEE Trans. Multimed. **21**(7), 1724–1736 (2018)
3. Wang, H., Yang, Y., Liu, B.: GMC: graph-based multi-view clustering. IEEE Trans. Knowl. Data Eng. **32**(6), 1116–1129 (2019)
4. Wang, C.D., Lai, J.H., Philip, S.Y.: Multi-view clustering based on belief propagation. IEEE Trans. Knowl. Data Eng. **28**(4), 1007–1021 (2015)
5. Zhang, C., et al.: Generalized latent multi-view subspace clustering. IEEE Trans. Pattern Anal. Mach. Intell. **42**(1), 86–99 (2018)
6. Tao, H., Hou, C., Yi, D., Zhu, J.: Unsupervised maximum margin incomplete multi-view clustering. In: Zhou, Z.-H., Yang, Q., Gao, Y., Zheng, Yu. (eds.) ICAI 2018. CCIS, vol. 888, pp. 13–25. Springer, Singapore (2018). https://doi.org/10.1007/978-981-13-2122-1_2
7. Liang, N., Yang, Z., Li, Z., Xie, S.: Co-consensus semi-supervised multi-view learning with orthogonal non-negative matrix factorization. Inf. Process. Manag. **59**(5), 103054 (2022)
8. Liang, N., Yang, Z., Li, Z., Xie, S., Su, C.Y.: Semi-supervised multi-view clustering with graph-regularized partially shared non-negative matrix factorization. Knowl.-Based Syst. **190**, 105185 (2020)
9. Guo, J., Ye, J.: Anchors bring ease: an embarrassingly simple approach to partial multi-view clustering. In: Proceedings of the AAAI Conference on Artificial Intelligence, vol. 33, pp. 118–125 (2019)
10. Ou, Q., Wang, S., Zhou, S., Li, M., Guo, X., Zhu, E.: Anchor-based multiview subspace clustering with diversity regularization. IEEE Multimed. **27**(4), 91–101 (2020)
11. Sun, M., et al.: Scalable multi-view subspace clustering with unified anchors. In: Proceedings of the 29th ACM International Conference on Multimedia, pp. 3528–3536 (2021)
12. Li, M., Xia, J., Xu, H., Liao, Q., Zhu, X., Liu, X.: Localized incomplete multiple kernel k-means with matrix-induced regularization. IEEE Trans. Cybernet. (2021)
13. Liu, X.: Incomplete multiple kernel alignment maximization for clustering. IEEE Trans. Pattern Anal. Mach. Intell. (2021)

14. Liu, X., et al.: Multiple kernel k k-means with incomplete kernels. IEEE Trans. Pattern Anal. Mach. Intell. **42**(5), 1191–1204 (2019)
15. Liang, N., Yang, Z., Xie, S.: Incomplete multi-view clustering with sample-level auto-weighted graph fusion. IEEE Trans. Knowl. Data Eng. (2022)
16. Zhong, G., Pun, C.M.: Improved normalized cut for multi-view clustering. IEEE Trans. Pattern Anal. Mach. Intell. **44**(12), 10244–10251 (2021)
17. Kang, Z., Lin, Z., Zhu, X., Xu, W.: Structured graph learning for scalable subspace clustering: from single view to multiview. IEEE Trans. Cybernet. **52**(9), 8976–8986 (2021)
18. Li, Z., Tang, C., Zheng, X., Liu, X., Zhang, W., Zhu, E.: High-order correlation preserved incomplete multi-view subspace clustering. IEEE Trans. Image Process. **31**, 2067–2080 (2022)
19. Zheng, Y.B., Huang, T.Z., Zhao, X.L., Jiang, T.X., Ji, T.Y., Ma, T.H.: Tensor n-tubal rank and its convex relaxation for low-rank tensor recovery. Inf. Sci. **532**, 170–189 (2020)
20. Shi, S., Nie, F., Wang, R., Li, X.: Multi-view clustering via nonnegative and orthogonal graph reconstruction. IEEE Trans. Neural Netw. Learn. Syst. (2021)
21. Xu, J., Ren, Y., Li, G., Pan, L., Zhu, C., Xu, Z.: Deep embedded multi-view clustering with collaborative training. Inf. Sci. **573**, 279–290 (2021)
22. Zhang, C., Fu, H., Hu, Q., Zhu, P., Cao, X.: Flexible multi-view dimensionality co-reduction. IEEE Trans. Image Process. **26**(2), 648–659 (2016)
23. Fei-Fei, L., Fergus, R., Perona, P.: Learning generative visual models from few training examples: an incremental Bayesian approach tested on 101 object categories. In: 2004 Conference on Computer Vision and Pattern Recognition Workshop, pp. 178–178. IEEE (2004)
24. Liu, S., et al.: Efficient one-pass multi-view subspace clustering with consensus anchors. In: Proceedings of the AAAI Conference on Artificial Intelligence, vol. 36, pp. 7576–7584 (2022)
25. Wang, S., et al.: Fast parameter-free multi-view subspace clustering with consensus anchor guidance. IEEE Trans. Image Process. **31**, 556–568 (2021)
26. Li, Z., Tang, C., Liu, X., Zheng, X., Zhang, W., Zhu, E.: Tensor-based multi-view block-diagonal structure diffusion for clustering incomplete multi-view data. In: 2021 IEEE International Conference on Multimedia and Expo (ICME), pp. 1–6. IEEE (2021)
27. You, C., Robinson, D., Vidal, R.: Scalable sparse subspace clustering by orthogonal matching pursuit. In: Proceedings of the IEEE Conference on Computer Vision and Pattern Recognition, pp. 3918–3927 (2016)

Deep Learning-Empowered Unsupervised Maritime Anomaly Detection

Lingxuan Weng[1], Maohan Liang[2], Ruobin Gao[3], and Zhong Shuo Chen[4]($\boxtimes$)

[1] School of Computer Science and Artificial Intelligence,
Wuhan University of Technology, Wuhan, China
[2] Department of Civil and Environmental Engineering,
National University of Singapore, Singapore, Singapore
[3] School of Civil and Environmental Engineering,
Nanyang Technological University, Singapore, Singapore
[4] School of Intelligent Finance and Business,
Xi'an Jiaotong-Liverpool University, Suzhou, China
Zhongshuo.Chen@xjtlu.edu.cn

Abstract. Automatically detecting anomalous vessel behaviour is an extremely crucial problem in intelligent maritime surveillance. In this paper, a deep learning-based unsupervised method is proposed for detecting anomalies in vessel trajectories, operating at both the image and pixel levels. The original trajectory data is converted into a two-dimensional matrix representation to generate a vessel trajectory image. A wasserstein generative adversarial network (WGAN) model is trained on a dataset of normal vessel trajectories, while simultaneously training an encoder to map the trajectory image to a latent space. During anomaly detection, the vessel trajectory image is mapped to a hidden vector by the encoder, which is then used by the generator to reconstruct the input image. The anomaly score is computed based on the residuals between the reconstructed trajectory image and the discriminator's residuals, enabling image-level anomaly detection. Furthermore, pixel-level anomaly detection is achieved by analyzing the residuals of the reconstructed image pixels to localize the anomalous trajectory. The proposed method is compared to autoencoder (AE) and variational autoencoder (VAE) model, and experimental results demonstrate its superior performance in anomaly detection and pixel-level localization. This method has substantial potential for detecting anomalies in vessel trajectories, as it can detect anomalies in arbitrary waters without prior knowledge, relying solely on training with normal vessel trajectories. This approach significantly reduces the need for human and material resources. Moreover, it provides valuable insights and references for trajectory anomaly detection in other domains, holding both theoretical and practical importance.

Keywords: Anomaly detection · Vessel trajectory · Generative
Adversarial Network · Deep learning · Automatic identification system

L. Weng and M. Liang—Equal contribution.

1 Introduction

Maritime safety and security rely heavily on effective maritime situational awareness (MSA) systems supported by surveillance and tracking technologies such as the automatic identification system (AIS) [1]. Surveillance operators are tasked with identifying and predicting potential conflicts, collisions, abnormal vessel behavior, and suspicious activities among a vast number of vessels traversing vast sea areas. However, with the increasing size of vessel and the complexity of maritime environments, the task of monitoring and identifying abnormal vessel behavior becomes challenging for operators.

Maritime anomaly detection involves detecting and flagging abnormal behavior in maritime activities or vessel movements [2]. Various studies have been conducted to assist operators in detecting and responding to abnormal situations in maritime surveillance systems. Three main categories of methods for maritime anomaly detection can be identified [3]: data-driven approaches, rule-based approaches, and hybrid approaches. Data-driven approaches focus on learning patterns and behaviors from historical vessel movement data, involving normalcy extraction and anomaly detection. Rule-based approaches identify anomalies based on specific rules or characteristics associated with abnormal behavior. Hybrid approaches combine data-driven and rule-based methods, utilizing machine learning techniques and predefined rules or event definitions. The growth of maritime traffic and the development of AIS have advanced monitoring systems for preventing accidents and detecting illegal activities. AIS data contains a wealth of vessel movement information, and deep learning techniques have emerged as powerful tools in anomaly detection [4]. These techniques autonomously learn patterns and representations from extensive datasets, enabling the detection of anomalies without explicit rule definition. Deep learning approaches offer advantages in capturing intricate relationships and dependencies within AIS data, automatically extracting complex features and representations. This enhances the effectiveness of anomaly detection in the maritime domain, which requires flexible and adaptable methods to detect diverse and dynamic anomalies. However, filtering out abnormal vessel trajectories from large amounts of data is a highly challenging task that requires extensive manual effort. Moreover, it is worth noting that abnormal vessel trajectories may not consistently exhibit abnormal behavior throughout their entire duration.

To overcome the above challenges, this paper proposes a deep learning-empowered unsupervised approach for maritime anomaly detection. The proposed model first generates the vessel trajectory into a vessel trajectory image to ensure the scale consistency of the trajectory. Then, this paper utilizes the WGAN [5] to learn the behavior patterns of normal vessel trajectories. Finally, the residuals of the reconstructed vessel trajectory and the original vessel trajectory image are used to calculate the anomaly score. Our approach does not rely on predefined rules or expert knowledge but instead learns from the data itself, allowing for adaptive and flexible anomaly detection. The main contributions can be summarized as

- The method introduces a novel approach that utilizes deep learning techniques, specifically a WGAN model and an encoder, for unsupervised anomaly detection in vessel trajectories. By generating vessel trajectory images and utilizing image reconstruction and residual analysis, it enables both image-level and pixel-level anomaly detection, providing a comprehensive solution.
- The method offers significant potential for anomaly detection in vessel trajectories without the need for a priori knowledge or annotated anomaly trajectory data. By training only on normal vessel trajectories, it overcomes the challenges associated with collecting and annotating anomaly data, thereby reducing the reliance on human and material resources. This contributes to the efficiency and practicality of implementing the method in real-world scenarios.
- Experimental results demonstrate that the proposed method outperforms alternative methods in terms of anomaly detection and pixel-level localization. This indicates that the method achieves optimal results in accurately identifying and localizing anomalies in vessel trajectories, enhancing the effectiveness of anomaly detection in maritime surveillance systems.

2   Related Work

In this section, we will give a brief overview of maritime anomaly detection. The research on maritime anomaly detection can be broadly categorized into three main categories based on their data processing strategies: data-driven approaches, rule-based approaches, and hybrid approaches [6]. Thus, this paper will introduce the three types of maritime anomaly identification methods respectively.

2.1   Data-Driven Approaches for Maritime Anomaly Detection

Data-driven approaches have been widely used in maritime anomaly detection to identify abnormal patterns and behaviors in vessel data. These methods leverage the power of clustering algorithms to group similar vessel trajectories together, allowing the identification of deviations from the normal clusters as anomalies. Ristic et al. [7] introduced an adaptive Kernel density estimation (KDE) method combined with particle filters to predict vessel positions based on the derived density. However, the popularity of clustering has increased due to its favorable performance and ease of implementation. Zhen et al. [8] employed the k-medoids algorithm for vessel trajectory clustering and Bayesian networks for anomaly detection. Li et al. [9] proposed a multi-step clustering approach using dynamic time warping (DTW) distance, principal components analysis (PCA), and k-medoids algorithm for robust AIS trajectory clustering. One limitation of k-medoids is the challenging determination of the number of clusters, and it may not capture trajectories deviating from the traffic pattern. Density-based clustering, such as DBSCAN, has gained attention due to its ability to recognize noise and derive arbitrarily shaped clusters [10]. Yan et al. [11] applied density-based clustering for identifying stops and moves in tracking points, while Liu et al. [12] enhanced DBSCAN with non-spatial attributes for extracting normal vessel trajectory patterns.

2.2   Rule-Based Approaches for Maritime Anomaly Detection

Rule-based methods have also been widely employed for maritime anomaly detection due to their interpretability and ease of implementation. These methods involve defining a set of predefined rules or thresholds based on expert knowledge or domain-specific guidelines. For instance, Laxhammar et al. [13] proposed a rule-based approach to identify abnormal vessel behaviors by defining rules related to speed, course, and proximity to restricted areas. Similarly, Lei et al. [14] developed a rule-based system to detect illegal vessel activities based on predefined rules related to vessel behaviors and patterns.

2.3   Hybrid Approaches for Maritime Anomaly Detection

In recent years, there has been a shift towards hybrid approaches, such as machine learning and deep learning, which can automatically learn patterns from data and adapt to new anomalies without relying on predefined rules. Karatacs et al. [15] employ deep learning to predict vessel trajectories compared to real movements for abnormal vessel detection. Nguyen et al. [16] propose and investigate the Sequential Hausdorff Nearest-Neighbor Conformal Anomaly Detector (SHNN-CAD) and the discords algorithm for online learning and sequential anomaly detection in trajectories. Challenges include the availability of labeled anomaly data for training, the interpretation and explainability of deep learning models, and the computational requirements for processing large-scale maritime data. However, the promise of deep learning in uncovering hidden patterns and detecting novel anomalies holds great potential for improving maritime surveillance and enhancing maritime safety and security.

3   Fast GAN-Based Anomalous Vessel Trajectory Detection

In this work, the original vessel trajectory is first remapped into an informative trajectory image. Subsequently, unsupervised learning using generative adversarial networks (GAN) is employed to capture the distribution of normal vessel trajectory images and generate trajectory images with normal trajectory features. An encoder is then trained to extract latent space features from the input image, which are utilized as input for the GAN. Vessel trajectory anomaly detection is achieved by calculating the loss between the original and generated images and deriving an anomaly score. The flow of the pixel-level unsupervised vessel trajectory anomaly detection method based on deep learning proposed in this paper is depicted in Fig. 1.

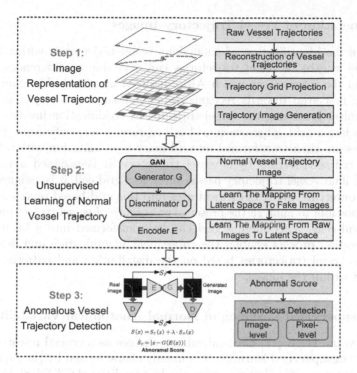

Fig. 1. The flowchart of our proposed deep learning-empowered unsupervised anomaly detection method.

3.1 Problem Definition

To better understand trajectory anomaly detection, it is imperative to first delineate key concepts: vessel trajectory points, vessel trajectories, and anomalous trajectories.

Definition 1. *Vessel Trajectory Points: Vessel trajectory points $P_i = \{lat_i, lon_i, t_i\}$ acquired through AIS equipment, where lat_i, lon_i and t_i in P_i representing the latitude, longitude, and timestamp respectively.*

Definition 2. *Vessel Trajectory: Vessel trajectory $T = \{P_1, P_2, \cdots, P_n\}$ is a sequence of vessel trajectory points in chronological order, where $P_i (1 \leq i \leq n)$, $n \in \{1, 2, \cdots, N\}$ denotes the n-th timestamp point and N is the length of the vessel trajectory T. The length may vary for different vessel trajectories.*

Definition 3. *Anomalous Trajectories: Anomalous trajectory refers to a vessel's movement pattern or path that deviates significantly from the expected or normal behavior, indicating potential irregularities, suspicious activities, or safety concerns in maritime navigation. It involves the identification of abnormal vessel movements, such as abrupt changes in speed, unusual route deviations, suspicious stops, or other anomalous behaviors that may indicate illegal activities, accidents, security threats, or operational abnormalities.*

3.2   Generation of Vessel Trajectory Images

To achieve the identification of anomalous vessel trajectories within a unified region, the lengths of vessel trajectories can vary due to differences in vessel speeds and the intervals at which AIS data is transmitted. Therefore, in this study, the first step towards recognizing anomalous vessel trajectories in the region involves gridifying the vessel trajectories. Gridification involves dividing the region into a grid, where each grid cell represents a specific area. The vessel trajectories are then transformed into a binary representation within this grid. If a grid cell is traversed by a vessel trajectory, it is assigned a value of 1; otherwise, if no vessel trajectory passes through a grid cell, it is assigned a value of 0. By converting the vessel trajectories into this grid-based representation, it becomes possible to analyze the presence or absence of vessel trajectories within specific grid cells. This one-hot grid can be transformed into a binary image. This generation process enables the subsequent identification and detection of anomalous vessel trajectories based on the distribution and patterns of vessel movement within the image.

3.3   Unsupervised Learning of Normal Anatomical Variability

In recent years, GAN [17] have gained prominence as a crucial research area in the field of unsupervised learning. It is mainly characterized by the capability to model an unknown distribution indirectly. In a traditional GAN [18], the generator and discriminator are trained using the Jensen-Shannon divergence (JSD) or the Kullback-Leibler divergence (KLD) as the objective function. However, these divergence metrics suffer from issues like mode collapse and vanishing gradients, which can make training unstable and result in poor sample quality. WGAN [19] addresses these issues by using the Wasserstein distance, also known as the Earth Mover's distance, as the objective function. The Wasserstein distance provides a more meaningful and stable measure of the difference between the true data distribution and the generated distribution. By minimizing the Wasserstein distance, WGAN encourages the generator to generate samples that are closer to the true data distribution.

The structure of WGAN is inspired by the idea of game theory and consists of two opposing models: generators (G) and discriminators (D). The generator's purpose is to capture the underlying distribution of real data and produce realistic new samples to deceive the discriminator, while the discriminator's role is to differentiate between real and generated samples. The architecture of our WGAN is visually illustrated in Fig. 2. In WGAN, both G and D can be implemented by the neural network. The input of G is a random vector z obeying a certain distribution p_z, and the output of G can be considered as a sample in the distribution p_g. Assume that the distribution of real data is p_{data}, and the generative adversarial network is trained only on the real data set. The generator G learns a mapping function that approximates the distribution of the real data. D is a binary classifier that determines whether the input data comes from real or generated samples, and its input consists of both real and generated

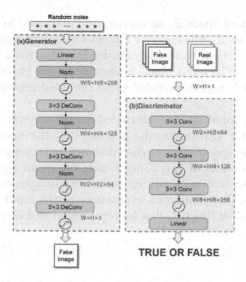

Fig. 2. The visual architecture of the Generative Adversarial Network (GAN) in our deep learning-based unsupervised learning method, comprising the generator and discriminator. The generator and discriminator are primarily implemented using convolutional and deconvolutional layers, respectively.

samples. The optimization objective of the GAN is to simultaneously maximize the discriminator's parameters and minimize the generator's parameters. The objective function is defined as follows.

$$\min_{G}\max_{D} V(D,G) = \mathop{E}_{x \sim p_{data}(x)} [\log G(x)] \\ + \mathop{E}_{x \sim p_z(x)} [\log(1 - D(G(z)))] \tag{1}$$

The proposed anomaly detection method involves two training steps: WGAN training and encoder training based on the trained WGAN. Both the WGAN and the encoder are trained using normal vessel trajectory images ($x = x_k \in \chi$) [20]. During testing, both normal and abnormal vessel trajectory images (y_m) along with their corresponding binary labels ($a_m \in 0, 1$) are used as test data.

In the first training process, a WGAN is trained to learn a nonlinear mapping from the latent space ($\mathcal{Z}$) to the vessel trajectory image space (χ). The WGAN generates normal images that capture the characteristics of the latent variables in $\mathcal{Z}$. The training involves optimizing both the generator (G) and the discriminator (D). The generator learns the distribution of normal images, while the discriminator evaluates the similarity between generated images and the distribution of normal images. Input noise from a potential space of dimension d is used to train the generator, aiming to generate data that closely resembles real data and deceives the discriminator. Once training is complete, the fixed weights of the generator and discriminator are used for subsequent encoder training and anomaly scoring.

To obtain the latent space $\mathcal{Z}$, an encoder (E) is trained to map vessel trajectory images from χ to $\mathcal{Z}$. The encoder follows a standard convolutional autoencoder (AE) design, with a trainable encoder E and a fixed generator WGAN serving as a decoder. The input image is encoded by the encoder to generate the latent vector z, which is then decoded using the fixed generator to reconstruct the image.

Additionally, the input and reconstructed images are evaluated separately by the fixed discriminator. During encoder training, only the encoder parameters are optimized, while the generator and discriminator parameters remain constant. The loss function consists of two components: the mean squared error (MSE) residual loss between the input and reconstructed images, and the MSE residual loss between the discriminator's evaluations of the input and reconstructed images. The overall loss function can be represented as follows:

$$L\left(x\right) = \frac{1}{n} \cdot \|x - G\left(E\left(x\right)\right)\|^2 + \frac{\lambda}{n_d} \cdot \|f\left(x\right) - f\left(G\left(E\left(x\right)\right)\right)\|^2, \qquad (2)$$

where the discriminator features of the intermediate layer $f\left(\cdots\right)$ are taken as the statistics for a given input, n_d is the dimension of the intermediate feature representation, and λ is the weighting factor.

3.4   Detection of Anomalous Vessel Trajectory

Anomaly behavior detection can be categorized into picture-level and pixel-level detection. Picture-level anomaly detection aims to determine whether the entire trajectory exhibits any anomalies. On the other hand, pixel-level anomaly detection focuses on identifying specific segments of the trajectory that contain anomalies. To detect vessel anomaly trajectories at the image level, we calculate the discrepancy between the input image and its reconstructed counterpart to derive the anomaly score. The quantification formula for the anomaly score corresponds to the specific definition of the loss function used in encoder training, as shown in the equation:

$$S\left(x\right) = S_T\left(x\right) + \lambda \cdot S_D\left(x\right), \qquad (3)$$

where $S_T\left(x\right) = \frac{1}{n} \cdot \|x - G\left(E\left(x\right)\right)\|^2$, $S_D\left(x\right) = \frac{1}{n_d} \cdot \|f\left(x\right) - f\left(G\left(E\left(x\right)\right)\right)\|^2$, λ is the weighting factor.

The distribution of anomaly fractions in vessel trajectory images typically follows a consistent pattern: Anomalous input images tend to yield higher anomaly scores, while normal input images tend to yield lower anomaly scores. Since the model is trained exclusively on normal images, it is capable of generating reconstructions that bear visual resemblance to the input image and align with the distribution of normal trajectory images. The degree of dissimilarity between the reconstructed image and the input image directly correlates with the level of abnormality exhibited by the input image. A normal input trajectory image displays minimal deviation from its reconstructed image, resulting in a lower

Table 1. The geographical information and vessel trajectory statistics in two different water areas used in the experiment.

Water Area	Time Span	Number of Trajectories	Longitude Span	Latitude Span
Chengshan Cape	Jul, 1–31, 2018	9269	122.58–123.18	37.14–37.78
Tianjin Port	Jul, 1–31, 2018	3156	117.46–118.24	38.71–39.21

anomaly score. Conversely, an anomalous image exhibits significant deviation from its corresponding reconstruction, leading to a higher anomaly score.

Moreover, for pixel-level anomaly detection, we utilize the absolute value of the pixel residuals to localize anomalies, as defined by:

$$\dot{S}_T\left(x\right) = \left|x - G\left(E\left(x\right)\right)\right| \tag{4}$$

4 Experiments

In this section, experiments will be conducted on various waters to validate the methodology proposed in this paper. Firstly, a brief description will be provided regarding the basic information concerning the study waters and the data utilized in this paper. Subsequently, detailed descriptions will be given regarding the evaluation indicators and the comparison benchmarks. Finally, experiments will be performed in real scenarios to demonstrate the superior performance of the deep learning-based unsupervised vessel trajectory anomaly detection method, both quantitatively and qualitatively.

4.1 Data Set and Baselines

In this paper, the datasets consists of AIS data collected from shore-based base stations for two waters (i.e., Tianjin Port (TJP) and Chengshan Cape (CSC)). Table 1 provides the geographic information and statistics of the dataset. To ensure temporal consistency of the AIS data, the vessel trajectories were interpolated using the cubic spline interpolation method, with a fixed time interval of 20 s between adjacent data points. Subsequently, the waters were discretized into a finite grid using the projection method. Specifically, in our experiments, a 66 × 62 grid was used to divide the study waters onto which the trajectories were projected. To generate abnormal vessel trajectories, we randomly selected 50% of the normal vessel trajectories. Subsequently, we introduced random partial trajectory elimination and partial trajectory flipping. This process created an abnormal trajectory dataset with issues such as missing and discontinuous trajectories. We employed the above-processed anomalous trajectory dataset and the remaining normal trajectory dataset as the test set for quantitative and qualitative evaluation. These datasets were excluded from the WGAN and encoder training process. To provide a fair evaluation of unsupervised deep learning-based anomaly detection methods for vessel trajectories, we compare

Table 2. The image-level anomaly detection performance of AE, VAE, and our proposed deep learning-based unsupervised anomaly detection method measured based on various evaluation metrics.

CSC	Accuracy	F1	Recall	AUC
AE	0.8826	0.8708	0.7942	0.9163
VAE	**0.9033**	**0.8972**	0.8458	0.9121
Ours	0.8885	0.8833	**0.8647**	**0.9404**

TJG	Accuracy	F1	Recall	AUC
AE	0.8700	0.8547	0.8010	0.9182
VAE	0.8367	0.8256	0.8056	0.8886
Ours	**0.8933**	**0.8847**	**0.8539**	**0.9441**

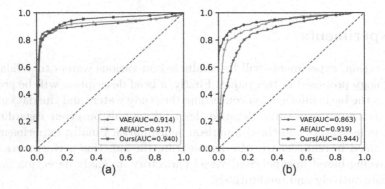

(a)   (b)

Fig. 3. Evaluation of image-level anomaly detection accuracy in two different water areas (a) CSC, (b) TJP.

their performance with anomaly detection using autoencoders (AE) and variant autoencoders (VAE). In addition, we used Accuracy, Recall, and F1 as evaluation metrics, which are very popular in the validation of anomaly detection models.

4.2   Experimental Evaluation

In this section, we will present the results of the quantitative evaluation, including the ROC curves and AUC values, as well as a detailed comparison of evaluation metrics such as Recall, F-score, and Accuracy among the three methods. Additionally, we will discuss the distribution of anomaly scores obtained from the anomaly detection results. Following the quantitative evaluation, a qualitative comparison of pixel-level anomaly localization performance will be conducted.

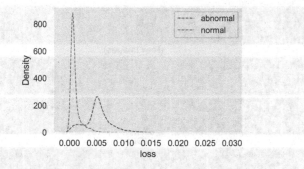

Fig. 4. The distribution of abnormal scores for vessel trajectories in the test set.

Quantitative Evaluation. The corresponding Receiver Operating Characteristic (ROC) curves and Area Under the Curve (AUC) values for the three methods are shown in Fig. 3. Detailed comparison of evaluation metrics (AUC, Recall, F-score, and Accuracy) among the three methods is presented in Table 2. The metrics were calculated the ROC curve, a common measure for estimating the optimal threshold for anomaly detection [21]. The results indicate that our trajectory anomaly detection method achieves the highest recall and AUC on the CSC dataset, surpassing the other two models to some extent. Although Accuracy and F-score are not the best, they are close to the top-performing models. On the TJG dataset, our method achieves the best performance on all metrics, demonstrating its effectiveness in vessel trajectory anomaly detection.

The distribution of anomaly scores $S(x)$, obtained from the anomaly detection results of normal and abnormal trajectory images in the test set, is shown in Fig. 4. There is a noticeable difference in the concentration of anomaly scores between normal and anomalous images, confirming the discriminative ability of our proposed method in distinguishing between the two. These findings strongly support the quantitative evaluation results presented in Table 2.

Qualitative Evaluation. We conducted a qualitative comparison of the pixel-level anomaly localization performance between our proposed method, AE, and VAE. The results are illustrated in Fig. 5. A subset of normal and anomalous vessel trajectories from the test set were selected. Each method was applied for anomaly detection, and the reconstructed results of the original trajectory images and the anomaly localization results were presented. The AE method tends to generate overly smooth images, leading to misidentification of incomplete shorter abnormal trajectories. VAE, on the other hand, does not fit the input images well, resulting in misclassifications of normal trajectories as abnormal. In terms of pixel-level anomaly localization performance, our method exhibits high confidence in identifying anomalous trajectories and achieves greater precision in localizing anomalous segments at the pixel level compared to AE and VAE. These results indicate that our method outperforms other alternatives not only

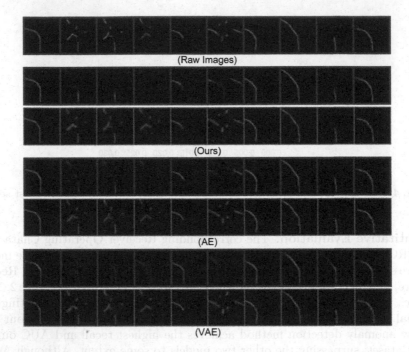

(Raw Images)

(Ours)

(AE)

(VAE)

Fig. 5. Comparison of pixel-level anomaly trajectory localization accuracy. The first row shows the input image of the vessel trajectory to be detected, starting with the second row we show the generated image on one row and the pixel-level anomaly trajectory localization results on the next row: Proposed method (rows 2 and 3), AE (rows 4 and 5) and VAE (rows 6 and 7).

in image-level anomaly detection performance but also in qualitative anomaly localization at the pixel level.

5   Conclusions

In conclusion, we propose a deep learning-based unsupervised method for detecting anomalies in vessel trajectories at both the image and pixel levels. The method first generates a vessel trajectory image by mapping the original vessel trajectory into a two-dimensional matrix. Then, we train a WGAN model on a dataset of normal vessel trajectories, and an encoder is trained to map the trajectory image to a hidden space. During anomaly detection, the encoder maps the vessel trajectory image to the hidden vector, which is then utilized by the generator to reconstruct the input image. The anomaly score is calculated based on the residuals of the reconstructed trajectory image and the residuals of the discriminator to achieve image-level anomaly detection. Furthermore, pixel-level anomaly detection is achieved by localizing the anomaly trajectory based on the residuals of the reconstructed image pixels. We have compared the proposed method with AE and VAE with experimental results showing that the method in this paper achieves the optimal anomaly detection and pixel-level localization results, superior to other alternative methods. The method has considerable potential for anomaly detection of vessel trajectories, allowing anomaly detection of vessel trajectories in arbitrary waters without any a priori knowledge, and can be implemented only by training on normal vessel trajectories in the waters. This approach provides an effective way of overcoming the difficulty of collecting and annotating vessel anomaly trajectory data, thus greatly reducing human and material resources. It also provides valuable insights and references for trajectory anomaly detection in other domains and holds both theoretical and practical importance.

References

1. Kazemi, S., Abghari, S., Lavesson, N., Johnson, H., Ryman, P.: Open data for anomaly detection in maritime surveillance. Exp. Syst. Appl. **40**(14), 5719–5729 (2013)
2. Liang, M., Zhan, Y., Liu, R.W.: MVFFNet: multi-view feature fusion network for imbalanced ship classification. Pattern Recogn. Lett. **151**, 26–32 (2021)
3. Liang, M., Liu, R.W., Li, S., Xiao, Z., Liu, X., Lu, F.: An unsupervised learning method with convolutional auto-encoder for vessel trajectory similarity computation. Ocean Eng. **225**, 108803 (2021)
4. Liang, M., Liu, R.W., Zhan, Y., Li, H., Zhu, F., Wang, F.Y.: Fine-grained vessel traffic flow prediction with a spatio-temporal multigraph convolutional network. IEEE Trans. Intell. Transp. Syst. **23**(12), 23694–23707 (2022)
5. Salimans, T., Goodfellow, I., Zaremba, W., Cheung, V., Radford, A., Chen, X.: Improved techniques for training GANs. In: Advances in Neural Information Processing Systems, vol. 29 (2016)
6. Forti, N., d'Afflisio, E., Braca, P., Millefiori, L.M., Willett, P., Carniel, S.: Maritime anomaly detection in a real-world scenario: ever given grounding in the Suez Canal. IEEE Trans. Intell. Transp. Syst. **23**(8), 13904–13910 (2021)
7. Ristic, B., La Scala, B., Morelande, M., Gordon, N.: Statistical analysis of motion patterns in AIS data: anomaly detection and motion prediction. In: 2008 11th International Conference on Information Fusion, pp. 1–7. IEEE (2008)
8. Zhen, R., Riveiro, M., Jin, Y.: A novel analytic framework of real-time multi-vessel collision risk assessment for maritime traffic surveillance. Ocean Eng. **145**, 492–501 (2017)
9. Li, H., Liu, J., Liu, R.W., Xiong, N., Wu, K., Kim, T.: A dimensionality reduction-based multi-step clustering method for robust vessel trajectory analysis. Sensors **17**(8), 1792 (2017)
10. Pallotta, G., Vespe, M., Bryan, K.: Vessel pattern knowledge discovery from AIS data: a framework for anomaly detection and route prediction. Entropy **15**(6), 2218–2245 (2013)
11. Yan, W., Wen, R., Zhang, A.N., Yang, D.: Vessel movement analysis and pattern discovery using density-based clustering approach. In: 2016 IEEE International Conference on Big Data (Big Data), pp. 3798–3806. IEEE (2016)
12. Liu, B., de Souza, E.N., Matwin, S., Sydow, M.: Knowledge-based clustering of ship trajectories using density-based approach. In: 2014 IEEE International Conference on Big Data (Big Data), pp. 603–608. IEEE (2014)

13. Laxhammar, R., Falkman, G.: Online learning and sequential anomaly detection in trajectories. IEEE Trans. Pattern Anal. Mach. Intell. **36**(6), 1158–1173 (2014)
14. Lei, P.R.: A framework for anomaly detection in maritime trajectory behavior. Knowl. Inf. Syst. **47**(1), 189–214 (2016)
15. Karataş, G.B., Karagoz, P., Ayran, O.: Trajectory pattern extraction and anomaly detection for maritime vessels. IoT **16**, 100436 (2021)
16. Nguyen, D., Vadaine, R., Hajduch, G., Garello, R., Fablet, R.: GeoTrackNet - a maritime anomaly detector using probabilistic neural network representation of AIS tracks and a contrario detection. IEEE Trans. Intell. Transp. Syst. **23**(6), 5655–5667 (2021)
17. Goodfellow, I.J., et al.: Generative adversarial networks (2014)
18. Radford, A., Metz, L., Chintala, S.: Unsupervised representation learning with deep convolutional generative adversarial networks. arXiv preprint arXiv:1511.06434 (2015)
19. Arjovsky, M., Chintala, S., Bottou, L.: Wasserstein generative adversarial networks. In: International Conference on Machine Learning, pp. 214–223. PMLR (2017)
20. Schlegl, T., Seeböck, P., Waldstein, S.M., Schmidt-Erfurth, U., Langs, G.: Unsupervised anomaly detection with generative adversarial networks to guide marker discovery. In: Niethammer, M., et al. (eds.) IPMI 2017. LNCS, vol. 10265, pp. 146–157. Springer, Cham (2017). https://doi.org/10.1007/978-3-319-59050-9_12
21. Fluss, R., Faraggi, D., Reiser, B.: Estimation of the Youden Index and its associated cutoff point. Biometrical J. J. Math. Meth. Biosci. **47**(4), 458–472 (2005)

Hazardous Driving Scenario Identification with Limited Training Samples

Zhen Gao[1], Liyou Wang[1], Jingning Xu[1]($\boxtimes$), Rongjie Yu[2], and Li Wang[3]

[1] School of Software Engineering, Tongji University, Shanghai 201804, China
2131479@tongji.edu.cn
[2] Key Laboratory of Road and Traffic Engineering of the Ministry of Education,
Tongji University, Shanghai 201804, China
[3] College of Electronic and Information Engineering,
Tongji University, Shanghai 201804, China

Abstract. Extracting hazardous driving scenarios from naturalistic driving data is essential for creating a comprehensive test scenario library for autonomous driving systems. However, due to the sporadic and low-probability nature of hazardous driving events, the available number of hazardous driving scenarios collected within a short period is limited, resulting in insufficient scenario coverage and a lack of diverse training samples. Such limited samples also lead to the poor generalization ability and weak robustness of the hazardous driving scenario identification model. To address these challenges, we propose a method to augment the limited driving scenario data through generative adversarial networks. By integrating DIG (Discriminator gradIent Gap) and APA (Adaptive Pseudo Augmentation) techniques into the original GAN framework, the quality of the generated data is enhanced. Benefiting from the advantages of augmented samples, we can leverage a more sophisticated ResNet architecture for feature extraction from compressed dashcam videos called motion profiles to identify hazardous driving scenarios. By incorporating the augmented samples into the training set, the AUC of the proposed hazardous driving scenario identification model is improved by 4% and surpassed the existing state-of-the-art methods.

Keywords: Hazardous driving scenario · Limited sample modeling · Generative adversarial network · Data augmentation · ResNet

1 Introduction

With advancements in autonomous driving and intelligent connected vehicle technology, a substantial amount of naturalistic driving data has been collected and stored. Moreover, for operating vehicles, there is a growing trend of mandatory camera installations to monitor vehicle operations for improving traffic safety. For private cars, the installation of dashcams is also becoming increasingly common. Dashcams serve to capture driving scenarios and determine responsibilities in the event of an accident. All these in-car dashcam videos are accumulating

© The Author(s), under exclusive license to Springer Nature Singapore Pte Ltd. 2024
B. Luo et al. (Eds.): ICONIP 2023, CCIS 1967, pp. 203–214, 2024.
https://doi.org/10.1007/978-981-99-8178-6_16

and occupy a large amount of storage space. These dashcam videos can be categorized into non-conflict driving scenarios, which hold limited significance, and hazardous driving scenarios, which play a crucial role in various areas, including the analysis of accident causation, the assessment of the driver's safe driving ability, driver's safety education, and the safety evaluation of autonomous driving systems. Therefore, it is essential to develop an accurate and efficient algorithm for identifying hazardous driving scenarios from the massive dashcam videos.

Current state-of-the-art models for recognizing hazardous driving scenarios primarily rely on deep learning approaches [1–4]. However, due to the infrequent occurrence of accidents, there is a limited number of collected dangerous driving scenarios in a finite period of time, which usually leads to model overfitting. Therefore, there is a need for the exploration of data augmentation techniques to address the inadequacy and diversity shortage issue of limited training samples.

Besides, the lack of sample adequacy and diversity also leads to the problem of limited model complexity. Complex DNN models are more prone to overfitting problems compared to lightweight CNNs, which has led to the frequent use of lightweight models [5] to alleviate the overfitting problem in hazardous driving scenario identification tasks, which helps mitigate the overfitting phenomenon to some extent, but also sacrifice the excellent representational capacity of complex models.

In summary, it is necessary to study the data augmentation method for driving scenarios to solve the problem of insufficient sample adequacy and diversity. Secondly, on the basis of augmented data, the complexity of the model network structure could be appropriately increased to strengthen the characterization ability and then improve the model performance. This paper aims to fill these gaps by presenting a GAN (Generative Adversarial Networks) based method to perform driving scenario data augmentation and construct a ResNet based model to identify hazardous driving scenarios. The main work and contributions of this paper are:

- A driving scenario data augmentation method under limited data constraint is proposed, based on the original GAN, regularization is then performed for the loss of Discriminator, and augmented samples are added during training for improving the consistency of distribution between generated and real data.
- A ResNet-based model for evaluating hazardous driving scenarios is proposed: a dashcam video is compressed into a motion profile by pixel averaging, and a ResNet network is used for feature extraction of motion profiles to evaluate accident risks during driving.
- The data augmentation method proposed in this paper is applied to improve the quality of the training samples, and the augmented model achieves SOTA (state-of-the-art) results compared to the existing methods.

The rest of the paper is organized as follows. The related work is discussed in Sect. 2. In Sect. 3, the description of the dataset is given in details. Data augmentation and model architecture are given in Sect. 4. Section 5 describes the experiment and the performance evaluation results. Conclusions and discussions are presented in Sect. 6.

2 Related Work

2.1 Hazardous Driving Scenario Identification

When faced with a hazardous driving scenario, drivers usually adopt risk avoidance manoeuvres such as sharp brakes and sharp turns. Before the availability of in-vehicle camera equipment, research was usually based on vehicle kinematic data combined with threshold judgments to identify hazardous driving scenarios [6–8]. Some studies developed a machine learning model to identify the near crash [9–12]. However, due to the lack of interactive information between vehicles, the accuracy of these methods is often not high.

Deep neural network-based methods have made significant progress in identifying hazardous driving scenarios from dashcam videos. Recent research [1,2] mainly utilized object detection methods such as SSD [13] and YOLO [14] to detect the positions of interacting traffic participants, which are then used to calculate indicators like TTC (Time-To-Collision) to determine the risk of traffic accidents. Yu [3] proposed an end-to-end method to automatically extract the location features of traffic participants directly from the network over multiple frames of dashboard video, and then calculate the probability of accident. In summary, compared to TTC thresholding-based approaches, video-based approaches are more data-dependent and the sample size and quality have great influences on the accuracy of the model.

2.2 Motion Profile Based Models

Existing video-based algorithms primarily use object detection and tracking algorithms or feature extraction networks to process frame-by-frame images [13,14], which can be computationally demanding and hinder real-time performance. Furthermore, the reliance on single-frame feature extraction fails to consider the continuity of video streams, leading to potential issues such as missing or skipping frames in the detection results, especially in challenging lighting conditions, or complex traffic scenarios. To improve real-time performance and model accuracy, some studies [15,16] have proposed utilizing motion profiles to express the conflicts between traffic participants in dashcam videos.

The vehicle motion is presented as a continuous curve in the motion profile. The width change of the curve represents the relative distance change between the vehicle in front and the ego vehicle, and the differentiation of the curve width can reflect the rate of change of the relative distance, i.e., TTC. Because the interacting traffic participants in the motion profile are difficult to identify, Gao [5,17] proposed a CNN network on the motion profile to extract the trajectory features of interacting traffic participants to determine the driving risk, which achieved real-time performance, but the model accuracy is still lacking compared with the algorithms to detect traffic participants on each single frame [2,4]. Motion profiles can reflect the motion trajectories of traffic participants in the driving scenario comprehensively. But due to the loss of object shape in the process of motion profile generation, they usually lack semantic information of objects.

Although the large compression of video information saves processing time, it is more difficult for the model to identify and extract the trajectory features of the interacting traffic participants directly from motion profiles, which requires more diverse training samples, and more effective feature extraction methods to achieve better performance.

3   Data Operation

The dashcam videos are obtained from an operating vehicle company which installed dashcam equipment on all operating vehicles, and filtered continuous in-vehicle forward videos according to deceleration, with video FPS (Frames Per Second) of 4 and resolution of 760 × 368. The videos of 8 s before and 4 s after the *deceleration* < −0.5 g moment for a total of 12 s were taken as samples. Two experts jointly labeled the videos (non-conflict driving scenario and hazardous driving scenario), and the third expert negotiated the judgment when a difference of opinion was encountered. Since too limited data on bicycle-vehicle conflicts and human-vehicle conflicts were filtered out and too few types of side conflicts were collected, these conflicts were not considered anymore. Only the largest number of vehicle-vehicle longitudinal conflicts were studied in this paper.

A total of 546 samples were collected, of which 200 were hazardous driving scenarios. The driving scenarios covered different time periods (including day and night), different weather (sunny, rainy, cloudy), and different road facilities (highways, expressways, city roads, and rural roads).

3.1   Motion Profile Generation

We compress the dashcam video into a image called motion profile. We then transform the problem from a video-based modeling task into an image-based modeling task. The motion profile image contains the compressed information of traffic participants and environment. The motion profile generation algorithm is as follows: First, take the upper, middle, and lower rectangular regions I_i^f, I_i^m, I_i^c of the dashcam video frame image $I_i(1 \leq i \leq n)$, which correspond to the far, middle, and close range area from the ego car. The image is then compressed, and pixel averaging is performed along the vertical direction for each rectangular region I_i^f, I_i^m, I_i^c to generate pixel averaging lines L_i^f, L_i^m, L_i^c. Finally, the pixel averaging lines L_i^f, L_i^m, L_i^c are arranged according to the sequence of video frames to form three motion profiles MP^f, MP^m, MP^c for far, middle, and close range.

$$
\begin{cases}
MP^f = [L_1^f, L_2^f, L_3^f...L_n^f] \\
MP^m = [L_1^m, L_2^m, L_3^m...L_n^m] \\
MP^c = [L_1^c, L_2^c, L_3^c...L_n^c]
\end{cases}
\tag{1}
$$

The effect of generating 3 motion profiles from the dashcam video for far, middle, and close range is shown schematically in Fig. 1.

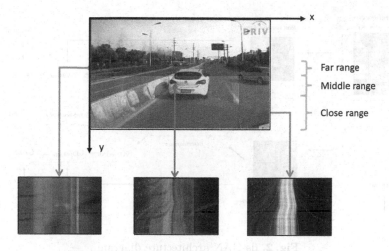

Fig. 1. Schematic of motion profile generation

4   Methodology

Real-time hazardous driving scenario evaluation algorithm mainly faces two problems of insufficiently diverse samples and the difficulty of extracting information about the location changes of traffic participants.

4.1   Data Augmentation

Instead of implementing data augmentation on driving videos, we propose to compress the continuous dashcam video over a period of time into a motion profile firstly, and then perform data augmentation based on the motion profile. Here we introduce how to complete the data augmentation of driving scenarios by GAN-based data generation method.

Data augmentation based on GAN [18] is a trend in recent years. In order to generate samples with specified categories, CGAN [19] adds classification labels to the GAN to guide the process of data generation. To improve the quality of generated data, WGAN [20] which optimizes the loss function, has been proposed. In this paper, we choose to use SNGAN [21] for motion profile data augmentation. SNGAN uses spectral parametrization to solve the 1-Lipschitz problem, which is difficult to be solved by WGAN, to further improve the quality of generated data.

Since the number of hazardous driving scenarios is too limited, we borrowed the method of GAN data augmentation under limited samples, and used the DIG (Discriminator gradIent Gap) [22] canonical term to avoid Discriminator overfitting to limited samples of hazardous driving scenarios, so that the convergence speed of Discriminator is stabilized, meanwhile, the APA [23,24] (Adaptive Pseudo Augmentation) idea is used to augment the training samples of GAN by taking generated fake samples as real samples to deceive Discriminator and improve its stability during the training process.

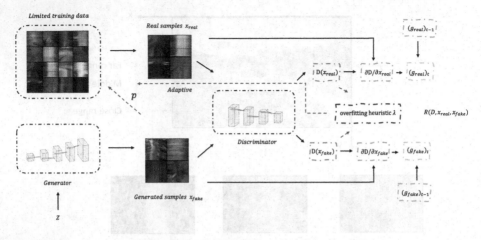

Fig. 2. ds-GAN architecture diagram

We propose a driving scenario GAN (ds-GAN) to augment the motion profile. ds-GAN's architecture is shown in Fig. 2. ds-GAN adopts the main architecture of SNGAN with the loss function as in Eq. 2 and 3.

$$L_G = \mathbb{E}_{z \sim p(z)} \left[\log D(G(z)) \right] \tag{2}$$

$$L_D = \mathbb{E}_{x \sim p_{\text{data}}(x)} \left[\log D(x) \right] - \mathbb{E}_{z \sim p(z)} \left[\log(1 - D(G(z))) \right] \tag{3}$$

To address the issue of the Discriminator's learning rate imbalance between real data and fake data, gradient parametric constraints are added to ds-GAN as in Eq. 4. This issue is addressed by introducing the loss function of the post-Discriminator as in Eq. 5. At the same time, the fake samples produced from the Generator are added to the real samples by data augmentation to deceive the Discriminator.

$$L_{DIG} = \left(\left\| \frac{\partial D}{\partial x_{\text{real}}} \right\|_2 - \left\| \frac{\partial D}{\partial x_{\text{fake}}} \right\|_2 \right)^2, \tag{4}$$

$$\tilde{L}_D = L_D + \lambda_{DIG} \cdot L_{DIG} \tag{5}$$

Algorithm 1: Enhanced model training using ds-GAN

Input: Training Dataset $\mathbb{D}$.
Output: enhanced model f_{da}.

1: **repeat**
2:     Sample mini-batch m images x_{real} from $\mathbb{D}$;
3:     Generate samples $x_{fake} \leftarrow G(x_{real})$;
4:     Calculate overfitting factor λ_{apa} and introduction probability p;
5:     $x'_{real} \leftarrow APA\_ADJUST(x_{real}, p)$;
6:     $output \leftarrow R(D, x'_{real}, x_{fake})$;
7:     Calculate L_{DIG} by $output$;
8:     Update the discriminator by descending $\tilde{L}_D = L_D + \lambda_{DIG} \cdot L_{DIG}$;
9:     Update the generator by descending L_G;
10: **until** ds-GAN training converges;
11: **repeat**
12:     Sample mini-batch m paired images x_{real} and labels y from $\mathbb{D}$;
13:     Generate augmentation samples: $x_{fake} \leftarrow G(x)$;
14:     $L_{real} \leftarrow CE(f_{da}(x_{real}), y)$;
15:     $L_{fake} \leftarrow CE(f_{da}(x_{fake}), y)$;
16:     Update the data augmentation model f_{da} by descending
       $L_{DA} = \lambda_{real}L_{real} + \lambda_{fake}L_{fake}$;
17: **until** model training converges;

4.2 Model Architecture

The hazardous driving scenario identification model is designed based on the ResNet network architecture, and the scenario is classified into two categories: non-conflict scenario with label 0 and hazardous scenario with label 1, and the displacement features of the interacting traffic participants are extracted through the ResNet network, with the pseudo code for network training shown as Algorithm 1.

5 Experiments and Results

5.1 Experiment Settings

Experiment Environment. The deep learning model training and evaluation were implemented on a workstation equipped with NVIDIA Tesla K40c GPU and Intel Core i7 processors with PyTorch framework.

Model Metrics. In the experiment, we used AUC (Area Under the Curve), accuracy, precision, recall, and F1 score as metrics to evaluate the performance of hazardous driving scenario identification model [5]. In addition, we measured the quality of GAN generated samples using FID (Fréchet Inception Distance) [22] in this paper.

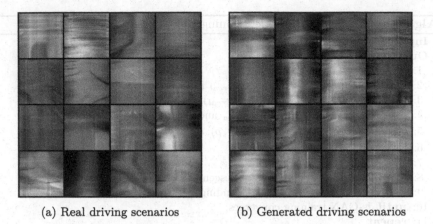

(a) Real driving scenarios   (b) Generated driving scenarios

Fig. 3. Different driving scenarios comparison

5.2   Effect of Data Augmentation

The data augmentation effect of the proposed ds-GAN is presented here. The randomly selected real driving scenarios samples are shown in Fig. 3a, and the generated driving scenarios samples are shown in Fig. 3b. It can be intuitively seen from Fig. 3b that the generated fake driving scenarios closely resemble the original real scenarios, demonstrating a high-quality data augmentation.

Table 1. Ablation experiments on data augmentation

Model	FID ↓ (Non-conflict)	FID ↓ (Hazardous)	FID ↓ (Both)
ds-GAN	**77.09**	**133.09**	**83.65**
ds-GAN w/o APA	102.82	155.03	98.13
ds-GAN w/o (DIG & APA)	124.66	175.10	129.82

Our ds-GAN is optimized using DIG and APA strategies, and the effect of these two optimization strategies is tested here. The training process of GAN with different strategies is shown in Fig. 4. It can be seen that the FID index can be effectively reduced and stabilized after DIG and APA optimization are adopted. From Table 1, in the case of dangerous driving scenarios, non-conflict driving scenarios, and a mixture of both scenarios, ds-GAN with both DIG as well as APA strategy achieves the best results. Moreover, the quality of the generated hazardous driving scenarios is not as good as the non-conflict scenarios due to the less data and more scattered distribution of the dangerous driving scenarios.

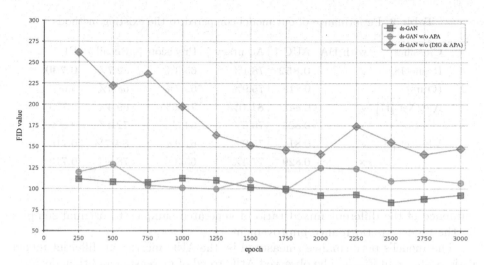

Fig. 4. FID change during training process of GAN with different strategies

5.3   Comparison with Existing Methods

Here our method is compared with the existing SOTA research methods, which used AlexNet [17] and Lightweight CNN [5] for feature extraction of motion profiles. The comparison of the model accuracy metrics is shown in Table 2. From the table, we can see that (1) the AUC of the model is improved by 4% with data augmentation, which indicates that data augmentation can effectively improve the accuracy of the model; (2) although there is no uniform improvement in precision and recall from the experimental results, after data enhancement, different performance models can be selected according to the actual need for precision and recall; (3) compared with the Lightweight CNN and AlexNet architectures used in the existing literature, we use the ResNet18 model architecture to evaluate hazardous driving scenarios and achieved SOTA results.

DA (Data augmentation) has shown that more complex models yield a more substantial improvement in performance. Specifically, the lightweight CNN model has a simpler layer structure compared to the other two network models, which is suitable for feature extraction tasks with less data and simpler samples, therefore, from the results, on the one hand, the performance of the lightweight CNN model decreases after data augmentation, on the other hand, AlexNet [17] outperforms the Lightweight CNN model [5], and ResNet performs even better than AlexNet [5], which emphasizes the significance of utilizing sophisticated neural network architectures in conjunction with data augmentation techniques.

5.4   Impact of Mixed Sample Ratio on Model Performance

The quantity of generated driving scenario samples appended to the training set has an impact on model performance. In this analysis, we delve into the

Table 2. Performance of our model compared to the existing methods

Classifier	with DA	AUC ↑	Accuracy ↑	Precision ↑	Recall ↑	F1 ↑
Resnet18	✓	**0.852**	78.1%	65.3%	82.7%	**0.730**
(Ours)	×	0.819	75.9%	65.7%	68.4%	0.670
AlexNet [17]	✓	0.842	**81.0%**	**75.6%**	69.4%	0.723
	×	0.823	75.5%	64.0%	72.4%	0.679
VK-Net [5]	✓	0.814	75.9%	61.9%	**84.7%**	0.716
	×	0.829	77.4%	65.3%	78.8%	0.713

influence of the different mixed ratio of generated samples to original samples on the model's performance improvement.

The model's performance (measured by the AUC metric) at different mixed ratio is shown in Fig. 5. The observed AUC trend of increasing and then decreasing as the mixed ratio increases can be explained by the interplay between data diversity and over-reliance. Initially, incorporating more generated samples enhances the model's ability to generalize and identify hazardous driving scenarios, leading to an increase in AUC. However, as the mixed ratio continues to rise, the model may become overly dependent on the generated samples, resulting in decreased performance and a decline in AUC. Therefore, there exists an optimal mixed ratio, as demonstrated by the highest AUC of 0.8517 at a mixed ratio of 0.6, where the model benefits from the augmented data without being overly influenced by it.

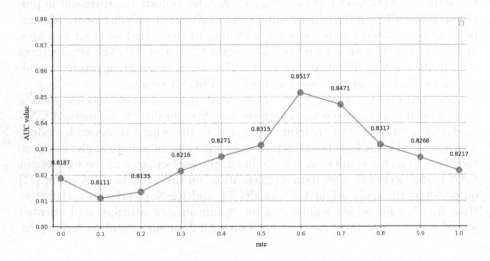

Fig. 5. AUC change under different proportion of generated samples

6 Conclusion and Future Work

In this work, we propose an approach to augment limited driving scenario data by incorporating DIG and APA techniques into the original GAN framework, thereby improving the quality of generated scenarios. Additionally, we introduce a method that utilizes ResNet for feature extraction from compressed motion profiles to identify hazardous driving scenarios. Through incorporating the generated samples into the training process of the hazardous driving scenario identification model, we achieve a 4% improvement in the model's AUC and attain the state-of-the-art performance.

In future work, we suggest two potential directions for improvement. Firstly, incorporating semantic information into motion profiles can enhance the model's performance by adding contextual details and object semantics. Secondly, exploring more complex model architectures, such as Transformer models, may lead to further advancements in identifying hazardous driving scenarios.

Acknowledgement. This work was sponsored by the National Key R&D Program of China (No. 2022ZD0115600), the Natural Science Foundation of Shanghai (No. 23ZR1465300 & No. 21ZR1465100), the Career Development Fund for Young Women Employees of Tongji University, the Fundamental Research Funds for the Central Universities (No. 22120220658) and the National Natural Science Foundation of China (No. 61702374).

References

1. Taccari, L., Sambo, F., Bravi, L., et al.: Classification of crash and near-crash events from dashcam videos and telematics. In: 2018 21st International Conference on Intelligent Transportation Systems (ITSC), pp. 2460–2465. IEEE (2018)
2. Ke, R., Cui, Z., Chen, Y., et al.: Lightweight edge intelligence empowered near-crash detection towards real-time vehicle event logging. IEEE Trans. Intell. Veh. **8**, 2737–2747 (2023)
3. Yu, R., Ai, H., Gao, Z.: Identifying high risk driving scenarios utilizing a CNN-LSTM analysis approach. In: 2020 IEEE 23rd International Conference on Intelligent Transportation Systems (ITSC), pp. 1–6. IEEE (2020)
4. Yamamoto, S., Kurashima, T., Toda, H.: Identifying near-miss traffic incidents in event recorder data. In: Lauw, H.W., Wong, R.C.-W., Ntoulas, A., Lim, E.-P., Ng, S.-K., Pan, S.J. (eds.) PAKDD 2020. LNCS (LNAI), vol. 12085, pp. 717–728. Springer, Cham (2020). https://doi.org/10.1007/978-3-030-47436-2_54
5. Gao, Z., Xu, J., Zheng, J.Y., et al.: A lightweight VK-net based on motion profiles for hazardous driving scenario identification. In: 2021 IEEE 23rd International Conference on High Performance Computing & Communications (HPCC), pp. 908–913. IEEE (2021)
6. Bagdadi, O.: Assessing safety critical braking events in naturalistic driving studies. Transp. Res. F Traffic Psychol. Behav. **16**, 117–126 (2013)
7. Jeong, E., Oh, C., Kim, I.: Detection of lateral hazardous driving events using in-vehicle gyro sensor data. KSCE J. Civ. Eng. **17**, 1471–1479 (2013)
8. Perez, M.A., Sudweeks, J.D., Sears, E., et al.: Performance of basic kinematic thresholds in the identification of crash and near-crash events within naturalistic driving data. Accid. Anal. Prev. **103**, 10–19 (2017)

9. Katrakazas, C., Quddus, M., Chen, W.H.: A new integrated collision risk assessment methodology for autonomous vehicles. Accid. Anal. Prev. **127**, 61–79 (2019)
10. Yan, L., Huang, Z., Zhang, Y., et al.: Driving risk status prediction using Bayesian networks and logistic regression. IET Intel. Transp. Syst. **11**(7), 431–439 (2017)
11. Arvin, R., Kamrani, M., Khattak, A.J.: The role of pre-crash driving instability in contributing to crash intensity using naturalistic driving data. Accid. Anal. Prev. **132**, 105226 (2019)
12. Osman, O.A., Hajij, M., Bakhit, P.R., et al.: Prediction of near-crashes from observed vehicle kinematics using machine learning. Transp. Res. Rec. **2673**(12), 463–473 (2019)
13. Ning, C., Zhou, H., Song, Y., et al.: Inception single shot multibox detector for object detection. In: 2017 IEEE International Conference on Multimedia & Expo Workshops (ICMEW), pp. 549–554. IEEE (2017)
14. Redmon, J., Divvala, S., Girshick, R., et al.: You only look once: unified, real-time object detection. In: Proceedings of the IEEE Conference on Computer Vision and Pattern Recognition, pp. 779-788 (2016)
15. Kilicarslan, M., Zheng, J.Y.: Direct vehicle collision detection from motion in driving video. In: 2017 IEEE Intelligent Vehicles Symposium (IV), pp. 1558–1564. IEEE (2017)
16. Kilicarslan, M., Zheng, J.Y.: Predict vehicle collision by TTC from motion using a single video camera. IEEE Trans. Intell. Transp. Syst. **20**(2), 522–533 (2018)
17. Gao, Z., Liu, Y., Zheng, J.Y., et al.: Predicting hazardous driving events using multi-modal deep learning based on video motion profile and kinematics data. In: 2018 21st International Conference on Intelligent Transportation Systems (ITSC), pp. 3352–3357. IEEE (2018)
18. Goodfellow, I., Pouget-Abadie, J., Mirza, M., et al.: Generative adversarial networks. Commun. ACM **63**(11), 139–144 (2020)
19. Mirza, M., Osindero, S. Conditional generative adversarial nets. arXiv preprint arXiv:1411.1784 (2014)
20. Arjovsky, M., Chintala, S., Bottou, L.: Wasserstein generative adversarial networks. In: International Conference on Machine Learning, pp. 214–223. PMLR (2017)
21. Miyato, T., Kataoka, T., Koyama, M., et al.: Spectral normalization for generative adversarial networks. arXiv preprint arXiv:1802.05957 (2018)
22. Fang, T., Sun, R., Schwing, A.: DigGAN: discriminator gradIent gap regularization for GAN training with limited data. arXiv preprint arXiv:2211.14694 (2022)
23. Jiang, L., Dai, B., Wu, W., et al.: Deceive D: adaptive pseudo augmentation for GAN training with limited data. Adv. Neural. Inf. Process. Syst. **34**, 21655–21667 (2021)
24. Karras, T., Aittala, M., Hellsten, J., et al.: Training generative adversarial networks with limited data. Adv. Neural. Inf. Process. Syst. **33**, 12104–12114 (2020)

Machine Unlearning with Affine Hyperplane Shifting and Maintaining for Image Classification

Mengda Liu, Guibo Luo, and Yuesheng Zhu[✉]

Shenzhen Graduate School, Peking University, Beijing, China
mdliu@stu.pku.edu.cn, {luogb,zhuys}@pku.edu.cn

Abstract. Machine unlearning enables a well-trained model to forget certain samples in the training set while keeping its performance on the remaining samples. Existing algorithms based on decision boundary require generating adversarial samples in input space in order to obtain the nearest but incorrect class labels. However, due to the nonlinearity of the decision boundary in input space, multiple iterations are needed to generate the adversarial samples, and the generated adversarial samples are affected by the bound of noise in adversarial attack, which greatly limits the speed and efficiency of unlearning. In this paper, a machine unlearning method with affine hyperplane shifting and maintaining is proposed for image classification, in which the nearest but incorrect class labels are directly obtained with the distance from the point to the hyperplane without generating adversarial samples for boundary shifting. Moreover, knowledge distillation is leveraged for boundary maintenance. Specifically, the output of the original model is decoupled into remaining class logits and forgetting class logits, and the remaining class logits is utilized to guide the unlearn model to avoid catastrophic forgetting. Our experimental results on CIFAR-10 and VGGFace2 have demonstrated that the proposed method is very close to the retrained model in terms of classification accuracy and privacy guarantee, and is about 4 times faster than Boundary Shrink.

Keywords: Machine Unlearning · Data Privacy · Information Security

1 Introduction

One person can meet around a thousand people in a day. But for most of them, we will forget as time goes by. However, when it comes to machine learning models, things become different. Making the model forget certain samples is a notoriously non-trivial problem. This leads to the risk of the stored training sample information in the model being stolen. Attackers can obtain the model's training samples through model inversion attack [5] or determine if a sample is in the training set through membership inference attack [15]. Recently, with the public's attention on privacy issues, more and more countries or regions have

© The Author(s), under exclusive license to Springer Nature Singapore Pte Ltd. 2024
B. Luo et al. (Eds.): ICONIP 2023, CCIS 1967, pp. 215–227, 2024.
https://doi.org/10.1007/978-981-99-8178-6_17

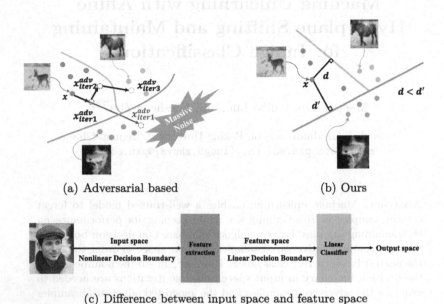

(a) Adversarial based   (b) Ours

(c) Difference between input space and feature space

Fig. 1. The difference between our method and the previous method: our method directly calculates the distance between samples and the decision boundary, while the previous method needed to generate adversarial samples, which may cause wrong results and be time-consuming.

passed legislation to protect users' personal data, such as the GDPR [18] in the EU and the California Consumer Privacy Act [13], which advocate for the right of users information to be forgotten-requiring companies to delete users' personal data upon request, whether in the database or in the model trained based on user data. In this context, the concept of "machine unlearning" has been emerged, explain the concept, like the model seems like not used the data from the beginning.

The optimal way to achieve forgetting is to retrain the initialized model with the original dataset, from which specific samples have been deleted. However, as the training data continues to increase, the cost of simple retraining is not acceptable. Therefore, a series of methods for accelerating unlearning have been proposed. And it can be divided into two parts: Exact unlearning and Approximate unlearning.

For Exact unlearning methods, they aims to accelerate the speed of a random initialization model's training on the remaining samples. SISA [2] is the first pioneer working in exact unlearning by dividing the original dataset into multiple disjoint shard datasets and training the model incrementally on each shard dataset. When a user's deletion request comes in, only the model corresponding to the shard dataset where the forgetting data is located needs to be retrained, thus achieving acceleration of retraining. Amnesiac Unlearning [7] records the gradients of the training data during training and subtracts the gradient update of the specific batch when a deletion request comes in. However, these methods either require saving the relevant training state during the training process or aggregating the trained models on multiple shard datasets as the final output, which greatly affects the real-time performance of the model.

While for Approximate unlearning methods, they trying to updates the parameters of a pretrained model, by adding noise to the model parameter space [6] or label [4,7]. Fisher forgetting [17] uses the Fisher Information Matrix to approximate the reduction of the Hessian matrix during forgetting, reducing the time and space costs of forgetting. However, the algorithm's computation cost is proportional to the square of the sample size, which has no advantage compared to the time required for retraining even on small datasets. Unlike parameter perturbation methods, Boundary Unlearning [4] proposes forgetting by shrinking and expanding the decision boundary and only using unlearning samples to finetune the model. which greatly accelerates the unlearning process compared to unlearning in the parameter space. Boundary shrink obtains the nearest but incorrect class labels by adversarial attacks in the input space, assigns labels to unlearning samples, and retrains the model. Boundary expanding assigns new labels to forgetting samples and deletes the new classifier after finishing finetuning. However, both shrinking and expanding suffer from the degradation of the remaining samples.

To find out the reason, we conducted a simple analysis of the unlearning process of Boundary Shrink. Specifically, Boundary Shrink obtains nearest but incorrect labels through adversarial attacks [14,19], and these incorrectly obtained labels can reflect the distance from the samples to the decision boundary to some extent. Due to the strong generalization of neural networks, unlearning based on the adversarial attack to obtain labels on decision boundaries is more consistent with the output of the model after retraining than using random labels. However, since adversarial sample generation is highly affected by the bound of noise, the adversarial samples generated by the model may not obtain the correct nearest labels, which causes the model to fail to unlearn samples properly. For example, in Fig. 1(a), the original sample belongs to the deer, and the nearest label should be horse. But due to the addition of excessive noise, the generated adversarial sample does not fall within the decision region of the horse. In addition, catastrophic forgetting of neural networks means that only using forgetting samples to finetune the model may cause the model to forget knowledge of the remaining samples, which leads to performance decline in the remaining classes.

In order to bypass the generation of adversarial examples to obtain corresponding labels, we simply decouple deep neural networks into two major parts: nonlinear feature extractors and affine classifiers, the inputs of which correspond to the model's input space and feature space, respectively. For the input space, due to the nonlinearity of the feature extractor, the decision boundary is also nonlinear with respect to the original input sample. However, for the feature space, since the classifier does not contain nonlinear components, the decision boundary for samples in the feature space is an affine hyperplane. Figure 1(c) shows the difference between input space and feature space. Therefore, we can directly obtain the corresponding labels by calculating the distance of the sample to each decision boundary, thus bypassing the generation of adversarial examples. Figure 1(b) shows how to calculate the distance between boundaries and samples in feature space. Moreover, based on the distance of the samples to the decision plane in the feature space, we can obtain the relationships between the samples and any class, making it possible to forget multiple classes compared to the Boundary Shrink, which can only obtain the relationship between the adversarial example class and the original class.

In order to cope with the performance decline in the remaining classes while ensuring the retention of knowledge of the remaining samples, we use knowledge distillation [9] to transfer knowledge of the remaining samples between the original model and the unlearning model.Specifically, we decompose the output of the original model into output for forgetting classes and output for remaining classes, and we supervise the output of the unlearning model on the remaining classes using knowledge of the remaining classes only.

2   Method

2.1   Preliminaries

In this paper, we mainly focus on class unlearning. For the K classification problem, a dataset $D = (x_i, y_i)_{i_1}^{N}$ is given, with N samples. For the ith sample, $x_i \in R$ corresponds to the input space, and $y_i \in \{1, \cdots, K\}$ corresponds to the label space. D_f represents the samples that the model needs to forget, and $D_f \subseteq D$, while the remaining samples are represented by $D_r = D/D_f$. The goal of unlearning is to make the model trained on D forget the knowledge on D_f while keeping the information learned on D_r unchanged.

2.2   Affine Hyperplane Unlearning

Due to the strong generalization ability of DNN networks, assuming that the logits of forgetting samples follow a uniform distribution on remaining classes or assuming that the mean of predictions for inputs belonging to the deleted class is equal to the mean of predictions of the remaining classes [1] may not be the optimal way. According to [4], most of the forgetting samples are predicted as specific classes, instead of random classes. Since nearest class labels can reflect the similarity between samples to some extent, assigning nearest but incorrect class labels to forgetting samples is crucial for boundary unlearning.

Proposition 1. *The distance from the original samples to the nearest decision boundary in the feature space is the lower bound of the distance between the original samples and adversarial samples in the feature space.*

Proof. Assume x and x_{adv} are the original sample and the corresponding adversarial sample in the feature space, where x corresponds to the label y and x_adv

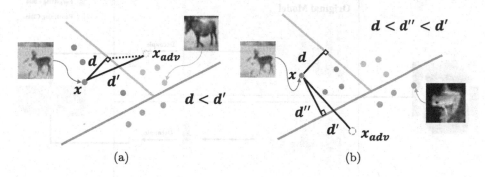

Fig. 2. The case of y_{nbi} and y_{adv}. (a) $y_{nbi} = y_{adv}$. (b) $y_{nbi} \neq y_{adv}$.

corresponds to the label y_{adv}. The distance between x and x_{adv} in the feature space is d, and the distance from x to the nearest decision boundary is d'. The class on the other side of the decision boundary closest to x is y_{nbi}.

If $y_{adv} = y_{nbi}$, since the distance from the sample to the decision boundary on the decision boundary is greater than the distance from the sample point to the other side of the decision boundary, we have $d' \leq d$.

If $y_{adv} \neq y_{nbi}$, since the shortest distance from the original sample to the hyperplane in the feature space is d', the distance from the sample point to the other decision boundary is $d'' \geq d'$, and because $d'' \leq d$, therefore $d' \leq d$.

Therefore, the distance based on the decision boundary is the lower bound of the distance between the original samples and adversarial samples in the feature space, which means that the labels obtained based on adversarial attacks may not be the nearest but incorrect class labels, Fig. 2 shows the case of the y_{adv} and y_{nbi}.

For multi-classification tasks, models usually use the softmax classifier. According to [12], the softmax classifier for K classification tasks can be seen as a combination of $\frac{K(K-1)}{2}$ binary classifiers. The decision boundary between classes can be obtained by $W_i - W_j$. Therefore, we can calculate the distance between each input sample and the decision boundary of another class in feature space to obtain the nearest but incorrect label. Given a feature vector $f \in R^D$, where D is the dimension of the vector, the forgetting class classifier W_i, an arbitrary remaining class classifier W_j, and their corresponding biases b_i and b_j, the distance between a sample point in feature space and the decision plane can be expressed as:

$$d_{ij} = \frac{|(W_i f_i + b_i) - (W_j f_j + b_j)|}{\sqrt{(W_i - W_j)^2}}. \tag{1}$$

Subsequently, we can select the class of the nearest but incorrect labels y_{nbi} for the sample point to its closest decision boundary.

$$y_{nbi} = argmin(d_{ij}). \tag{2}$$

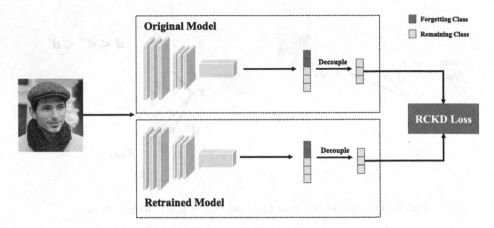

Fig. 3. Remaining Class Knowledge Distillation.

Same as [4], we achieve boundary shrink by assigning y_{nbi} to forgetting samples to unlearn the decision boundary. We then finetune the model based on the incorrect labels, and the loss function can be represented as follows:

$$\mathcal{L}_{cls} = CE(x_i, y_{nbi}),\qquad(3)$$

where CE is the cross entropy loss.

Note that, compared to [4], we do not need to repeat the process of generating adversarial examples multiple times and then using them as inputs to obtain labels through forward propagation. This greatly improves the speed of machine unlearning. Moreover, compared to obtaining labels through adversarial attacks, labels obtained based on decision boundary can achieve model parameter updates at a higher learning rate. Intuitively, a higher learning rate means that the distribution of model parameters changes more from the original model, so the model's forgetting of data is more complete.

2.3   Affine Hyperplane Maintaining

Simply considering using forgetting samples to finetune the model may lead to catastrophic forgetting of the remaining samples. In that case, Li et al. [10] used knowledge distillation to transfer knowledge between different neural network architectures, while LwF [11] used knowledge distillation to avoid catastrophic forgetting of old data when training models on new datasets.

However, using knowledge distillation is suboptimal because the model output also includes information about forgetting samples. Recently, motivated by Decoupled Knowledge Distillation (DKD) [20], which separates specific classes through Target Class Knowledge Distillation (TCKD) and Non-Target Class Knowledge Distillation (NCKD) by decoupling the original model output, In light of this, we decoupled the output of the original model into Forgetting Class Knowledge Distillation (FCKD) and Remaining Class Knowledge Distillation

Table 1. Single class unlearning on CIFAR-10 and Lacuna-10 dataset.

Dataset	Method	$A_{D_{rt}}$	$A_{D_{ft}}$	A_{D_r}	A_{D_f}	A_{MIA}
CIFAR-10	Original	100	100	85.07	83.30	64.30
	Retrain	100	0.00	85.56	0.00	52.90
	Negative Gradient	98.67	8.60	81.17	7.00	61.40
	Random Label	99.50	9.84	82.29	8.00	64.90
	Boundary Shrink	99.16	7.04	82.04	5.70	60.70
	Boundary Expanding	99.62	9.96	82.56	8.30	64.10
	Ours	**99.96**	**0.40**	**85.36**	**0.00**	59.30
Lacuna-10	Original	100	100	97.44	99.00	54.00
	Retrain	100	0.00	97.22	0.00	43.00
	Negative Gradient	97.75	7.00	94.08	7.29	44.00
	Random Label	97.92	9.50	94.08	9.38	50.00
	Boundary Shrink	97.94	3.00	94.64	3.12	**43.00**
	Boundary Expanding	97.83	7.55	94.11	8.00	46.00
	Ours	**99.94**	**0.00**	**96.65**	**0.00**	**43.00**

(RCKD). Figure 3 shows the Remaining Class Knowledge Distillation. Since the knowledge of the class that has not been forgotten can reflect the relationship between classes, using RCKD can help maintain the decision boundary of classifiers for the remaining samples. Specifically, for any training sample x_i, the classification probabilities predicted by the model can be represented as $p = [p_1, \cdots, p_c]$, where p_i is the probability of the i-th class, C_r is the number of remaining classes. We decompose the model predictions into remaining class predictions and forgetting class predictions. Since we only need the output results of the remaining classes, the probability of the i-th remaining class is:

$$p_i = \frac{exp(z_i)}{\sum_{j-1}^{c_r} exp(z_j)}, \tag{4}$$

Where z_i represents the logits of the i-th remaining class.

Note that, compared to the original DKD, we removed the logits of the forgetting classes from the model outputs before performing softmax to prevent information from the forgetting classes from being transferred to the unlearned model.

Finally, based on KL divergence, we implement knowledge distillation for the remaining classes. For the original model output p_i^t and the model output after finetuning on the original model p_i^s, RCKD can be represented as:

$$\mathcal{L}_{RCKD} = KL(p_i^t, p_i^s). \tag{5}$$

3   Experiments

3.1   Experimental Settings

Datasets. We conducted experiments on CIFAR-10 and VGGFace2 datasets. CIFAR-10 includes 60,000 images, of which 50,000 are used as training set and the remaining 10,000 as test set. VGGFace2 is a large-scale face recognition dataset that contains nearly 331 million images of 9,131 individuals, Similarly to [6], we randomly selected 10 and 100 classes for experimentation, which we named Lacuna-10 and Lacuna-100 respectively. The classes in Lacuna-10 and Lacuna-100 do not overlap, and each class consists of 500 randomly selected images. We used the first 400 images of each class for training and the remaining 100 for testing.

Baselines. We compare our results with the methods that only finetune on D_r for fair comparisons.
Retrain: Reinitialize the model and retrain the model.
Negative Gradients: Update model parameters in the direction of gradient ascent.
Random Labels: Assign random labels to the remaining classes on forgetting classes and finetune the model.
Boundary Shrink: Using adversarial attack to obtain wrong labels and assign to the unlearning samples.
Boundary Expanding: Assign new labels to forgetting classes, and after training, delete the classifier for new classes.

Evaluation Metrics. We evaluate the performance of the model in terms of both utility and privacy guarantees, and a good model should have similar performance in both aspects compared to the retrained model.

For utility guarantee, we use four accuracy metrics to evaluate them: the remaining samples accuracy for the training set $A_{D_{rt}}$, the forgetting samples accuracy for the training set $A_{D_{ft}}$, the remaining samples accuracy for the testing set A_{D_r}, and the forgetting samples accuracy for the testing set A_{D_f}.

For privacy guarantee, we use membership inference attacks to determine if the forgetting samples in the training set are recognized as being in the training set by a shallow model. We use logistic regression as the shallow model and use the loss of sample as input. The success rate of the membership inference attack can be represented as A_{MIA}.

Implementation Details. The experiments were conducted using one NVIDIA GeForce RTX 2080Ti GPU. For CIFAR-10, we conducted single class unlearning experiment based on the All-CNN [16]. For the original model, we trained it for 30 epochs using SGD with a learning rate of 0.01 and a batch size of 64. For the retrain model, we finetune the original model using incorrect labels for 10 epochs with a learning rate of 0.001. For the VGGFace2 [3] dataset, we conducted experiments using ResNet-18 [8]. For Lacuna-10, we conducted single class unlearning experiment, while for Lacuna-100, we conducted multiple class unlearning experiment to evaluate the algorithm's unlearning performance in more realistic scenarios. The training parameters were kept consistent with

Table 2. Multiple class unlearning on Lacuna-100 dataset.$\#y_f$ denote number of forgetting classes.

$\#y_f$	Metrics	Original	Retrain	Ours
25	$A_{D_{rt}}$	96.37	97.28	97.10
	$A_{D_{ft}}$	94.11	0.00	0.00
	A_{D_r}	77.69	82.54	81.26
	A_{D_f}	76.04	0.00	0.00
	A_{MIA}	59.27	51.28	54.52
50	$A_{D_{rt}}$	95.01	96.81	96.09
	$A_{D_{ft}}$	96.60	0.00	0.00
	A_{D_r}	75.74	82.76	81.62
	A_{D_f}	78.82	0.00	0.02
	A_{MIA}	59.50	49.54	51.64
75	$A_{D_{rt}}$	94.62	98.56	98.33
	$A_{D_{ft}}$	96.20	0.00	0.00
	A_{D_r}	74.56	86.96	87.06
	A_{D_f}	78.19	0.00	0.00
	A_{MIA}	59.23	49.40	52.79

those used in the CIFAR-10 experiments. In the case of the privacy guarantee of class unlearning, we exclude the loss of the remaining samples and consider whether the shallow model can distinguish between forgetting samples in the training set or in the test set based on the loss.

3.2   Experimental Results

For single class unlearning, we compared our experimental results with the following methods: Retrain, Random Labels, Negative Gradient, Boundary Shrink and Boundary Expanding. Due to the poor results of the above methods, we compare our results only with the Retrain Model in the subsequent experiments. Note that during experiments, we used the same network structure and kept hyperparameters the same in [4].

Utility Guarantee. We first report the accuracy metrics of our method on CIFAR-10 and Lacuna-10 datasets. Because our method can obtain the nearest but incorrect class labels, the model tends to misclassify samples in a more natural way, which leads to a better utility guarantee. As shown in Table 1, in the single class unlearning experiment, our method achieves close to zero accuracy for the forgetting samples. On CIFAR-10 dataset, the accuracy drops by 6.64% and 5.70% respectively, for the forgetting samples in the training and test sets compared with Boundary Shrink. For the Lacuna-10 dataset, our method achieves a drop in accuracy of 3.00% and 3.12% respectively, for the forgetting samples in the training and test sets compared with the SOTA. Compared to other methods, our method maintains similar accuracy to the retrained model

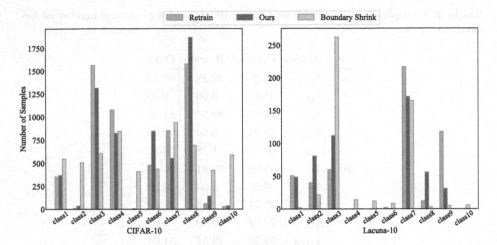

Fig. 4. Model output on CIFAR-10 and Lacuna-10 datasets.

in the remaining samples for both datasets, with only a slight drop of 0.2% on CIFAR-10 and 0.57% on Lacuna-10. Table 2 shows the result of the multi-class unlearning experiment. We can clearly see that our method achieves accuracy consistent with the retrained model. When unlearning 75 classes, our method even outperforms the retrained model by 0.1% in accuracy for the remaining classes.

Privacy Guarantee. We then report the privacy protection performance of the algorithm. From Table 1, we can see our method is closest to the retrained model in terms of the success rate of membership inference attacks, with a difference of 6.4% on CIFAR-10 and equal on Lacuna-10. This indicates that the output properties of our model are basically consistent with the retrained model. Table 2 shows the results of unlearning multiple classes, and we can also see that our method is very close to the retrained model in terms of the success rate of membership inference attacks. We further compared the model output for the forgetting samples with Boundary Shrink and found that our method's output for forgetting samples was more close to the retrained model. Since our method generates the nearest but incorrect class labels, which are retrained models mostlikely to misclassify, our method's output is closer to the retrained model. Figure 4 shows our method achieves the best forgetting effect for forgetting samples.

Unlearning Speed. Unlearning speed is also a very important indicator for evaluating algorithms. We compare our results with Boundary Shrink and Boundary Expanding with 10 epoch training. Figure 5 shows a comparison of the unlearning speed between our method and other methods. Since our method does not need to generate adversarial samples through multiple iterations, it can be comparable to Boundary Expanding in terms of unlearning time. For Boundary Shrink, our method is about 4 times faster than Boundary Shrink.

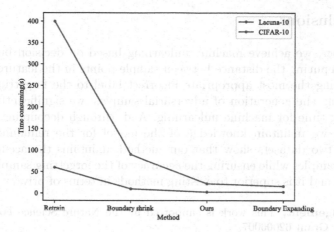

Fig. 5. Unlearning speed on CIFAR-10 and Lacuna-10 datasets.

Table 3. Ablation study on CIFAR-10 dataset.

Variant	Candidate				$A_{D_{rt}}$	$A_{D_{ft}}$	A_{D_r}	A_{D_f}	A_{MIA}
	y_{adv}	y_{nbi}	KD	RCKD					
No.1		✓			99.86	0.60	84.39	**0.00**	**59.30**
No.2	✓				99.16	7.04	82.04	5.70	60.70
No.3		✓	✓		99.92	41.08	84.51	39.90	69.30
No.4		✓		✓	**99.96**	**0.40**	**85.36**	**0.00**	**59.30**

3.3 Ablation Study

Effectiveness of Affine Hyperplane Unlearning. By removing the RCKD term from the loss function, we compared our method with the nearest erroneous label obtained through adversarial attacks. From Table 3 first line and second line, it can be seen that our method still outperforms existing methods, with a difference of 0.7% and 2.35% on the remaining classes of the training set and test set, respectively, and is also 1.4% better than Boundary Shrink in privacy guarantee.

Effectiveness of Affine Hyperplane Maintaining. We further evaluated the role of RCKD in the loss function. Table 3 last two lines show that using the entire model output for knowledge distillation resulted in preserving a significant amount of knowledge about forgetting samples from the training set, with 41.08% accuracy on forgetting samples in the training set, while RCKD successfully reduced the model's accuracy on forgetting samples in the training set and test set.

4   Conclusions

In this paper, we achieve machine unlearning based on decision boundary by directly computing the distance between sample points in the feature space and then assigning the most appropriate incorrect label to the forgetting samples. By bypassing the generation of adversarial samples, we significantly speed up the training time for machine unlearning. And, through decoupling knowledge distillation, we maintain knowledge of the model for the remaining samples. Results on two datasets show that our method maintains the accuracy of the remaining samples while ensuring the accuracy of the forgetting samples is as low as possible, and it is superior to existing methods in terms of privacy protection.

Acknowledgement. This work is supported by the Nature Science Foundation of China under Grant 62006007.

References

1. Baumhauer, T., Schöttle, P., Zeppelzauer, M.: Machine unlearning: linear filtration for logit-based classifiers. Mach. Learn. **111**(9), 3203–3226 (2022)
2. Bourtoule, L., et al.: Machine unlearning. In: 2021 IEEE Symposium on Security and Privacy, pp. 141–159 (2021)
3. Cao, Q., Shen, L., Xie, W., Parkhi, O.M., Zisserman, A.: VGGFace2: a dataset for recognising faces across pose and age. In: 2018 13th IEEE International Conference on Automatic Face & Gesture Recognition, pp. 67–74 (2018)
4. Chen, M., Gao, W., Liu, G., Peng, K., Wang, C.: Boundary unlearning: rapid forgetting of deep networks via shifting the decision boundary. In: Proceedings of the IEEE/CVF Conference on Computer Vision and Pattern Recognition, pp. 7766–7775 (2023)
5. Fredrikson, M., Jha, S., Ristenpart, T.: Model inversion attacks that exploit confidence information and basic countermeasures. In: Proceedings of the 22nd ACM SIGSAC Conference on Computer and Communications Security, pp. 1322–1333 (2015)
6. Golatkar, A., Achille, A., Soatto, S.: Eternal sunshine of the spotless net: selective forgetting in deep networks. In: Proceedings of the IEEE/CVF Conference on Computer Vision and Pattern Recognition, pp. 9304–9312 (2020)
7. Graves, L., Nagisetty, V., Ganesh, V.: Amnesiac machine learning. In: Proceedings of the AAAI Conference on Artificial Intelligence, pp. 11516–11524 (2021)
8. He, K., Zhang, X., Ren, S., Sun, J.: Deep residual learning for image recognition. In: Proceedings of the IEEE Conference on Computer Vision and Pattern Recognition, pp. 770–778 (2016)
9. Hinton, G.E., Vinyals, O., Dean, J.: Distilling the knowledge in a neural network. arXiv preprint arXiv:1503.02531 (2015)
10. Li, K., Yu, R., Wang, Z., Yuan, L., Song, G., Chen, J.: Locality guidance for improving vision transformers on tiny datasets. In: Avidan, S., Brostow, G., Cissé, M., Farinella, G.M., Hassner, T. (eds.) ECCV 2022. LNCS, vol. 13684, pp. 110–127. Springer, Cham (2022). https://doi.org/10.1007/978-3-031-20053-3_7
11. Li, Z., Hoiem, D.: Learning without forgetting. IEEE Trans. Pattern Anal. Mach. Intell. **40**(12), 2935–2947 (2017)

12. Moosavi-Dezfooli, S.M., Fawzi, A., Frossard, P.: DeepFool: a simple and accurate method to fool deep neural networks. In: Proceedings of the IEEE Conference on Computer Vision and Pattern Recognition, pp. 2574–2582 (2016)
13. Pardau, S.L.: The California consumer privacy act: towards a European-style privacy regime in the united states. J. Tech. L. Pol'y **23**, 68 (2018)
14. Rosenberg, I., Shabtai, A., Elovici, Y., Rokach, L.: Adversarial machine learning attacks and defense methods in the cyber security domain. ACM Comput. Surv. (CSUR) **54**(5), 1–36 (2021)
15. Shokri, R., Stronati, M., Song, C., Shmatikov, V.: Membership inference attacks against machine learning models. In: 2017 IEEE Symposium on Security and Privacy, pp. 3–18 (2017)
16. Springenberg, J.T., Dosovitskiy, A., Brox, T., Riedmiller, M.: Striving for simplicity: the all convolutional net. arXiv preprint arXiv:1412.6806 (2014)
17. Tarun, A.K., Chundawat, V.S., Mandal, M., Kankanhalli, M.: Fast yet effective machine unlearning. IEEE Trans. Neural Netw. Learn. Syst. 1–10 (2023)
18. Voigt, P., Von dem Bussche, A.: The EU general data protection regulation (GDPR). In: A Practical Guide, 1st Ed., vol. 10(3152676), pp. 10–5555. Springer, Cham (2017)
19. Xu, H., Li, Y., Jin, W., Tang, J.: Adversarial attacks and defenses: frontiers, advances and practice. In: Proceedings of the 26th ACM SIGKDD International Conference on Knowledge Discovery & Data Mining, pp. 3541–3542 (2020)
20. Zhao, B., Cui, Q., Song, R., Qiu, Y., Liang, J.: Decoupled knowledge distillation. In: Proceedings of the IEEE/CVF Conference on Computer Vision and Pattern Recognition, pp. 11953–11962 (2022)

An Interpretable Vulnerability Detection Framework Based on Multi-task Learning

Meng Liu[1], Xiaohui Han[1,2,3(✉)], Wenbo Zuo[1,2], Xuejiao Luo[1], and Lei Guo[4]

[1] Key Laboratory of Computing Power Network and Information Security, Ministry of Education, Shandong Computer Science Center (National Supercomputer Center in Jinan), Qilu University of Technology (Shandong Academy of Sciences), Jinan, China
hanxh@sdas.org
[2] Shandong Provincial Key Laboratory of Computer Networks, Shandong Fundamental Research Center for Computer Science, Jinan, China
[3] Quan Cheng Laboratory, Jinan, China
[4] Shandong Normal University, Jinan, China

Abstract. Vulnerability detection (VD) techniques are critical to software security and have been widely studied. Many recent research works have proposed VD approaches built with deep learning models and achieved state-of-the-art performance. However, due to the black-box characteristic of deep learning, these approaches typically have poor interpretability, making it challenging for analysts to understand the causes and mechanisms behind vulnerabilities. Although a few strategies have been presented to improve the interpretability of deep learning models, their outputs are still difficult to understand for those with little machine learning knowledge. In this study, we propose IVDM, an **I**nterpretable **V**ulnerability **D**etection Framework **B**ased on **M**ulti-task Learning. IVDM integrates the VD and explanation generation tasks into a multi-task learning mechanism. It can generate explanations of the detected vulnerabilities in the form of natural language while performing the VD task. Compared with existing methods, the explanations outputted by IVDM are easier to understand. Moreover, IVDM is trained based on a large-scale pre-trained model, which brings it the cross-programming-language VD ability. Experimental results conducted on both a dataset collected by ourselves and public datasets have demonstrated the effectiveness and rationality of IVDM.

Keywords: Vulnerability Detection · Multi-task Learning · Interpretability · Black-box

1 Introduction

Hidden vulnerabilities in software can lead to system exploitation by attackers and cause serious security problems. The existence of software vulnerabilities has brought significant challenges to the security of the system. Although researchers have developed various techniques and tools to aid in vulnerability detection

B. Luo et al. (Eds.): ICONIP 2023, CCIS 1967, pp. 228–242, 2024.
https://doi.org/10.1007/978-981-99-8178-6_18

(VD) and mitigation [1,2], the number of vulnerabilities disclosed each year in the Common Vulnerabilities and Exposures (CVE) continues to rise. It still needs much more effort to promote the performance of VD.

Early VD techniques relied on reviewers' understanding of the code and their accumulated discernment experience, which was a time-consuming and labor process [3]. Moreover, as the size and complexity of the code increased, these approaches that solely rely on human judgment become ineffective. Subsequently, approaches based on machine learning have gradually become the focus of academia and industry. The VD techniques based on machine learning analyzes code automatically and detects whether there are vulnerabilities in the code. However, these techniques still require experts to define a set of vulnerability features for constructing source code representations [4].

With the successful application of deep learning models in various fields in recent years, end-to-end approaches based on deep learning models have become a new trend. Deep learning-based approaches mainly utilize neural networks such as Convolutional Neural Network (CNN) and Long Short-Term Memory network (LSTM) to automatically learn code features [5,6], alleviating the need for expert-defined features. Even in the face of some vulnerability code samples with inconspicuous features, these approaches can extract effective feature information and detect vulnerability codes. However, due to the black-box nature of deep learning models, their outputs often lack interpretability, making it challenging for analysts to understand the causes and mechanisms behind vulnerabilities based on the results. To enhance the interpretability of VD, some researches try to determine the critical parts within each code that are most relevant to the vulnerabilities using attention mechanisms [7]. However, this type of explanation still presents certain difficulties for non-machine learning experts. Therefore, to better understand code vulnerabilities and make explanations more understandable, it is necessary to develop an intuitive and interpretable VD tool.

In this study, we propose IVDM, an InterpretableVulnerability Detection Framework Based on Multi-task Learning. IVDM integrates the VD and explanation generation tasks into a multi-task learning mechanism. It can generate explanations of the detected vulnerabilities in the form of natural language while performing the VD task. Through such explanations, programmers can gain a better understanding of the existing vulnerabilities in the code, thereby helping them in vulnerability remediation. Moreover, IVDM is built based on the large-scale pre-trained model UniXcoder [8], which captures the semantic information of the code without considering programming language or syntax variations. Therefore, IVDM is widely adaptable and has cross-programming-language VD ability.

The main contributions of this paper are as follows:

(1) We propose a multi-task learning framework IVDM, which can generate explanations in natural language form. Compared with existing methods, IVDM uses natural language to explain the reasons and mechanisms behind vulnerabilities. This explanation is easier to understand by programmers who with little machine learning knowledge.
(2) We design a VD fine-tuning strategy built on a large-scale pre-trained model to construct IVDM. On the one hand, this large-scale pre-training model has satisfactorily text generation ability due to massive text training. On the other hand, it captures the semantic correlation between different language codes and can detect vulnerabilities in multiple programming languages. This eliminates the need to maintain models for each programming language and greatly reduces associated costs.
(3) We constructed a dataset containing multiple programming languages, and evaluated IVDM on the constructed dataset and public datasets. Experimental results show that IVDM outperforms the baseline methods. It can accurately detect whether there are vulnerabilities in the code and provide human-readable interpretation of the detection results.

The remainder of this paper is organized as follows. Section 2 reviews the related prior work. Section 3 introduces some preliminaries, and Sect. 4 details our proposed method. Section 5 describes the experimental setup, evaluation metrics, and analysis of experimental results. Section 6 concludes the paper.

2 Related Work

With the rapid development of machine learning technology, more and more researchers apply machine learning to the field of VD. Kronjee et al. [9] applied data flow analysis techniques to extract features from code samples. These features are then used to train various machine learning classifiers to detect SQL injection (SQLi) and cross-site scripting (XSS) vulnerabilities. Ren et al. [10] used the characteristics of buffer overflow vulnerabilities to construct a model based on decision tree algorithm, which can perform VD on the code at the function level. These methods require experts to define features and use traditional machine learning models such as support vector machines and naive bayes to detect vulnerabilities.

In recent years, the emergence of deep learning approaches offers a solution that reduces the reliance on manually defined features. These approaches have stronger feature learning capabilities. Dam et al. [11] utilized LSTM network to learn syntactic and semantic features of Java code for VD. The VulDeepecker model proposed by Li et al. [12] detects whether there are loopholes and the location of the loopholes in the C language program at the slice level. Li et al. [13] proposed an improved model SySeVR, which focuses on how to obtain code representations containing syntactic and semantic information for application in VD. Zhou et al. [14] introduced Devign, which uses abstract syntax tree (AST) as its backbone and encodes program control and data dependencies at different levels into a joint graph with heterogeneous edges. Then, a gated graph recurrent layer is employed to learn composite graph node features for detecting vulnerable code. The DeepWukong proposed by Cheng et al. [15] first performs program slicing to extract fine-grained but complex semantic features. Then, it utilizes graph convolutional networks to learn code features for VD.

Although different neural network models with various structures in deep learning can learn vulnerability features from a large number of codes, their decision-making processes often lack interpretability. Some researchers have proposed two different interpreters, PGExplainer and GNNExplainer. They provide explanations for neural networks by outputting subgraphs that play a crucial role in the model's predictions [16,17]. In order to make the VD results interpretable, Yan et al. [7] employed an attention mechanism to focus the model's attention on code lines that are critical for VD. These key code lines can provide a certain level of explanation. Zou et al. [18] adopt a custom heuristic search to get the important tokens that are used to generate some vulnerability rules. These rules are used to explain the model's decision process. Li et al. [19] introduced the IVDetect tool, which utilizes the graph neural network interpreter GNNExplainer to provide explanations for function-level VD results.

Differently, IVDM generates natural language explanations for detection results, which is more intuitive and readable. UniXcoder has learned rich language representation and the ability to generate text after pre-training. By fine-tuning UniXcoder, IVDM can capture the semantic context related to the vulnerability and extract information related to the characteristics and mechanism of the vulnerabilities. Furthermore, we did not focus on a specific programming language, but considered a wider range of programming languages. IVDM is trained based on a large-scale pre-trained model, which brings it the cross-programming-language VD ability.

3   Methodology

3.1   Overview of Our Detection Model

We choose UniXcoder as the initial model of the encoder, and further fine-tune it to improve performance, for extracting feature representations for each code sample. Our IVDM model's workflow is illustrated in Fig. 1. First, the tokenization pipeline converts the raw code into sequential representations, and then sends them to the encoder for feature extraction. After encoding, they are input into the classifier and decoder respectively. In the following sections, we will provide detailed explanations of each step.

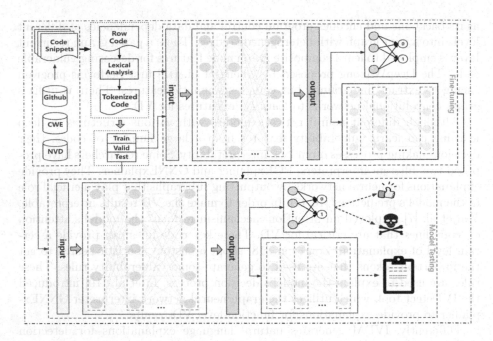

Fig. 1. The overall architecture of IVDM. IVDM uses the collected code snippets for fine-tuning, and then uses the fine-tuned model for vulnerability detection and interpretation of the detection results.

3.2 Encoder: Features Extraction

First, we obtain token representation P = [CLS],$s_1,s_2,...s_n$,[SEP] for each sample. [CLS] is a special token, and its corresponding output vector serves as the feature representation for the sequence. s_n represents the n-th token in the code sequence, and [SEP] is used to separate different sequence segments. Next, convert the data into a form that the model can learn and put it into the model. Specifically, each token is mapped to a corresponding word vector representation. The process is shown in equation (1) :

$$Q = tokenizerID(P) \tag{1}$$

Finally, this results in a two-dimensional tensor H^0 with a shape of $|Q| \times d_h$. Here, $|Q|$ represents the length of the input sequence Q, and d_h represents the dimension of the hidden layer.

The UniXcoder model applies 12 Transformer layers on the input vectors to generate hidden states $H^N = \{h_0{}^N, h_1{}^N, ..., h_{n-1}{}^N\}, N \in [1, 12]$. Each layer consists of an identically structured transformer that uses a multi-head self-attention operation and a feed-forward layer on the previous layer's output. To calculate the output of the multi-head self-attention for the l-th transformer layer, we follow the procedure outlined below:

$$Q_i = H^{l-1}W_i{}^Q, K_i = H^{l-1}W_i{}^K, V_i = H^{l-1}W_i{}^V \tag{2}$$

$$head_i = softmax(\frac{Q_i K_i^T}{\sqrt{d_k}} + M)V_i \tag{3}$$

$$\tilde{X} = concat(head_1, head_2, ..., head_u)W_i^O, \tag{4}$$

where the output $H^{l-1} \in \mathbb{R}^{|Q| \times d_h}$ of the previous layer is linearly mapped to triplets of queries, keys, and values using the model parameter $W_i^Q, W_i^K, W_i^V \in \mathbb{R}^{d_h \times d_k}$, d_k is the dimension of the header, u is the number of headers and $M \in \mathbb{R}^{|Q| \times |Q|}$ is a mask matrix to control the context a token can attend to when computing its contextual representation.

Afterwards, the output of the multi-head self-attention is processed through feed-forward layers and normalization layers to generate the final contextual representation, i.e.:

$$H^l = LN(H^{l-1} + LN(FFN(\tilde{X}^l))), \tag{5}$$

where H^{l-1} is the output matrix of the $(l-1)$-th layer, FFN represents a two-layer fully connected neural network, with the first layer activated by the ReLU function and the second layer without any activation function. LN refers to the operation of layer normalization. Finally, we obtain the output $H^{12} = \{h_1^{12}, h_1^{12}, ..., h_n^{12}\}$ of the encoder at the last transformer layer, where h_i represents the encoder representation of the i-th token in the input sequence.

3.3   Objective Task

Task 1: Code Vulnerability Detection. The first task is to use a classifier to detect whether the input code contains vulnerabilities. First, we select the feature representation corresponding to the [CLS] token from the features extracted by the encoder. Then, we input this feature representation into a classifier composed of fully connected layers and a sigmoid function to generate the final prediction result from y'. i.e.:

$$y' = \sigma(W_c h_1^{12} + b_c), \tag{6}$$

where W_c and b_c are the weight matrix and bias vector of the linear layer, and $\sigma(\cdot)$ represents the sigmoid activation function.

Task 2: Vulnerability Semantics Generation. The second task is to generate explanations of the detected vulnerabilities in the form of natural language through the decoder. After obtaining the feature representation extracted by the encoder, the decoder utilizes these context vectors to generate the predicted output of the target word. We adopt Transformer as the model's decoder. Specifically, the decoder utilizes a feed-forward neural network to transform each context vector to generate a predicted output probability distribution for each word. In the Transformer decoder, each context vector is processed as input through multiple attention mechanisms and feed-forward neural network layers. Through the combination of these layers, the decoder is able to capture the semantic and syntactic information in the input sequence and generate the corresponding predicted output probability distribution. We can obtain the following equation:

$$P(y_i|y_1, y_1, ..., y_{i-1}, x) = softmax(W_s \cdot g_i + b_s), \tag{7}$$

where y_i represents the i-th word in the target sequence, $\{y_1, y_2, ..., y_{i-1}\}$ represents the sequence of the first to $(i-1)$-th words in the target sequence, x represents the source sequence. The generated state g_i represents the state vector at the i-th step of the decoder and is used to maintain information about the source code snippet and previously generated words. W_s and b_s are the weight matrix and bias vector of the feed-forward neural network, and the softmax function converts the output into a probability distribution.

3.4   Training

The training process of IVDM is based on the fine-tuning of the UniXcoder model. Fine-tuning is a technique for supervised training on the basis of pre-trained models, which enables IVDM to adapt to the specific requirements of the task by performing gradient updates on a specific task with labeled code data. We randomly select a part of the data from the whole dataset as the training set to ensure the representativeness and diversity of the training data. During the fine-tuning process, by jointly optimizing the VD and explanation generation tasks, the IVDM model is able to better learn the vulnerability features and semantic representations in the code. After training, the IVDM model can achieve better performance, improve the accuracy of VD and generate high-quality vulnerability explanations.

The loss function for Task 1 is defined as:

$$L_2 = -\frac{1}{N} \sum_i^N (y_i \log(y_i') + (1 - y_i) \log(1 - y_i')), \tag{8}$$

where N represents the number of samples, $y_i \in \{0, 1\}$ represents the true label of sample i, and $y_i' \in \{0, 1\}$ represents the predicted label of sample i.

The loss function of task Task 2 is defined as:

$$L_1 = -\frac{1}{N} \sum_{i=1}^N \sum_{j=1}^m \log p(y_i^{(i)}), \tag{9}$$

where N represents the number of samples, m represents the length of the comment sequence for each sample, and $p(y_i^{(i)})$ represents the probability of generating the j-th word.

The total loss function of Task 1 and Task 2 is defined as:

$$Loss = \alpha L_1 + (1 - \alpha)L_2, \tag{10}$$

where the hyperparameter $\alpha \in [0, 1]$ is used to control the balance between task losses.

4   Evaluation and Analysis

4.1   Dataset

We constructed two datasets named **dataset I** and **dataset II**. We collected these datasets from multiple sources, including the National Vulnerability Database (NVD), CVE database, and open-source projects hosted on GitHub. These datasets cover various types of vulnerability samples, aiming to provide a comprehensive and representative vulnerability dataset. This allows for more accurate assessment and validation of IVDM and baseline performance and effectiveness. The upper part of Table 1 gives the details of **dataset I**, which contains a total of 7814 samples at the function level. The lower part of Table 1 gives the details of **dataset II**, which contains a total of 6663 samples at the function level. In addition to including our self-constructed datasets, we also use two publicly available datasets, **dataset III** [13] and **dataset IV** [23]. Through these publicly available datasets, we can compare IVDM with other research works and validate the generality and effectiveness of our method. **Dataset III** collected 32,531 non-vulnerable samples from more than 1,000 popular versions of 6 open source projects, and 457 vulnerable samples from the NVD and CVE. **Dataset IV** collected 1591 open source C/C++ programs from the NVD, of which 874 are vulnerable samples; 14000 C/C++ programs were collected from the Software Assurance Reference Dataset (SARD), of which 13906 programs are vulnerable (i.e. "bad" or "mixed"). "Bad" means the code sample has the vulnerability, and "Mixed" means the code sample contains a patched version of the vulnerability but is also available.

Evaluation Metrics: For VD task, we adopted standard evaluation metrics for detection tasks to evaluate the performance of IVDM. These metrics are Recall, Precision, F1 and Accuracy. For explanation generation task, we use BLEU-4 to evaluate the semantic generation quality of IVDM and other baseline models.

Table 1. Details of dataset I and dataset II

Datasets	Language	Number	Total
Dataset I	Java	4007	
	Python	583	
	C/C++	864	
	Go	2000	7814
	PHP	157	
	Jsp	98	
	HTML	105	
Dataset II	C	6663	6663

4.2   Experimental Results

A. Comparison of Detection Performance between IVDM and Baselines

We conduct experiments on IVDM and baselines using **dataset I**, which includes three traditional machine learning methods and six deep learning methods. We compared their performance on precision, recall, accuracy, and F1, and the experimental results as shown in Table 2. Compared with machine learning models, IVDM achieved an average improvement of 21.3% in accuracy and 45.9% in recall. This significant improvement indicates that IVDM has better performance and effect in VD task than traditional machine learning models. Several methods based on deep learning have improved in various aspects of performance. For example, the BiLSTM [28] and GNN-ReGVD [29] methods achieved accuracies of 95.8% and 96.5%, respectively. However, IVDM outperformed these methods in four detection metrics. IVDM has a recall of 98.8%, indicating that IVDM can discover and capture as many vulnerability samples as possible, reducing the false negative rate. This is crucial for identifying vulnerabilities in systems and taking timely remedial measures. Additionally, **dataset I** is composed of multiple programming languages, including Java, Python, C/C++, Go, PHP, Jsp, and HTML. We found that none of the three traditional machine learning models performed satisfactorily on this multilingual dataset. This result may be due to significant differences in syntactic and semantic features between different programming languages. Traditional machine learning models struggle to fully capture and understand the differences between these languages. In contrast, IVDM directly analyzes the source code, focusing on the characteristics of the code rather than its variations. The cross-programming-language detection ability of IVDM makes it better adapt to the multilingual environment.

Table 2. Performance comparison of the IVDM model with baselines on dataset I.

Model	Precision	Accuracy	Recall	F1
SVM [26]	80.6	77.9	51.4	62.7
NB [24]	61.0	69.6	40.7	48.8
RF [25]	92.6	85.9	66.4	77.3
TextCNN [27]	76.6	84.5	81.6	79.0
BiLSTM [28]	95.8	95.8	92.5	94.1
VulDeeLocator [12]	88.5	74.6	95.7	79.0
VulBERTa(CNN) [22]	94.2	94.8	90.2	97.7
VulBERTa(MLP) [22]	44.2	57.7	48.3	43.2
GNN-ReGVD [29]	97.7	96.2	92.01	94.8
IVDM(ours)	**99.6**	**99.0**	**98.8**	**99.4**

We then conduct experiments comparing IVDM with rule-based analysis tools on **dataset II**. We choose two popular tools, FlawFinder and RATS [30], and the experimental results are presented in Fig. 2. According to the comparison results, the accuracy of FlawFinder is 79.3%, and the accuracy of RATS is 72.2%. The IVDM demonstrates absolute superiority over both tools in terms of precision and accuracy. This result is understandable since traditional rule-based analysis tools rely on expert-defined rules or patterns for VD. However, there are various types of vulnerabilities in practical applications, and the characteristics of each vulnerability are different. Since it is difficult for experts to define all patterns of all vulnerabilities, the detection performance of these rule-based tools is limited, resulting in relatively low precision. It is worth noting that **dataset II** only contains the C programming language, and the two rule-based tools, FlawFinder and RATS, only achieve recall of 86% and 87%. This further validates the challenges traditional tools face when detecting vulnerabilities. IVDM not only has excellent ability in the environment where multiple programming languages can be detected, but also shows excellent performance in the case of a single programming language.

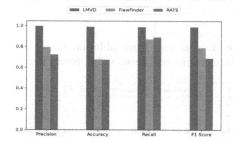

Fig. 2. Comparison of performance with two rule-based detection tools.

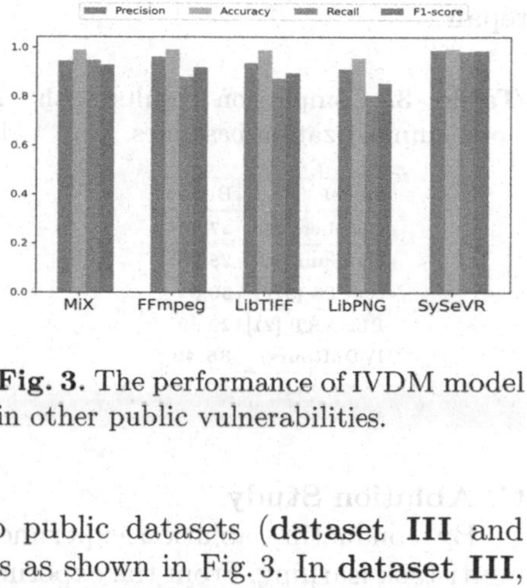

Fig. 3. The performance of IVDM model in other public vulnerabilities.

IVDM conducted experiments on two public datasets (**dataset III** and **dataset IV**), and the experimental results as shown in Fig. 3. In **dataset III**, we selected the data of three projects (FFmpeg, LibTIFF, and LibPNG) in the original paper for experiments. We also combined the data from these three projects into a single dataset called Mix. In addition, **dataset IV** contains 126 types of vulnerabilities, each type is uniquely identified by a common vulnerability enumeration identifier. The experimental results show that all indicators of IVDM on the **dataset IV** have reached more than 95%. It means that IVDM can better capture the characteristics and patterns of vulnerabilities, and effectively detect potential vulnerabilities. On the unbalanced dataset of **dataset III**, IVDM can also achieve 96% accuracy. This shows that IVDM can effectively deal with imbalanced datasets and correctly identify vulnerability samples. The reason for this phenomenon is that IVDM adopts a multi-task learning method in the training process, so that it can make full use of the information in the data set and has a strong generalization ability.

B. Interpretability Comparison between IVDM and Baselines

We compare the generative performance of IVDM with several baselines on **dataset I**, and the experimental results as shown in Table 3. The BLEU metric is a commonly used method for automatic machine translation evaluation and has also been applied to other natural language processing tasks. In the experimental results, the IVDM performed best on the BLEU-4 score. This means that compared with the baselines, IVDM can generate explanations that are closer to the standard answer words and generate natural language explanations related to real vulnerabilities. IVDM improved BLEU-4 scores by an average of 6.58% compared to baselines that simply performed summarization. This improvement shows that IVDM can accurately capture the semantic characteristics of vulnerabilities. IVDM can better learn and understand vulnerability-related features and contextual information by fine-tuning VD and explanation generation tasks. Therefore, IVDM can accurately describe and explain code vulnerabilities in natural language. Compared with other studies on VD interpretability, natural language description provides more specific and easier-to-understand vulnerability information, which can help improve the efficiency and accuracy of vulnerability repair.

Table 3. Comparison results with code summarization baselines.

Model	Bleu-4
Codebert [20]	27.53
GypSum [32]	28.26
codeT5 [31]	30.64
PLBART [21]	29.39
IVDM(ours)	**35.49**

Table 4. Analysis from ablation study on IVDM, experimental results of different tasks.

Model	Precision	Accuracy	Recall	F1	Bleu-4
Task 1	97.2	97.3	95.9	96.5	–
Task 2	–	–	–	–	30.28
IVDM(ours)	**99.6**	**99.0**	**98.8**	**99.4**	**35.49**

C. Ablation Study

By conducting ablation experiments, we demonstrated the effectiveness of multi-task learning. From the experimental results in Table 4, it can be observed that IVDM performs the best when simultaneously performing both VD and vulnerability semantics generation tasks, outperforming the single-task models. However, when conducting the VD or vulnerability semantics generation task separately, there is a certain degree of performance decline. In summary, multi-task learning is crucial for IVDM to improve the performance of vulnerability detection and vulnerability semantic generation.

D. Case Study

We evaluate the accuracy and clarity of the generated vulnerability description through specific examples. We provide three examples of vulnerability semantic descriptions generated by IVDM and the baseline methods, along with the ground truth. As shown in Fig. 4 that the information generated by Code-BERT has an excellent grammatical structure, but it does not describe the vulnerability. The code contains a vulnerability in the password verification section, but CodeBERT generates the information "print sensitive information." Code-BERT does not focus on vulnerabilities in the code, but instead generates a functional summary of the current code. In Fig. 5, due to the short length of the code, there is not enough information to generate meaningful and useful descriptions. This leads to the generated natural language descriptions by Code-BERT being meaningless and unreadable. Although the description generated by CodeT5 complies with language rules, it fails to capture the cause of the vulnerability accurately. While IVDM does not generate a perfect explanation either, it still successfully captures the key point of the vulnerability "NullPointerException." Compared to CodeBERT and CodeT5, the IVDM model can express vulnerability information more accurately (highlighted in red in the generated description) and provide intuitive and clear descriptions.

```
public boolean VerifyAdmin(String password) {
if (password.equals("68af404b513073584c4b6f22b6c63e6b")) {
System.out.println("Entering Diagnostic Mode...");
return true,
}
System.out.println("Incorrect Password!");
return false;
```
Ground truth:It is likely that an attacker will be able to read the key and compromise the system.
CodeBERT:This function prints sensitive information, which may cause serious security issues.
CodeT5 : It is probable that an attacker will be able to fetch the passwords and compromise the system.
IMVD:It is likely that an attacker will be able to read the key and disrupt the system.

```
private User user;
public void someMethod() {
...
String username = user.getName();
}
```
Ground truth:An uninitialized field in a Java class is used in a seldomcalled method, which would cause a NullPointerException to be thrown.
CodeBERT:to define user variables. View variable Calls
CodeT5:The user did not define the variable correctly here, which can result in unexpected values.
IMVD:It does not properly initialize variables, which can cause null pointer exceptions.

Fig. 4. A case study of a code snippet using a hard-coded cryptographic key.

Fig. 5. A case study of a code snippet on NullPointerException.

5   Conclusion

In this study, we propose IVDM, a framework that combines VD and explanation generation tasks. The explanations generated by IVDM exhibit clear semantic structure and contextual coherence, enabling accurate portrayal of the nature of vulnerabilities and their potential impacts in the code. IVDM provides a new perspective on interpretability in VD. Additionally, IVDM leverages large-scale pre-training models to learn feature representations of code samples, enabling efficient VD and intuitive explanation of the results. Extensive experiments conducted on multiple datasets demonstrate the effectiveness of IVDM in addressing syntax and semantic differences between different programming languages. Experimental results demonstrate that IVDM produces a better explanation ability. Through multi-task learning, IVDM can generate explanations better than summary generation models alone. The simultaneous mutual promotion of VD and generate explanations tasks further improves the performance of the model. At the same time, IVDM can not only detect vulnerabilities efficiently, but also can cope with the challenges of multiple programming languages, providing strong support for cross-programming-language VD.

Acknowledgements. This work was supported in part by the Shandong Provincial Natural Science Foundation of China under Grant ZR2022MF295 and Grant ZR2022MF257, in part by the Pilot Project for Integrated Innovation of Science, Education and Industry of Qilu University of Technology (Shandong Academy of Sciences) under Grant 2022JBZ01-01, in part by the Fundamental Research Promotion Plan of Qilu University of Technology (Shandong Academy of Sciences) under Grant 2021JC02020, and in part by the Joint Open Project of Shandong Computer Society and Provincial Key Laboratory under Grant SKLCN-2021-03.

References

1. Cao, S., Sun, X., Bo, L., Wei, Y., Li, B.: Bgnn4vd: constructing bidirectional graph neural-network for vulnerability detection. Inf. Softw. Technol. **136**, 106576 (2021)
2. Wartschinski, L., Noller, Y., Vogel, T., Kehrer, T., Grunske, L.: VUDENC: vulnerability detection with deep learning on a natural codebase for python. Inf. Softw. Technol. **144**, 106809 (2022)
3. Hin, D., Kan, A., Chen, H., Babar, M.A.: LineVD: statement-level vulnerability detection using graph neural networks. In: Proceedings of the 19th International Conference on Mining Software Repositories, pp. 596–607 (2022)
4. Napier, K., Bhowmik, T., Wang, S.: An empirical study of text-based machine learning models for vulnerability detection. Empir. Softw. Eng. **28**(2), 38 (2023)
5. Sun, H., et al.: VDSimilar: vulnerability detection based on code similarity of vulnerabilities and patches. Comput. Secur. **110**, 102417 (2021)
6. Wu, Y., Zou, D., Dou, S., Yang, W., Xu, D., Jin, H.: VulCNN: an image-inspired scalable vulnerability detection system. In: Proceedings of the 44th International Conference on Software Engineering, pp. 2365–2376 (2022)
7. Yan, G., Chen, S., Bail, Y., Li, X.: Can deep learning models learn the vulnerable patterns for vulnerability detection? In: 2022 IEEE 46th Annual Computers, Software, and Applications Conference (COMPSAC), pp. 904–913. IEEE (2022)
8. Guo, D., Lu, S., Duan, N., Wang, Y., Zhou, M., Yin, J.: UniXcoder: unified cross-modal pre-training for code representation. In: Muresan, S., Nakov, P., Villavicencio, A. (eds.) Proceedings of the 60th Annual Meeting of the Association for Computational Linguistics (Volume 1: Long Papers), ACL 2022, Dublin, Ireland, 22–27 May 2022, pp. 7212–7225. Association for Computational Linguistics (2022). https://doi.org/10.18653/v1/2022.acl-long.499
9. Kronjee, J., Hommersom, A., Vranken, H.: Discovering software vulnerabilities using data-flow analysis and machine learning. In: Proceedings of the 13th International Conference on Availability, Reliability and Security, pp. 1–10 (2018)
10. Ren, J., Zheng, Z., Liu, Q., Wei, Z., Yan, H.: A buffer overflow prediction approach based on software metrics and machine learning. Secur. Commun. Netw. **2019** (2019)
11. Dam, H.K., Tran, T., Pham, T., Ng, S.W., Grundy, J., Ghose, A.: Automatic feature learning for predicting vulnerable software components. IEEE Trans. Software Eng. **47**(1), 67–85 (2018)
12. Li, Z., et al.: VulDeePecker: a deep learning-based system for vulnerability detection. In: 25th Annual Network and Distributed System Security Symposium, NDSS 2018, San Diego, California, USA, 18–21 February 2018, The Internet Society (2018)

13. Li, Z., Zou, D., Xu, S., Jin, H., Zhu, Y., Chen, Z.: SySeVR: a framework for using deep learning to detect software vulnerabilities. IEEE Trans. Dependable Secure Comput. **19**(4), 2244–2258 (2021)
14. Zhou, Y., Liu, S., Siow, J., Du, X., Liu, Y.: Devign: effective vulnerability identification by learning comprehensive program semantics via graph neural networks. In: Advances in Neural Information Processing Systems, vol. 32 (2019)
15. Cheng, X., Wang, H., Hua, J., Xu, G., Sui, Y.: DeepWukong: statically detecting software vulnerabilities using deep graph neural network. ACM Trans. Softw. Eng. Methodol. (TOSEM) **30**(3), 1–33 (2021)
16. Luo, D.: Parameterized explainer for graph neural network. Adv. Neural. Inf. Process. Syst. **33**, 19620–19631 (2020)
17. Ying, Z., Bourgeois, D., You, J., Zitnik, M., Leskovec, J.: GNNExplainer: generating explanations for graph neural networks. In: Advances in Neural Information Processing Systems, vol. 32 (2019)
18. Zou, D., Zhu, Y., Xu, S., Li, Z., Jin, H., Ye, H.: Interpreting deep learning-based vulnerability detector predictions based on heuristic searching. ACM Trans. Softw. Eng. Methodol. (TOSEM) **30**(2), 1–31 (2021)
19. Li, Y., Wang, S., Nguyen, T.N.: Vulnerability detection with fine-grained interpretations. In: Proceedings of the 29th ACM Joint Meeting on European Software Engineering Conference and Symposium on the Foundations of Software Engineering, pp. 292–303 (2021)
20. Feng, Z., et al.: CodeBERT: a pre-trained model for programming and natural languages. In: Cohn, T., He, Y., Liu, Y. (eds.) Findings of the Association for Computational Linguistics: EMNLP 2020, Online Event, 16–20 November 2020. Findings of ACL, vol. EMNLP 2020, pp. 1536–1547. Association for Computational Linguistics (2020). https://doi.org/10.18653/v1/2020.findings-emnlp.139
21. Ahmad, W.U., Chakraborty, S., Ray, B., Chang, K.: Unified pre-training for program understanding and generation. In: Toutanova, K., et al. (eds.) Proceedings of the 2021 Conference of the North American Chapter of the Association for Computational Linguistics: Human Language Technologies, NAACL-HLT 2021, 6–11 June 2021, pp. 2655–2668. Association for Computational Linguistics (2021). https://doi.org/10.18653/v1/2021.naacl-main.211
22. Hanif, H., Maffeis, S.: VulBERTa: simplified source code pre-training for vulnerability detection. In: 2022 International Joint Conference on Neural Networks (IJCNN), pp. 1–8. IEEE (2022)
23. Lin, G., et al.: Cross-project transfer representation learning for vulnerable function discovery. IEEE Trans. Industr. Inf. **14**(7), 3289–3297 (2018)
24. Webb, G.I., Keogh, E., Miikkulainen, R.: Naïve bayes. Encyclopedia Mach. Learn. **15**, 713–714 (2010)
25. Rigatti, S.J.: Random forest. J. Insur. Med. **47**(1), 31–39 (2017)
26. Noble, W.S.: What is a support vector machine? Nat. Biotechnol. **24**(12), 1565–1567 (2006)
27. Chen, Y.: Convolutional neural network for sentence classification. Master's thesis, University of Waterloo (2015)
28. Van Houdt, G., Mosquera, C., Nápoles, G.: A review on the long short-term memory model. Artif. Intell. Rev. **53**, 5929–5955 (2020)
29. Nguyen, V.A., Nguyen, D.Q., Nguyen, V., Le, T., Tran, Q.H., Phung, D.: ReGVD: revisiting graph neural networks for vulnerability detection. In: Proceedings of the ACM/IEEE 44th International Conference on Software Engineering: Companion Proceedings, pp. 178–182 (2022)

30. Guo, W., Fang, Y., Huang, C., Ou, H., Lin, C., Guo, Y.: HyVulDect: a hybrid semantic vulnerability mining system based on graph neural network. Comput. Secur. **121**, 102823 (2022)
31. Wang, Y., Wang, W., Joty, S., Hoi, S.C.: Codet 5: identifier-aware unified pre-trained encoder-decoder models for code understanding and generation. arXiv preprint arXiv:2109.00859 (2021)
32. Wang, Y., Dong, Y., Lu, X., Zhou, A.: Gypsum: learning hybrid representations for code summarization. In: Proceedings of the 30th IEEE/ACM International Conference on Program Comprehension, pp. 12–23 (2022)

Co-GAN: A Text-to-Image Synthesis Model with Local and Integral Features

Lulu Liu, Ziqi Xie, Yufei Chen(✉), and Qiujun Deng

College of Electronic and Information Engineering, Tongji University,
Shanghai 201804, China
yufeichen@tongji.edu.cn

Abstract. Text-to-Image synthesis is a promising technology that generates realistic images from textual descriptions by deep learning model. However, the state-of-the-art text-to-image synthesis models often struggle to balance the overall integrity and local diversity of objects with rich details, leading to unsatisfactory generation results of some domain-specific images, such as industrial applications. To address this issue, we propose Co-GAN, a text-to-image synthesis model that introduces two modules to enhance local diversity and maintain overall structural integrity respectively. Local Feature Enhancement (LFE) module improves the local diversity of generated images, while Integral Structural Maintenance (ISM) module ensures that the integral information is preserved. Furthermore, a cascaded central loss is proposed to address the instability during the generative training. To tackle the problem of incomplete image types in existing datasets, we create a new text-to-image synthesis dataset containing seven types of industrial components, and test the effects of various existing methods based on the dataset. The results of comparative and ablation experiments show that, compared with other current methods, the images generated by Co-GAN incorporate more details and maintain the integrity.

Keywords: Text-to-Image synthesis · Adversarial generation model · Cross modality · Deep learning

1 Introduction

Text-to-image generation is a cross-modal task that combines textual processing with visual generations [1–3], and simultaneously boosts the research in both two fields [4,5]. Accordingly, the research on text-to-image generation has gained increasing attention [6,7]. It can be widely cited in visual image retrieval [8], the medical field [9], computer-aided design [10], and so on.

Complete structure is the cornerstone of an image, and rich and diverse details can improve the authenticity of the image. Image presentation is very important, we investigate the CAXA industrial resource pool [11] and find that

Supported by the National Key Research and Development Plan (2020YFB1712301).

image-displayed resources contribute to higher views and downloads, as shown in Table 1. From the 4th column in Table 1, we find that the averaged downloads with a sharp decrease come from those resources that lack visual image sensory attractions. Seriously, due to the lack, the user downloads interest has drastically decreased from 75.96 to 17.68. The absence of images is obviously detrimental to the dissemination of resources, thus, it is necessary to introduce text-to-image generation technology to compensate for the appealing of missing and rich-detailed visual images.

Table 1. Survey Of CAXA Industrial Resource Pool

Resource Type	Ratio	Average Views	Average Downloads
With image	44.4%	692.5	75.96
Without image	55.6%	628.44	17.68

However, images produced by the existing methods are usually only semantically consistent with the real in terms of the overall structure, while poor in the diversity for local details, which is caused by the textual description features in these methods are not fully exploited. The existed text-to-image generation works are normally based on constant textual information during the model training [12] while neglecting the dynamic adjustment and selection for the textual guidance, which results in a lack of diversity in local features. In terms of the overall structure, these methods rely heavily on the image features generated in the initial stage [13]. When the initial generation is not good enough and the structure is broken, the final results will also perform poorly. Hence, in our method, both the aforementioned local diversity and the overall structural integrity are considered.

To sum up, we propose Co-GAN that generates semantically consistent images with the input textual guidance using local and integral cross-modal semantic features, based on a deeper exploration of textual description guidance. And our main contributions are highlighted as follows.

- We propose a Local Feature Enhancement (LFE) module that leverages semantic information at the word level to dynamically select and supplement local diversities for the generation.
- We propose an Integral Structural Maintenance (ISM) module to obtain integral information on visual features in different network layers, to restrict the structural consistency of generated images.
- We propose a cascaded central loss to address the issue of discriminator gradient penalty failure during the adversarial generation training.
- To address the lack of relevant datasets, we create an industrial component text-image dataset collected from the Traceparts platform [14].

2   Related Work

GAN-based methods can simultaneously guarantee the quality of the generations within a tolerable computation time. Usually, GAN based structure consists of two basic components, known as generator G and discriminator D [4]. G is to generate samples as real and to deceive D, while D tries to identify the authenticity between the real and the generation. Before the relevant cross-modal generation work [15] is proposed, text representation and image generation are two independent research branches. Later, more works contribute to the cross-modal generation. StackGAN [16] employs a two-stacked structure to achieve the goal of text-to-image generation. And StackGAN++ [17] improves the two-stage into a tree-like multi-level structure. Based on the aforementioned work, AttnGAN [18] incorporates fine-grained word-level text features into the generation process. MirrorGAN [19] constrains the bi-direction semantic consistency from both text-to-image and image-to-text. These stacked generations are easily affected by the initial generator. Several works try to reduce the dependence between the final result and the initial generation through the use of additional structure [20] or the design of a loss function [21]. Until the proposal of DF-GAN [22], a new single-stage structure is founded. VLM-GAN [23], RAT-GAN [24], and SD-GAN [25] are both in this form. To better utilize conditionally textual information, SSA-GAN [26] jointly trains the text encoder with G and D.

Based on the single-stage generative backbone, our method both considers the dynamically further selection of the word-level textual information and the design of a reasonable loss function to help GAN generate local diversity detailed images more stably. Different from those existed works, which stabilize the training or explore the textual information with additional extra networks, the deeper use of textual information is simply achieved by a difference mask matrix jointly with the attention mechanism in our method, thereby, avoiding to aggregate the training complexity of the network.

3   Method

3.1   Overview

The overview of our model is presented in Fig. 1. Our model is mainly composed of a generator G, a discriminator D, and a pretrained text encoder. Additionally, local feature enhancement module (LFE) and integral structure maintenance module (ISM) are proposed and incorporated.

The pretrained text encoder encodes the text descriptions into a sentence feature s and word features w, separately. And G mainly consists of several generative blocks (GBlock) [22], which is based on the textual encoding constraints to perform up-samplings and generate images as real as possible to deceive D. And D conducts down-samplings through DownBlocks [22] to identify the possibility of an image being true.

Notably, in our method, LFE and ISM are proposed to enhance the local diversities and balance the global integrity for the generations. LFE dynamically supplements the input s for the next GBlock generation's t with more word-level information w, *i.e.*, $F_{LFE} : (s, w, t) \rightarrow s\prime$. Besides, ISM extracts the integral information (θ, ρ) from different visual features t_i, $F_{ISM} : (t_i \mid s\prime) \rightarrow (\theta, \rho)$ to collaborate with loss function to limit the overall integrity of the generations. Finally, the proposed cascaded central loss narrows the integral distance between the generated features t_g and the real features t_r over the integral information (θ, ρ).

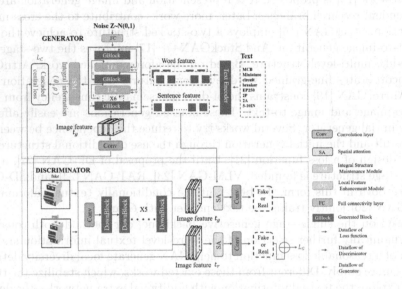

Fig. 1. Overview of Co-GAN

3.2   Local Feature Enhancement Module

The fixed and constant textual conditions cannot bring sufficient diversity changes to the generation. To enrich the diversity in local details, we introduce the LFE module, which dynamically learns and selects additional textual features used for guidance enhancement in different generation stages. The calculation of LFE can divide into two parts. As shown in Fig. 2, the first part on the left is to obtain the difference matrix M, and the second part on the right is to adopt the attention mechanism. It takes three inputs: the word-level textual feature w, the sentence-level textual feature s, and the intermediate image feature t.

The first part for LFE is to get the difference mask matrix M, which characterizes the differences across the visual and textual modality. To obtain global information t_{avg} in the visual modality, we apply the global averaged pooling $GAP(\cdot)$ on the generated visual features t. And the difference δ across two modalities is calculated by $\delta = t_{avg} - s\prime$. Then, based on δ, we can get the difference mask M by:

$$M_{i,j} = \begin{cases} 1 & \delta_{i,j} > 0 \\ 0 & \delta_{i,j} \le 0 \end{cases} \tag{1}$$

where $\delta_{i,j}$ means the ith row and the jth column in δ, and M is the binary matrix with the same shape as δ.

Subsequently, as shown on the right in Fig. 2. We spatially expand the global averaged image features t_{avg} into the word-level spatial dimension and obtain the expanded $t\_expanding$, so does the difference matrix M to be $M\_expanding$. Later, $t\_expanding$ is used as key vectors $K(t\_expanding)$. The Hadamard product result between $M\_expanding$ and word-level textual features w works as query vector $Q(M\_expanding \odot w)$. And w is served as value vector $V(w)$. Through the attention mechanism in formula (2), we acquire the additional information and finally get the enhanced $s\prime$ in formula (3).

$$\text{Attention} = \text{softmax} \left(\frac{Q_{M\_\text{expanding} \odot w} K^T_{t\_\text{expanding}}}{\sqrt{d_{K_{t\_\text{expanding}}}}} \right) \odot V_w \tag{2}$$

$$s' = s + \text{Attention} \odot w \tag{3}$$

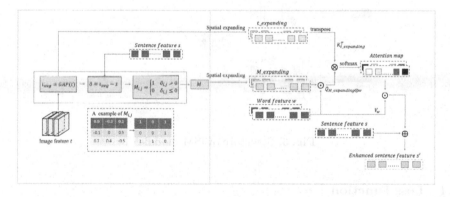

Fig. 2. Structure of LFE module

Compared with ordinary (K, Q, V) attention mechanisms, LFE calculates the matrix M ahead booting attention similarities. Different from applying attention straightly to visual features, LFE pays more attention to textual enhancement through the generated visual base in M. The textual additions selected by visual features relimits to the next stage of visual image generation in GBlock, which therefore builds a mutual cyclic constraint across these two modalities. And the generation of visual features is entirely learned by the GAN network itself within the guidance, and without any extra manual modifications.

3.3   Integral Structural Maintenance Module

For textual descriptions within the same semantics, we hope the generated results to keep consistent and invariant in integral performance, and to perceive diversities in the local. However, continuously diverse additions from the word-level fracture the integral information of the generations as the deepening of the network. Moreover, these various enhancements compared to the invariant sentence increase learning difficulties for the network and slow down the training process.

To align the integrity of visual features in different generation stages, ISM calculates the mean and deviation as the integral information from fragmentation due to the varying enhanced guidance. More details for ISM are shown in Fig. 3. ISM consists of seven Single-scale Generation Transfer Modules (SGTM), each SGTM is connected followed a GBlock [22]. The single-scale feature generated by different GBlocks is fed into SGTM to transform the visual shape into the same scale, as $255 \times 1 \times 1$. Then, the seven transformed features are combined to calculate the aggregated integral information mean and deviation (θ, ρ). In all, ISM converts intermediate image features t into a uniform dimension and combines deep and shallow visual features by averaging their integral information pooled values. Later, this integral information is collected to compute the cascaded central loss, and details will be described in Sect. 3.4.

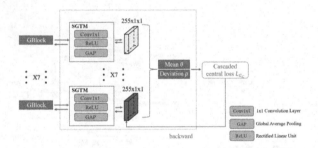

Fig. 3. Structure of ISM module

3.4   Loss Function

For stable network learning, we propose a cascaded central loss to eliminate the bias in integrity caused by local diversity enhancements. Formally, we measure the integral information (θ, ρ) distance between the generated t_g and the real image features t_r. And in Co-GAN, we jointly use the following two losses: (1) matching aware gradient penalty MA_GP [22]; (2) cascaded central loss.

Matching Aware Gradient Penalty. MA_GP is the constraint implemented on D to help the network converge, which penalizes the real images within the matching textual description.

Cascaded Central Loss. However, limited by the initial capabilities of the generator G, there is a clear difference between the visual generation and the real, which helps D to identify the authenticity of the generations a lot. Thereby the loss L_D converges quickly about zero. Additionally, according to MA_GP, G fails to receive the gradient optimization from D and goes to vanish as well. We define the occurrence of this phenomenon as discriminator gradient penalty failure. In this case, it becomes impossible to train an effective G. Thus, the cascaded central loss L_C is proposed as follows:

$$L_C = \ln\left(1 + e^{\sum_{i=1}^{d}\|\theta_r - \theta_g\|}\right) + \ln\left(1 + e^{\sum_{i=1}^{d}\|\rho_r - \rho_g\|}\right) \tag{4}$$

where θ_r with ρ_r is from integral information extracted by ISM on real images, θ_g with ρ_g is from the generation, and d denotes the dimension of visual features.

L_C is directly applied to G and helps to converge over integral information between the generated and the real. Overall, in Co-GAN we simultaneously use MA_GP and L_C to form independent regularizations respectively on D and G. The discriminator loss L_D used in Co-GAN is the same one in MA_GP, and the generator loss L_{G_C} is obtained as a combination from L_G in MA_GP with a weighted L_C, where α is the coefficient weight for L_C.

$$\begin{aligned}
L_D = &- E\left[\min\left(0, -1 + D\left(t_r \mid s\right)\right)\right] \\
&- (1/2)E[\min(0, -1 - D(G(z \mid s), s))] \\
&- (1/2)E\left[\min\left(0, -1 - D\left(t_r \mid s\right)\right)\right] \\
&+ k\left[\left(\|\nabla_{t_r}D(t_r \mid s)\| + \|\nabla_e D\left(t_r \mid s\right)\|\right)^p\right]
\end{aligned} \tag{5}$$

$$L_G = -E[D(G(z \mid s) \mid s)] \tag{6}$$

$$L_{G_C} = L_G + \alpha L_C \tag{7}$$

4 Experiments and Discussion

4.1 Set up

Datasets. Most of the existed text-to-image generation work has been studied on COCO [27] or OXFORD-102 [28] dataset. However, these datasets mainly consist of natural images that have no strict requirements for the rigor of the integral structure, even blurry natural image structure can be tolerated. However, in reality, there are also images with extremely high requirements for boundary integrity and clarity, such as industrial workpieces, for those fuzzy boundaries will greatly limit the functional representation. Therefore, we create a dataset collected from the industrial software platform TraceParts with text-image pairs of real data. According to the shape of industrial components, we divide the dataset into the following categories: Gear-shaped, Handle-shaped, Knob-shaped, Rectangular, Hat-shaped, Wheel-shaped, and Irregular-shaped. An example of our dataset is shown in Fig. 4.

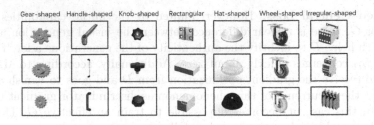

Fig. 4. A simple example for our dataset

Settings. Our method experiments on Pytorch 1.10.1 with a single RTX3060 GPU and CUDA 11.3. The Generator (G) and Discriminator (D) are alternately and iteratively updated with Adam's optimization algorithm. The learning rate for G is set to 0.0001 and the rate for D is set to 0.0004. Limited by the computational power of our GPU, the batch size is set as 24.

4.2 Experiments and Discussion

Ablation of LFE Module. As shown in Table 2, the averaged IS score [29] and FID [30] values of the generations are both improved with the addition of LFE, And we also conduct enlarge visualizations of local details with LFE as shown in Fig. 5, from up to down each row is followed by the enlarged view of its local details.

Table 2. Ablation of LFE module

Method	DF-GAN [22]	DF-GAN [22] + LFE module
IS	3.63	3.65
FID	10.14	9.22

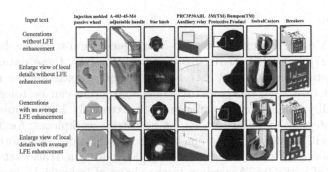

Fig. 5. Ablation results of LFE module

The IS score measures the quality and diversity of the generations and the higher the IS score the richer the diversity of generations. similarly, the FID value is used to evaluate the similarity between the generations and the real visual features, the lower the FID value means the better the quality of the generated images.

Ablation of ISM Module. As shown in Table 3, we compare the quantitative results for respectively applying ISM to DF-GAN [22] and ours. Without ISM module, the discriminator gradient penalty failure occurs in our training process and we cannot obtain a convergently experimental result. The visualizations of ISM are shown in Fig. 6, to observe the boundary integrity of generations more intuitively, we extract corresponding boundaries below each generation in the 2nd and the 4th row.

Table 3. Ablation of ISM module

Method	DF-GAN [22]	DF-GAN [22] + ISM module	LFE + ISM module
IS	3.63	3.63	3.67
FID	10.14	10.12	8.74

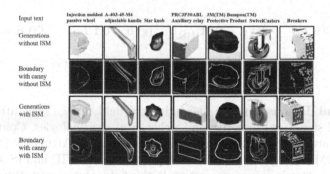

Fig. 6. Ablation results of ISM module (row 1 to row 4 from up to down)

Ablation of Hyperparameter α. The loss trend for Generator and Discriminator with different hyperparameters α are shown from Fig. 7(a) to 7(h). Specifically, as defined in MA_GP, L_G (which means $\alpha = 0$ in L_{G_C}) accepts inverse gradient optimization from L_D to improve its generation ability. As shown in Fig. 7(a) and 7(e), due to the weak generation ability of G in th initial training stage, the distribution of generated images differs significantly from that of real images. D can easily perform discriminative tasks, with L_D close to zero, and cannot provide effective generation optimization guidance to G through gradient backpropagation, resulting in drastic changes in L_G and turbulence in model training. However, with the balance of α between L_G and L_C, G can still obtain effective generation guidance, and narrow the gap between the initial training generations and the real through the integral information, even if the discriminator gradient penalty failure occurs. Accordingly, due to the reduction of the difference between the real and the generations, the difficulties of discriminative tasks for D has correspondingly increased. Based on the difficulties, the gradient of L_D can continuously update and L_D can converge more smoothly, avoiding the illusion of rapid convergence of L_D due to strong discrimination ability in the initial training stage.

After comparisons, for the stability of the model training, we set $\alpha = 0.1$ as the coefficient weight for L_C in L_{G_C}.

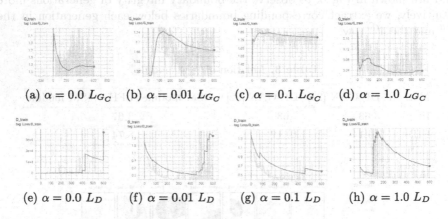

(a) $\alpha = 0.0$ L_{G_C} (b) $\alpha = 0.01$ L_{G_C} (c) $\alpha = 0.1$ L_{G_C} (d) $\alpha = 1.0$ L_{G_C}

(e) $\alpha = 0.0$ L_D (f) $\alpha = 0.01$ L_D (g) $\alpha = 0.1$ L_D (h) $\alpha = 1.0$ L_D

Fig. 7. Ablation loss of Hyperparameter α

Results and Discussion. In the industrial field, image generation requires more overall structural integrity and richer local morphologies. Compared with state-of-the-art methods including StackGAN++ [17], AttnGAN [18], Mirror-GAN [19], DM-GAN [20], and DF-GAN [22] on metrics FID and IS, our method can perform better in image generation in this field, the quantitatives are shown in Table 4, and the visualizations are shown in Fig. 8. With the help of LFE, Co-GAN can fully utilize the supplement of word-level textual description features and introduce more diversity enhancements in the dynamic generation process for local details, which increases the authenticity of the generations and improves the IS score. Meanwhile, the combination of ISM and L_{G_C} narrows the integral information gap between the generated visual features and the real, which improves the similarity between the generations and the real images and results in the decrease in FID value. However, almost all experimental methods are not accurate enough in generating complex gear-shaped images while the generations on simple rectangle-shaped images all perform realistic.

Table 4. Quantitative comparisons of different methods

Method	Metric	Gear-shaped	Handle-shaped	Knob-shaped	Rectangular	Hat-shaped	Wheel-shaped	Irregular-shaped	Average
StackGAN++ [17]	FID	20.47	18.25	18.10	8.28	17.40	20.76	20.71	17.71
	IS	3.19	3.46	3.37	3.62	3.35	3.24	3.50	3.39
AttnGAN [18]	FID	20.97	19.74	18.29	8.74	19.77	20.20	20.88	18.37
	IS	3.24	3.43	3.39	3.67	3.41	3.25	3.48	3.41
MirrorGAN [19]	FID	16.61	11.40	12.13	8.02	11.98	12.16	13.10	12.20
	IS	3.25	3.52	3.40	3.77	3.42	3.49	3.44	3.47
DM-GAN [20]	FID	15.90	10.12	11.14	7.96	11.69	10.04	11.06	11.13
	IS	3.37	3.48	3.52	3.77	3.43	3.51	3.42	3.50
DF-GAN [22]	FID	15.79	8.69	10.14	7.89	9.05	8.93	10.35	10.12
	IS	3.48	3.66	3.64	3.80	3.61	**3.63**	3.59	3.63
Ours	FID	**11.23**	**8.07**	**7.81**	**7.33**	**8.43**	**8.81**	**7.61**	**8.74**
	IS	**3.48**	**3.69**	**3.66**	**3.82**	**3.64**	3.61	**3.79**	**3.67**

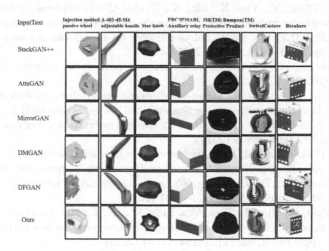

Fig. 8. Generated results of different methods

5 Conclusion

To solve the problem that local details are less diverse and integral information is fractured in cross-modal text-to-image generations, we proposed Co-GAN with two novel modules and a loss. LFE module introduces more fine-grained word-level textual features into textual guidance to enhance the diversity of local details in a simple way. And ISM module helps preserve integral information of visual generation features. At last, cascaded central loss stabilizes the training process. According to the above experiments, the proposed method can stably generate images that are semantically consistent with the input textual descriptions, and keep the overall structural integrity and local details diverse, which provides a guarantee for the automatic cross-modal generation and completion task for the visual missing resources. Based on the existing experiments, future research will try to explore the diverse local details in more complex and trivial image features.

Acknowledgements. This work was supported by the National Key Research and Development Plan (2020YFB1712301).

References

1. Van Den Oord, A., Kalchbrenner, N., Kavukcuoglu, K.: Pixel recurrent neural networks. In: International Conference on Machine Learning, pp. 1747–1756. PMLR (2016)
2. Agnese, J., Herrera, J., Tao, H., Zhu, X.: A survey and taxonomy of adversarial neural networks for text-to-image synthesis. Wiley Interdisc. Rev. Data Min. Knowl. Discov. **10**(4), e1345 (2020)
3. Yang, X., Chen, Y., Yue, X., Lin, X., Zhang, Q.: Variational synthesis network for generating micro computed tomography from cone beam computed tomography. In: 2021 IEEE International Conference on Bioinformatics and Biomedicine (BIBM), pp. 1611–1614. IEEE (2021)
4. Goodfellow, I., et al.: Generative adversarial networks. Commun. ACM **63**(11), 139–144 (2020)
5. Ho, J., Jain, A., Abbeel, P.: Denoising diffusion probabilistic models. In: Advances in Neural Information Processing Systems, vol. 33, pp. 6840–6851 (2020)
6. Dinh, L., Sohl-Dickstein, J., Bengio, S.: Density estimation using real NVP. arXiv preprint arXiv:1605.08803 (2016)
7. Kingma, D.P., Dhariwal, P.: Glow: generative flow with invertible 1×1 convolutions. In: Advances in Neural Information Processing Systems, vol. 31 (2018)
8. He, P., Wang, M., Tu, D., Wang, Z.: Dual discriminative adversarial cross-modal retrieval. Appl. Intell. **53**(4), 4257–4267 (2023)
9. Xu, L., Zhou, S., Guo, J., Tian, W., Tang, W., Yi, Z.: Metal artifact reduction for oral and maxillofacial computed tomography images by a generative adversarial network. Appl. Intell. **52**(11), 13184–13194 (2022)
10. Du, W., Xia, Z., Han, L., Gao, B.: 3D solid model generation method based on a generative adversarial network. Appl. Intell. 1–26 (2022)
11. CAXA-gongyeyun. http://www.gongyeyun.com/SoftService/ResourceDownList
12. Singh, V., Tiwary, U.S.: Visual content generation from textual description using improved adversarial network. Multimed. Tools Appl. **82**(7), 10943–10960 (2023)
13. Mao, F., Ma, B., Chang, H., Shan, S., Chen, X.: Learning efficient text-to-image synthesis via interstage cross-sample similarity distillation. Sci. China Inf. Sci. **64**, 1–12 (2021)
14. TraceParts. https://www.traceparts.cn/zh
15. Reed, S., Akata, Z., Yan, X., Logeswaran, L., Schiele, B., Lee, H.: Generative adversarial text to image synthesis. In: International Conference on Machine Learning, pp. 1060–1069. PMLR (2016)
16. Zhang, H., et al.: StackGAN: text to photo-realistic image synthesis with stacked generative adversarial networks. In: Proceedings of the IEEE International Conference on Computer Vision, pp. 5907–5915 (2017)
17. Zhang, H., et al.: Stackgan++: realistic image synthesis with stacked generative adversarial networks. IEEE Trans. Pattern Anal. Mach. Intell. **41**(8), 1947–1962 (2018)
18. Xu, T., et al.: AttnGAN: fine-grained text to image generation with attentional generative adversarial networks. In: Proceedings of the IEEE Conference on Computer Vision and Pattern Recognition, pp. 1316–1324 (2018)

19. Qiao, T., Zhang, J., Xu, D., Tao, D.: MirrorGAN: learning text-to-image genera-
    tion by redescription. In: Proceedings of the IEEE/CVF Conference on Computer
    Vision and Pattern Recognition, pp. 1505–1514 (2019)
20. Zhu, M., Pan, P., Chen, W., Yang, Y.: DM-GAN: dynamic memory generative
    adversarial networks for text-to-image synthesis. In: Proceedings of the IEEE/CVF
    Conference on Computer Vision and Pattern Recognition, pp. 5802–5810 (2019)
21. Liao, K., Lin, C., Zhao, Y., Gabbouj, M.: DR-GAN: automatic radial distortion
    rectification using conditional GAN in real-time. IEEE Trans. Circuits Syst. Video
    Technol. **30**(3), 725–733 (2019)
22. Tao, M., Tang, H., Wu, F., Jing, X.Y., Bao, B.K., Xu, C.: DF-GAN: a simple and
    effective baseline for text-to-image synthesis. In: Proceedings of the IEEE/CVF
    Conference on Computer Vision and Pattern Recognition, pp. 16515–16525 (2022)
23. Cheng, Q., Wen, K., Gu, X.: Vision-language matching for text-to-image synthesis
    via generative adversarial networks. IEEE Trans. Multimed. (2022)
24. Ye, S., Wang, H., Tan, M., Liu, F.: Recurrent affine transformation for text-to-
    image synthesis. IEEE Trans. Multimed. (2023)
25. Ma, J., Zhang, L., Zhang, J.: SD-GAN: saliency-discriminated GAN for remote
    sensing image super resolution. IEEE Geosci. Remote Sens. Lett. **17**(11), 1973–
    1977 (2019)
26. Liao, W., Hu, K., Yang, M.Y., Rosenhahn, B.: Text to image generation with
    semantic-spatial aware GAN. In: Proceedings of the IEEE/CVF Conference on
    Computer Vision and Pattern Recognition, pp. 18187–18196 (2022)
27. Lin, T.-Y., et al.: Microsoft COCO: common objects in context. In: Fleet, D.,
    Pajdla, T., Schiele, B., Tuytelaars, T. (eds.) ECCV 2014. LNCS, vol. 8693, pp.
    740–755. Springer, Cham (2014). https://doi.org/10.1007/978-3-319-10602-1_48
28. Nilsback, M.E., Zisserman, A.: Automated flower classification over a large number
    of classes. In: 2008 Sixth Indian Conference on Computer Vision, Graphics & Image
    Processing, pp. 722–729. IEEE (2008)
29. Heusel, M., Ramsauer, H., Unterthiner, T., Nessler, B., Hochreiter, S.: GANs
    trained by a two time-scale update rule converge to a local NASH equilibrium.
    In: Advances in Neural Information Processing Systems, vol. 30 (2017)
30. Salimans, T., Goodfellow, I., Zaremba, W., Cheung, V., Radford, A., Chen, X.:
    Improved techniques for training GANs. In: Advances in Neural Information Pro-
    cessing Systems, vol. 29 (2016)

Graph Contrastive ATtention Network for Rumor Detection

Shaohua Li[ID], Weimin Li[✉][ID], Alex Munyole Luvembe[ID], and Weiqin Tong[ID]

School of Computer Engineering and Science, Shanghai University,
Shanghai 200444, China
{flowingfog,wmli,luvembe,wqtong}@shu.edu.cn

Abstract. Detecting rumors from the vast amount of information in online social media has become a formidable challenge. Rumor detection based on rumor propagation trees benefits from crowd wisdom and has become an important research method for rumor detection. However, node representations in such methods rely on limited label information and lose a lot of node information when obtaining graph-level representations through pooling. This paper proposes a novel rumor detection model called Graph Contrastive ATtention Network (GCAT). GCAT adopts a graph attention model as the encoder, applies graph self-supervised learning without negative label pairs as an auxiliary task to update network parameters, and combines multiple pooling techniques to obtain the graph-level representation of the rumor propagation tree. To verify the effectiveness of our model, we conduct experiments on two real-world datasets. The GCAT model outperforms the optimal baseline algorithms on both datasets, proving the effectiveness of the proposed model.

Keywords: Rumor detection · Rumor propagation tree · Graph contrastive learning

1 Introduction

The spread of rumors in social media has become a significant problem that can lead to real-world consequences, and it has attracted widespread attention from researchers [9,10,14]. Rumor detection is, therefore, a crucial task in today's information age. Rumor detection is a research field that automatically identifies false information or rumors in online social media platforms such as Twitter, Facebook, and Weibo. The area has gained significant attention in recent years due to the increasing prevalence of false information in social media and its impact on individuals, organizations, and society.

The dissemination of information in social networks has received much attention from researchers [11,12,23]. The spread of true and false rumors on social media elicited different responses, forming a rumor propagation structure [15] that can be used for automatic rumor identification. Detection methods based on user interactions show stronger robustness in the face of deliberately fabricated content. With the development of graph neural networks and their excellent performance in structured data processing, graph neural networks [8] have achieved widespread applications in a number of fields including rumor detection [7,9,19,22]. Some researchers applied graph neural network methods to aggregate user feedback information, and achieved good results in the task of false news detection [9]. BiGCN [1] applies top-down and bottom-up graph convolutional networks to obtain sequence features of rumor depth propagation and structural features of breadth dispersion.

Faced with the high cost of annotation, some methods [9,13,18] attempt to reduce reliance on label information through contrastive learning. GACL [18] perceives the differences between different classes of conversation threads through contrastive learning while employing an adversarial feature transformation module to generate conflicting samples and extract event-invariant features. Lin et al. [13] introduced an adversarial contrastive learning framework designed for low-resource scenarios. However, in supervised classification tasks, contrastive learning methods based on positive and negative pairs struggle to effectively handle cases where different samples belong to the same class. Different samples of the same class are treated as negative pairs, pushing each other far apart in the embedding space.

This paper proposes a novel Graph Contrastive ATtention Network (GCAT) for rumor identification. GCAT benefits from the crowd and takes the embedding of user history information and source claim as the initial features of the rumor propagation tree, adopts the graph attention model as the aggregated neighborhood information, and applies graph self-supervised learning without negative label pairs as auxiliary tasks to update network parameters. To obtain a graph-level representation as the final embedding of the statement, we integrate multiple pooling methods to give downstream classifiers a more vital ability to distinguish different rumor propagation graphs. Therefore, the main contributions of this paper can be summarized as follows:

- We adopt graph self-supervised learning without negative label pairs as an auxiliary task to model the structure of rumor propagation, reducing the dependence on limited label information;
- When obtaining graph-level representations, we integrate multiple pooling methods so that the downstream classifier has a stronger ability to distinguish different rumor propagation trees;
- We conduct experiments on two real-world datasets, and the proposed model leads both in terms of Accuracy and F1 value;
- The proposed model also achieves good performance on the early detection task with less user participation;

2   Preliminary Knowledge

We define rumors as a set of statements $C = \{c_1, c_2, \cdots, c_n\}$, where each statement c_i contains the source post s_i and the response to it, i.e., $c_i =$

$\{s_i, r_{i1}, r_{i2}, \cdots, r_{i,m}, G_i\}$, and each r_{i*} is the response to the source post s_i, and G_i refers to the structure of response cascade related to C_i. In the propagation structure $G_i = (V_i, E_i)$, the root node stands for the rumor source s_i, while the other nodes correspond to responses from users either directly responding to the root node or participating in the discussion. Edges in E_i are constructed based on the reply relationships within the discussion.

Using $X_i = [x_1^i, \cdots, x_{n_i-1}^i]$ to denote features of nodes in claim c_i, which includes embeddings for both the source rumor and participating users responds. The goal of rumor identification can be represented as a supervised classification problem, learning a classifier f from declarations with user participants:

$$f : (X_i, G_i) \rightarrow Y_i \qquad (1)$$

where Y_i stands for the ground-truth label of C_i, which usually takes one of two categories: false rumor or true rumor.

3   Graph Attentive Self-supervised Learning Model

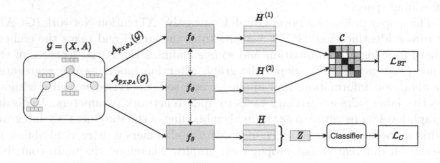

Fig. 1. Architecture of GCAT. GCAT includes two components: a graph contrastive learning network and a rumor identification network, and the graph encoders share parameters.

As shown in Fig. 1, we propose an end-to-end rumor identification model that combines the attention mechanism and graph self-supervised learning, called the Graph Contrastive ATtention Network (GCAT). GCAT comprises a rumor recognition network and a graph self-supervised learning network, which share the graph encoder.

3.1   Rumor Identification Network

In the rumor identification network, we take the embeddings of declared content and user feedback as initial features and adopt a graph neural network (GNN) with attention mechanism to aggregate features from neighbor nodes in the rumor propagation tree, and then the graph-level representations obtained by fusion of multiple pooling methods on node representations. Finally, these representations are sent to an MLP classifier to judge the authenticity of rumors.

Node Representation Update. Combining user feedback and rumor propagation structure can significantly improve the performance of rumor detection [6]. In GCAT, we take embeddings of the original claim and user feedback as initial features for nodes and update node representations through a weighted aggregation process over the rumor propagation tree.

The attention mechanism learns the weights for aggregating features from neighboring nodes [3,20]. We denote the hidden features as $h = h_1, h_2, \cdots, h_N$, $h_i \in R_d$, where d is the dimension of features for each node. The attention layer produces a new set of node features $h' = \{h'_1, h'_2, \cdots, h'_N\}, h'_i \in R'_d$ as its output. The attention weight a_{ij} is calculated according to the following formula:

$$\alpha_{ij} = \frac{exp(\mathbf{a}^T LeakyReLU(\mathbf{W}[h_i||h_j]))}{\sum_{k \in \mathcal{N}_i} exp(\mathbf{a}^T LeakyReLU(\mathbf{W}[h_i||h_k]))} \tag{2}$$

where N_i is the neighbor of node i, $W \in R^{d' \times d}$ is a trainable matrix for transforming node attributes, $a \in \mathbb{R}^{2d'}$ is the weight vector used to obtain the attention coefficient, and $||$ represents the splicing operation.

We replace the adjacency matrix with the attention matrix and compute a linear combination of the neighbor node features to update the representation of a node:

$$h'_i = \sigma \left(\sum_{j \in \mathcal{N}_i} \alpha_{ij} W h_j \right) \tag{3}$$

where σ represents the nonlinear activation function.

Fusion Pooling. In this context, the goal is to incorporate more user response information into determining the veracity of a rumor statement by considering the entire statement, including user feedback. A powerful readout operation is needed to extract a graph-level representation to achieve this. However, a single pooling operation often fails to extract enough information from the representations of different nodes. We combine two commonly used pooling operations, mean pooling and max pooling, to obtain the graph-level representation of a claim:

$$Z = FC(Concat(Max(H, G), Mean(H, G))) \tag{4}$$

The process involves concatenating the maximum pooling and average pooling of hidden representations H over the graph G. Subsequently, a fully connected layer is applied to combine the process of mean pooling and max pooling, resulting in graph-level representations Z of statements that incorporate user participants.

Classifier. Through fusion pooling, we can obtain the graph-level representations $Z \in \mathbb{R}^{n \times d'}$ of the rumor propagation structure. We take Z as the input to the classifier and generate the prediction results $\hat{Y}$:

$$\hat{Y} = Softmax(FC(Z)) \tag{5}$$

3.2 Graph Contrastive Learning Network

To effectively deal with the problem of insufficient data, we also use contrastive learning between different augmented views as an auxiliary task to train a graph encoder.

Graph Augmentation. We apply a combined augmentation of edge removal and node feature masking on the input rumor propagation tree to generate two views $\mathcal{G}^{(1)}$ and $\mathcal{G}^{(2)}$. When performing edge removal, we remove edges according to the generated mask of size $|E|$, and the mask elements are sampled from the Bernoulli distribution $\mathcal{B}(1 - p_A)$. Regarding masking node features, we employ a similar scheme to generate masks of size k sampled from the Bernoulli distribution $\mathcal{B}(1 - p_X)$.

The encoder network $f_\theta(\cdot)$ takes an augmented graph as input and computes representation for each node in the graph. These two enhanced views $\mathcal{G}^{(1)}$, $\mathcal{G}^{(2)}$ pass through the same encoder and get two embedding matrices $H^{(1)}$ and $H^{(2)}$.

Loss Function. We employ a contrastive learning approach to encourage the cross-correlation matrices to approximate a diagonal matrix for different hidden representations $H^{(1)}$ and $H^{(2)}$ of different views, which is inspired by [21] and [2]. We first normalize the embedding matrix along the batch dimension and then compute the empirical cross-correlation matrix $C \in R^{d \times d}$. The cross-correlation matrix is calculated based on the network embedding, and the single element C_{ij} is obtained as follows:

$$C_{ij} = \frac{\sum_b h_{b,i}^{(1)} h_{b,j}^{(2)}}{\sqrt{\sum_b (h_{b,i}^{(1)})^2} \sqrt{\sum_b (h_{b,j}^{(2)})^2}} \tag{6}$$

where b is the batch index, and i, j are the embedding indices, the obtained C is a square matrix with a size equal to the dimensionality of the network's output.

The loss function $\mathcal{L}_{CL}$ compares C with the identity matrix and optimizes them to be as close as possible. It consists of two parts: (I) an invariant term and (II) a redundant reduction term, where the first term enforces the diagonal element C_{ii} equal to 1, and the second term will optimize the off-diagonal elements C_{ij} to zero, which will result in uncorrelated components of the embedding vector.

$$\mathcal{L}_{CL} = \sum_i (1 - C_{ii})^2 + \lambda_1 \sum_i \sum_{j \neq i} C_{ij}^2 \tag{7}$$

where $\lambda_1 = \frac{1}{d}$ defines the trade-off between invariance and redundancy reduction when optimizing the overall loss function. This loss aims to reduce redundancy in self-supervised learning, and it is conceptually simple and easy to implement without needing large batches or stop-gradient operations.

3.3 Model Fusion

To effectively update the parameters of the rumor recognition network, especially its graph encoder, we fuse the graph contrastive learning network and the rumor recognition network through joint training of loss functions. To do this, we take the weighted sum of these losses as the objective function:

$$\mathcal{L} = \mathcal{L}_{CE} + \lambda_2 * \mathcal{L}_{CL} \tag{8}$$

where λ_2 is empirically set to 0.1, and the losses of these two parts are in the same order of magnitude.

4 Experiments

To verify the effectiveness of the proposed model, we compare GCAT with several recently proposed algorithms, especially those based on rumor propagation paths on two real datasets. In addition, we also compare the performance of different models on the early detection task to understand the actual application value of the model. Finally, to clarify the contribution of different components and features to the rumor identification task, we conduct ablation experiments on GCAT.

4.1 Datasets

We conduct experiments on data obtained through FakeNewsNet [17], which consists of true and false news data checked on two fact-checking sites, Politifact and Gossipcop, and the social feeds of engaged users on Twitter. The statistics of the dataset are shown in the Table 1:

Table 1. Dataset statistics.

	Politifact	Gossipcop
#graphs	314	5464
#true	157	2732
#fake	157	2732
#total nodes	41,050	314,262
#total edges	40,740	308,798
#Avg. nodes per graph	131	58

For the Politifact and Gossipcop datasets, we adopt the pre-trained Bert model [4] to encode text features. The content of a news item is encoded as a token sequence with a maximum length of 512. The classification tag [CLS] inserted at the beginning of each input sequence collects information about the entire sequence and is used for the classification task. When it comes to encoding the characteristics of users participating in interactions, we captured nearly 200 historical tweets for each user. The maximum sequence length of tweets is set to 16, and the average of different tweet representations is considered user preference.

4.2   Baselines

We compare the proposed model with algorithms proposed in recent years, especially those based on rumor propagation paths. The baseline models compared in our experiments are as follows:

- CSI [16] combines text and user response features, including a capture module that captures the time-varying characteristics of a user's response to a given document, a scoring model that scores user engagement, and an ensemble module.
- GCNFN [15] generalizes convolutional neural networks to graphs, allowing the fusion of various features such as content, user profiles, interactions, and rumor propagation trees.
- GNN-CL [6] applies continuous learning to progressively train GNNs to achieve balanced performance on existing and new datasets.
- BiGCN [1] proposes a bidirectional graph convolutional network that aggregates the features of adjacent nodes in the rumor propagation tree in both top-down and bottom-up directions.
- UPFD [5] combines various signals such as news text features and user endogenous preference rumor propagation features by jointly modeling news content and rumor propagation tree.

We add two additional baselines that directly classify word2vec and BERT source content embeddings, where we employ a 3-layer fully connected network and apply DropOut to prevent overfitting.

Many baseline methods utilize additional information, and to ensure fair comparisons, we implement the baseline algorithms only with features such as news content, user profiles, user preference embeddings, and rumor propagation trees. We apply the same batch size (64) and hidden layer dimension (64) for different methods. In the division of the data set, the training set, validation set, and test set account for 70%, 10%, and 20%, respectively.

4.3   Experimental Results and Analysis

The performance of GCAT and baseline models on two real datasets is shown in Table 2.

Both $Word2Vec + MLP$ and $Bert + MLP$ abandon the rumor propagation structure and only utilize news content features. These two models' performance

Table 2. Prediction Performance.

Model	Politifact				Gossipcop			
	Acc.	Prec.	Rec.	F1	Acc.	Prec.	Rec.	F1
Word2Vec	0.8438	0.9224	0.8530	0.8359	0.7706	0.8065	0.7174	0.7696
BERT	0.8594	0.9309	0.8387	0.8551	0.7550	0.7567	0.7482	0.7504
CSI	0.8281	0.8947	0.8293	0.8181	0.9159	0.9290	0.9003	0.9153
GCNFN	0.8594	0.8810	0.9024	0.8458	0.9406	0.9339	0.9498	0.9400
GNN-CL	0.7188	0.8750	0.5833	0.7176	0.9333	0.9202	0.9424	0.9325
BiGCN	0.8906	0.9381	0.8537	0.8843	0.9598	0.9437	0.9596	0.9596
UPFD	0.8594	0.9444	0.8293	0.8533	0.9707	0.9713	0.9885	0.9705
GACL	0.8672	0.8947	0.8226	0.8665	0.9013	0.9217	0.8810	0.8994
GCAT	**0.9375**	**0.9655**	**0.9032**	**0.9373**	**0.9799**	**0.9791**	0.9802	**0.9795**

is close to several other methods on the smaller Politifact dataset but worse on the larger Gossipcop dataset. We believe this is because fake news in the Gossipcop dataset is more deceptive and cannot be effectively identified by simple text features, requiring the help of crowd wisdom. The CSI model considers user participation, but the user representation is not updated with the message-passing process, and the effect is not as good as the model based on graph neural networks. The poor performance of $GNN - CL$ on the Politifact dataset is because the user profile contains insufficient information. $BiGCN$ extracts more structural information from the Politifact dataset, with fewer but deeper rumor propagation trees, and achieves better results. On the Gossipcop dataset with more rumor propagation trees and shallow propagation depth, UPFD achieves higher accuracy and F1 value by applying more features after sufficient training.

It can be seen that the proposed $GCAT$ model outperforms all metrics on both datasets, especially the smaller Politifact. $GCAT$ combines news contents and user feedback to obtain initial embeddings, applies a graph neural network with attention to aggregate neighbor features, and takes graph self-supervised learning without negative samples as an auxiliary task to train the graph encoder. What's more, we incorporate multiple pooling methods when obtaining graph-level representations. GCAT effectively fuses multiple features, retains more node information during the pooling process, and performs better.

To further assess the effectiveness of GCAT, we plotted ROC curves as shown in Fig. 2. The results indicate that the model performs well on both datasets, with a ROC score of 0.96 on the Politifact dataset and an even higher ROC score of 0.99 on the Gossipcop dataset. This demonstrates that the dataset can effectively distinguish between samples of different categories.

4.4   Early Detection

Early detection of rumors is significant for intervening in spreading false rumors and reducing the adverse effects. In the early stage of rumor propagation, there

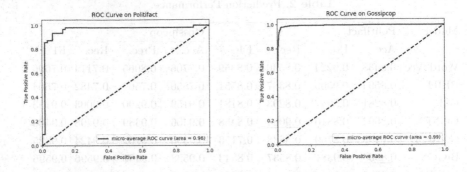

Fig. 2. The ROC curve of GCAT on Politifact and Gossipcop datasets.

are few participating users, which will affect the performance of models benefit from crowd wisdom. To evaluate the performance of the proposed model on this task, we simulate the propagation of rumors at different stages by limiting the number of participating nodes (from 0 to 50) in chronological order.

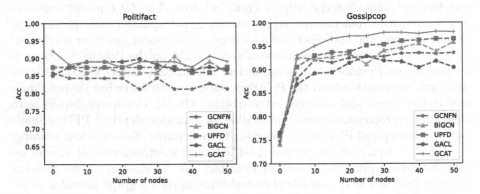

Fig. 3. The effect of the method based on message propagation varies with the number of participating nodes

As shown in Fig. 3, we compare the performance of the rumor propagation-based model for early detection on Politifact and Gossipcop. The input only contains news content when the number of node participants is 0. On the Politifact dataset, GCAT is significantly better than the baseline methods and can benefit from the increase of participating nodes. When no nodes are involved in the large-scale Gossipcop dataset, all models except GNNCL using the profile are ineffective, indicating that the news content corresponding to the Gossipcop dataset is more deceptive. GCAT improves steadily with the increase of the number of nodes on Acc and F1 and can make compelling predictions in the case of a few participating nodes, which shows the effectiveness of the proposed model in the early identification task.

4.5 Ablation Study

We conducted ablation experiments to clarify the contribution of different features and components to the model. The experimental results are shown in Table 3.

- In $GCAT_{-user}$, we disregard user participation and solely rely on textual information for classification, the model is essentially equivalent to BERT.
- In $GCAT_{-mean}$, we remove mean pooling when obtaining graph-level representations. While in $GCAT_{-max}$, we removed max pooling when obtaining a graph-level representation.
- In $GCAT_{-ssl}$, we remove the self-supervised contrastive learning network in GCAT.
- GCAT contains all the above features and functional components with good results.

Table 3. Prediction performance on Politifact and Gossipcop.

Model	Politifact		Gossipcop	
	ACC	F1	ACC	F1
$GCAT_{-user}$	0.8594	0.8551	0.7495	0.7463
$GCAT_{-mean}$	0.9375	0.9374	0.9735	0.9730
$GCAT_{-max}$	0.8906	0.8906	0.9625	0.9619
$GCAT_{-ssl}$	0.8906	0.8906	0.9771	0.9766
$GCAT$	**0.9375**	**0.9373**	**0.9799**	**0.9791**

On the smaller Politifact dataset, $GCAT_{-user}$ achieves better results without applying news features for representation enhancement than on Gossipcop. We speculate this is because concatenating graph-level representations with news features introduces more parameters, and the smaller dataset makes the model insufficiently trained. A comparison of the model variants obtained by removing the two pooling methods shows that max pooling is a more efficient way to obtain graph-level representations under the current task. However, a comparison with the final model using both pooling methods shows that better results can be achieved by combining multiple pooling methods. In $GCAT_{-ssl}$, after removing the self-supervised network, the parameter update of the graph encoder lacks auxiliary information, and the model performance degrades. The GCAT model, with all features and components, achieves the best results.

5   Conclusion and Future Work

In this article, we propose a novel rumor detection method for online social networks based on user participation called Graph Contrastive ATtention Network (GCAT). GCAT applies weighted attention to aggregate neighbor features in the rumor propagation tree and takes graph contrastive self-supervised learning without negative samples as the auxiliary task to help update network parameters. When obtaining graph-level representations, we apply multiple pooling methods to ensure that the final graph-level representation retains more node information. Experimental results on two real-world datasets show that our proposed GCAT model outperforms baseline methods in both Accuracy and F1. We also simulated early detection of rumors to verify the practical value of the model. Additionally, we build several variants of GCAT to compare the contributions of different components and features to the model. In the future, we will consider more effective graph representation learning methods to model rumor propagation paths and further explore the problem of rumor detection under complex conditions such as multimodality and few labels.

Acknowledgements. This work was supported in part by the National Key Research and Development Program of China (grant number 2022YFC3302601), the High-Performance Computing Center of Shanghai University, and the Shanghai Engineering Research Center of Intelligent Computing System (2252600) for providing the computing resources.

References

1. Bian, T., et al.: Rumor detection on social media with bi-directional graph convolutional networks (2020)
2. Bielak, P., Kajdanowicz, T., Chawla, N.V.: Graph Barlow Twins: a self-supervised representation learning framework for graphs. arXiv preprint arXiv:2106.02466 (2021)
3. Brody, S., Alon, U., Yahav, E.: How attentive are graph attention networks? In: International Conference on Learning Representations (2022)
4. Devlin, J., Chang, M.W., Lee, K., Toutanova, K.: BERT: pre-training of deep bidirectional transformers for language understanding. arXiv preprint arXiv:1810.04805 (2018)
5. Dou, Y., Shu, K., Xia, C., Yu, P.S., Sun, L.: User preference-aware fake news detection. In: Proceedings of the 44nd International ACM SIGIR Conference on Research and Development in Information Retrieval (2021)
6. Han, Y., Karunasekera, S., Leckie, C.: Graph neural networks with continual learning for fake news detection from social media. CoRR abs/2007.03316 (2020)
7. Hu, W., et al.: Spatio-temporal graph convolutional networks via view fusion for trajectory data analytics. IEEE Trans. Intell. Transp. Syst. **24**, 4608–4620 (2022). https://doi.org/10.1109/TITS.2022.3210559
8. Kipf, T.N., Welling, M.: Semi-supervised classification with graph convolutional networks. arXiv preprint arXiv:1609.02907 (2016)

9. Li, S., Li, W., Luvembe, A.M., Tong, W.: Graph contrastive learning with feature augmentation for rumor detection. IEEE Trans. Comput. Soc. Syst. 1–10 (2023). https://doi.org/10.1109/TCSS.2023.3269303. Conference Name: IEEE Transactions on Computational Social Systems

10. Li, W., Guo, C., Liu, Y., Zhou, X., Jin, Q., Xin, M.: Rumor source localization in social networks based on infection potential energy. Inf. Sci. **634**, 172–188 (2023)

11. Li, W., Li, Z., Luvembe, A.M., Yang, C.: Influence maximization algorithm based on gaussian propagation model. Inf. Sci. **568**, 386–402 (2021)

12. Li, W., Zhong, K., Wang, J., Chen, D.: A dynamic algorithm based on cohesive entropy for influence maximization in social networks. Expert Syst. Appl. **169**, 114207 (2021)

13. Lin, H., Ma, J., Chen, L., Yang, Z., Cheng, M., Guang, C.: Detect rumors in microblog posts for low-resource domains via adversarial contrastive learning. In: Findings of the Association for Computational Linguistics: NAACL 2022, pp. 2543–2556. Association for Computational Linguistics, Seattle, United States (2022)

14. Luvembe, A.M., Li, W., Li, S., Liu, F., Xu, G.: Dual emotion based fake news detection: a deep attention-weight update approach. Inf. Process. Manage. **60**(4), 103354 (2023)

15. Monti, F., Frasca, F., Eynard, D., Mannion, D., Bronstein, M.M.: Fake news detection on social media using geometric deep learning (2019)

16. Ruchansky, N., Seo, S., Liu, Y.: CSI: a hybrid deep model for fake news detection. In: Proceedings of the 2017 ACM on Conference on Information and Knowledge Management, pp. 797–806. CIKM 2017, Association for Computing Machinery, New York, NY, USA (2017)

17. Shu, K., Mahudeswaran, D., Wang, S., Lee, D., Liu, H.: FakeNewsNet: a data repository with news content, social context and dynamic information for studying fake news on social media. arXiv preprint arXiv:1809.01286 (2018)

18. Sun, T., Qian, Z., Dong, S., Li, P., Zhu, Q.: Rumor detection on social media with graph adversarial contrastive learning. In: Proceedings of the ACM Web Conference 2022, pp. 2789–2797. WWW 2022, Association for Computing Machinery, New York, NY, USA (2022)

19. Tao, Y., Wang, C., Yao, L., Li, W., Yu, Y.: Item trend learning for sequential recommendation system using gated graph neural network. Neural Comput. Appl. 1–16 (2021)

20. Veličković, P., Cucurull, G., Casanova, A., Romero, A., Lio, P., Bengio, Y.: Graph attention networks. In: International Conference on Learning Representations (2018)

21. Zbontar, J., Jing, L., Misra, I., LeCun, Y., Deny, S.: Barlow twins: self-supervised learning via redundancy reduction. In: Proceedings of the 38th International Conference on Machine Learning. Proceedings of Machine Learning Research, vol. 139, pp. 12310–12320. PMLR (2021)

22. Zhang, C., Li, W., Wei, D., Liu, Y., Li, Z.: Network dynamic GCN influence maximization algorithm with leader fake labeling mechanism. IEEE Trans. Comput. Soc. Syst. 1–9 (2022). https://doi.org/10.1109/TCSS.2022.3193583

23. Zhou, X., Li, S., Li, Z., Li, W.: Information diffusion across cyber-physical-social systems in smart city: a survey. Neurocomputing **444**, 203–213 (2021)

E³-MG: End-to-End Expert Linking via Multi-Granularity Representation Learning

Zhiyuan Zha, Pengnian Qi, Xigang Bao, and Biao Qin[✉]

School of Information, Renmin University of China, Beijing, China
{zhazhiyuan99,pengnianqi,baoxigang,qinbiao}@ruc.edu.cn

Abstract. Expert linking is a task to link any mentions with their corresponding expert in a knowledge base (KB). Previous works that focused on explicit features did not fully exploit the fine-grained linkage and pivotal attribute inside of each expert work, which creates a serious semantic bias. Also, such models are more sensitive to specific experts resulting from the isolationism for class-imbalance instances. To address this issue, we propose **E³-MG** (**E**nd-to-**E**nd **E**xpert Linking via **M**ulti-**G**ranularity Representation Learning), a unified multi-granularity learning framework, we adopt a cross-attention module perceptively mining fine-grained linkage to highlight the expression of masterpieces or pivotal support information and a multi-objective learning process that integrates contrastive learning and knowledge distillation method is designed to optimize coherence between experts via document-level coherence. E³-MG enhances the representation capability of diverse characteristics of experts and demonstrates good generalizability. We evaluate E³-MG on KB and extern datasets, and our method outperforms existing methods.

Keywords: Entity Linking · Contrastive Learning · Knowledge Distillation

1 Introduction

Given a document, expert linking aims to extract the mentions of experts and associate these experts with the corresponding entities in the academic KB. Recently, with the flood of unreliable generated content entering the web, quick and accurate identification of experts can well assist users to improve efficiency. Generative paradigms are powerless in generating accurate academic information, and knowledge bases serve as support for reliable and accurate information in the academic domain [1], which makes expert linking more important.

However, Experts are often ambiguous, how to accurately identify the names of experts is one of the important issues due to a large number of experts' name

---

Z. Zha and P. Qi—Equal contribution.

© The Author(s), under exclusive license to Springer Nature Singapore Pte Ltd. 2024
B. Luo et al. (Eds.): ICONIP 2023, CCIS 1967, pp. 268–280, 2024.
https://doi.org/10.1007/978-981-99-8178-6_21

duplication and social media always using abbreviated names daily. For example, "L Wang", maybe means "Wang Liang", "Wang Lun". A basic paradigm for expert linking representation learning is to use the publication of experts for identification and to build representations using the individual information of the papers. The traditional approach uses feature engineering [2,3] to analyze features, but such limited features lose a lot of information.

The graph method and reinforcement learning [4–6] such as CONNA requires a large amount of data labeling and tedious calculations, so it is costly and labor-intensive to update this fast information, and previous work [4,7] which does not apply to this scenario where daily updates are required, this paradigm requires a dedicated and extensive manual labeling design. Chen [8] proposes a contrastive and adversarial learning framework. But this work averages the representation of each paper which does not effectively use the connections of the paper and expert, resulted in a large number of missing semantics and unbalanced semantic expressions.

To address the above issues,we perform enhancements at different granularities based on a constrastive learning approach to achieve fast alignment on new samples, reducing the requirement for manual annotation. We propose a cross-attention module to enhance the mining of key attributes and the fine-grained linkage contained within different pairs of paper and solve the semantic deficiencies brought by the previous isolated processing way. Also, previous schemes can lead to problems that can be expressed sensitively to specific experts due to the imbalance of the sample. Some experts are more distinctive due to the number of papers and their long involvement in the same field, which would make them easier to identify and to build representations, we propose expert coherence via distillation learning that experts with small samples of papers got significant improvement. Compare graph methods or annotated external knowledge requires a large number of manual annotation and pre-training, and our method ensures that the model can be rapidly deployed for updates.

To this end, in an academic knowledge base named AMiner [9], we propose $\mathrm{E^3}$-MG, a unified multi-granularity learning framework, as shown in Fig. 1, our approach alleviates the above challenges and experiments are conducted to demonstrate the effectiveness of our proposed method.

The main contributions of this work can be summarized as follows:

We propose an innovative method for expert linking, which first unify cross-attention and knowledge distillation with contrastive learning, jointly performed as a multi-objective learning process, high similarity and coherence can be favored by the generated embeddings.

We first use cross-attention to exploit fine-grained linkage inside of expert and mining expert coherence with knowledge distillation, effectively capturing fine-grained semantics, and significantly relieving the specific expert sensitivity due to the imbalance class.

With by conducting experiments on KB and two extern datasets, The results show that our $\mathrm{E^3}$-MG significantly outperforms existing methods in all datasets.

2   Related Work

2.1   Name Entity Linking

Named entities require higher accuracy than normal entity linking tasks, [10–12] propose various forms of name entity linking tasks for different scenarios of learning tasks such as multimodal,and academic expert name entity linking is one of the most important tasks, and [8,13,14] propose effective schemes including graph neural network, contrastive learning, feature engineering, and other methods, as a prominent approach in this field.

2.2   Representation Learning

For retrieval tasks based on similarity modeling, representation learning is the most important foundation and core [15], not only applied to single text entity linking, including other tasks [16], representation learning mapping real entities in the mathematical space to perform mathematical operations. In [17], it was confirmed that the supervised contrast learning approach is more suitable for enhancing the representation of optimized text. And [18] has shown that knowledge distillation methods can improve the representation, and it can be used for different purposes in another retrievel scene such as entity linking [19,20].

3   Methodology

3.1   Problem Definitions

Expert Entities Definitions. While each expert in a KB is composed of his published works. Formally, we denote each expert p from knowledge base premise as a set of papers $E_p = \left\{M_p^1, \cdots, M_p^l\right\}$, M^i is the published literature of E_p, l is the number of samples in the KB. A paper q or k from expert p is composed of a set of support information of this paper, such as title, keywords, and venue. Here is the encoder h (e.g.mBERT), q and k means the different paper from an expert, we use [CLS] + support information + [SEP] as the h input to obtain the preliminary embedding $M_p^{(q)} = h(q)$, $M_p^{(k)} = h(k)$ from the same author p, $M_p^{(q)}$ denote current representation of a single paper instance, so we set m_i indicates a piece of support information. m and n are processed to the same length to compute the attention from one batch.

$$\mathbf{M}_p^{(q)} = \text{MLP}(\text{CLS}(q)) = \left\{\mathbf{m}_i^{(q)} \mid \mathbf{m}_i^{(q)} \in \mathbb{R}^k, i = 1, 2, \cdots, m\right\}, \qquad (1)$$

where CLS(q) indicates the CLS token embedding of BERT, k is the dimension of the encoder's hidden state. we define a set of randomly sampled papers I_p of p from AMiner [9], i.e. a sample I_p can be formulated as follows: $I_p = \left\{M_p^1, \cdots, M_p^L\right\}$, $I_p \in E_p$,I_p^+ is sampled from the same expert, while a negative one I_p^- is from a different expert. where $p_i \in p$ and L are the maximal number of sampled papers for each instance. Here we assume that any expert from an external source can be linked with certainty in AMiner, and verify the assumption.

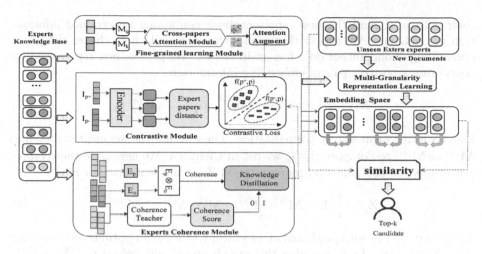

Fig. 1. Overview of Unify E^3-MG

3.2 Fine-Grained Linkage via Cross Attention

To better exploit fine-grained linkage between papers, we need to enable the highlights expression of masterpieces or pivotal attributes. We used a cross-attention mechanism to make the key information centered for better clustering. We first introduce the attention matrix in the author's internal composition vector $\mathbf{C} \in \mathbb{R}^{m \times n}$ of the token level. Each element $\mathbf{C}_{i,j} \in \mathbb{R}$ indicates the impact between the i-th token of paper M^q and the j-th token of paper M^k, two random articles from the same author, Our attentional module is set as follows:

$$\mathbf{C}_{i,j} = \mathbf{P}^T \tanh \left(\mathbf{W} \left(\mathbf{m}_i^{(q)} \odot \mathbf{m}_j^{(k)} \right) \right), \tag{2}$$

where $\mathbf{W} \in \mathbb{R}^{d \times k}, \mathbf{P} \in \mathbb{R}^d$, and $\odot$ denotes the elementwise production operation. Then the attentive matrix could be formalized as:

$$\mathbf{c}_i^{(q)} = \mathrm{softmax} \left(\mathbf{C}_{i,:} \right), \mathbf{c}_j^{(k)} = \mathrm{softmax} \left(\mathbf{C}_{:,j} \right), \tag{3}$$

$$\mathbf{m}_i^{(q)'} = \mathbf{M}^{(k)} \cdot \mathbf{c}_i^{(q)}, \mathbf{m}_i^{(k)'} = \mathbf{M}^{(q)} \cdot \mathbf{c}_j^{(k)} \tag{4}$$

Next, we obtain the attention weight of each paper sample relative to other paper samples based on the attention matrix. From this, we use attentional weighting to fuse the representations of these samples to get fine-grained linkage and pivotal support information.

$$\mathbf{m}_i^{(q)''} = \left[\mathbf{m}_i^{(q)}; \mathbf{m}_i^{(q)'}; \mathbf{m}_i^{(q)} - \mathbf{m}_i^{(q)'}; \mathbf{m}_i^{(q)} \odot \mathbf{m}_i^{(q)'} \right], \tag{5}$$

$$\tilde{\mathbf{m}}_i^{(q)} = \mathrm{ReLU} \left(\mathbf{W}_i^{(p)} \mathbf{m}_i^{(q)''} + \mathbf{b}_i^{(q)} \right), \tag{6}$$

where $[\cdot; \cdot; \cdot; \cdot]$ represents the concatenation operation. $\mathbf{m}_i^{(q)} - \mathbf{m}_i^{(q)'}$ indicates the difference between the original and after attention, and $\mathbf{m}_i^{(q)} \odot \mathbf{m}_i^{(q)'}$ represents their relevance. And the new representations containing fine-grained information are obtained. We expect that such operations could help cluster the paper embedding of one author on a fine-grained level. We get the new representation containing a positive pair of embedding information:

$$\hat{\mathbf{M}}^{(q)} = \text{Layer Norm} \left(\tilde{\mathbf{m}}_1^{(q)}, \tilde{\mathbf{m}}_2^{(q)}, \ldots, \tilde{\mathbf{m}}_m^{(q)} \right), \tag{7}$$

LayerNorm(.) means layer normalization. Then we use the augment embedding $\mathbf{M}^{(q)} = \hat{\mathbf{M}}^{(q)}$, to compose the I_p we list in Chap. 3.4. We aggregate these representations and the pair-level representation $\mathbf{Z}$ for the sentence pair is

$$\mathbf{Z}_p = \left[\hat{\mathbf{M}}^{(q)}; \hat{\mathbf{M}}^{(k)}; \hat{\mathbf{M}}^{(q)} - \hat{\mathbf{M}}^{(k)}; \hat{\mathbf{M}}^{(q)} \odot \hat{\mathbf{M}}^{(k)} \right] \tag{8}$$

Therefore, we use softmax-based cross-entropy to form an optimization goal for the classification of which author this sample specifically belong to. This step is joint training with the following loss function:

$$\mathcal{L}^{CE} = \text{CrossEntropy}(\mathbf{W}\mathbf{Z_p} + \mathbf{b}, p), \tag{9}$$

where $\mathbf{W}$ and $\mathbf{b}$ are trainable parameters. $\mathbf{Z}_p$ is the fine-grained linkage representation from the cross-attention module and p is the correct author.

3.3   Expert Coherence via Knowledge Distillation

The obtained models tend to be more sensitive to experts with more papers [4], limiting the generalization ability of the models. So we propose the following operation based on the method of knowledge distillation, we set all mentions $A = \{E_p, E_o, \cdots, E_z\}$. The Experts-Network Coherence Teacher Model follows a cross-encoder architecture to improve the ability to capture the coherence of experts, we set COH to indicate coherence between experts.

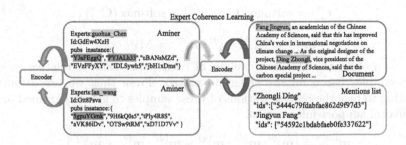

Fig. 2. Case of Expert Coherence Learning

Expert Coherence Distillation. First, we use the same initial text encoder as in Sect. 3.1, then we use BERT as a cross-encoder in the positive sample acquisition by them to determine the overall coherence between the two experts, similar to building an overall judgment over the experts to measure the relationship between the experts, and $COH = 0$ means unrelated and $COH = 1$ means closely connected collaborators are considered. Our score calculation scheme is as follows:

$$COH_{p,o} = \sigma(W^T \text{BERT}^{Tch}([CLS, E_p, SEP, E_o])) \qquad (10)$$

With such a scheme, we can capture the conference between entities on a document level from A, where $[CLS, E_p, SEP, E_o]$ means the concatenation of experts embedding with CLS and SEP token padded; we take the final layer's hidden state corresponding to CLS as the output of BERT. $W (\in \mathbb{R}^{d\times 1})$ is the linear projection which reduces the output vector to a real value, and $\sigma(\cdot)$ is the sigmoid activation. The ability of this coherence teacher model comes from the supporting information from the KB and new context. The coherence score $COH_{q,k}$ for the mentions q k. Then, E^3-MG is learned to imitate the teacher's prediction with the inner product of the query and ad embedding, we set E^3-MG$(E_p)=\hat{E}_p$, E^3-MG$(E_o)=\hat{E}_o$ where the following loss is minimized:

$$\mathcal{L}^{COH}(E_p, E_o) = \min. \sum_q \sum_a \left\| COH_{E_p, E_o} - \left\langle \hat{E}_p, \hat{E}_o \right\rangle \right\| \qquad (11)$$

In this place, $\langle \cdot \rangle$ indicates the inner product operator. Therefore, there is an internal correlation among authors who worked together or created excellent work together, which is crucial for our study of the cooperation information between different expert entities. Figure 2 is an example.

3.4   Multi-granularity Representation Learning Framework

Inspired by the retrieval scheme [8,21], a dynamic contrastive learning is proposed. For the anchor instance I_p, the positive counterpart. Given a set of triplets $\left\{ \left(I_p, I_p^+, I_p^- \right) \right\}$, the triplet loss function is defined as:

$$L^{CL} = \sum_{\left(I_p, I_p^+, I_p^- \right)} \max \left\{ 0, m + f\left(p, p^- \right) - f\left(p, p^+ \right) \right\} \qquad (12)$$

Given a set of experts $\{p\}$ on KB, we aim at pre-training an expert encoder h and a metric function $f : \{I_p, I_{p'}\} \to \{y\}$ to infer the alignment label between p and p', we use the metric function to measure that $y = 1$ implies p and p' are identical and 0 means irrelevant, where m is a margin, we fine-tune h and f on the external source $\{\tilde{e}\}$ such that h and f are transferred to unseen external sources, then we acquire a set of paper embeddings $I_p = \left\{ \mathbf{M}_p^l \right\}_{l=1}^L$ and $I_{p'} = \left\{ \mathbf{M}_{p'}^j \right\}_{j=1}^L$ respectively. Then we compute a similarity matrix $\mathbf{A}$ between I_p and $I_{p'}$. Each element $\alpha_{lj} = \left\| \mathbf{M}_p^l - \mathbf{M}_{p'}^j \right\|_2^2$ is calculated by the normalized euclidean distance between the m-th paper in I_p and n-th paper in $I_{p'}$. Here we transform α_{lj} into a K-dimensional distribution to extract the similarity with mean μ_k and variance σ_k implying how likely α_{mn} corresponds to the k-th similarity pattern:

$$\mathbf{K}\left(\alpha_{lj}\right) = \left[K_1\left(\exp\left[-\frac{\left(\alpha_{lj} - \mu_1\right)^2}{2\sigma_1^2}\right]\right), \cdots, K_K\left(\exp\left[-\frac{\left(\alpha_{lj} - \mu_k\right)^2}{2\sigma_k^2}\right]\right)\right],$$

(13)

then we sum up over rows to represent the similarities between m-th paper in I_p and all the papers in $I_{p'}$, and sum up over columns to represent the similarity between I_p and $I_{p'}$, After all, we apply an MLP layer to obtain the similarity score:

$$f\left(p, p'\right) = \text{MLP}\left(\sum_{l=1}^{L} \log \sum_{j=1}^{L} \mathbf{K}\left(\alpha_{lj}\right)\right)$$

(14)

Compared to the static scheme of embedding a single point [4], For the multi-granularity features of the expert linking problem samples, we carry over to use a dynamic relative distance triple loss.

E^3-MG is developed to select experts E_p from the entire KB, the set of mentions $S_D = \{s^1, \cdots, s^N\}$, If E^3-MG traversing to the mentions s^i currently. The above objective is formulated by the following equation:

$$\max \sum_{E_p} f(s^i, E_p) + \sum_{i \neq j}^{n} \text{COH}(s^i, s^j),$$

(15)

$$\text{s.t. COH}(s^i, s^j) \geq \varepsilon, \forall s^i, s^j \in D,$$

(16)

where ε is the threshold of the Document-level coherence score. Figure 3 is a case of this scenario.

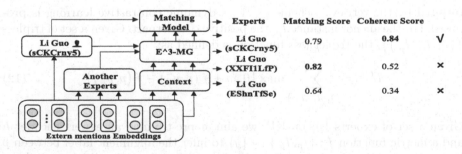

Fig. 3. Case of Expert Linking

For ease of optimization, The cross-attention mechanism is used to learn joint representations of paper pairs, we make the following relaxation to the above objective:

$$\mathcal{L}^{E^3-MG} = \mathcal{L}^{CL} + \alpha * \mathcal{L}^{COH} + \beta * \mathcal{L}^{CE}, \tag{17}$$

where α, β is the positive value trading off the importance between fine-grained and document-level entity coherence.

4 Experiment

In this section, we evaluated our proposed model on three datasets and validated the effectiveness of our approach by ablation experimental enrichment, etc. We first evaluate the representation capability by author identification and paper clustering on AMiner. Also to verify the performance improvement of the unseen expert, we evaluate the external task by linking external experts, i.e., linking unknown names in the news, experts, from LinkedIn.com to AMiner experts.

4.1 Datasets

This is the whoiswho dataset proposed by [22], which is a reliable benchmark in expert identification and has been widely used. Here are some details about these datasets: **AMiner**: A complex dataset with 421 names, 45187 experts and 399255 English papers, and 18 average expert candidates. **Chinese News**: The dataset includes 20,658 reliable news, 1,824 names, and 8.79 average expert candidates from these articles from Chinese technical websites. **LinkedIn**: This dataset was constructed from 50,000 LinkedIn homepages in English, of which 1,665 homepages and 4.85 average expert candidates were associated with the AMiner.

4.2 Experiment and Analysis

For the experimental part, we performed the following setup, we used Python 3.8.10, and PyTorch 1.14.0, and used two NVIDIA TITAN RTX with 48 GB memory size. Here we follow [8] setting, using pre-training and adversarial fine-tuning, the baseline mode shown in the table. We set the number of epoch rounds to 20 and limited by the resources, 9 we set the size of the negative sampling and paper sampling and L = 6 is set in the process of paper sampling.

Preprocess. For example, names "L Wang", including "Wang Lan", and "W Lan". Given an external expert to be linked, we choose the AMiner experts with similar names as candidates. Adjust the position of the last name and first name, or abbreviate the abbreviation, as well as use just the last name. Also, we filter the experts with less than 6 papers to ensure that too few paper authors affect the sampling effect.

Table 1. PC on AMiner

Model	P-Pre	P-Rec	P-F1
Louppe [2]	0.609	0.605	0.607
Zhang [7]	0.768	0.551	0.642
G/L-Emb [4]	**0.835**	0.640	0.724
Unsupervised [8]	0.332	0.591	0.425
CODE [8]	0.724	0.789	0.755
E³-MG	0.735	**0.812**	**0.772**

Table 2. AI on Aminer

Model	HR@1	HR@3	MRR
GBDT [23]	0.873	0.981	0.927
Camel [24]	0.577	0.737	0.644
HetNetE [25]	0.582	0.759	0.697
CONNA [6]	**0.911**	0.985	**0.949**
CODE-pre [8]	0.898	0.964	0.934
E³-MG-pre	0.905	**0.988**	0.946

Paper Clustering (PC). Given papers with the same author, a set of papers is asked to return to identify which papers with the same name belong to. We use hierarchical agglomerative clustering algorithm with embedding by model, and use pairwise Precision, Recall, and F1-score (P-Pre, P-Rec, and P-F1) [4] to evaluate the performance of the models. Here are the baseline models, **Louppe** [2], in this work, the similarity between papers is measured using a de-metric similarity based on manual features. And multiple types of expert paper graphs are constructed to learn different features in **Zhang** [7] work. **G/L-Emb** [4] learns paper embedding on an external paper knowledge network, so it gets a good score in accuracy. **CODE** [8] used the paradigms of contrastive learning and adversarial learning from which the features of the paper were extracted. As described in Table 1, Compared to the previous best results, our model improves 0.023 on Recall and 0.017 on F1 value, which demonstrate the effectiveness (Figs. 4 and 5).

Author Identificaton (AI). Author Identification is a task to assign a new paper to the right expert, This task tests the learning effect of representation learning on papers and candidate experts and returns candidate results by similarity. Here are some of the previous baseline models, **GBDT** [23] is a feature-engineering model, **Camel** [24] use title and keywords as support information to one-hot embedding, **HetNetE** [25] used each paper's author, affiliations, and venues append to the title and keywords. **CONNA** [6] are built between multiple attribute learning to establish the linkage of marker embeddings, and

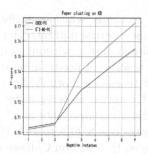

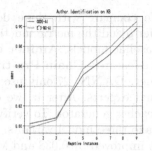

Fig. 4. Effect of NS on PC   **Fig. 5.** Effect of NS on AI

requires a large amount of data labeling and tedious calculations. Since our model requires much less external data and manual annotation, Table 2 shows that we got slightly worse on HR@1, but after improving by 0.024 on HR@3 score.

External Expert Linking. We compared our model with these baseline models: As described in Table 3, HR@K and MRR are also used as metrics to evaluate the ranked AMiner candidates. HR@K (HR@K, K = 1,3) to measure the proportion of correctly assigned experts. And we set HR@K to 1,3,5 in our experiments to measure the MRR of correctly assigned experts following previous settings. We use different serial training approaches to verify the effect of our multi-granularity representation learning paradigm.

Table 0. Performance of External Expert Linking to AMiner

External Sources	News			LinkedIn		
	HR@1	HR@3	MRR	HR@1	HR@3	MRR
Unsupervised-CODE	0.329	0.731	0.559	0.805	0.963	0.886
CODE(AMiner-Only)	0.737	0.927	0.837	0.897	0.982	0.940
E³-MG(AMiner-Only)	0.746	0.931	0.843	0.899	0.981	0.942
CODE(Chain Pre-training)	0.739	0.927	0.839	0.895	0.978	0.939
E³-MG(Chain Pre-training)	0.780	0.944	0.865	0.905	0.986	0.946
CODE(DANN)	0.743	0.928	0.842	0.901	0.983	0.943
E³-MG(DANN)	0.786	0.947	0.869	0.911	0.983	0.948
CODE	0.753	0.936	0.848	0.904	0.987	0.945
E³-MG	**0.791**	**0.958**	**0.873**	**0.917**	**0.990**	**0.954**
E³-MG w/o L^{CE}	0.777	0.949	0.864	0.908	0.983	0.946
E³-MG w/o L^{COH}	0.783	0.944	0.866	0.910	0.980	0.941

We compare with three domain adaptation methods which transfer the encoder g and the metric function f from KB to the external datasets. Here are the baseline models, **Aminer only** extracts domain-agnostic without domain-private features. **Chain Pre-training** [26] chains a series of pre-training stages together. **DANN** [27]only use a shared generator to encode both AMiner and external experts. After it, we coded AMiner and external experts using only one teacher generator. To verify the efficacy of different components in E³-MG, we make three model variants. Compared to the previous SOTA model, HR@1 scores improved by 0.038 and MRR average improved by 0.025 on Chinese news dataset, HR@1 scores improved by 0.013 and MRR average improved by 0.009 on Linkedin dataset, thanks to our multi-granular modeling approach, which taps into different levels of information.

Ablation Study. To investigate the effectiveness of different modules, we perform comparisons between full models and their ablation methods. The results are shown in Table 3, we use the same training paradigm, As shown in table w/o L^{CE} drops 0.012 HR@1 scores. And w/o L^{COH} drops 0.008 HR@1 scores shows that coherence and fine-grained linkage is a practical solution, and such a multi-granularity training paradigm is effective.

5   Conclusion

We propose **E³-MG**, a unified multi- granularity learning framework performed as a multi-objective learning process, It is possible to acquire both high similarity and coherence by the learning paradigm, exploit fine-grained linkage inside of an expert with cross-attention and mining expert coherence with knowledge distillation, effectively capturing fine-grained semantics and significantly relieving the specific expert sensitivity due to the imbalance of class. In the future, we will use it in more scenarios, such as multimodal name entity linking.

References

1. Mallen, A., Asai, A., Zhong, V., Das, R., Hajishirzi, H., Khashabi, D.: When not to trust language models: investigating effectiveness and limitations of parametric and non-parametric memories," ArXiv, vol. abs/2212.10511 (2022)
2. Louppe, G., Al-Natsheh, H.T., Susik, M., Maguire, E.J.: Ethnicity sensitive author disambiguation using semi-supervised learning. In: International Conference on Knowledge Engineering and the Semantic Web (2015)
3. Tang, J., Fong, A.C.M., Wang, B., Zhang, J.: A unified probabilistic framework for name disambiguation in digital library. IEEE Trans. Knowl. Data Eng. **24**, 975–987 (2012)
4. Zhang, Y., Zhang, F., Yao, P., Tang, J.: Name disambiguation in aminer: clustering, maintenance, and human in the loop. In: Proceedings of the 24th ACM SIGKDD International Conference on Knowledge Discovery & Data Mining (2018)
5. Trivedi, R.S., Sisman, B., Dong, X., Faloutsos, C., Ma, J., Zha, H.: Linknbed: multi-graph representation learning with entity linkage, ArXiv, vol. abs/1807.08447 (2018)

6. Chen, B., et al.: CONNA: addressing name disambiguation on the fly. IEEE Trans. Knowl. Data Eng. **34**, 3139–3152 (2019)
7. Zhang, B., Hasan, M.A.: Name disambiguation in anonymized graphs using network embedding. In: Proceedings of the 2017 ACM on Conference on Information and Knowledge Management (2017)
8. Chen, B., et al.: Code: contrastive pre-training with adversarial fine-tuning for zero-shot expert linking. In: Proceedings of the AAAI Conference on Artificial Intelligence, vol. 36, no. 11, pp. 11 846–11 854 (2022)
9. Wan, H., Zhang, Y., Zhang, J., Tang, J.: AMiner: search and mining of academic social networks. Data Intell. **1**, 58–76 (2019)
10. Sun, W., Fan, Y., Guo, J., Zhang, R., Cheng, X.: Visual named entity linking: a new dataset and a baseline. In: Conference on Empirical Methods in Natural Language Processing (2022)
11. Joko, H., Hasibi, F.: Personal entity, concept, and named entity linking in conversations. In: Proceedings of the 31st ACM International Conference on Information & Knowledge Management, Atlanta, GA, USA, 17–21 October 2022, Hasan, M.A., Xiong, L., (eds.) ACM, pp. 4099–4103 (2022). https://doi.org/10.1145/3511808.3557667
12. Galperin, R., Schnapp, S., Elhadad, M.: Cross-lingual UMLS named entity linking using UMLS dictionary fine-tuning. In; Findings of the Association for Computational Linguistics: ACL 2022, Dublin, Ireland, 22–27 May 2022, Muresan, S., Nakov, P., Villavicencio, A. (eds.) Association for Computational Linguistics, pp. 3380–3390 (2022). https://doi.org/10.18653/v1/2022.findings-acl.266
13. Park, S.-S., Chung, K.-S.: CONNA: configurable matrix multiplication engine for neural network acceleration. Electronics **11**(15), 2373 (2022)
14. Zhang, H., Chen, Q., Zhang, W., Nie, M.: HSIE: improving named entity disambiguation with hidden semantic information extractor. In: 2022 14th International Conference on Machine Learning and Computing (ICMLC) (2022)
15. Bengio, Y., Courville, A.C., Vincent, P.: Representation learning: a review and new perspectives. IEEE Trans. Pattern Anal. Mach. Intell. **35**, 1798–1828 (2012)
16. Adjali, O., Besançon, R., Ferret, O., Borgne, H.L., Grau, B.: Multimodal entity linking for tweets. In: Advances in Information Retrieval, vol. 12035, pp. 463–478 (2020)
17. Li, S., Hu, X., Lin, L., Wen, L.: Pair-level supervised contrastive learning for natural language inference. In: ICASSP 2022–2022 IEEE International Conference on Acoustics, Speech and Signal Processing (ICASSP), pp. 8237–8241 (2022)
18. Zhou, X., et al.: Multi-grained knowledge distillation for named entity recognition. In: North American Chapter of the Association for Computational Linguistics (2021)
19. Jia, B., Wu, Z., Zhou, P., Wu, B.: Entity linking based on sentence representation. Complex **2021**, 8 895 742:1–8 895 742:9 (2021)
20. Zhang, J., et al.: Uni-retriever: towards learning the unified embedding based retriever in bing sponsored search. In: Proceedings of the 28th ACM SIGKDD Conference on Knowledge Discovery and Data Mining (2022)
21. Xiong, C., Dai, Z., Callan, J., Liu, Z., Power, R.: End-to-end neural ad-hoc ranking with kernel pooling. In: Proceedings of the 40th International ACM SIGIR Conference on Research and Development in Information Retrieval (2017)
22. Chen, B., et al.: Web-scale academic name disambiguation: the who is who benchmark, leaderboard, and toolkit, ArXiv, vol. abs/2302.11848 (2023)
23. Li, J., Liang, X., Ding, W., Yang, W., Pan, R.: Feature engineering and tree modeling for author-paper identification challenge. In: KDD Cup 2013 (2013)

24. Zhang, C., Huang, C., Yu, L., Zhang, X., Chawla, N.: Camel: content-aware and meta-path augmented metric learning for author identification. In: Proceedings of the 2018 World Wide Web Conference (2018)
25. Chen, T., Sun, Y.: Task-guided and path-augmented heterogeneous network embedding for author identification. In: Proceedings of the Tenth ACM International Conference on Web Search and Data Mining (2016)
26. Logeswaran, L., Chang, M.-W., Lee, K., Toutanova, K., Devlin, J., Lee, H.: Zero-shot entity linking by reading entity descriptions, ArXiv, vol. abs/1906.07348 (2019)
27. Ganin, Y., et al.: Domain-adversarial training of neural networks, ArXiv, vol. abs/1505.07818 (2015)

TransCenter: Transformer in Heatmap and a New Form of Bounding Box

Deqi Liu[1], Aimin Li[2(✉)], Mengfan Cheng[3], and Dexu Yao[3]

[1] Key Laboratory of Computing Power Network and Information Security, Ministry of Education, Shandong Computer Science Center, Qilu University of Technology (Shandong Academy of Sciences), Jinan, China
[2] Shandong Engineering Research Center of Big Data Applied Technology, Faculty of Computer Science and Technology, Qilu University of Technology (Shandong Academy of Sciences), Jinan, China
lam@qlu.edu.cn
[3] Shandong Provincial Key Laboratory of Computer Networks, Shandong Fundamental Research Center for Computer Science, Jinan, China
{10431210366,10431210601}@stu.qlu.edu.cn

Abstract. In current heatmap-based object detection, the task of heatmap is to predict the position of keypoints and its category. However, since objects of the same category share the same channel in the heatmap, it is possible for their keypoints to overlap. When this phenomenon occurs, existing heatmap-based detectors are unable to differentiate between the overlapping keypoints. To address the above issue, we have designed a new heatmap-based object detection model, called TransCenter. Our model decouples the tasks of predicting the object category and keypoint position, and treats object detection as a set prediction task. We use a label assignment strategy to divide the predicted sets into positive and negative samples for training. The purpose of this is to allow different objects to have their own heatmap channel without sharing with other, thereby completely eliminating the occurrence of overlapping. To make the model easier to learn, we leverage the characteristic that heatmaps can reduce the solution space, proposed a novel approach for predicting bounding boxes. We use the encoder-decoder structure in transformers, treat the prediction of bounding boxes as an encoding task, use the form of a heatmap to represent the position and size. Then, we treat category prediction and offset prediction of the bounding box as decoding tasks, where the offset prediction is outputted through regression.

Keywords: Heatmap · Keypoint · Object Detection · Keypoint Overlap

1 Introduction

In current object detection algorithms, there are two types of output formats for prediction. One is to directly output specific coordinates, which is based on regression [1–6]. The other is to output a Gaussian heatmap of object keypoints, which is based on heatmap

[7–9], the output is shown in Fig. 1. The second method is actually more like a classification task. Initially, the heatmap-based method was often used in the field of human pose estimation [10, 11]. Later, as object detection algorithms developed, people applied the heatmap-based method to object detection. One of the most intuitive advantages of this method is that the model does not require the construction of complex anchors, because each pixel in the heatmap can be approximated as an anchor [12–14] in a sense.

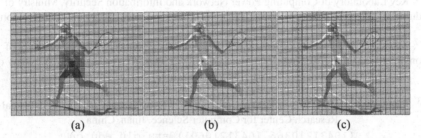

(a)   (b)   (c)

Fig. 1. Representation of bounding boxes in heatmap-based models. (a) Illustrates how a heatmap represents the position of an keypoint. (b) Represents the regression of width and height at that location. (c) Represents the offset of keypoint.

Unlike regression-based approaches, directly output the 2D coordinates of an object is an extremely nonlinear process [15]. It is also a more challenging form of supervised learning, as the network has to independently convert spatial positions into coordinates. The heatmap-based method utilizes the explicit rendering of the Gaussian heatmap, allowing the model to learn the output target distribution by learning a simple filtering method that filters the input image into the final desired Gaussian heatmap [16]. This greatly simplifies the learning difficulty of the model and is very consistent with the characteristics of convolutional. Furthermore, the regression-based method has a faster training and inference speed and can achieve end-to-end full-differentiation training, but they are prone to overfitting and have poor generalization ability. Compared with the regression, the heatmap-based method specifies the learned distribution, which is more robust for various situations (occlusion, motion blur, truncation, etc.) than it. Additionally, the heatmap-based method can explicitly suppress the response at non-keypoints.

After analyzing the strengths and weaknesses of both output forms, we decided to integrate them in our model. Our model employs CNN as the backbone to extract low-level features, and then utilizes Transformer to capture global dependencies at a higher level. We assign the task of predicting bounding boxes to the encoding layer of the Transformer. Unlike any other forms of bounding box prediction, we approximate both the position and size prediction of bounding boxes as a classification task, using the form of Gaussian heatmap for output. This approach significantly reduces the solution space and makes it easier for the network to learn. However, relying solely on the prediction of bounding boxes through the encoding layer is not accurate enough, as the limitations of the heatmap result in the position coordinates and sizes being quantized, leading to a significant error when mapping the predicted bounding boxes to the original image. To reduce the prediction error, we not only classify the object categories in the decoding layer, but also use regression to more accurately predict the position and size offsets of the bounding boxes, thus achieving more precise localization.

In current heatmap-based models, the task of the heatmap is not only to predict the position of keypoints, but also to predict its category. The heatmap has K channels, which is equal to the number of categories in the dataset, each channel is responsible for predicting different category. When multiple objects of the same category share one channel, the overlap of their keypoints is inevitable, Fig. 2 shows the overlap of key-points. This is why larger heatmap sizes lead to better detection in these models. Larger heatmap can preserve more feature information and ensure sufficient distances between the keypoints of different objects, enabling the model to distinguish them effectively. In our model, we implemented a simple solution to this problem. By decoupling the task of predicting keypoint position from its category prediction, we were able to assign one heatmap exclusively to each object, ensuring that each keypoint corresponded to a unique location on the heatmap, and the constraint of heatmap size on the model is alleviated.

Fig. 2. The figure above shows the situation when the geometric center of an object is used as a keypoint. As the model gets deeper, the downsampling rate of the image increases, which may cause different object keypoints to appear in the same cell.

As we have decoupled the task of object classification from that of position prediction, in order to establish accurate association between the objects and their bounding boxes, we have drawn inspiration from DETR [17]. We consider an image as a set, with the objects in the image being the items in this set. Each item is composed of a category label and a bounding box. Our model predicts a fixed-size set, with the set size N being much larger than the number of objects present in the image. During the training process, we match each predicted object with every object in the ground-truth set, generating a matching cost for each pair of predicted and ground-truth objects. We then use these matching costs to assign positive and negative samples, and train the model accordingly.

We summarize our contributions as follows:

- We proposed a novel form of predicting bounding boxes, which is different from any previous output forms. We approximate the prediction of the location and size of the bounding box as a classification problem, using the response values at each position in the heatmap to determine the specific location and size of the bounding box. This approach can narrow down the solution space and make the network easier to learn.

- Unlike other heatmap-based object detectors, we decoupled the category prediction from the keypoint position prediction. This allows each object to have its own heatmap, thus eliminating the overlapping keypoints problem that arises from multiple objects sharing the same heatmap channel.

2 Related Work

2.1 Heatmap in Human Pose Estimation

The heatmap-based approach has become the mainstream method in this field. This approach trains the model to learn a Gaussian probability distribution map by rendering each point in the ground-truth as a Gaussian heatmap. The network output consists of K heatmaps, corresponding to K keypoints, and the final estimation is obtained by using argmax or soft-argmax to locate the point with the highest value.

2.2 Heatmap in Object Detection

The most mainstream approach in the object detection field is still obtaining the position and size of the bounding box through direct regression. However, many heatmap-based object detection models have emerged so far. These models can be broadly categorized into two types. The first type outputs one keypoint of the object through a heatmap, such as CenterNet [9], which considers the geometric center of the object as the keypoint, and then obtains the precise size of the object through regression. The second type predicts multiple different keypoints of the object. Then, through some matching method, the keypoints belonging to the same object are associated together to determine the specific position and size of the bounding box, thus avoiding direct regression of coordinates and size. Representatives of this type of method include CornerNet [7], which determines the bounding box by predicting the two diagonal points of an object, and ExtremeNet [18], which uses five heatmaps to predict the four extreme points and central region of an object, etc. However, since heatmaps can only predict a rough position, such methods still require regression to obtain position offsets of keypoints for more precise adjustments.

2.3 Transformer with Heatmap

The Transformer was originally proposed by Vaswani et al. [19] and was initially applied to the field of natural language processing. In recent years, the Transformer has also gained significant attention in the field of computer vision [20, 21]. Sen Yang et al. [22] applied the Transformer to the task of heatmap prediction, and only used the encoder. They believed that pure heatmap prediction is simply an encoding task, and that the Transformer-based keypoint localization method is consistent with the interpretability of activation maximization [23]. Up to now, there are very few methods that use the Transformer for heatmap prediction tasks, and most of them combine the Transformer with regression techniques. Therefore, this novel combination of the Transformer with heatmap prediction is a bold attempt for us.

3 Method

3.1 Model Structrue

Usually, heatmap-based models choose to use HourglassNet [24] to produce high-resolution feature maps, as this network structure is capable of capturing and integrating information at all scales of the image. However, our network uses a lighter backbone, ResNet-50 [25], instead. We pass the extracted low-level features to a Transformer to obtain a more advanced feature representation. Our Transformer consists of an encoder and a decoder. The main task of the encoder is to perform a rough prediction of the bounding box, including its position and size. The main task of the decoder is to predict the object category and adjust the bounding box, which includes position offsets and size offsets. The overall structure of the model is shown in Fig. 3.

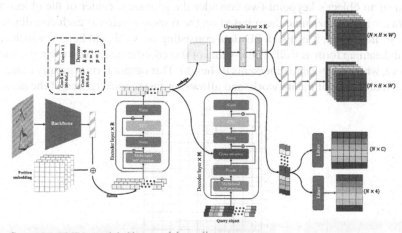

Fig. 3. Structure of the model. Our model predicts a set of size N, where each item in the set consists of a position heatmap, a size heatmap, a category, and offsets to ensure that each object has its own individual heatmap.

The low-level features extracted by the CNN are first compressed in channel dimension via a convolutional layer with a kernel size of 1, and then sent to the encoder of Transformer. In the encoder, since the feature maps output by the backbone is flattened into a 1D sequence of pixels, the Transformer can calculate the correlation between each pixel and all other pixels of the feature maps. The encoder consists of several encoding layers, each composed of multi-head self-attention and a FFN (feedforward neural network). A normalization module follows each module. The output of the encoding layer is fed into a continuous upsampling operation [26] before being fed into the bounding box prediction head, which is composed of convolutional layers. In the first two layers of the prediction head, larger-sized convolutional kernels are used to aggregate information from the feature map. Finally, a convolutional kernel with a size of 1 is used to obtain the position and size of bounding boxes, which are output as Gaussian heatmaps.

The output of the encoder is also fed into the decoder. The decoder consists of multiple decoding layers, each composed of multi-head self-attention, cross-attention, and FFN.

Like the encoding layers, there is also a normalization module following each module in the decoder. The output of the decoder is then separately sent to the classification head and offset prediction head, both of which are composed of linear layers.

3.2 Bounding Box in Heatmap Form

In current mainstream object detection models, the predicted form of bounding boxes are usually in the form of numerical values, which is a common approach in regression-based models. In heatmap-based models, however, only the coordinate of the bounding box is output as heatmap, while the size of the bounding box is obtained as specific values through regression. In our experiments, we have demonstrated that the size of the bounding box can also be obtained in the form of heatmap, as shown in Fig. 4. We can view the heatmap as a 2D coordinate system with limited width and height, and for the position of an object's keypoint (we consider the geometric center of the object as the keypoint), we can determine them based on the response value at each coordinate. For the size of the object, it is also a 2D data consisting of width and height, which can be output in heatmap form as well. The x-value of this coordinate can represent the width of the object, while the y-value represents its height. This output format greatly reduces the prediction difficulty of the network, and allows for faster convergence of the network.

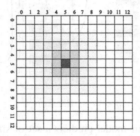

Fig. 4. Represent the bounding box in the form of a heatmap. The darker the color on the heatmap, the higher the response value. If the image represents a position heatmap, the horizontal axis denotes the x-coordinate of the keypoint, and the vertical axis symbolizes the y-coordinate. Consequently, the keypoint is at the (5, 5) coordinates. Alternatively, if the image represents a size heatmap, the horizontal axis signifies the object's width, while the vertical axis represents its height, thus, the size can be expressed as (5, 5).

Predicting bounding boxes using heatmaps can be approximated as a classification task. In this case, the coordinates in the heatmap can be considered as "categories". As the decreases of size, during the initial model learning phase, the probability of "guessing" the correct "category" increases. Thus, the size of heatmap determine the lower limit of the model. Similarly, due to the heatmap's constraints, coordinates can only appear as integers. Thus, when a bounding box is mapped back to its original-sized image after downsampling, it inevitably results in quantization errors. A smaller solution space also imposes limitations on the model's upper limit. To effectively compensate for the generated errors, we predict the offsets of the position and size through a regression method following the decoding layer.

3.3 Offset

In heatmap-based models, the ground-truth representation of offsets is illustrated as in Eq. 1.

$$\begin{cases} x - \lfloor x \rfloor \\ y - \lfloor y \rfloor \\ w - \lfloor w \rfloor \\ h - \lfloor h \rfloor \end{cases} \tag{1}$$

In this formula, x and y represent center coordinates, while w and h denote width and height. However, such a representation may not be suitable for our model, as during the training process, the offset loss struggles to decrease significantly. After comparing the format of labels for offsets in CenterNet [9], we found that, offset regression is performed for specific locations. However, using Eq. 1 to calculate offsets is not effective in representing spatial positions. Therefore, we have chosen to use Eq. 2 to create the labels for offsets.

$$\begin{cases} (x - \lfloor x \rfloor) * e^{(\lfloor x \rfloor / S_w)} \\ (y - \lfloor y \rfloor) * e^{(\lfloor y \rfloor / S_h)} \\ (w - \lfloor w \rfloor) * e^{(\lfloor w \rfloor / S_w)} \\ (h - \lfloor h \rfloor) * e^{(\lfloor h \rfloor / S_h)} \end{cases} \tag{2}$$

In the formula, S_w and S_h represent the width and height of the heatmap. We multiply the offset with the coordinates and size, thus incorporating spatial information into offset. The reason for using exp is that after quantization, the coordinates and sizes may become 0. The division by the size of heatmap is for normalization.

3.4 Label Assignment

Due to the fact that the predicted set is far greater than the ground-truth set, we need to divide the predicted set into positive and negative samples, with the number of positive equaling the number of ground-truth items. We use the Hungarian algorithm to assign labels, and the cost matrix will be constructed using classification cost, L1 cost, and GIOU cost. When computing the L1 and GIOU costs between predicted results and ground-truth, we first map these three components (position heatmap, size heatmap, and offset) to the image space to obtain specific bounding boxes. Subsequently, by utilizing the Hungarian algorithm, we perform a one-to-one pairing match between the predicted boxes and ground-truth boxes. The reason for selecting this holistic approach for calculation is because if we separately calculate the costs for these three components, the weightings of each component are not easy to balance, which could lead to a certain part dominates the allocation of samples.

3.5 Loss Function

For the calculation of the losses, we did not convert the output into bounding boxes as in calculating costs, but instead calculated the losses for each component of the model separately.

For the prediction of categories, we defined the output format of the network as (B, N, C + 1), where B is the batch size, N is the fixed size of the set, and C is the number of categories. We set C + 1 categories in total, with the additional one defined as the background. In order to avoid the interference of a large number of background classes, we set the weight of the background to 0.1. For the calculation of classification loss, we chose to use binary cross-entropy function, as shown in Eq. 3.

$$
L = \frac{-1}{pos + neg * 0.1} \sum_i^N
\begin{cases}
[y_i \log(\hat{y}_i) + (1 - y_i) \log(1 - \hat{y}_i)] & \text{if } y_i \text{ not background} \\
[y_i \log(\hat{y}_i) + (1 - y_i) \log(1 - \hat{y}_i) * 0.1] & \text{otherwise}
\end{cases}
\tag{3}
$$

where *pos* signifies the quantity of positive samples, and *neg* signifies the quantity of negative samples, $pos + neg = N$.

When calculating the loss of the heatmap, instead of equally penalizing negative locations, we reduce the penalty given to negative locations within a radius of the positive location. This is because even if a negative bounding box is close enough to its ground-truth, it can still result in a bounding box that overlaps sufficiently with the ground-truth box. In our model, we use Gaussian heatmaps for both the position heatmaps and the size heatmaps when labeling ground-truth values. The outputs of these two heatmaps are both formatted as (B, N, H, W), where N is a fixed set size and H, W are the size of heatmaps. We use Gaussian Focal Loss to calculate the losses, all channels are involved in the calculation, as shown in Eq. 4.

$$
L = \frac{-1}{pos + neg * 0.1} \sum_{n=1, y=1, x=1}^{N, H, W}
\begin{cases}
(1 - p_{nyx})^\alpha \log(p_{nyx}) & \text{if } g_{nyx} = 1 \\
(1 - g_{nyx})^\beta (p_{nyx})^\alpha \log(1 - p_{nyx}) & \text{otherwise}
\end{cases}
\tag{4}
$$

where H and W are the size of heatmap, α and β are two hyperparameters, we use $\alpha = 2$ and $\beta = 4$. p_{nyx} is the prediction value in (x, y), and the weight of penalty g_{nyx} at location (x, y) is calculated based on the Gaussian radius r, as shown in Eq. 5.

$$
g_{nyx} = \exp(\frac{(x - \hat{x})^2 + (y - \hat{y})^2}{-2\varphi^2})
\tag{5}
$$

where $(\hat{x}, \hat{y})$ denote the positive coordinates, and $x \in [\hat{x} - r, \hat{x} + r], y \in [\hat{y} - r, \hat{y} + r], \varphi$ is an object size-adaptive standard deviation, default $\varphi = \frac{2r+1}{6}$.

When calculating the offset loss, we use the SmoothL1 loss. Unlike the calculation for heatmaps, we only select positive samples for calculation, because in the offset values, 0 represents a distance, while in heatmaps, 0 represents "none". These have completely different properties, and it is meaningless to calculate the offset for a non-existent bounding box.

4   Experiment

Our experiments were carried out on the PASCAL VOC 2007 + 2012 [28] dataset. On the PASCAL VOC dataset, we used 17K labeled images from the entire dataset for training and 2K labeled images for validation. Training was conducted using a single A100 GPU. Experimental results demonstrate that the model is effective in distinguish objects with overlapping centers because each object has its own independent heatmap, as shown in Fig. 5.

Fig. 5. When the centers of the objects overlap, the heatmap channels of the two objects are independent of each other, so it does not hinder the model discrimination.

We set the input image dimensions to 512×512 and chose ResNet-50 as the backbone. After downsampling, the size of feature map fed into the transformer is 16×16. The output from the encoding layer undergoes upsampling via transposed convolution, resulting in our final heatmap size of 64×64. We counted the number of parameters and Flops of other heatmap-based models, as shown in Table 1. In contrast, the number of parameters and Flops of our model are far less than them. We compared our model's detection performance with other models, demonstrating that our model achieves good detection results while having significantly fewer parameters and computation requirements than other models, as shown in Table 2.

We find that these models use HourglassNet to extract the underlying features. The advantage of HourglassNet is that it can output a high-resolution feature map, capture and integrate the information of all scales of the image, but the cost is that it needs to pay a huge amount of calculations and parameters, as shown in Table 3. In addition, these models also use keypoint pooling to improve the detection effect, which also requires a lot of computing resources. In contrast, our model is much more lightweight.

Table 1. Our model has far fewer parameters and Flops than other heatmap-based models.

Model	Flops	Params
CornerNet [7]	452.96G	201.04M
CentripetalNet [27]	491.70G	205.76M
CenterNet [9]	292.70G	191.25M
TransCenter 16 × 16 (Our)	25.07G	45.79M
TransCenter 32 × 32 (Our)	28.81G	46.84M
TransCenter 64 × 64 (Our)	43.74G	47.89M
TransCenter 128 × 128 (Our)	103.46G	48.94M

Table 2. AP comparison of our model with other heatmap-based models.

Model	$AP_{0.5:0.95}$	$AP_{0.5}$	$AP_{0.75}$	AP_S	AP_M	AP_L
CenterNet	54.4	78.2	59.5	16.7	34.1	63.9
CornerNet	55.9	72.3	59.3	12.7	35.6	64.7
CentripetalNet	57.2	77.0	61.0	26.1	37.5	65.6
TransCenter (Our)	**53.7**	**77.6**	**58.7**	**17.4**	**32.3**	**64.9**

Table 3. Parameters and Flops of each module.

Model	Flops	Params
HourglassNet [24]	234.522G	187.7M
ResNet-50 [25]	20.366G	23.508M
CornerPooling [7]	25.3G	1.542M
CenterPooling [27]	39.812G	2.427M

4.1 Lower Limit of Model

In the experiment, we adjusted the size of the heatmap to 16 × 16, 32 × 32, 64 × 64 and 128 × 128. After many comparative experiments, we find that the lower limit of our model is much higher than other models. Our model has reached 44.9 (16 × 16), 42.3 (32 × 32), 40.5 (64 × 64) and 38.2 (128 × 128) mAP0.5 in the initial rounds of training, as shown in Fig. 6. Two points can be seen from this set of data, first, the lower limit of the model is inversely proportional to the size of the heatmap. Secondly, our model only needs less time cost to achieve a relatively satisfactory detection effect. At present, the evaluation indicators of the model are all aimed at the upper detection limit of the model, but we believe that a higher lower limit of the model can make more trade-offs between time cost, equipment cost and detection effect.

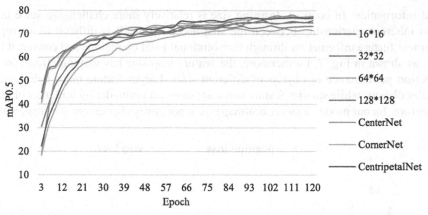

Fig. 6. It can be seen from the curve that mAP0.5 of our model is much higher than other models in the early training period.

4.2 Upper Limit of Model

Table 4 shows the detection effect of the model with different scales of heatmaps. We find that the size of the heatmap is not necessarily proportional to the upper limit of the model. Because the size affects the model in many ways. As the heatmap size decrease, the position and size prediction become easier. However, this does not necessarily translate to better model performance. Smaller size makes position and size predictions easier, but also rough. Consequently, the role of offset becomes much more apparent. Suppose the input image size is 512×512, and the output heatmaps are 16×16. In this case, a 0.5 offset maps to $0.5/16 \times 512 = 16$ pixels in the original image. As the size increase, the position and size prediction become more difficult. Conversely, the impact of offset on the prediction results will diminish.

Table 4. We conducted several comparative experiments and proved that the heatmap size of 64×64 has the best detection results. Moreover, the Flops and the number of parameters at this size are only 15% and 25% of CenterNet's respectively. Compared to CentripetalNet, Flops has only 8.9% of it and 23.3% of its parameters.

Model	$AP_{0.5:0.95}$	$AP_{0.5}$	$AP_{0.75}$
TransCenter(16×16)	50.5	73.5	56.3
TransCenter(32×32)	52.4	75.1	57.7
TransCenter(64×64)	53.7	77.6	58.7
TransCenter(128×128)	52.8	75.8	59.1

In addition to this, the heatmap's learning capabilities for position and size differ. Regarding position learning, an object's feature information generally gathers at its location, making convolution operations well-suited since their role is to aggregate local information. In contrast, learning size is relatively more challenging since there is no inherent relationship between the size and position. It is difficult to aggregate complete feature information through convolutional local operations and convert it into size, as shown in Fig. 7. Furthermore, the feature map size has varying effects on the detection performance of objects of different sizes. Larger feature maps help capture smaller objects, while smaller feature maps are more accommodating for larger objects. Therefore, for our model, a larger heatmap size is not always better, nor is a smaller size.

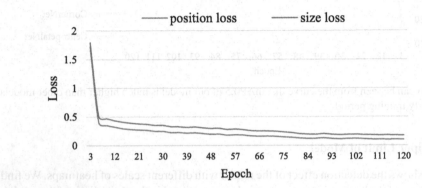

Fig. 7. The position loss decreases faster than the size loss, which fully demonstrates that the model's ability to learn size and position is not the same.

5 Conclusion

We proposed a novel method to predict the position and size of the bounding box in the form of heatmap, so as to greatly reduce the solution space, which is more conducive to the learning of the model, and also improves the prediction lower limit of the model. We decouple the position prediction task from the category prediction task, thus thoroughly solving the problem of keypoint overlap in heatmap-based models. Although the current experimental results are not enough to reach the level of SOTA, but compared with other heatmap-based models, we have fewer parameters and less computation, and this cost is completely acceptable. We will continue this research direction and continue to optimize our model to achieve better detection results.

Acknowledgment. This work was supported by the Key R&D Plan of Shandong Province, China (No.2021CXGC010102).

References

1. Redmon, J., Divvala, S., Girshick, R., Farhadi, A.:You only look once: unified, real-time object detection, In: 2016 IEEE Conference on Computer Vision and Pattern Recognition (CVPR), Las Vegas, NV, USA, 2016, pp. 779–788, doi: https://doi.org/10.1109/CVPR.2016.91

2. Redmon, J. Farhadi, A.:YOLO9000: better, faster, stronger, In: 2017 IEEE Conference on Computer Vision and Pattern Recognition (CVPR), Honolulu, HI, USA, 2017, pp. 6517–6525, doi: https://doi.org/10.1109/CVPR.2017.690

3. Redmon, J., Farhadi, A.: YOLOv3: An Incremental Improvement." ArXiv abs/1804.02767 (2018): n. pag

4. Bochkovskiy, A., et al.: YOLOv4: Optimal speed and accuracy of object detection. ArXiv abs/2004.10934 (2020): n. pag

5. Girshick, R., Donahue, J., Darrell, T., Malik, J.: Rich feature hierarchies for accurate object detection and semantic segmentation, In: 2014 IEEE Conference on Computer Vision and Pattern Recognition, Columbus, OH, USA, 2014, pp. 580–587, doi: https://doi.org/10.1109/CVPR.2014.81

6. He, K., Zhang, X., Ren, S., Sun, J.: Spatial pyramid pooling in deep convolutional networks for visual recognition. IEEE Trans. Pattern Anal. Mach. Intell. 37(9), 1904–1916 (2015). https://doi.org/10.1109/TPAMI.2015.2389824

7. Law, H., Deng, J.: CornerNet: detecting objects as paired keypoints. Int. J. Comput. Vis. 128, 642–656 (2020)

8. Duan, K., Bai, S., Xie, L., Qi, H., Huang Q. Tian, Q.: CenterNet: keypoint triplets for object detection, In: 2019 IEEE/CVF International Conference on Computer Vision (ICCV), Seoul, Korea (South), 2019, pp. 6568–6577, doi: https://doi.org/10.1109/ICCV.2019.00667

9. Zhou, X., Koltun, V., Krähenbühl, P.: Tracking Objects as Points. In: Vedaldi, A., Bischof, H., Brox, T., Frahm, J.-M. (eds.) ECCV 2020. LNCS, vol. 12349, pp. 474–490. Springer, Cham (2020). https://doi.org/10.1007/978-3-030-58548-8_28

10. He, K., Gkioxari, G., Dollar, P., Girshick, R.: Mask R-CNN. In Int. Conf. Comput. Vis., pp 2961–2969, 2017

11. Kreiss, S., Bertoni, L., Alahi, A., Paf, P.: Composite fields for human pose estimation. In IEEE Conf. Comput. Vis. Pattern Recog., pages 11977–11986, 2019

12. Ren, S., He, K., Girshick, R., Sun, J.: Faster R-CNN: towards real-time object detection with region proposal networks. IEEE Trans. Pattern Anal. Mach. Intell. 39(6), 1137–1149 (2017). https://doi.org/10.1109/TPAMI.2016.2577031

13. Girshick, R B.: Fast R-CNN.In: 2015 IEEE International Conference on Computer Vision (ICCV), pp .1440–1448 (2015)

14. Dai, J., Li, Y., He, K., Sun. J.:2016. R-FCN: object detection via region-based fully convolutional networks. In Proceedings of the 30th International Conference on Neural Information Processing Systems (NIPS'16). Curran Associates Inc., Red Hook, NY, USA, pp. 379–387

15. Nibali, Aiden et al. Numerical Coordinate Regression with Convolutional Neural Networks." ArXiv abs/1801.07372 (2018): n. pag

16. Jin, H., Liao, S., Shao, L.: Pixel-in-Pixel Net: towards efficient facial landmark detection in the wild. Int. J. Comput. Vision 129(12), 3174–3194 (2021). https://doi.org/10.1007/s11263-021-01521-4

17. Carion, N., Massa, F., Synnaeve, G., Usunier, N., Kirillov, A., Zagoruyko, S.: End-to-End Object Detection with Transformers. In: Vedaldi, A., Bischof, H., Brox, T., Frahm, J.-M. (eds.) ECCV 2020. LNCS, vol. 12346, pp. 213–229. Springer, Cham (2020). https://doi.org/10.1007/978-3-030-58452-8_13

18. Zhou, X., et al. Bottom-up object detection by grouping extreme and center points.In: 2019 IEEE/CVF Conference on Computer Vision and Pattern Recognition (CVPR), pp 850–859 (2019)

19. Vaswani, A., et al.: Attention is all you need. In Proceedings of the 31st International Conference on Neural Information Processing Systems (NIPS'17). Curran Associates Inc., Red Hook, NY, USA, pp 6000–6010 (2017)

20. Ramachandran, P., Parmar, N., Vaswani, A., Bello, I., Levskaya, A., Shlens, J. Stand-alone self-attention in vision models. In NIPS, pp 68–80, 2019

21. Bello, I., Zoph, B., Le, Q., Vaswani, A., Shlens. J.: Attention augmented convolutional networks. In ICCV, pp. 3286–3295, 2019
22. Yang, S., Quan, Z., Nie, M., Yang, W.:TransPose: keypoint localization via transformer,In: 2021 IEEE/CVF International Conference on Computer Vision (ICCV), Montreal, QC, Canada, 2021, pp. 11782–11792, doi: https://doi.org/10.1109/ICCV48922.2021.01159
23. Erhan, D., Bengio, Y., Courville, A., Vincent, P.: Visualizing higher-layer features of a deep network. Tech. Rep. Univ. Montreal **1341**(3), 1 (2009)
24. Newell, A., Yang, K., Deng, J.: Stacked Hourglass Networks for Human Pose Estimation. In: Leibe, B., Matas, J., Sebe, N., Welling, M. (eds.) ECCV 2016. LNCS, vol. 9912, pp. 483–499. Springer, Cham (2016). https://doi.org/10.1007/978-3-319-46484-8_29
25. He, K., Zhang, X., Ren, S., Sun, J.: Deep residual learning for image recognition. IEEE Conf. Comput. Vis. Pattern Recogn. (CVPR) **2016**, 770–778 (2016). https://doi.org/10.1109/CVPR. 2016.90
26. Zeiler, M.D., Krishnan, D., Taylor, G.W., Fergus, R.: Deconvolutional networks. IEEE Comput. Soc. Conf. Comput. Vis. Pattern Recogn. **2010**, 2528–2535 (2010). https://doi.org/10.1109/CVPR.2010.5539957
27. Dong, Z., Li, G., Liao, Y., Wang, F., Ren, P., Qian, C.: CentripetalNet: pursuing high-quality keypoint pairs for object detection. IEEE/CVF Conf. Comput. Vis. Pattern Recogn. (CVPR) **2020**, 10516–10525 (2020). https://doi.org/10.1109/CVPR42600.2020.01053
28. Everingham, M., Gool, L.V., Williams, C.K.I., Winn, J., Zisserman, A.: The pascal visual object classes (VOC) challenge. Int. J. Comput. Vis. **88**(2), 303–338 (2010). https://doi.org/10.1007/s11263-009-0275-4

Causal-Inspired Influence Maximization in Hypergraphs Under Temporal Constraints

Xinyan Su[1,2], Jiyan Qiu[1,2], Zhiheng Zhang[3], and Jun Li[1(✉)]

[1] Computer Network Information Center,
Chinese Academy of Sciences, Beijing, China
[2] University of Chinese Academy of Sciences, Beijing, China
{suxinyan,qiujiyan,lijun}@cnic.cn
[3] Institute for Interdisciplinary Information Sciences,
Tsinghua University, Beijing, China
zhiheng-20@mails.tsinghua.edu.cn

Abstract. Influence Maximization is a significant problem aimed to find a set of seed nodes to maximize the spread of given events in social networks. Previous studies are contributing to the efficiency and online dynamics of basic IM on classical graph structure. However, they lack an adequate consideration of individual and group behavior on propagation probability. This can be attributed to inadequate attention given to node **I**ndividual **T**reatment **E**ffects (ITE), which significantly impacts the probability of propagation by dividing the sensitive attributes of nodes. Additionally, current research lacks exploration of temporal constraints in influence spreading process under higher order interference on hypergraphs. To fill these two gaps, we introduce two sets of basic assumptions about the impact of ITE on the propagation process and develop a new diffusion model called the Latency Aware Contact Process on Causal Independent Cascading (LT-CPCIC) under time constraints on hypergraphs. We further design Causal-Inspired Cost-Effective Balanced Selection algorithm (CICEB) for the proposed models. CICEB first recovers node ITE from observational data and applies three types of debiasing strategies, namely DebiasFormer, DebiasCur and DebiasInteg, to weaken the correlation between the propagation effects of different pre- and post-nodes. Finally, we compare CICEB with traditional methods on two real-world datasets and show that it achieves superior effectiveness and robustness.

Keywords: Influence Maximization · Hypergraphs · Causal Inference

1 Introduction

Social networks have witnessed a booming in the propagation of information in recent years. This phenomenon has given rise to a surge of interest in the laws of rumor spreading [17], viral marketing [20], products promotion [3], etc.

© The Author(s), under exclusive license to Springer Nature Singapore Pte Ltd. 2024
B. Luo et al. (Eds.): ICONIP 2023, CCIS 1967, pp. 295–308, 2024.
https://doi.org/10.1007/978-981-99-8178-6_23

Consequently, it has become essential to address the question of how to achieve maximum impact. In other words, how can we ensure that a message is delivered to the largest possible audience or that the maximum number of individuals accept a suggestion from propagators? This problem is known as Influence Maximization (IM), which seeks to find a set of K nodes to maximize the influence of a particular event under certain diffusion model [8].

Recent studies primarily focus on exploring the online dynamics [11], efficiency [21] and fairness [1] of IM within basic graph structures. However, current research faces two key challenges, one internal in structure and the other external on assumptions. Internally, it is necessary to consider IM with more complex relationships between individuals due to the insufficient representation power on relationships of only two nodes connected. But, there is still much that needs to be done in modeling complex interactions among individuals and their belonging group units in a network. Externally, most IM scenarios make assumptions about individual behavior relating to the diffusion process. However, in practice, imposing different outer prior-conditions can result in significant variation in individual choices, which impacts the diffusion process. Unfortunately, existing algorithms have not adequately accounted for these differences, and thus may fall short in accurately modeling diffusion dynamics. Besides, IM problems in real-world settings often require timely resolution due to the time-sensitive nature of events involved. Given the recent research focus on time-related IM [4,15], it is necessary to impose strict time constraints when considering the aforementioned challenges to ensure alignment with relevant contextual restrictions. To illustrate, suppose a primary school aims to educate all students on the harms of smoking within a week. However, the school faces the constraint of limited speakers and devices such that only a selected few students can be educated. Additionally, small groups exist among the students, and the school intends to educate them by finding key nodes to spread the key message. Furthermore, it is assumed that the students have never received similar education before, and the lecture contains a video showing the physical damage caused by smoking, potentially eliciting a shock response. Each student will have a different stress response before and after receiving the lecture, and those experiencing greater mood swings will exhibit a relatively more negative attitude towards spreading the message, potentially affecting the overall optimal solution. A similar problem can arise with individuals who are more sensitive to price in medicine sales. Thus, it is critical to consider the above two challenges under time constraints when developing effective resolutions for IM-related problems.

Previous studies have attempted to address the two challenges, but with certain limitations. To address the internal structure problem, research on community influence maximization has explored relationships within communities. These studies focus on identifying core nodes in each community as a representation of the overall solution [5,7].This approach significantly differs from our objectives, which aim to provide a global solution while considering relationships within more complex cross-clusters. Moreover, IM in Hypergraphs models complex relationships efficiently but faces difficulties in representing the propagation

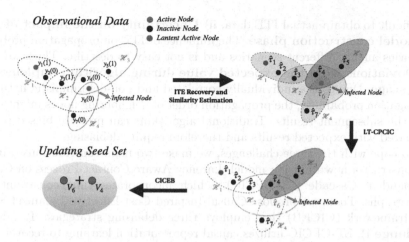

Fig. 1. The process of CICEB under LT-CPCIC. We select the red-colored node as the current active node and restore the ITE values for all nodes. Next, the hyperedge with the most similar ITE values is chosen to be propagated to proceed the propagation process. We then use CICEB to select the next seed. (Note that during this process, ITE values serve two purposes: 1) as a self-representation for each node, where each node tends to propagate towards nodes with similar ITE values, and 2) influencing the propagation probability). In this case, hyperedge $\mathcal{H}_3$ (in dark pink) with the highest similarity is chosen to be propagate. (Color figure online)

process at the group level of the network, particularly under time constraints [22]. As for the external interference problem, learning-based methods incorporating outer factors affecting nodes, with the goal of learning the individual diffusion pattern or node embedding in place of specific propagation processes, have been proposed in previous research [12,16]. Nevertheless, it is worth noting that the learned characteristics may not necessarily reflect the impact of external interventions in real situations because they only correlate with the impact and cannot causally affect it. Thus, inaccurate impact estimation could indirectly impact the quality of the outputs and model transferability.

To conclude, addressing the above mentioned challenges demands not only modeling the propagation process related to the group's internal and external relationships but also focusing on changes in propagation propensity among various individuals subject to specific interventions, or in other words, the Individual Treatment Effect(ITE), which has a great impact on propagation. Existing models primarily leverage global propagation probability modeling or no modeling at all, which poses a challenge in accurately depicting the change of propensity. Therefore, incorporating ITE to model the propagation probability of various individuals in the propagation process may significantly enhance effectiveness. However, creating an algorithm under the influence of ITE poses three significant challenges: i) **Data loss in the data collection phase**. While collecting observable data, only one state is obtainable under external treatment, making it difficult to obtain actual ITE data. ii) **Uncertainty on the impact of ITE in model construction phase**. The influence of ITE on propagation probability varies among different scenarios and is not easy to describe. Thus, we have iii) **Deviations from the expected value during algorithm implementation stage**. Due to the individually correlated and non-deterministic nature of propagation probability, the propagation order of neighboring nodes might influence the subsequent results. Traditional algorithms can produce biased results compared with expected results and therefore require debiasing.

To cope with the three challenges, we make two main contributions. Firstly, we construct a new diffusion model, Latency Aware Contact Process on Causal Independent Cascade (LT-CPCIC), which incorporates time constraints on hypergraphs. Then, we propose Causal-Inspired Cost-Effective Balanced Selection framework (CICEB) that employs three debiasing strategies. To address **challenge i)**, LT-CPCIC utilizes causal representation learning to recover individual treatment effects (ITE) of each node from observed data. As for **challenge ii)**, the influence of each node is estimated based on the positive or negative effects of ITE on individual propagation intentions during its propagation process. Lastly, to settle **challenge iii)**, the algorithm removes interference from the selection process of each seed node. With different probabilities, it takes into account the effect of back-and-forth propagation orders between neighboring nodes. Figure 1 displays an overview of the entire process. While our probabilistic model, accounting for individual correlations, produces more realistic results than alternative approaches, it exhibits a drawback of overparameterization. Nevertheless, our method demonstrates low sensitivity to hyperparameters in experiments.

Three main contributions of our paper can be summarized as follows: **1)** We rethink the problem of individual relevance in propagation probabilities. Unlike previous parameter-free methods and deep learning strategies, we present two reasonable assumptions on how an individual's ITE can impact their propagation willingness in real-world scenarios. **2)** We propose a novel propagation model LT-CPCIC to estimate influence of nodes under temporal constraints and design a corresponding influence maximization algorithm CICEB, which is aligned with our problem settings. **3)** We experimentally validate the effectiveness and robustness of our proposed framework on two real-world datasets.

2   Related Work

In this section, we will give a brief review of influence maximization and causal representation learning.

2.1   Influence Maximization

Influence Maximization (IM) is algorithmically recognized first in literature [8]. It aims at finding a K-size seed set S to maximize the influence under certain diffusion model P on a graph $\mathcal{G}$, which can be formalized as: $\sigma(S) =$

$\arg max_{S \subseteq \mathcal{V} \wedge |S|=K} \sigma(S_0)$, where $\mathcal{G} = (\mathcal{V}, \mathcal{E})$, where $\sigma(\cdot)$ is the influence function. Current research on IM can be divided into two groups: classical IM and context-aware IM. The former can be classified into four types: simulation-based method [8,10], sketch-based method [21], heuristic method and learning-based method [15]. However, classical IM assumes an unlimited time length, which is impractical for most real-life scenarios that require time-sensitivity. In contrast, context-aware IM analyzes different factors' impact, where the most crucial type is time-constraint IM. Previous studies mostly focus on both discrete and continuous time models on traditional graphs and neglect hypergraphs. Additionally, there is a lack of research on the impact of individual and group behavior.

2.2   Causal Representation Learning

Causal learning essentially models interference effects via the process of recovering missing data from observational data [18]. In our study, we focus on the outcome of nodes ($eg : n_0$) under treatment or non-treatment, which can be represented as $y_0(n_0)$ and $y_1(n_0)$, respectively. However, only one outcome can be observed, making the recovery of the other one a challenge that falls under the domain of causal learning. Two strategies exist for estimating the missing outcome: weight-based methods and representation-based methods. Studies have shown that deep learning-based methods bring about increased robustness and explainability [6,9]. Thus, our paper adopts the representation-based approach using deep learning-based models.

3   Problem Formulation

We present a brief description of notations in Table 1, then we define the problem formally.

We define the problem as follows. Given a hypergraph $\mathcal{G}(\mathcal{V}, \mathcal{E})$, nodes spread under diffusion model D with a time bound T. A chosen initial seed set S_0 is given at the beginning of the propagation, and the final result is represented by S_K. Three types of node states during propagation are: active, latent active and inactive, where the number of activated nodes of seeds are noted by the influence function σ_T Here, diffusion model D is LT-CPCIC we proposed. With the impact of each node ITE τ on the diffusion process, our objective is to identify a seed set S_T of size K that maximizes the spread of nodes within time bound T starting from S_0, which is formulated as:

$$\arg\max\{\sigma_T(S_0)\}, s.t. |S_0| = K. \tag{1}$$

Then we introduce two main assumptions below.

Assumption 1 *(Basic ITE settings)*

1) Bounded ITE: The ITE is constraint to a constant L, which can be expressed as $\max_{v \in \mathcal{V}} |\tau_v| \leq L$.

Table 1. Notations in our paper.

Symbol	Descriptions
$\mathcal{G}(\mathcal{V}, \mathcal{E})$	hypergraph (nodes $\mathcal{V}$, hyperedges $\mathcal{E}$)
m, n	the indices of nodes and hyperedges
X_i	the covariate of node i
$\mathcal{N}_e$	the set of nodes in hyperedges $e \in \mathcal{E}$
$\mathcal{H}_i$	the set of hyperedges containing node i
$P_{i,j}, P_{i,h_{iq}}$	actual diffusion probability from node i to node j/ one of its hyperedges $h_{iq} \in \mathcal{H}_i$
p_i^{lat}	distribution of propagation latency of node i
$p_{i,j}$	probability node i activates node j under no external assumptions
$Sim_{i,h_{iq}}$	the similarity of node i and hyperedge $h_{iq} \in \mathcal{H}_i$
S_i	the active nodes in each iteration (finial seed set: S_K)
$y_i(1), y_i(0)$	the potential outcome of each node i
$\tau_i, \hat{\tau}_i$	the true/estimated ITE of each node i

2) *Consistency and unconfoundedness: The intervention level is the same for all individuals, and potential outcomes for each individual do not depend on the intervention status of other individuals. Specifically, if the environmental summary function is noted as $o_i = D(\mathcal{G}, T_{-i}, X_{-i})$, then we have, $y_i(1), y_i(0) \perp\!\!\!\perp \{T_i, o_i\} \mid x_i$.*

3) *Expressiveness of the environmental summary function [14]: The potential outcomes $y_i(1), y_i(0)$ can be determined given x_i as well as the environmental function.*

Assumption 2 *(External impact on propagation process)*

1) *The individual's own characteristic variables do not influence the propagation itself, and only one external treatment is considered in this context*

2) *Each node prefers to propagate to hyperedges that are similar in degree to the extent to which they were affected before and after the intervention, which is referred to ITE.*

3) *The impact of ITE on individual propagation probability can be expressed as two types, denoted as ϕ. The first type of impact is positive and is power-law correlated with ITE. Specifically, it can be expressed as $\phi^P = \alpha e^{ite_i - L_0}$, where L_0 is a hyperparameter and α is the ratio of influence. The second one is negative and can be notated as $\phi^N = \alpha e^{-ite_i - L_0}$.*

4   Proposed Framework

In this part, we first introduce the proposed propagation model LT-CPCIC, which describes the diffusion process in the presence of external intervention. We then present our designed framework of a causal-inspired influence maximization algorithm called CICEB. This algorithm utilizes the LT-CPCIC model to estimate influence, with reference to Assumptions 1 and 2.

4.1   Latency Aware Contact Process on Causal Independent Cascading Model

To accurately quantify the propagation power of nodes on hypergraphs under the influence of individual ITE with time constraints, we propose the Latency Aware Contact Process on Causal Independent Cascading Model(LT-CPCIC). In this model, LT indicates the activating latency feature of the diffusion process referring to [13], while CP represents the way messages are passed to specific groups, which aligns with the information spreading routines follow [19]. Based on the problem settings described in Sect. 3, we define the state of nodes as $St_v \in \{AC, INAC, PAC\}$, where AC represents active, $INAC$ denotes inactive and PAC signifies pending active. An active node will activate inactive nodes in groups (hyperedge) that have the most similar ITE level, which is measured by Sim. The activated nodes at time step t will first become pending active, then transmit to active after a time decay $\delta(t)$, where $\delta(t) \sim p_u^{lat}$. Due to the temporal sequence, when one node is activated simultaneously, it will get the minimum activating time decay. The detailed process is illustrated as follows:

– **Step 1.** Initially at time step 0, the state of all nodes in the active set S_a are AC and the rest nodes are $INAC$ state. The time limit of the model is denoted as T.

– **Step 2.** At time step t, each node i in S_a first selects one hyperedge in its hyperedge set $\mathcal{H}_i$. Each edge $h_{iq} \in \mathcal{H}_i$ has a probability $P_{i,h_{iq}}$ of being chosen. Then, the node i activates $INAC$ nodes j in the chosen hyperedge with a probability $P_{i,j}$. For the successfully activated nodes: if their state is $INAC$, they change to the PAC state with an activating time decay $\delta(t)$, sampled from $P_i^{lat}(t)$, where $i \in S_a$. Their real activating time is stored as $t + \delta(t)$. Otherwise, if the nodes are already in the PAC state, their activating time is updated to the minimum of $t + \delta(t)$.

– **Step 3.** Check nodes with PAC state and change their state to AC if their activating time is t, then put them into S_a set.

– **Step 4.** Repeat procedure **Step2**, **Step3** until time step arrives T.

Based on Assumption 2, the probability $P_{i,h_{iq}}$ for choosing a hyperedge, as well as the probability $P_{i,j}$ for node propagation, are determined by the group ITE level and individual ITE level, respectively. These probabilities are defined by Eqs. 2, 3 and 4, where β_0, λ_0, α_0 are adjustable parameters, and L is constant. It should be noted that the correlation between $P_{i,j}$ and node ITE can be either positive or negative, depending on the specific settings, and is determined by the $sign(.)$ function in this case.

$$P_{i,h_{iq}} = \frac{Sim_{i,h_{iq}}}{\sum_{h_{il} \in \mathcal{H}_i} Sim_{i,h_{il}}} \qquad (2)$$

$$Sim_{i,h_{iq}} = \beta_0(\hat{\tau}_i - \mu_i) + \lambda_0 \sqrt{\frac{\sum_{v_j \in \mathcal{N}_i}(\hat{\tau}_{v_j} - \mu_i)^2}{|\mathcal{N}_i|}}, \mu_i = \frac{\sum_{v_j \in \mathcal{N}_i} \hat{\tau}_{v_j}}{|\mathcal{N}_i|} \qquad (3)$$

$$P_{i,j} = \alpha_0 \exp\left(sign(\hat{\tau}_i) - L\right) + (1 - \alpha_0)p_{i,j} \qquad (4)$$

To elaborate, we give a specific spreading example in Fig. 1.

4.2   Causal-Inspired Cost-Effective Balanced Selection Algorithm

We propose the CICEB framework to identify the most influential seeds. This framework consists of three strategies aimed at balancing the biased outcomes caused by multiple probabilities in different diffusion chains. We employ a simulation-based approach as the underlying strategy due to the limitations on the a priori assumptions of the propagation model and theoretical guarantees of the results. Based on the general algorithm CELF [10], CICEB calculates the influence of each node using the LT-CPCIC model described in Sect. 4.1. It ensures theoretical guarantees with a monotonous and submodular σ_T, which represents our strategies. The details of CICEB are provided in Algorithm 1. In lines 8–16 of Algorithm 1, each node is updated with the largest relative marginal gain under the balanced strategy. The three balancing solutions are presented in Algorithm 2. Let the current seed set be denoted as S_t. The first strategy, *DebiasFormer*, aims to balance the deviation from the expected influence caused by the ITE personalized probability of the parent node. It computes the mean value of the marginal gain of each new node under the current seed set S_t and the marginal gain under the seed set in the previous step S_{t-1}, as shown in lines 3–6. The second solution, *DebiasCur*, calculates the mean value of the marginal gain of each new node under the current seed set and the marginal gain under the current seed set with an additional randomly selected node r, in order to eliminate the bias introduced by the node's selection in the next step, as presented in lines 8–11. The third strategy, *DebiasInteg*, integrates both of the aforementioned solutions and computes the mean of the three values, as indicated in lines 12–17.

5   Experiments

5.1   Experimental Settings

Datasets. To evaluate the performance of the proposed algorithms, we conducted experiments on two realistic hypergraph datasets: the contact-primary-school dataset and the Email-Eu dataset [2]. The contact-primary-school dataset, referred to as "Contact" for brevity, consists of a sequence of simplices representing hyperedges where students contact each other at specific timestamps. The Email-Eu dataset records email communications in chronological order. The problem settings for the Contact dataset are discussed in Sect. 1, while for the Email-Eu dataset, our objective is to identify key individuals who can maximize the spread of a shocking news within a given time limit. In this context, the treatment refers to receiving a lecture or not for the Contact dataset and knowing the news for the Email-Eu dataset, where ITE represents the degree of fear after receiving the treatment. We assume that the feature variable X_i for each node i varies across diverse groups and simulate them as follows:

$$X_i \sim \sum_{l=1}^{L} \omega_l \mathcal{N}(\mu_l, I), \tag{5}$$

where ω_l is the proportionality coefficient and μ_l is a constant.

Algorithm 1. Causal-Inspired Cost-Effective Balanced Selection (CICEB)

Require: Hypergraph $\mathcal{G}(\mathcal{V}, \mathcal{H})$, size of the seed set K, influence estimation function $\sigma(.)$, debiasing strategy type Ty.

1: Initialization: $S_0 = \emptyset$, $MargDic = \{\}$, $k = 0$.
   /* Get the marginal gain to $MargDic$ */
2: **for** $v \in \mathcal{V}$ **do**
3:    $MargDic[v] = \sigma(\{v\})$
4: **end for**
   /* Sort $MargDic$ in decreasing order of value */
5: **Sort** $(MargDic, Reverse = True)$
6: **for** $|S_0| < K$ **do**
7:    $check = False$
8:    **while** not $check$ **do**
9:       $cur =$**Top**$(MargDic)$
       /*Re-compute marginal gain*/
10:      $MargDic[cur] = $**StrategyFunc**$(S_0, cur, Ty)$
11:      **Sort** $(MargDic, Reverse = True)$
       /*Check if previous top node stays on top after each sort*/
12:      **if** $cur ==$**Top**$(MargDic)$ **then**
13:         $check = True$
14:      **end if**
15:      $S_0 = S_0 \cup \{cur\}$
16:    **end while**
17: **end for**
Ensure: The deterministic seed set S_0 with $|S_0| = K$.

Algorithm 2. StrategyFunc

Require: Hypergraph $\mathcal{G}(\mathcal{V}, \mathcal{H})$, influence estimation function $\sigma(.)$, current seed set S_t, current estimating node cur, debiasing strategy type Ty

1: Initialization: S_t, $MargDic = \{\}$, $k = 0$.
2: **ITE recovery**.
3: **if** $Ty == DebiasFormer$ **then**
4:     $a = \sigma(S_t \cup \{cur\}) - \sigma(S_t)$
5:     $b = \sigma(S_{t-1} \cup \{cur\}) - \sigma(S_{t-1})$
6:     $res = mean(a, b)$
7: **else if** $Ty == DebiasCur$ **then**
8:     $a = \sigma(S_t \cup \{cur\}) - \sigma(S_t)$
9:     $r = RandomChoice(\mathcal{V}, 1)$
10:     $b = \sigma(S_t \cup \{cur\} \cup \{r\}) - \sigma(S_t \cup \{r\})$
11:     $res = mean(a, b)$
12: **else if** $Ty == DebiasInteg$ **then**
13:     $a = \sigma(S_t \cup \{cur\}) - \sigma(S_t)$
14:     $b = \sigma(S_{t-1} \cup \{cur\}) - \sigma(S_{t-1})$
15:     $r = RandomChoice(\mathcal{V}, 1)$
16:     $c = \sigma(S_t \cup \{cur\} \cup \{r\}) - \sigma(S_t \cup \{r\})$
17:     $res = mean(a, b, c)$
18: **end if**
Ensure: The debiased marginal gain res of node cur.
Function ITE recovery:
Require: feature X_i, treatment T_i and environmental information T_{-i} of each node i in Hypergraph $\mathcal{G}(\mathcal{V}, \mathcal{E})$.
19: Balance $\{X_1, \cdots, X_{|\mathcal{V}|}\}$ through weighting strategy follow [14]
20: $P_v = ENV(X_v, \mathcal{G}, T_{-v}, t_v) for v \in \mathcal{V}$
21: Rebalance (P_v) for all node $v \in \mathcal{V}$
22: $y_v(1) = \text{MLP1}(Z_v, P_v)$
23: $y_v(0) = \text{MLP0}(Z_v, P_v)$
24: $\tau_v = y_v(1) - y_v(0)$ for all node $v \in \mathcal{V}$
Ensure: The individual causal effect τ_v on each node.

Baselines and Parameter Settings. We set Round $= 20$ and K $= 15$, each set of experiment running for 10 times and compare the total performance of the results. We choose traditional CELF and random selection strategy as the baselines to evaluate our three strategies. And other parameters are set as: $\alpha_0 = \beta_0 = \lambda_0 = 0.5$, $p_0 = 0.01$. Our proposed algorithm as well as diffusion framework are performed with python 3.8 on Ubuntu 18.04 system.

5.2 Results and Analysis

In this section, we present and analyze the results obtained from our experiments conducted on the two datasets. Our analysis focuses on the following two key aspects:

Effectiveness: Performance Comparison and Ablation Analysis. To compare the performance of three methods, we initially obtain seeds using various methods. Subsequently, we estimate the final number of influenced nodes and the degree of randomness with a increasing sum of seeds. The results are presented in Fig. 4 and Fig. 3, and we can summarize them into three main conclusions. Firstly, from a comprehensive perspective, it is evident that DebiasCur outperforms both the random method and the traditional CELF method. Secondly, DebiasInteg demonstrates the second-best performance, while Debias-Former shows slightly lower performance than the random method in the Contact but significantly higher performance than random selection in Email-Eu. Lastly, in terms of algorithmic result stability, DebiasCur exhibits the least randomness. The maximum and minimum values increase smoothly as more seeds are considered, in contrast to the random method. In conclusion, our method shows substantial overall improvement compared to the traditional CELF method, confirming its efficacy. Additionally, the ablation analysis demonstrates the superiority of our method as we diminish the key balancing part of CICEB, which leads it to degenerate into CELF.

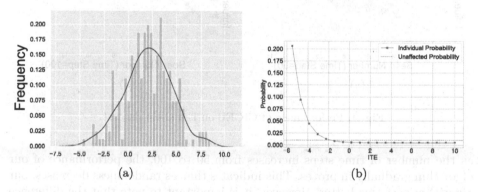

Fig. 2. a) ITE Distribution of Contact dataset. b) Variation of Individual Probability with ITE. The distribution of ITE itself approximates to a Gaussian distribution, while the actual individual propagation probability exhibits an exponential decrease trend as ITE increases. This aligns with the parameter settings specified in our main text.

Robustness: Analysis on Propagation Probability and Time Step Parameters. Figure 2(a) presents a model description of the ITE distribution on Contact, while Fig. 2(b) depicts the corresponding probability distribution. Observations from this study reveal that, in our experimental setting, personalized propagation probabilities show fluctuations that negatively correlate with ITE. This is in contrast to constant probability values that remain unaffected by external interference. These findings demonstrate the crucial role of our algorithm in bias correction and provide an explanation for its superior performance over traditional algorithms. Furthermore, we compare the performance of our algorithm with 50 and 100 time steps. The results show that

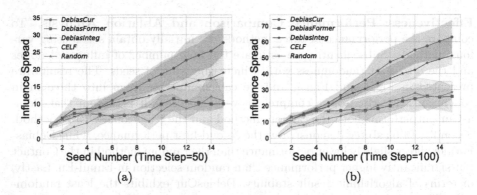

Fig. 3. Performance of CICEB on contact-primary-school dataset.

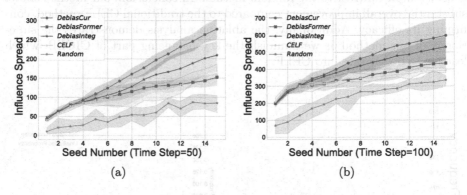

Fig. 4. Performance of CICEB on Email-Eu dataset.

as the number of time steps increases from 50 to 100, the performance of our algorithm gradually improves. This indicates that as randomness decreases, our algorithm performs better. However, it is important to note that the difference in performance between the two cases is small, suggesting the robustness of our algorithm. Besides, we add noise to ITE simulation results to further validate the robustness across variations in the parameter with the variation of P. More details can be find in https://github.com/suxinyan/CIIM-EXP/.

6   Conclusions and Future Work

We introduce a new propagation model LT-CPCIC that captures the time-constrained propagation process influenced by Individual Treatment Effects (ITE). In contrast to traditional approaches that focus on feature-based propagation modeling, our model explicitly considers the impact of ITE on individual propagation probabilities. Based on this new model, we design the CICEB algorithm for the first time, which incorporates three different strategies. Our experimental results demonstrate promising overall performance, with the DebiasCur algorithm exhibiting the best results.

Moving forward, we propose several promising future directions: 1) Provide theoretical proofs to support the guarantee of our algorithm. 2) Investigate the impact of parameter variations on the robustness of practical algorithms. 3) Explore the reasons underlying the insufficient performance of the DebaisFormer algorithm (we speculate that it may be related to the interference caused by varying sizes of parent nodes in the propagation chain).

References

1. Becker, R., D'angelo, G., Ghobadi, S., Gilbert, H.: Fairness in influence maximization through randomization. J. Artif. Intell. Res. **73**, 1251–1283 (2022)
2. Benson, A.R., Abebe, R., Schaub, M.T., Jadbabaie, A., Kleinberg, J.: Simplicial closure and higher-order link prediction. Proc. Natl. Acad. Sci. (2018)
3. Domingos, P., Richardson, M.: Mining the network value of customers. In: Proceedings of the Seventh ACM SIGKDD International Conference on Knowledge Discovery and Data Mining, pp. 57–66 (2001)
4. Erkol, Ş, Mazzilli, D., Radicchi, F.: Effective submodularity of influence maximization on temporal networks. Phys. Rev. E **106**(3), 034301 (2022)
5. He, Q., et al.: CAOM: a community-based approach to tackle opinion maximization for social networks. Inf. Sci. **513**, 252–269 (2020)
6. Hernan, M.A., Robins, J.M.: Causal Inference: What If. Chapman Hill/CRC, Boca Raton (2020)
7. Huang, H., Shen, H., Meng, Z., Chang, H., He, H.: Community-based influence maximization for viral marketing. Appl. Intell. **49**, 2137–2150 (2019)
8. Kempe, D., Kleinberg, J., Tardos, É.: Maximizing the spread of influence through a social network. In: Proceedings of the Ninth ACM SIGKDD International Conference on Knowledge Discovery and Data Mining, pp. 137–146 (2003)
9. Kilbertus, N., Rojas Carulla, M., Parascandolo, G., Hardt, M., Janzing, D., Schölkopf, B.: Avoiding discrimination through causal reasoning. In: Advances in Neural Information Processing Systems, vol. 30 (2017)
10. Leskovec, J., Krause, A., Guestrin, C., Faloutsos, C., VanBriesen, J., Glance, N.: Cost-effective outbreak detection in networks. In: Proceedings of the 13th ACM SIGKDD International Conference on Knowledge Discovery and Data Mining, pp. 420–429 (2007)
11. Li, S., Kong, F., Tang, K., Li, Q., Chen, W.: Online influence maximization under linear threshold model. In: Advances in Neural Information Processing Systems, vol. 33, pp. 1192–1204 (2020)
12. Ling, C., et al.: Deep graph representation learning and optimization for influence maximization. In: International Conference on Machine Learning, pp. 21350–21361. PMLR (2023)
13. Liu, B., Cong, G., Xu, D., Zeng, Y.: Time constrained influence maximization in social networks. In: 2012 IEEE 12th International Conference on Data Mining, pp. 439–448. IEEE (2012)
14. Ma, J., Wan, M., Yang, L., Li, J., Hecht, B., Teevan, J.: Learning causal effects on hypergraphs. In: Proceedings of the 28th ACM SIGKDD Conference on Knowledge Discovery and Data Mining, pp. 1202–1212 (2022)
15. Ma, L., et al.: Influence maximization in complex networks by using evolutionary deep reinforcement learning. IEEE Trans. Emerg. Top. Comput. Intell. (2022)

16. Panagopoulos, G., Malliaros, F.D., Vazirgianis, M.: Influence maximization using influence and susceptibility embeddings. In: Proceedings of the International AAAI Conference on Web and Social Media, vol. 14, pp. 511–521 (2020)
17. Pei, S., Makse, H.A.: Spreading dynamics in complex networks. J. Stat. Mech: Theory Exp. **2013**(12), P12002 (2013)
18. Schölkopf, B., et al.: Toward causal representation learning. Proc. IEEE **109**(5), 612–634 (2021)
19. Suo, Q., Guo, J.L., Shen, A.Z.: Information spreading dynamics in hypernetworks. Phys. A **495**, 475–487 (2018)
20. Tang, S.: When social advertising meets viral marketing: sequencing social advertisements for influence maximization. In: Proceedings of the AAAI Conference on Artificial Intelligence, vol. 32 (2018)
21. Tang, Y., Shi, Y., Xiao, X.: Influence maximization in near-linear time: a martingale approach. In: Proceedings of the 2015 ACM SIGMOD International Conference on Management of Data, pp. 1539–1554 (2015)
22. Zhu, J., Zhu, J., Ghosh, S., Wu, W., Yuan, J.: Social influence maximization in hypergraph in social networks. IEEE Trans. Netw. Sci. Eng. **6**(4), 801–811 (2018)

Enhanced Generation of Human Mobility Trajectory with Multiscale Model

Lingyun Han[✉]

School of Software Engineering, Tongji University, Shanghai, China
henlinyun@tongji.edu.cn

Abstract. Over the past three years, the COVID-19 pandemic has highlighted the importance of understanding how people travel in contemporary urban areas in order to produce high-quality policies for public health emergencies. In this paper, we introduce a multiscale generative model called *MScaleGAN* that generates human mobility trajectories. Unlike existing models where both location and time were discretized, resulting in generated results that were concentrated on certain points, *MScaleGAN* can produce trajectories with higher detail for better capturing urban road systems spatially and human behaviors' irregularity temporally. Experimental results show that our method generates better performance than previous models based on distribution similarities of individual and collective metrics compared with real GPS trajectories. Furthermore, we study the application of *MScaleGAN* on COVID-19 spread simulation and find that the spreading process under generated trajectories is similar to that under real data.

Keywords: Human activity recognition · Generative models · Learning for big data

1 Introduction

Understanding human mobility is crucial for urban planning [1], disaster management [17], and epidemic modeling [7]. However, due to privacy concerns and limited availability, obtaining urban-scale mobility trajectories in reality is almost impossible. To further study human mobility behaviors, methods and models that generate realistic mobile trajectories have become practical solutions and have received increasing attention.

Based on certain simplified assumptions of human mobility, previous works modeled individuals' mobility as Markov chains, in which transitional probabilities were calculated for location generation [16]. Following the success of generative models in computer vision [8], recent work has introduced standard CNN-based GANs to produce trajectory images [13] or used attention mechanisms to generate predictions of discrete locations [7]. However, the majority of existing work includes certain steps to discretize spatial-temporal points due to memory limitations or other practical considerations, resulting in generated results

© The Author(s), under exclusive license to Springer Nature Singapore Pte Ltd. 2024
B. Luo et al. (Eds.): ICONIP 2023, CCIS 1967, pp. 309–323, 2024.
https://doi.org/10.1007/978-981-99-8178-6_24

with intrinsic problems of limited precision and coarse detail. For example, one discretized location can include two distinct regions in some applications, such as epidemic modeling, where inside or outside certain walls can produce heavy differences, but two areas may be discretized into one block due to resolution problems. In these types of situations, previous approaches have failed to produce reliable trajectories, and further improvements are needed.

To produce more realistic mobility trajectories, we propose an improved deep generative model called *MScaleGAN*, which follows recent works [7,24] and includes many improvements. The most important feature is a secondary neural network generator G_d that enhances the coarse output of the main generator G_c with more detail. The main generator G_c is a self-attention-based sequential modeling network that is augmented with prior knowledge of road system structure and human mobility regularities. The discriminator D is a CNN-based model that is also enhanced with prior knowledge of human mobility to evaluate the quality of the whole trajectory that results from both generator G_c and G_d. Finally, the three parts are combined within a Wasserstein GAN-GP [9] framework, including a reward feedback stage inspired by [26] to bridge generator and discriminator. Moreover, a pre-train phase is included to further improve training efficiency and final performance.

Our contributions can be summarized as follows:

- **Multiscale Generation:** We propose an improved multiscale model called *MScaleGAN* to generate human mobility trajectories with enhanced detail in both spatial and temporal domains.
- **Improved Training:** Compared with existing works, *MScaleGAN* includes a rescheduled training process that improves the training efficiency and stability of convergence.
- **Application Reliability:** The generated results of *MScaleGAN* have more similarity with real input data in both objective and subjective metrics. These results are further validated with the application of COVID-19 spread simulation and achieve a consistent spreading process compared with real data.

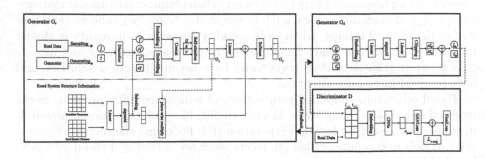

Fig. 1. The *MScaleGAN* framework for human mobility trajectory generation.

2 Related Work

With the proliferation of mobile devices, especially smartphones, location-based services are ubiquitous in contemporary life. The huge data generated by these services has introduced extensive research on human mobility. Within this vast research field, trajectory generation is a challenging task. There are two main approaches to this problem: sequence-based and image-based.

In sequence-based approaches, trajectory generation can be treated as a sequence prediction task, similar to text generation in natural language processing. In early works, a single trajectory sequence was treated as Markov chains, and then Hidden Markov models were introduced to add extra hidden states [25]. Following the success of GANs, researchers have tried to generate synthetic trajectories via adversarial methods. Feng et al. [7] introduced a self-attention-based GAN that integrated prior knowledge with a model-free framework. Choi et al. [4] tried to generate trajectories via generative adversarial imitation learning (GAIL), in which individuals and their locations were regarded as agents and action outcomes, and then the generation process was modeled by a Markov decision process and solved by imitation learning. Wei et al. [21] proposed an ImIn-GAIL model to simulate driving trajectories from sparse real-world data for solving the data sparsity challenges in behavior learning. Xu et al. [23] designed a two-stage generative model called DeltaGAN which can generate a fine-grained timestamp for each location to capture temporal irregularity more precisely. Wei et al. [22] modeled state transition in trajectories by learning the decision model and integrating the system dynamics with GAIL. As a result, this method can generate long-term movement patterns with more fidelity. Feng et al. [6] introduce DeepMove framework, which is an attentional RNN for mobility prediction from sparse and lengthy trajectories. In DeepMove, historical trajectories are handled by an attention mechanism to extract mobility patterns, while a GRU handles current trajectories. Yuan et al. [27] presented a novel framework based on generative adversarial imitation learning(GAIL), which can capture the spatiotemporal dynamics underlying trajectories by leveraging neural differential equations. However, sequence-based generation approaches have some problems in that they tend to capture the spatial properties in trajectories improperly, such as distances and radius of gyration.

On the other hand, image-based approaches can well depict the spatial features of trajectories. In these approaches, the tessellated trajectory is usually treated as an image for training a generative model. Then, random noises are inputted to generate new trajectories. Ouyang et al. [13] regarded each cell point as a specific location and two-channel images were used to represent the duration and timestamp of the visits. For repetitive visits, more channels were used in the generated images. Song et al. [15] used a four-layer deep convolutional GAN (DCGAN) to generate trajectories on a mobile phone dataset. Wang et al. [19] introduced an image-based two-stage GAN model to generate large-scale trajectories. Cao and Li [2] designed a TrajGen model. A DCGAN was first trained to generate location images. Then, the Harris corner detector was used to extract locations from the generated images. Finally, a sequence-to-sequence model was trained to assign temporal information to locations. Jiang et al. [11] propose a two-stage generative adversarial framework to generate the continuous trajectory on the road network, which built the generator under the human mobility hypothesis of the A* algorithm to learn the human mobility behavior and used a two-stage generation process to overcome the weak point of the existing stochastic generation process. Wang et al. [20] introduced a model which leveraged an important factor neglected by the literature -the underlying road properties and decomposed trajectory into map-matched road sequences with temporal information and embeds them to encode spatiotemporal features to generate trajectory sequences with judiciously encoded features. Generally, the main disadvantage of this approach is that ordinal dependency in the location of trajectories is ignored. For example, repetitive visits to one location are almost indescribable due to a single pixel per location in the image.

Despite the differences between sequence-based and image-based approaches, existing works have included certain steps to discretize spatial-temporal points and assume that precision deviation is acceptable. This paper aims to alleviate this problem with the multiscale model in order to achieve better fine-grained results.

3 Methods

Figure 1 illustrates the proposed *MScaleGAN* framework, which includes a main generator G_c and a secondary generator G_d along with an overall discriminator D. The following sections will explain each component in more detail.

3.1 Problem Definition and Basic Setting

Human mobility data consists of spatial-temporal trajectories $S = \{x_1, x_2, ..., x_N\}$ in which each x_i represents a tuple $\langle l_i, t_i \rangle$ as a visiting record data, with t_i denoting the ith timestamp and l_i denoting the location (lat, lon) of the record. Because it is almost impossible to model the distribution $P(S)$, particularly for long sequences, the common simplification is to factorize the joint probability as follows: $P(S) = P(x_1)P(x_2|x_1) \prod_{t=3}^{n} P(x_t|x_{1:t-1})$. This means treating the generating method as a sequential process. In most related works, GPS coordinates are discretized into a grid of M×M sizes by truncating after certain digits after the decimal point of GPS records in the data pre-processing phase. However, in this work, GPS coordinates are kept untouched as inputs.

3.2 Generator

As shown in Fig. 1, there are two pieces of generators working successively to generate trajectories. The former is the coarse-scale generator G_c, which generates discretized predictions of locations. The latter is the detail-scale generator G_d, which takes samples of G_c outputs as inputs, combined with other information, to generate more fine-grained detail.

Coarse Scale Generator. Because the coarse-scale locations have a large proportion of contribution to the final quality of generated results and the fact that sequence generation is a difficult task, the coarse-scale generator G_c is the most complicated component of the whole model. It is a self-attention-based network that generates the trajectory, where the self-attention mechanism [18] is used to directly capture the correlation in the sequence and can better model complex and long-term patterns in human mobility compared with classic RNNs [6,12]. First, as a sequence generator that takes previous spatial-temporal points and predicts the next, the first inputted spatial-temporal point is sampled from real datasets or generated by the generator of the previous version. Each inputted spatial-temporal point $\langle l, t \rangle$ is then discretized and decomposed into a coarse part $\langle l', t' \rangle$ and a detailed part $\langle \Delta l, \Delta t \rangle$, which is useful for detail-scale generator G_d. For example, $\langle l, t \rangle$ is $\langle (39.906667, 116.397552), 1688390482 \rangle$, in which $(39.906667,116.397552)$ is (lat, lon) and 1688390402 is the timestamp. Then, the $\langle l', t' \rangle$ can be $< (39.9066, 116.3975), 16883904 >$ and $\langle \Delta l, \Delta t \rangle$ can be $< (0.000067, 0.000052), 82 >$. In practice, the discretizing method depends on the characteristics of the actual datasets and considers the limitation of the hardware. Next, $\langle l', t' \rangle$ are converted to embedding vectors respectively and concatenated into one embedding representation, which is the input of a standard multi-head attention network. The output of attention layer O_A is then post-processed by a linear fully connected layer.

To integrate the prior knowledge of the road system structure, there are two pre-computed relation tables, each with a size of $N \times N$, where N is the number of discretized locations l'. One table is for **transition frequency**, which means the frequency of two locations in the same trajectory inside the real GPS datasets. The other table is for **road system distance**, which is an augmented version of Euclidean distance, where only the discretized location visited by trajectories inside real datasets can be on the path to compute distance, and computationally it can be obtained by simply breadth-first searching (BFS) on the discretized grid map after masking out zero-visiting cells. When used for the generator, the two tables should be pre-processed with a linear layer and then apply a non-linear sigmoid activation function. Then using l' as an index to select only relevant data and fuse with outputs of attention layer O_A by piece-wise multiplication. The results are then added back to the main routine results. Finally, after applying the softmax function, the next location prediction O_c is computed.

Detail Scale Generator. The detail-scale generator G_d takes the output of G_c to produce detailed information. It is not necessary to be a sequence process; therefore, it is relatively simple compared with G_c. We use MLP-based architectures as G_d in this paper. The next location prediction O_c is used to sample the next location l'_n, which is combined with the first location l'_0 and the next time t'_n as the input of G_d. First, all inputs are converted to embedding vector representation and concatenated into one. Second, the vector representation passes a linear layer before applying the sigmoid function. Finally, after using another linear layer to reshape data, these results are clipped inside a certain range corresponding with a discretizing method to produce $\langle \Delta l_n, \Delta t_n \rangle$, which can be added to $\langle l'_n, t'_n \rangle$ for generating the final detailed next location $\langle l_n, t_n \rangle$.

3.3   Discriminator

The quality of trajectories can't be evaluated with only every single point; therefore, the discriminator must take the whole trajectory as an input. While executing the generator repeatedly can produce the whole trajectory as negative samples, the positive samples can simply be the real GPS trajectory for the input dataset. The discriminator is based on CNN in this paper. The whole trajectory first uses a spatial-temporal embedding layer to convert into a 2D feature matrix. Then several convolution layers followed by Max-pooling layers apply to the feature matrix as in the standard CNN process. Finally, flatten features from CNN are post-processed with a linear layer with the sigmoid activation function to produce the final result O_{pred}. The O_{pred} is essential for computing the loss function for model training, which will be described elaborately in the next section.

3.4   Model Training

The model training process consists of two stages: pre-training and adversarial training. During pre-training, G_c is trained using real GPS data as samples to predict the next location. Similarly, G_d is trained to output detailed information for discretizing real GPS trajectories as inputs. The pre-training task for D is to distinguish whether inputs are from the original real GPS trajectory or the fake trajectory generated by randomly shuffling the real one and adding noise.

In Algorithm 1, the adversarial model training process has been described in detail. There are some more explanations for some important parts of the algorithm. First, $G_c GanLoss$ for $loss_{G_c}$ is computed as $loss_{G_c} = -\sum selectedProb \odot reward$, where $\odot$ means the element-wise product, and $selectedProb$ means selected element of $prob$ with corresponding $samples$. L_{vavg} is the loss of mobility regularity of velocity, which reflects the fact that in certain areas, the velocity at which humans move should be within certain ranges, which can be formulated as follows.

$$L_{vavg} = \sum_{i}^{n} |v_i - v_{avg}(l_i, t_i)|$$

$$v_i = \frac{l_i - l_{i-1}}{t_i - t_{i-1}}$$

Where $v_{avg}(l, t)$ represents the average trajectory velocity at a certain location l during a certain time range t, which can be pre-computed from real GPS data. Second, the $G_d GanLoss$ for $loss_{G_d}$ is computed as $loss_{G_d} = \sum |detailedPred - \Delta samples| \odot reward$, where $\Delta samples$ represents the same method for discretizing out detailed parts in G_c to apply to target $samples$. The $Abs(detailedPred)$

is calculated as $Abs(\Delta l, \Delta t) = |\Delta l| + |\Delta t|$, which reflects the tendency that output detailed spatial-temporal points' distribution should be more concentrated around the center $\langle l', t' \rangle$.

4   Experiments

4.1   Datasets and Evaluation Metrics

- **Dataset.** This paper uses two real-world mobile datasets to measure the performance of our proposed framework. The first dataset [28] is the GeoLife dataset from Microsoft Research Asia (MSRA) which contains the real GPS trajectory data of 182 users in Beijing from April 2007 to August 2012, in which data is sampled every minute on average. The second dataset is the Porto dataset from the Kaggle trajectory prediction competition [5] which contains GPS movement trajectory data of 442 taxis from July 2013 to June 2014 with an average sampling period of 15 s. Table 1 lists the summary statistics of the preprocessed trajectory datasets used for this paper. For each trajectory dataset, we randomly split the whole dataset into three parts: training set, validation set, and test set in the ratio of 6:2:2.
- **Evaluation Metrics.** Following the common practice in previous work [7,13], several individual trajectory and geographical metrics are used to evaluate the distribution similarity (Jensen-Shannon divergence) between real and generated data, which include:
  1. **Distance:** the cumulative travel distance per trajectory within a day.
  2. **Duration:** the daily sum of stay duration of each visited location.
  3. **Radius:** the radius of gyration for a trajectory.
  4. **DailyLoc:** the number of unique visited locations in a trajectory.
  5. $P(r)$**:** the visiting probability of one location r
  6. $P(r, t)$**:** the visiting probability of one location r at time t

4.2   Analysis of Pre-training

As mentioned in Subsect. 3.4, pre-training MScaleGAN's discriminator should benefit convergence speed and stability. The architecture of the pre-trained discriminator model is almost identical to that of the final MScaleGAN model, except for the lack of prior knowledge for the human mobility regularity loss and a change to the last activation layer. The pre-trained model was given a simple supervised task of classifying real and fake samples. Fake data is created by swapping a random number of original r trajectories' positions with random positions.

Figure 2 shows the change in loss and accuracy when the discriminator is pre-trained. The results indicate that the discriminator can distinguish real samples with close to 90% accuracy, which is a good result because most of the fake samples differ from the original real samples by an average of one position change. After completing pre-training, we transfer the pre-trained model weights

---

Algorithm 1: Adversarial Model Training

$epoch \leftarrow 0$;
$G_c \leftarrow$ coarse scale generator;
$G_d \leftarrow$ detail scale generator;
$G_{copy} \leftarrow$ copy of G_c and G_d;
$D \leftarrow$ discriminator;
$inputs \leftarrow$ sampled spatial-temporal points as inputs;
while $epoch < maxEpoches$ **do**

    $samples \leftarrow sampling(G_c, G_d)$;
    $i \leftarrow 0$;
    $rewards \leftarrow EMPTY$;
    while $i < samplingNum$ **do**

        $l \leftarrow 0$;
        while $l < sequenceLength$ **do**

            $subSeq \leftarrow takeSubSeq(samples, l)$;
            $completeSeq \leftarrow supplementSeq(G_{copy}, subSeq)$;
            $qualityPred \leftarrow discriminatorPredict(D, completeSeq)$;
            $rewards \leftarrow accumulate(rewards, qualityPred)$;
            $l \leftarrow l + 1$;

        end
        $i \leftarrow i + 1$;

    end
    $prob \leftarrow generatingPredict(G_c, input)$;
    $detailedPred \leftarrow generatingDetail(G_d, prob)$;
    $loss_{G_c} \leftarrow G_cGanLoss(prob, samples, rewards)$;
    $loss_{G_c} \leftarrow loss_{G_c} + L_{vavg}(samples)$;
    $G_c \leftarrow optimizerUpdate(G_c, loss_{G_c})$;
    $loss_{G_d} \leftarrow G_dGanLoss(detailedPred, samples, rewards)$;
    $loss_{G_d} \leftarrow loss_{G_d} + Abs(detailedPred)$;
    $G_d \leftarrow optimizerUpdate(G_d, loss_{G_d})$;
    $G_{copy} \leftarrow updateParameters(G_{copy}, G_c, G_d)$;
    $t \leftarrow 0$;
    while $t < maxEpochesD$ **do**

        $samplesD \leftarrow sampling(G_c, G_d)$;
        $predD \leftarrow discriminatorPredict(D, samplesD)$;
        $loss_D \leftarrow NegativeLogLikelihoodLoss(predD)$;
        $D \leftarrow optimizerUpdate(D, loss_D)$;
        $t \leftarrow t + 1$;

    end
    $epoch \leftarrow epoch + 1$;

end

---

to MScaleGAN. The impact of the pre-trained MScaleGAN overall model can be observed from the loss graphs of each module of the model architecture in Fig. 2. The expected behavior of the discriminator loss (green line) is to converge to 0, which means that the discriminator judges real samples and fake samples

Table 1. Statistics of trajectory dataset

	Geolife-Beijing	Kaggle-Porto
total number of trajectory	945070	684084
average record points per trajectory	30.23	46.05
average distance(km)	8.87	9.74
average duration(hrs)	1.84	1.06

as equivalent and cannot distinguish between them. The loss of the generator (orange and blue lines) should converge to 0, indicating that the statistical distance between the original data and the generated data distribution is low. It is expected that pre-training MScaleGAN leads to a convergent and highly stable training process. Without pre-training, the model can never fully converge at almost all hyperparameter value settings. Therefore, a pre-trained version of MScaleGAN is used for the rest of this paper as it leads to significantly better results.

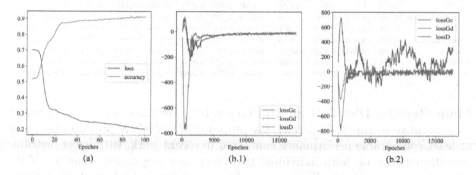

Fig. 2. Pre-train result. (a) Pre-training effect on accuracy and loss of Discriminator. (b) The effect of pre-training on the overall model of MScaleGAN. (1) With pre-training, (2) Without pre-training. (Color figure online)

4.3   Performance Comparison

Compared Methods. The baselines from Markov Models, Deep Prediction Models, and Deep Generative Models are used to compare performance.

- **Markov Models:** (1) *Markov* [16]: This model defines the state as the visited location and assumes the next location only depends on the current location. Then, the transition matrix can be constructed to acquire the first-order transition probability between locations. (2) *IO-HMM* [25]: Transition and emission models work together to maximize the likelihood of observed trajectories when initialization.

- **Deep Prediction Models:** (1) *GRUPred* [3]: Gated Recurrent Units are used to predict the next location after given historically visited locations. (2) *TransAutoencoder* [10]: The framework builds an encoder to extract information from historical data and feeds them to a decoder for reconstructing trajectories.
- **Deep Generative Models:** (1) *SeqGAN* [26]: Discrete location data is produced by combining Reinforcement Learning and GAN. (2) *MoveSim* [7]: The method combines GAN with the prior knowledge of human mobility regularity to improve performance. (3) *Ouyang GAN* [13] A GAN based on CNNs. (4) *TS-TrajGen* [11] A two-stage generative adversarial framework can generate the continuous trajectory on the road network

Table 2. Distribution comparison between real and generated trajectory data on two datasets. Lower values indicate more realistic results.

	Geolife-Beijing						Kaggle-Porto					
	Distance	Radius	Duration	DailyLoc	P(r)	P(r,t)	Distance	Radius	Duration	DailyLoc	P(r)	P(r,t)
Markov	0.0186	0.1475	0.0754	0.3155	0.4341	0.8194	0.0152	0.1453	0.0684	0.2946	0.4935	0.7582
IO-HMM	0.0164	0.0984	0.0094	0.0954	0.6112	0.8254	0.0163	0.0895	0.0082	0.0844	0.5894	0.7834
GRUPred	0.0194	0.1141	0.0121	0.1031	0.4334	0.8341	0.0173	0.1021	0.0102	0.0974	0.4845	0.7524
TransAutoencoder	0.0182	0.1031	0.0095	0.0984	**0.4241**	0.8214	0.0186	0.1104	0.0125	0.1012	_0.4721_	0.6904
SeqGAN	0.0171	0.0757	0.0085	0.0831	0.5814	0.8184	0.0153	0.0687	0.0075	0.0782	0.5023	_0.6849_
Ouyang GAN	0.0102	0.0632	0.0064	0.0723	0.5524	0.8043	0.0142	0.0701	0.0081	0.0704	0.4743	0.7389
MoveSim	0.0089	**0.0541**	_0.0021_	0.0641	0.5114	0.7882	_0.0071_	0.0641	**0.0017**	0.0588	0.4987	0.7054
TS-TrajGen	_0.0081_	0.0552	0.0024	**0.0594**	0.5225	_0.7642_	0.0075	**0.0594**	_0.0023_	_0.0564_	0.4853	0.6854
MScaleGAN	**0.0078**	_0.0549_	**0.0018**	_0.0605_	_0.4305_	**0.7532**	**0.0068**	_0.0611_	0.0025	**0.0552**	0.4628	**0.6781**

Main Results: Distribution Similarity. In Table 2 lists the performance of all generative methods. Generally, the proposed *MScaleGAN* produces better or at least comparable performance compared to recent work, with lower distribution discrepancy on both individual trajectory and geographical metrics. With dedicated designs for handling previously ignored detailed information, *MScaleGAN* generates trajectories that are close to reality, especially with the metrics of individual trajectories. On the other hand, the edge is a little narrower on the metrics of geographical metrics $P(r)$ and $P(r,t)$, which may be the result of deviations equalizing mutual influence statistically. The performance of each model on the Kaggle-Porto dataset is better than that on the Geolife-Beijing dataset to varying degrees. This difference may be due to the smaller data volume of the Kaggle-Porto dataset compared to the Geolife-Beijing dataset.

In Fig. 3 and Fig. 4, the qualitative performance of generative models is visualized by the distribution and popularity of visited locations of trajectories. The *MScaleGAN* not only can successfully capture the relative popularity of different locations in urban areas, such as main ring roads, highways, and busy hot spots but also provides much more realistic results in the detailed scale compared with previous work. It is worth noting that MScaleGAN can capture and generate corresponding trajectories regardless of the more regular road network structure in Beijing or the more naturally extended road system in Porto.

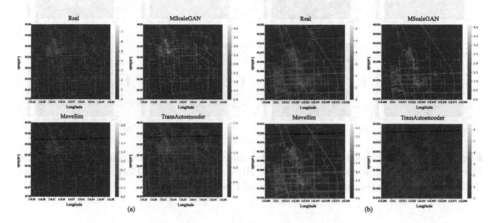

Fig. 3. Geographical visualization of real and generated trajectories of the Geolife-Beijing dataset on different scales. (a) The city-wise view of results, ranging within the 5th Ring Road of Beijing. (b) A more detailed view, focusing on the busiest Haidian District.

4.4 Ablation Study

In order to verify the effectiveness of each module of MScaleGan, this section conducts an ablation study on the detail-scale generator G_d, the regularity loss of human mobility $L_{v_{avg}}$, and the road network structure information (road info) on the Geolife-Beijing dataset. The results of the ablation study are shown in Table 3 and Fig. 6. Overall, the performance of the model degraded to varying degrees after removing each module, both for quantitative performance and qualitative visualization analysis. In particular, removing the detail-scale generator G_d will cause a significant degradation in generation accuracy, and removing the road info will result in the generation of a considerable number of trajectories that do not conform to the actual road network. Relatively speaking, removing the regularity loss of human mobility $L_{v_{avg}}$ has little effect on qualitative visualization results, but it will cause significant degradation in time-related quantitative metrics (Duration, DailyLoc, $P(r,t)$).

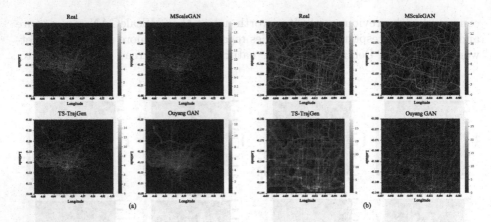

Fig. 4. Geographical visualization of real and generated trajectories of the Kaggle-Porto dataset on different scales. (a) The city-wise view of the Porto area. (b) A more detailed view, focusing on the city center of Porto.

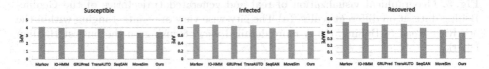

Fig. 5. Application of generated trajectory data to simulate the spreading process of the COVID-19 epidemic, measured by comparing with the simulation result of real data.

Table 3. The results of ablation study of some model modules on the Geolife-Beijing dataset

	Distance	Radius	Duration	DailyLoc	P(r)	P(r,t)
w/o G_d	0.0176	0.1675	0.0653	0.1856	0.5341	0.8193
w/o $L_{v_{avg}}$	0.0124	0.1044	0.1594	0.2554	0.4112	0.9824
w/o road info	0.0282	0.2125	0.1092	0.2023	0.8234	0.9103
Full Model	0.0078	0.0552	0.0018	0.0605	0.5041	0.7532

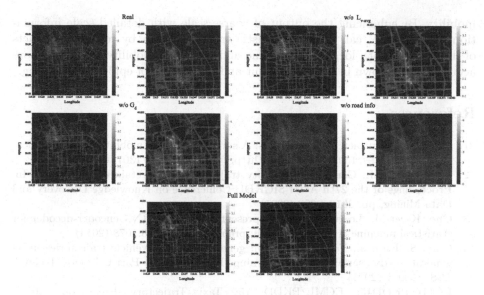

Fig. 6. Visual analysis of ablation study on the Geolife-Beijing dataset

4.5 Application: Epidemic Spreading Simulation

For analyzing the application of generated trajectories, we study the spreading of COVID-19 with generated data as input to the SIR model. Following recent work [14], for epidemic diffusion simulation in 7 days, the detailed setting is:

1. 12,000 individuals initialize as Susceptible(S) orInfected(I) with a proportion of $9:1$.
2. When an S person p_S goes within a certain range of an Infected and Spreading (IS) for a certain amount of time, then p_S immediately becomes Infected and Not Spreading (INS) with probability 0.5 at the exposure time t.
3. At $t + t_{IS}$ day in which $t_I S \sim N(5, 10)$,p_S becomes IS.
4. At $t + t_R$ day in which $t_R \sim N(12, 24)$,p_S becomes Isolated or Recovered(R).

Running the simulation with trajectory data and computing (mean) absolute percentage error ($MAPE$ or APE) between generated data and real data on the number of different categories (S, I, R): the number of S or I individuals calculated at the end of the 7th day, and the number of R is the daily number of individuals who become recovered from the 7th day until the day when the whole test population is recovered. As illustrated in Fig. 5, the *MScaleGAN* model can provide epidemic spreading study with the advantage of relatively small divergence in the resulting distribution of (S, I, R).

5 Conclusion

In order to further study human mobility behaviors, in this paper we introduce an improved generative model *MScaleGAN* to produce trajectories of human mobility. By enhancing the output of coarse scale with detailed scale information, *MScaleGAN* can produce realistic trajectory data by reducing artifacts of discretizing both locations and time and capturing mobility behaviors more accurately, as shown in our experiment and application of epidemic spreading.

References

1. Asgari, F., Gauthier, V., Becker, M.: A survey on human mobility and its applications. CoRR abs/1307.0814 (2013). http://arxiv.org/abs/1307.0814
2. Cao, C., Li, M.: Generating mobility trajectories with retained data utility. In: Proceedings of the 27th ACM SIGKDD Conference on Knowledge Discovery and Data Mining, pp. 2610–2620 (2021)
3. Cho, K., et al.: Learning phrase representations using RNN encoder-decoder for statistical machine translation. arXiv preprint arXiv:1406.1078 (2014)
4. Choi, S., Kim, J., Yeo, H.: Trajgail: generating urban vehicle trajectories using generative adversarial imitation learning. Transport. Res. Part C Emerg. Technol. **128**, 103091 (2021)
5. ECML/PKDD15: ECML/PKDD 15: Taxi Trajectory Prediction (2015). https://www.kaggle.com/competitions/pkdd-15-predict-taxi-service-trajectory-i/overview
6. Feng, J., et al.: DeepMove: predicting human mobility with attentional recurrent networks. In: Proceedings of the 2018 World Wide Web Conference, pp. 1459–1468 (2018)
7. Feng, J., Yang, Z., Xu, F., Yu, H., Wang, M., Li, Y.: Learning to simulate human mobility. In: Proceedings of the 26th ACM SIGKDD International Conference on Knowledge Discovery and Data Mining, pp. 3426–3433 (2020)
8. Goodfellow, I.J., et al.: Generative adversarial networks (2014). https://doi.org/10.48550/ARXIV.1406.2661, https://arxiv.org/abs/1406.2661
9. Gulrajani, I., Ahmed, F., Arjovsky, M., Dumoulin, V., Courville, A.C.: Improved training of Wasserstein GANs. In: Advances in Neural Information Processing Systems, vol. 30 (2017)
10. Hinton, G.E., Krizhevsky, A., Wang, S.D.: Transforming auto-encoders. In: Honkela, T., Duch, W., Girolami, M., Kaski, S. (eds.) ICANN 2011. LNCS, vol. 6791, pp. 44–51. Springer, Heidelberg (2011). https://doi.org/10.1007/978-3-642-21735-7_6
11. Jiang, W., Zhao, W.X., Wang, J., Jiang, J.: Continuous trajectory generation based on two-stage GAN. arXiv preprint arXiv:2301.07103 (2023)
12. Liu, Q., Wu, S., Wang, L., Tan, T.: Predicting the next location: a recurrent model with spatial and temporal contexts. In: Thirtieth AAAI Conference on Artificial Intelligence (2016)
13. Ouyang, K., Shokri, R., Rosenblum, D.S., Yang, W.: A non-parametric generative model for human trajectories. In: IJCAI, vol. 18, pp. 3812–3817 (2018)
14. Rambhatla, S., Zeighami, S., Shahabi, K., Shahabi, C., Liu, Y.: Toward accurate spatiotemporal COVID-19 risk scores using high-resolution real-world mobility data. ACM Trans. Spat. Algorithms Syst. (TSAS) **8**(2), 1–30 (2022)
15. Song, H.Y., Baek, M.S., Sung, M.: Generating human mobility route based on generative adversarial network. In: 2019 Federated Conference on Computer Science and Information Systems (FedCSIS), pp. 91–99. IEEE (2019)

16. Song, L., Kotz, D., Jain, R., He, X.: Evaluating location predictors with extensive Wi-Fi mobility data. ACM SIGMOBILE Mob. Comput. Commun. Rev. **7**(4), 64–65 (2003)
17. Song, X., Zhang, Q., Sekimoto, Y., Shibasaki, R., Yuan, N.J., Xie, X.: Prediction and simulation of human mobility following natural disasters. ACM Trans. Intell. Syst. Technol. (TIST) **8**(2), 1–23 (2016)
18. Vaswani, A., et al.: Attention is all you need. In: Advances in Neural Information Processing Systems, vol. 30 (2017)
19. Wang, X., Liu, X., Lu, Z., Yang, H.: Large scale GPS trajectory generation using map based on two stage GAN. J. Data Sci. **19**(1), 126–141 (2021)
20. Wang, Y., Li, G., Li, K., Yuan, H.: A deep generative model for trajectory modeling and utilization. Proc. VLDB Endow. **16**(4), 973–985 (2022)
21. Wei, H., Chen, C., Liu, C., Zheng, G., Li, Z.: Learning to simulate on sparse trajectory data. In: Dong, Y., Mladenić, D., Saunders, C. (eds.) ECML PKDD 2020. LNCS (LNAI), vol. 12460, pp. 530–545. Springer, Cham (2021). https://doi.org/10.1007/978-3-030-67667-4_32
22. Wei, H., Xu, D., Liang, J., Li, Z.J.: How do we move: modeling human movement with system dynamics. In: Proceedings of the AAAI Conference on Artificial Intelligence, vol. 35, pp. 4445–4452 (2021)
23. Xu, N., et al.: Simulating continuous-time human mobility trajectories. In: Proceedings of 9th International Conference on Learning and Representation, pp. 1–9 (2021)
24. Yin, J.: Learn to simulate macro-and micro-scopic human mobility. In: 2022 IEEE International Conference on Smart Computing (SMARTCOMP), pp. 198–199. IEEE (2022)
25. Yin, M., Sheehan, M., Feygin, S., Paiement, J.F., Pozdnoukhov, A.: A generative model of urban activities from cellular data. IEEE Trans. Intell. Transp. Syst. **19**(6), 1682–1696 (2017)
26. Yu, L., Zhang, W., Wang, J., Yu, Y.: SeqGAN: sequence generative adversarial nets with policy gradient. In: Proceedings of the AAAI Conference on Artificial Intelligence, vol. 31 (2017)
27. Yuan, Y., Ding, J., Wang, H., Jin, D., Li, Y.: Activity trajectory generation via modeling spatiotemporal dynamics. In: Proceedings of the 28th ACM SIGKDD Conference on Knowledge Discovery and Data Mining, pp. 4752–4762 (2022)
28. Zheng, Y., Xie, X., Ma, W.Y., et al.: Geolife: a collaborative social networking service among user, location and trajectory. IEEE Data Eng. Bull. **33**(2), 32–39 (2010)

SRLI: Handling Irregular Time Series with a Novel Self-supervised Model Based on Contrastive Learning

Haitao Zhang, Xujie Zhang, Qilong Han[✉], and Dan Lu

School of Computer Science and Technology,
Harbin Engineering University, Heilongjiang, China
hanqilong@hrbeu.edu.cn

Abstract. The advancement of sensor technology has made it possible to use more sensors to monitor industrial systems, resulting in a large amount of irregular, unlabeled time-series data. Consequently, a large volume of irregular and unlabeled time series data is produced. Learning appropriate representations for those series is a very important but challenging task. This paper presents a self-supervised representation learning model SRLI (Self-supervised Representation Learning for Irregularities). We use T-LSTM to construct the irregularity encoder block. Based on this, we design three data augmentation methods. First, the raw time-series data are transformed into different yet correlated views. Second, we propose a contrasting module to learn robust representations. Lastly, to further learn discriminative representations, we reconstruct the series and try to get the imputation values of the unobserved positions. Rather than in a two-stage manner, our framework can generate the instance-level representation for ISMTS directly. Experiments show that this model has good performance on multiple data sets.

Keywords: time series · representation learning · irregularly sampled data

1 Introduction

Time Series (TS) data is a series of data recorded in chronological order, such as sound, temperature, photoelectric signals, brain waves, and so on. Time series modeling plays a vital role in various industries, including financial markets, demand forecasting, and climate modeling, is one of the most challenging problems in data mining. Unlike image and text data, time series data in the real world are typically only recognized by domain experts, which makes the cost of

This work was supported by the National Key R&D Program of China under Grant No. 2020YFB1710200 and Heilongjiang Key R&D Program of China under Grant No. GA23A915.

labeling such data very high. To address this issue, self-supervised representation learning has gained considerable attention as it extracts effective representations from unlabeled data for downstream tasks. In the field of computer vision, self-supervised representation learning has shown promising results in various applications [10,18].

In real-world applications, time series data is often irregularly sampled multivariate time series (ISMTS). [16] Practical issues can lead to various types of irregularities caused by sensor failures, external forces in physical systems, sampling rate differences between devices, and limitations of the industrial environment. Irregular sampling poses a significant challenge to machine learning models, which typically assume fixed intervals with no missing data.

Existing methods for dealing with irregularly sampled data fall into two categories: 1) missing data-based perspective and 2) raw data-based perspective. The missing data-based perspective converts irregularly sampled time series data (ISMTS) into equally spaced data which results in missing data showing up. This method makes it harder to draw conclusions from the data and negatively impacts downstream performance. The raw data-based perspective utilizes the irregular data directly by incorporating varying time interval features into the model's input or designing models capable of processing different sampling rates.

The major contributions of this paper are summarized as follows:

- We proposed a Self-supervised Representation Learning model for Irregular Sampled Time-series(SRLI). Also, we introduce an data augmentation and adversarial strategies to improve the presentation capabilities of our model.
- We couple the imputation and prediction process together and optimize them simultaneously to generate representations of the ISMTS rather than in a two-stage manner. Additionally, our model can generate reliable predicted values for incomplete data locations.
- We conducted multiple types of experiments, and the results showed the effectiveness and generality of our model.

2 Related Work

Time series plays an important role in various industries. Developing universal representations for time series is a fundamental yet challenging problem. An effective method of representing time series data can facilitate comparison of similarity between sequences as well as aid in diverse data mining tasks. The representation method of time series data should be capable of significantly reducing data dimension and subsequently reconstructing the original data using the resulting representation. Additionally, the method must prioritize local or global features, be computationally efficient, and accommodate noise without significant performance degradation.

2.1 Self-supervised Learning for Time-Series

Recent advances in self-supervised learning originated from utilizing pretext tasks on images to acquire advantageous representations [6,10,14]. Contrastive learning promotes the unsupervised representation of data by reducing the distance of comparable data pairs in feature space, while maximizing other data pairs. To create these related data pairs, data augmentation is generally applied. As the current contrastive learning framework is primarily tailored toward image data, the augment methods used for image data, such as rotation and flip, are inadequate for time-series data. Therefore, there are still few augment methods for time-series data, include data interpolation, augmentation in time-frequency domain, decomposition-based methods, statistical probability models, generative models, and deep learning generative models. TimeCLR [15] combining the advantages of DTW and InceptionTime, presents a data augmentation framework centered on DTW. This framework can create a targeted phase shift and amplitude change phenomenon, as well as preserve the structure and feature information of the time series data. Different tasks may require different methods of augmentation, such as adversarial samples, unsupervised learning, reconstruction error, prediction error, time-frequency domain augmentation, etc. It is necessary to select the most suitable methods for a particular task, and to combine and adjust them appropriately to achieve better results.

Inspired by the success of contrastive learning, few works have recently leveraged contrastive learning for time series data. CPC [11] learned representations by predicting the future in the latent space and showed great advances in various speech recognition tasks. TS-TCC [4] propose a novel cross-view temporal and contextual contrasting modules to improve the learned representations for time-series data. TS2Vec [17] performs contrastive learning in a hierarchical way over augmented context views, which enables a robust contextual representation in an arbitrary semantic level. T-LOSS [5] performs instance-wise contrasting only at the instance level. TNC [13] encourages temporal local smoothness in a specific level of granularity.

2.2 Representation Learning on Irregular Sampled Time-Series

In a multivariate scenario, irregularity refers to the situation where observations from different sensors may not be temporally aligned. As shown in Fig. 1, we chose three key air quality indicators, namely pm2.5, co, and so2. The original data exhibited uneven time intervals between consecutive collections, which varied not only across different dimensions but also within the same dimension. This non-uniformity further complicates the analysis. Existing two ways can handle the irregular time intervals problem [12]: 1) Determining a fixed interval, treating the time points without data as missing data. 2) Directly modeling time series, seeing the irregular time intervals as information.

Methods from the missing data-based perspective can convert ISMTS into equally spaced data. However, the appearance of missing values can damage the temporal dependencies of sequences, making it infeasible to directly apply many existing models. To handle missing values in time series, numerous efforts have been made to develop models. The core problem of these methods is how to impute the missing data accurately. While common methods such as smoothing, interpolation [7], and so on. There are also deep learning based on recurrent neural networks [3] and adversarial generative networks [9].

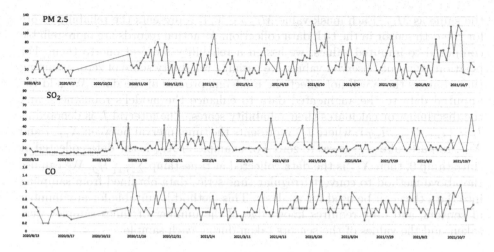

Fig. 1. Multivariate irregular sampled time-series.

End-to-end approaches process the downstream tasks directly based on modeling the time series with irregular intervals. Prior efforts [8]using the basic RNN model to cope with missing data in sequential inputs and the output of RNN being the final characteristics for prediction. Then, to improve this basic idea, they addressed the task of multilabel classification of diagnoses by given clinical time series and found that RNNs can make remarkable use of binary indicators for missing data, improving AUC, and F1 significantly. TE-ESN [4] propose a novel time encoding mechanism, the timestamp of each sampling is encoded as additional information input, and the data is modeled using an echo state network. RAINDROP [19] model dependencies between sensors using a graph structure where the model inputs each sample individually and updates the network with an attention mechanism.

The current methods used in deep learning still face several challenges, mainly evident in two key aspects: First, they depend on labeled data, and the insufficiency of labeled data restricts their performance. Second, the few methods available are not able to model multidimensional irregular time series.

3   Method

3.1   Problem Definition

Let $D = \{(X_i, M_i, I_i, y_i) | i = 1, ..., N\}$ denote an irregular time series dataset with N samples. Each sample contains total of T sets of data were successfully collected, each set contains data from at least one sensor or at most all K sensors. Every $X_i = \{X_{i,1}, X_{i,2}, X_{i,3}..., X_{i,T}\}$ is an irregular multivariate time series with it's corresponding mask set M_i, interval set I_i and label y_i. The data collected for the t-th time in i-th data will be recorded in $X_{i,t}$, which has dimension K × 1, the same as $M_{i,t}$. Each mask value $M_{i,t,d} \in [0,1]$ represents the reliability score for the d-th sensor in the t-th data collection. As actual records are more reliable compared to imputed values, we assign a reliability score of one to them. For positions without observed values, we assign a reliability score of zero, assuming that changes there will not affect the modeling process of the data. We add the imputed data to the augmented data to enhance the model's robustness, and the discriminator calculates their reliability scores. The interval I_i is denoted as $I_i = \{I_{i,1}, I_{i,2}, ...I_{i,T}\}$, where $I_{i,t}$ represents the time interval between $X_{i,t-1}$ and $X_{i,t}$. In particular, $I_{i,1}$ is set to zero. For example, if $X_{i,1}$ is the data collected at timestamp 0, then $X_{i,t}$ is the data collected at timestamp $\sum_{j=1}^{t} I_{i,j}$. We can use the above method to convert irregular time series data obtained from sampling into data that can be used for analysis. This method is suitable for any number of sensors, and only requires knowledge of the number of times samples were taken and the values of the sensors taken during each sample.

3.2   Model Architecture

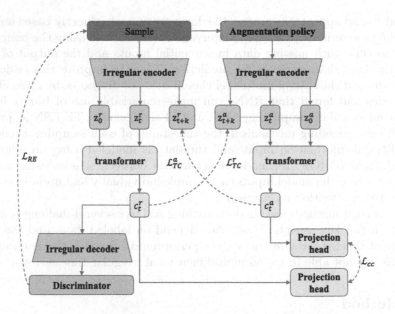

Fig. 2. Overall architecture of proposed SRLI model.

This section presents a comprehensive description of our proposed SRLI. As illustrated in Fig. 2, we employ irregular augmentation to generate data that closely approximates the distribution of the original data. The IrregPredict Contrasting module then explores the temporal features of the data through cross-prediction tasks to further maximize the similarity between the contexts. Our GenDiscrim module reconstructs the data and predicts values for unobserved locations, while a discriminator ensures that their distribution is similar to that of the authentic data. It should be acknowledged that we abandon the data augmentation module to allow the discriminator to evaluate the credibility of our augmented data. Instead, we use an encoder-decoder architecture for a few epochs until the discriminator training error meets the predefined threshold.

3.3   Irregular Augmentation

The effectiveness of contrastive learning heavily relies on data augmentation, which is utilized to maximize the similarity between different views of the same sample and minimize its similarity with other samples. As a result, developing appropriate data augmentation strategies is critical for the success of contrastive learning. Time series data have special properties that some existing data augmentation methods cannot adapt well to. Moreover, data augmentation methods introduce errors inevitable which increase the difficulty of data modeling.

Given a sample D with related M, I. We designed various data augmentation methods not only to explicitly convey our data representation to the network but also to enhance its robustness.

Mask Augmentation. We add gaussian noise with zero mean and per-element standard deviation to the data marked as unobserved value with mask 0. This augmentation method is designed to train the network to handle missing data with mask 0 explicitly. d_i is the observation we get at time i. Let $\epsilon \sim \mathcal{N}\left(\mathbf{0}, \sigma_c^2 \mathbf{I}\right)$, we have:

$$d_i^a = d_i + (1 - m_i)\epsilon \tag{1}$$

Interpolation Augmentation. We add new data between adjacent data points through interpolation, and changing the distance between the interpolated values and adjacent data points. Let d_t and d_{t+k} be adjacent data points, we have:

$$d_{t+j}^a = d_t + (d_{t+k} - d_t) * \frac{j}{k}, j \in (0, k) \tag{2}$$

Jitter Augmentation. We generate augmented data by applying translation and jittering on the original data, and adjusting the reliability score of the new data accordingly.

3.4   IrregTrans Contrasting

For a given data D with related mask M and interval I, the IrregTrans module uses the T-LSTM [2] module to generate representation z_{t+k} for each timestamp t. T-LSTM can take time interval and missing mask into consideration. Once the model reads the entire input sequence, the encoder can get the hidden state of the input by Eq. (3). We use two T-LSTM modules to model the influence of reliability mask and interval respectively, then a linear module is used to recombine the effects of mask and interval according to Fig. 3. After that, we use transformer to get the representation c_t of the overall context because of its efficiency and speed. We use linear function to map c_t to the same dimension of z_t. To predict future timesteps, then we use log-bilinear model that would preserve the mutual information between the input c_t and z_{t+k}.

$$
\begin{aligned}
C_{t-1}^S &= tanh(W_d C_{t-1} + b_d) \\
\hat{C}_{t-1}^S &= C_{t-1}^S * g(\Delta_t) \\
C_{t-1}^T &= C_{t-1} - C_{t-1}^S \\
C_{t-1}^* &= C_{t-1}^T + \hat{C}_{t-1}^S \\
f_t &= \sigma(W_f x_t + U_f h_{t-1} + b_f) \\
i_t &= \sigma(W_i x_t + U_i h_{t-1} + b_i) \\
o_t &= \sigma(W_o x_t + U_o h_{t-1} + b_o) \\
\tilde{C} &= tanh(W_c x_t + U_c h_{t-1} + b_c) \\
C_t &= f_t * C_{t-1}^* + i_t * \tilde{C} \\
h_t &= o * tanh(C_t)
\end{aligned}
\tag{3}
$$

In our approach, the augmentation data generates c_t^a and the raw data generates c_t^r. We use the augmentation data z_t^a to predict the future timesteps of the raw data c_{t+k}^r and vice versa. The objective of the contrastive loss is to reduce the dot product between the predicted representation and the true representation of other samples, while simultaneously increasing the dot product with other samples $\mathcal{N}_{t,k}$ in the minibatch. Accordingly, we calculate the two losses as follows:

$$
\mathcal{L}_{TC}^a = -\frac{1}{K} \sum_{k=1}^K \log \frac{\exp\left((\mathcal{W}_k\,(c_t^a))^T z_{t+k}^r\right)}{\sum_{n \in \mathcal{N}_{t,k}} \exp\left((\mathcal{W}_k\,(c_t^a))^T z_n^r\right)}
\tag{4}
$$

$$
\mathcal{L}_{TC}^r = -\frac{1}{K} \sum_{k=1}^K \log \frac{\exp\left((\mathcal{W}_k\,(c_t^r))^T z_{t+k}^a\right)}{\sum_{n \in \mathcal{N}_{t,k}} \exp\left((\mathcal{W}_k\,(c_t^r))^T z_n^a\right)}
\tag{5}
$$

3.5   GenDiscrim

Similar to the irregular encoder, we use T-LSTM to generate reconstructed data and constrain its distribution close to the real data with the discriminator. Eq. (6) and Eq. (7) show how the GenDiscrim module constrains the presentation result. First, we calculate the mean squared error loss (MSE) between the reconstructed data and the observed values in the original data, to evaluate the accuracy of the decoder. Then, we replace all non-observed values in the original data with simulated data generated by the decoder. Athen a discriminator is used to judge determine whether each data point is the original data, so that the output of the decoder is closer to the distribution of the original data.

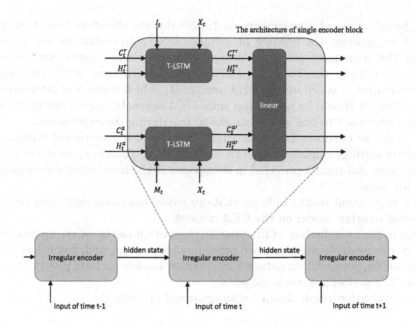

Fig. 3. Irregular encoder.

$$\mathcal{L}_{\text{rec}} = \frac{1}{N} \sum_{i=1}^{N} \|(X_i - f_{\text{dec}} (c_t^r)) \odot M_i\|_2^2 \tag{6}$$

$$\mathcal{L}_{dis} = -[E \underbrace{\log (D (X_{\text{real}}))}_{X_{\text{real}} \sim U} + E \underbrace{\log \left(1 - D \left(\hat{X}_{\text{pre}}\right)\right)}_{\hat{X}_{\text{pre}} \sim U}] \tag{7}$$

$$= -\frac{1}{N} \sum_{i=1}^{N} [M_i \odot \log (D (U_i)) + (1 - M_i) \odot \log (1 - D (U_i))]$$

4 Experiment

In this section, we evaluate the learned representations of SRLI on time series classification, forecasting.

4.1 Time Series Classification

Assigning labels to the entire time-series during classification requires instance-level representations that can be attained by performing max pooling over all timestamps. Subsequently, we utilize the same protocol as T-Loss [5] by training an SVM classifier with an RNF kernel on top of these instance-level representations to complete the classification.

A broad range of experiments on time-series classification have been performed to appraise the efficacy of instance-level representations when pitted against the other most advanced unsupervised time-series representation techniques, specifically the TS-TCC, TS2Vec, and conventional DTW techniques. This evaluation is based on the UEA archive [1], which consists of 30 multivariate datasets. It should be noted that since UEA is evenly spaced, our model sets all time intervals I to one and all masks to one during the experiment.

In order to ensure a fair comparison, we kept our experimental data and parameter settings consistent with those of TS2Vec. Therefore, we directly used the experimental results provided in that paper as the benchmark for comparing with our model.

Our experiment results indicate that our model has made significant progress compared to other model on the UEA dataset.

Due to the introduction of irregular augmentation methods, the optimization effect of our model is average when other models perform well, but it has shown significant improvement on datasets where other models perform poorly. Overall, our model's performance is more stable.

The evaluation result detail are summarized in Table 1.

4.2 ISMTS Forecasting

The goal of the time series forecasting task is to predict the next H observations $x_{t+1}, ..., x_{t+H}$ given the previous T observations $x_{t-T+1}, ..., x_t$. We utilize the instance-level representation R_i to predict future observations by training a linear regression model along with an L_2 norm penalty. The resulting model takes R_i as the input to directly forecast the future values $\hat{x}$. For the construction of irregular sampled data, we randomly eliminated 5% of data points from the original dataset. In the case of models that are not capable of directly modeling irregular data, we replaced the corresponding positions with zeros.

We compare the performance of SRLI and existing SOTAs on three public datasets, including three ETT datasets [20] and Electricity dataset. Follow previous works, we use MSE to evaluate the forecasting performance.

The evaluation results are shown in Table 2.

4.3 Forecasting on Industrial Dataset

To verify the performance of our model on real datasets, we collected air quality data from a region in China from August 2020 to December 2021. This includes PM2.5, PM10, SO2, NO2 and other data. Data sources are collected every minute for **Air quality M**. We calculated hourly and daily mean values respectively to obtain data sets **Air quality H** and **Air quality D**. Experiments were carried out on three data sets. We used 60% of them for training. The remaining 40% are used as validation sets and test sets, with 20% being used for each. In the experiment, through the observation data within 30 timesteps of the given model, various air indicators in the next 15, 30 and 60 timesteps are predicted

Table 1. Accuracy score of our method compared with those of other methods of unsupervised representation on UEA datasets. The representation dimensions are set to 320 for fair comparison.

Dataset	SRLI	TS2Vec	TS-TCC	DTW
ArticularyWordRecognition	**0.987**	**0.987**	0.953	0.976
AtrialFibrillation	**0.530**	0.200	0.267	0.200
BasicMotions	0.875	0.975	**1.00**	0.975
CharacterTrajectories	0.975	**0.995**	0.985	0.989
Cricket	0.900	0.972	0.917	**1.00**
DuckDuckGeese	0.56	**0.680**	0.380	0.600
EigenWorms	**0.852**	0.847	0.779	0.618
Epilepsy	0.957	**0.964**	0.957	0.946
ERing	**0.912**	0.874	0.904	0.133
EthanolConcentration	0.314	0.308	0.285	**0.323**
FaceDetection	**0.602**	0.501	0.544	0.529
FingerMovements	**0.635**	0.480	0.460	0.530
HandMovementDirection	**0.379**	0.338	0.243	0.231
Handwriting	0.475	**0.515**	0.498	0.286
Heartbeat	**0.753**	0.683	0.751	0.717
JapaneseVowels	0.936	**0.984**	0.930	0.949
Libras	**0.876**	0.867	0.822	0.870
LSST	0.468	**0.537**	0.474	0.551
MotorImagery	0.523	0.510	**0.610**	0.500
NATOPS	**0.922**	0.928	0.822	0.883
PEMS-SF	**0.81**	0.682	0.734	0.711
PenDigits	0.974	**0.989**	0.974	0.977
PhonemeSpectra	**0.356**	0.233	0.252	0.151
RacketSports	0.812	**0.855**	0.816	0.803
SelfRegulationSCP1	0.819	0.812	**0.823**	0.775
SelfRegulationSCP2	**0.604**	0.578	0.533	0.539
SpokenArabicDigits	0.973	**0.988**	0.970	0.963
StandWalkJump	**0.467**	**0.467**	0.333	0.200
UWaveGestureLibrary	0.895	**0.906**	0.753	0.903
InsectWingbeat	0.452	**0.466**	0.264	–
On the first 29 datasets				
AVG	**0.729**	0.712	0.682	0.650
Rank	2.00	2.138	2.966	2.931

Table 2. Accuracy score of our method compared with those of other methods of unsupervised representation on UEA datasets. The representation dimensions are set to 320 for fair comparison.

datasets		SRLI	TS2Vec	RAINDROP
ETTH1	24	**0.534**	0.599	0.613
	48	**0.505**	0.629	0.635
	128	**0.687**	0.755	0.733
ETTH2	24	**0.392**	0.398	0.402
	48	**0.356**	0.580	0.573
	128	**0.396**	1.901	1.352
ETTm1	24	0.437	**0.436**	0.448
	48	**0.434**	0.515	0.524
	128	**0.531**	0.549	0.557
Electricity	24	**0.278**	0.287	0.290
	48	**0.259**	0.307	0.315
	128	**0.271**	0.332	0.344
Avg		**0.564**	0.809	0.754

respectively. This setting is to verify the performance stability of our model in different length prediction tasks. The results indicate that our model is efficient in predicting in Air quality M and Air quality H, however, it performs poorly when it comes to predicting in Air quality D. The results details are shown in the table below (Table 3 and Fig. 4):

Table 3. Accuracy score of the forecasting task on the air quality dataset.

datasets	D	SRLI	TS2Vec	RAINDROP
Air quality M	15	**0.342**	0.357	0.366
	30	**0.358**	0.361	0.372
	60	**0.385**	0.396	0.392
Air quality H	15	**0.355**	0.369	0.372
	30	**0.379**	0.384	0.391
	60	**0.387**	0.403	0.405
Air quality D	15	0.384	0.382	0.397
	30	**0.396**	0.397	0.402
	60	0.412	**0.405**	0.408
Avg		**0.376**	0.383	0.389

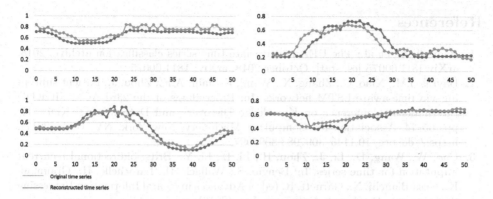

Fig. 4. The comparison result between the reconstructed data generated by SRLI and the original data in Air quality D.

5   Conclusion

We proposed an unsupervised representation learning method for multivariate irregular sampled time series named SRLI. The goal of SRLI is to extract structural and latent information from unlabeled data samples and produce high-quality representations via self-supervised learning. The results of our prediction and classification experiments shows that the generated representations are effective for various downstream tasks. It also does a good job of reconstructing the original data from its instance-level representation. While generating the reconstructed data, the decoder also generates imputation values for the missing locations. This not only proves the effectiveness of our model in extracting dependencies from time series, but also demonstrates the model's imputation accuracy for unobserved positions. In general our model encodes the series by using the mask matrix and the time interval vector, respectively, and potentially adjusts the distribution of the encoding using a predictive pseudo-task and mask discriminator. Our SRLI framework enables the feature extractor to obtain a high-quality representation in an end-to-end manner. Multiple types of experiments have validated the importance of each component of the SRLI model, showing that our model not only has comparable performance to the supervised model, but is also able to effectively handle high-dimensional irregular data. In the future, we will explore the possibility of applying the framework to more sparse and high-dimensional irregular time series, and further improve its performance using suitable time-series data augmentations.

References

1. Bagnall, A., et al.: The UEA multivariate time series classification archive, 2018. arXiv:1811.00075 [cs, stat], October 2018. arXiv: 1811.00075
2. Baytas, I.M., Xiao, C., Zhang, X., Wang, F., Jain, A.K., Zhou, J.: Patient subtyping via time-aware LSTM networks. In: Proceedings of the 23rd ACM SIGKDD International Conference on Knowledge Discovery and Data Mining. KDD '17, pp. 65–74. Association for Computing Machinery, New York, NY, USA (2017). https://doi.org/10.1145/3097983.3097997
3. Cao, W., Wang, D., Li, J., Zhou, H., Li, L., Li, Y.: Brits: bidirectional recurrent imputation for time series. In: Bengio, S., Wallach, H., Larochelle, H., Grauman, K., Cesa-Bianchi, N., Garnett, R. (eds.) Advances in Neural Information Processing Systems, vol. 31. Curran Associates, Inc. (2018)
4. Eldele, E., et al: Time-series representation learning via temporal and contextual contrasting. In: Zhou, Z. (ed.) Proceedings of the Thirtieth International Joint Conference on Artificial Intelligence, IJCAI 2021, Virtual Event / Montreal, Canada, 19–27 August 2021, pp. 2352–2359. ijcai.org (2021). https://doi.org/10.24963/ijcai.2021/324
5. Franceschi, J.Y., Dieuleveut, A., Jaggi, M.: Unsupervised scalable representation learning for multivariate time series. In: Advances in Neural Information Processing Systems, vol. 32. Curran Associates, Inc. (2019)
6. Gidaris, S., Singh, P., Komodakis, N.: Unsupervised representation learning by predicting image rotations. In: International Conference on Learning Representations (2018). https://openreview.net/forum?id=S1v4N2l0-
7. Kreindler, D.M., Lumsden, C.J.: The effects of the irregular sample and missing data in time series analysis. In: Nonlinear Dynamics, Psychology, and Life Sciences (2006)
8. Lipton, Z.C., Kale, D.C., Elkan, C., Wetzel, R.: Learning to diagnose with LSTM recurrent neural networks, March 2017. https://doi.org/10.48550/arXiv.1511.03677, arXiv:1511.03677 [cs]
9. Miao, X., Wu, Y., Wang, J., Gao, Y., Mao, X., Yin, J.: Generative semi-supervised learning for multivariate time series imputation. In: Thirty-Fifth AAAI Conference on Artificial Intelligence, AAAI 2021, Thirty-Third Conference on Innovative Applications of Artificial Intelligence, IAAI 2021, The Eleventh Symposium on Educational Advances in Artificial Intelligence, EAAI 2021, Virtual Event, 2–9 February 2021, pp. 8983–8991. AAAI Press (2021). https://ojs.aaai.org/index.php/AAAI/article/view/17086
10. Noroozi, M., Favaro, P.: Unsupervised learning of visual representations by solving jigsaw puzzles. In: Leibe, B., Matas, J., Sebe, N., Welling, M. (eds.) ECCV 2016. LNCS, vol. 9910, pp. 69–84. Springer, Cham (2016). https://doi.org/10.1007/978-3-319-46466-4_5
11. van den Oord, A., Li, Y., Vinyals, O.: Representation learning with contrastive predictive coding. arXiv preprint arXiv:1807.03748 (2018)
12. Sun, C., Hong, S., Song, M., Li, H.: A review of deep learning methods for irregularly sampled medical time series data. arXiv preprint arXiv:2010.12493 (2020)
13. Tonekaboni, S., Eytan, D., Goldenberg, A.: Unsupervised representation learning for time series with temporal neighborhood coding. In: 9th International Conference on Learning Representations, ICLR 2021, Virtual Event, Austria, 3–7 May 2021. OpenReview.net (2021). https://openreview.net/forum?id=8qDwejCuCN

14. Wang, J., Jiao, J., Liu, Y.-H.: Self-supervised video representation learning by pace prediction. In: Vedaldi, A., Bischof, H., Brox, T., Frahm, J.-M. (eds.) ECCV 2020. LNCS, vol. 12362, pp. 504–521. Springer, Cham (2020). https://doi.org/10.1007/978-3-030-58520-4_30

15. Yang, X., Zhang, Z., Cui, R.: TimeCLR: a self-supervised contrastive learning framework for univariate time series representation. Knowl.-Based Syst. **245**, 108606 (2022). https://doi.org/10.1016/j.knosys.2022.108606

16. Yoon, J., Zame, W.R., van der Schaar, M.: Estimating missing data in temporal data streams using multi-directional recurrent neural networks. IEEE Trans. Biomed. Eng. **66**(5), 1477–1490 (2019). https://doi.org/10.1109/TBME.2018.2874712

17. Yue, Z., et al.: Ts2vec: towards universal representation of time series. In: Proceedings of the AAAI Conference on Artificial Intelligence, vol. 36, pp. 8980–8987 (2022)

18. Zhang, R., Isola, P., Efros, A.A.: Colorful image colorization. In: Leibe, B., Matas, J., Sebe, N., Welling, M. (eds.) ECCV 2016. LNCS, vol. 9907, pp. 649–666. Springer, Cham (2016). https://doi.org/10.1007/978-3-319-46487-9_40

19. Zhang, X., Zeman, M., Tsiligkaridis, T., Zitnik, M.: Graph-guided network for irregularly sampled multivariate time series. In: The Tenth International Conference on Learning Representations, ICLR 2022, Virtual Event, 25–29 April 2022. OpenReview.net (2022)

20. Zhou, H., et al.: Informer: beyond efficient transformer for long sequence time-series forecasting. In: Proceedings of the AAAI Conference on Artificial Intelligence, vol. 35, no. 12, pp. 11106–11115, May 2021. https://doi.org/10.1609/aaai.v35i12.17325, https://ojs.aaai.org/index.php/AAAI/article/view/17325

Multimodal Event Classification in Social Media

Hexiang Wu, Peifeng Li[(⊠)], and Zhongqing Wang

School of Computer Science and Technology, Soochow University, Suzhou, China
20205227103@stu.suda.edu.cn, {pfli,wangzq}@suda.edu.cn

Abstract. Currently, research on events mainly focuses on the task of event extraction, which aims to extract trigger words and arguments from text and is a fine-grained classification task. Although some researchers have improved the event extraction task by additionally constructing external image datasets, these images do not come from the original source of the text and cannot be used for detecting real-time events. To detect events in multimodal data on social media, we propose a new multimodal approach which utilizes text-image pairs for event classification. Our model uses a unified language pre-trained model CLIP to obtain visual and textual features, and builds a Transformer encoder as a fusion module to achieve interaction between modalities, thereby obtaining a good multimodal joint representation. Experimental results show that the proposed model outperforms several state-of-the-art baselines.

Keywords: Multimodal event classification · Transformer · Image-text pair classification

1 Introduction

With the rapid development of mobile internet, social media has gradually become an important platform for people to obtain and publish information. Among the rich and diverse information on these platforms, a considerable amount of it revolves around events. Event extraction is one of the important tasks in natural language processing, which aims to detect event trigger words (i.e. event detection, ED) and event arguments (i.e. event argument extraction, EAE) from text. It classifies words in sentences based on sentence-level or lexical-level features. However, real-world information not only includes text, but also other modalities such as images that are specific to certain events.

To utilize other modalities such as images to help improve event extraction, previous researches [8–10] have used finely-grained image features that are consistent with the event types and event arguments in the corpus to perform multimodal event extraction. These research methods focus more on aligning modalities at a fine-grained level, but ignore other supplementary information that may be provided by images. Social media platforms such as Twitter often have a large number of comments or broadcasts about specific events during their occurrence, which usually consist of both text and images. These two modalities have a good correlation, as shown in Fig. 1. Therefore, this paper proposes

B. Luo et al. (Eds.): ICONIP 2023, CCIS 1967, pp. 338–350, 2024.
https://doi.org/10.1007/978-981-99-8178-6_26

Image

Text | Irma minor damage at univision's parking lot. |

Fig. 1. A image-text pair of destruction event from social media

a multimodal event classification task that determines the event category of a sample based on both text and image information.

Using large pre-trained models can learn good feature representations, and can still achieve good performance when transferred to tasks in other domains with smaller data volumes. Transfer learning has gradually become a common method in computer vision, natural language processing, and multi-modal deep learning. To achieve the fusion of pre-trained features from different modalities, current text-image-based multimodal methods are generally divided into two categories. The first category uses two separate unimodal pre-trained models to obtain features for each modality, and then performs multi-modal fusion before classification. The second category obtains features from the intermediate layers of each modality's unimodal model, and then uses a unified network for multimodal fusion. Previous work [1] has shown that an effective multimodal classification algorithm needs to consider both intra-modality processing and inter-modality interaction.

The core idea of this paper is to integrate good features obtained from each modality, and use a Transformer encoder to automatically achieve soft alignment between different modalities, thereby achieving effective feature extraction of individual modalities and interaction between modalities. To this end, this paper proposes a model called CLIP-TMFM (CLIP-Transformer for Multimodal Fusion Model) that uses a Transformer [4] encoder to effectively interact with the features of the vision language pre-trained model CLIP [3]. CLIP is pre-trained on a dataset of 400 million image-text pairs collected from the Internet through contrastive learning, and learns text and image features through contrastive learning between matched and unmatched text-image pairs during training. Therefore, the pre-trained text and image features obtained from CLIP actually have interaction between modalities. To fully exploit the pre-trained visual language model CLIP, CLIP-TMFM uses a Transformer encoder to learn cross-modal interaction between fine-grained features of text and images, and then focuses on important information in individual modal features to obtain good cross-modal representations. The experimental results on CrisisMMD [5] show that the proposed method outperforms several state-of-the-art baselines.

2   Related Work

In recent years, some novel methods have introduced other solution paradigms for event extraction tasks. Du et al. [14] used a question-answering approach to process event extraction tasks, inputting questions and original text together into a BERT model with custom question templates, and then extracting event triggers and arguments. Lu et al. [15] proposed a sequence-to-structure generation paradigm TEXT2EVENT, which uses the pre-trained generation model T5 [16] to generate target linear structures from input text. Due to the small size of event extraction datasets, the performance of generation methods usually lags behind state-of-the-art classification methods. Liu et al. [17] proposed a template-based, dynamic prefix-based generation method for event extraction tasks, achieving competitive results compared to a SOTA classification method ONEIE [13].

Events do not only exist in text, and some researchers have used image features as supplementary information to enhance event extraction. Zhang et al. [8] retrieve visual patterns from a background visual repository using external image-caption pairs to improve event extraction. Li et al. [9] proposed a weakly supervised training framework for multimodal event extraction using existing unimodal corpora, learning multi-modal common space embedding through structured representation and graph neural networks. Tong et al. [10] constructed an image dataset for event extraction benchmarks and proposed a dual recurrent multimodal model to conduct deep interactions between images and sentences for modality features aggregation. However, the above multimodal event extraction methods all use artificially constructed external image libraries to improve event extraction, and these images usually do not have a natural correspondence with the text data, but just select image features with similar semantics based on the event triggers and the event arguments.

3   Methodology

3.1   Task Definition

Let $S = \{p_1, p_2, ..., p_n\}$ represents a set of multimedia social media posts, where n is the number of posts, p_i is the i-th post. Let $p_i = \{post, img\}$, where $post$ is the text of the source post, img is an image attached to the source post. We need to learn a model $f : p_i \rightarrow Y(p_i \in S)$ to classify each post into the predefined event types Y.

3.2   Overview

This paper uses the pre-trained vision language model CLIP to extract features from text-image pairs, and constructs a fusion module based on Transformer encoder for multi-modal feature interaction. CLIP uses 400 million pairs of images and text collected from the Internet to obtain modal representations

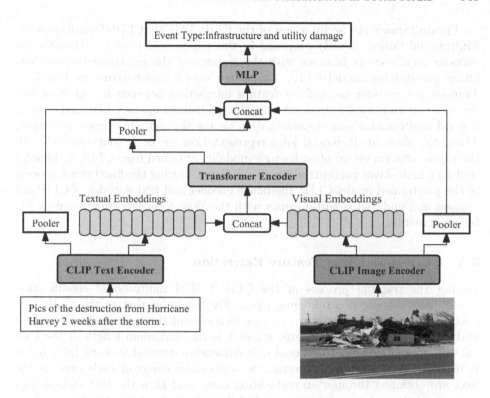

Fig. 2. The architecture of CLIP-TMFM.

of images and text through Contrastive Language-Image Pre-training, and it is a model with strong representation ability. CLIP chooses two different model architectures as image encoders. The first model is slightly modified based on the image model ResNet, where the main modification is to propose using attention pooling to replace the last average pooling in the ResNet [22] network structure. The second model is Vision Transformer (ViT) [23], and the main modification of CLIP on this model is to add an additional layer normalization before the Transformer. CLIP chooses Transformer as the text encoder, using [SOS] and [EOS] as the start and end lexical units of the text sequence, and the output of the final layer of the text encoder Transformer corresponding to the [EOS] lexical unit is used as the feature representation of the text, which will be mapped to the multi-modal embedding space.

Since its inception, the Transformer [4] model structure has had a profound impact on fields including natural language processing and computer vision, and many of the most advanced models today are based on the Transformer model structure. The image encoder used in the pre-trained CLIP model used in this paper adopts Vision Transformer [23] as the basic architecture, and the text encoder adopts Transformer encoder as the basic architecture.

Figure 2 shows the architecture of the CLIP-TMFM (CLIP-Transformer for Multimodal Fusion Model) proposed in this paper, which uses a Transformer encoder to effectively interact with the features of the pre-trained visual language pre-training model CLIP. This paper uses a fusion structure based on Transformer encoder to perform feature interaction between fine-grained features of text and images obtained from the pre-trained model CLIP, and obtains a good multi-modal joint representation vector through this fusion structure. Then, the above multi-modal joint representation vector is concatenated with the representation vector of each single modality obtained from CLIP, and finally sent to a multi-layer perceptron for classification. During the fine-tuning process of the pre-trained model CLIP, the image encoder and text encoder of CLIP are trained and updated simultaneously with the Transformer encoder used as the fusion structure.

3.3   CLIP-Based Text Feature Extraction

During the training process of the CLIP-TMFM multi-modal classification model, each sample is a text-image pair. For the input text to the CLIP text encoder, this paper uses $s = w_1, \cdots, w_L$ to represent the text after tokenization and padding to a uniform length, where L is the maximum length of the lexical unit sequence. After the lexical unit sequence converted to word table index is input to the CLIP text encoder, the word embeddings of each word in the text are obtained through an embedding layer, and then the text embeddings are passed through a multi-layer stacked Transformer encoder to obtain a fine-grained representation of the text. CLIP uses the representation corresponding to the [SOS] token added after the text in the hidden state of the last layer of the encoder as the text representation. And the position trained to represent the whole sentence is different from BERT [11], which chooses the 0th token(i.e. [CLS]) as the sentence representation. The text feature obtained through the CLIP text encoder is shown as follows:

$$T_h, t_p = \text{CLIP}_{\text{Text-Encoder}}(s), \tag{1}$$

where $T_h \in \mathbb{R}^{L \times d_T}$ is the fine-grained feature matrix corresponding to each lexical unit of the text; $t_p \in \mathbb{R}^{d_T}$ is the corresponding representation of [EOS] after the text in T_h, used to represent the entire sentence; L is the maximum number of lexical units in the text, and d_T is the dimension of the text feature.

3.4   CLIP-Based Image Feature Extraction

This paper uses a pre-trained CLIP with Vision Transformer (ViT) as the image encoder. ViT applies the Transformer structure to the field of computer vision and has achieved good performance on image classification tasks. ViT divides an input image $v \in \mathbb{R}^{C \times H \times W}$ into fixed-size patches, and then flattens each block to obtain the image patches embedding $v \in \mathbb{R}^{N \times (P^2 \cdot C)}$, where C, H, and

W are the number of channels, height, and width of the input image, respectively; (P, P) is the resolution of each image patch, and $N = HW/P^2$ is the number of patches. Then, a linear mapping is performed on $v \in \mathbb{R}^{N \times (P^2 \cdot C)}$ to obtain $\bar{v} \in \mathbb{R}^{N \times d_I}$, where d_I is the dimension of the image feature, which is consistent with the hidden layer dimension of the Transformer encoder in ViT. The Vision Transformer in the CLIP image encoder is structurally similar to the above ViT. The CLIP image encoder selects the output representation of the encoder corresponding to the 0th position in the image patches sequence as the representation vector of the entire image during the pre-training, which is similar to ViT. The computational process to obtain the global representation and fine-grained features from the CLIP Image encoder is shown as follows:

$$I_h, i_p = \text{CLIP}_{\text{Image-Encoder}}(v), \tag{2}$$

where $I_h \in \mathbb{R}^{N \times d_I}$ is the fine-grained feature matrix corresponding to the image block sequence; $i_p \in \mathbb{R}^{d_I}$ is the feature vector in the final hidden state of the encoder corresponding to the 0th position in the image block sequence, used to represent the entire image; N is the number of blocks the image is divided into, and d_I is the dimension of the image feature.

3.5   Transformer-Based Multimodal Fusion

We use a Transformer encoder as a multimodal fusion module to fully interact between fine-grained feature matrices from single modalities. The Transformer encoder can be a stack of multiple encoder blocks, each consisting of a multi-head self-attention layer (MSA) and a multi-layer perceptron layer (MLP). This paper mainly elaborates on the internal calculation process of using a single encoder block for multimodal fusion. The fine-grained feature matrices from single modalities are concatenated and input into the Transformer encoder, and then self-attention is applied to all features, including the text lexical unit sequence and image block sequence, to automatically focus on important information in the text and image and achieve interaction between modalities.

The Transformer encoder accepts fine-grained features from single modalities pre-trained by CLIP, concatenates the text feature T_h and the image feature I_h, and represents them as Z. The encoder first applies three linear layers to the feature matrix Z to obtain Z_q, Z_k, and Z_v. Then, it uses Z_q as a query to attend to Z_k, and obtains attention scores for different positions of the text sequence and image patches sequence through matrix operations and softmax layers. Finally, the attention scores are multiplied with Z_v to obtain the hidden state matrix of the encoder output, represented as Z_h. To obtain an effective multimodal fusion representation, we select the same positions where the CLIP text encoder and CLIP image encoder choose global representations in the pre-training setting, namely, the corresponding representations at the position of the [EOS] token of the text sequence and the first position of the image patches sequence in the latter half of the hidden state Z_h. These representations are used as the multimodal fusion representation obtained from the encoder. The above calculation process can be represented as follows:

$$Z_h = \text{MSA}(Z), \tag{3}$$

$$z_{pt} = \text{LN}(Z_h^E), \tag{4}$$

$$z_{pi} = \text{LN}(Z_h^L), \tag{5}$$

where $Z \in \mathbb{R}^{(L+N) \times d}$ is the feature matrix input to the multimodal fusion encoder, and $Z_h \in \mathbb{R}^{(L+N) \times d}$ is the final hidden state output by the encoder. $Z_h^E \in \mathbb{R}^d$ corresponds to the feature representation of the [EOS] symbol in the text sequence, and $Z_h^L \in \mathbb{R}^d$ corresponds to the first position of the image block sequence in the final hidden state, which represents the hidden state of the Lth position of the overall sequence of text and image. L is the maximum length of the text sequence. z_{pt} and z_{pi} correspond to the Pooler output of the Transformer encoder in Fig. 2, which are essentially obtained by applying a layer normalization layer (LN) to Z_h^E and Z_h^L.

3.6 Late Fusion Strategy

To utilize the high-level representations obtained from single-modal pre-trained models, we concatenates the text representation t_p obtained through the CLIP text encoder, the image representation i_p obtained through the CLIP image encoder, and the features z_{pt} and z_{pi} obtained through the multimodal fusion module. The resulting joint representation is then fed into a multi-layer perceptron (MLP) for event type classification. The MLP used in this paper adopts the same structural settings as the Vision Transformer, which mainly consists of two stacked linear layers, with a neural dropout layer added after each linear layer. The GELU activation function is used between the two linear layers to enable the MLP to have the ability of nonlinear learning.

This paper uses cross-entropy loss function as the target loss function of the event classifier. The cross-entropy error function on a mini-batch data containing n samples can be calculated as follows:

$$l(T, Y) = -\sum_{i=1}^n t_i \cdot log(y_i), \tag{6}$$

where T and Y are the true label sequence and predicted label sequence of the mini-batch sample, respectively. t_i represents the true label of the i-th sample, and y_i represents the label predicted by the event classifier for that sample.

4   Experimentation

4.1   Experimental Settings

This paper conducted experiments on the multimodal text-image classification dataset of CrisisMMD [5], which consists of user-generated content on social

Table 1. Dataset statistics of CrisisMMD

Dataset	Train	Dev	Test	Total
Informative	9601	1573	1534	12708
Humanitarian	6126	998	995	8079

media related to disaster events. The purpose of CrisisMMD is to detect information about real-world disaster events based on social media posts. Each sample in the dataset is a text-image pair, and some samples correspond to the same text information due to the one-to-many relationship between text and images in social media posts. This paper focuses on two sub-task datasets within Crisis-MMD. The first sub-task dataset is the informative task, which aims to determine whether a sample is related to a disaster event based on its text-image information. The second sub-task dataset is the humanitarian task, which identifies the humanitarian event category of a sample based on its text-image information, including five categories: infrastructure and utility damage, affected individuals, rescue volunteering or donation effort, other relevant information, and not humanitarian. The dataset division of CrisisMMD in this paper is consistent with the default setting [6], and the specific sample numbers for each division are shown in Table 1.

We adopt Precision (P), Recall (R), and F measure (F1) as evaluation metrics for assessing the results of multimodal event classification.

This paper uses the pre-trained CLIP, which uses the Transformer structure as the image encoder and text encoder, to obtain single-modal feature representations and fine-grained features. The version of CLIP pre-trained in this paper is $ViT\text{-}B/32$. The attention head number in the Transformer encoder of the multimodal fusion module of the proposed CLIP-TMFM model in this paper is 12. During the training process, Adam is used as the optimizer, and a total of 40 iterations are trained with a learning rate setting of $5e^{-6}$.

4.2 Baselines and Overall Performance

To verify the effectiveness of the proposed model CLIP-TMFM, this paper uses the following models as baseline systems for comparison: (1) **BERT** [11] is a language representation model based on the Transformer encoder. (2) **CLIP**$_{\text{Text}}$ and **CLIP**$_{\text{Image}}$ are the text encoder and image encoder in the pre-trained CLIP, respectively, and this paper uses them as single-modal comparison models for the proposed method. (3) **Concat**$_{\text{CLIP}}$ is a multimodal model that fuses the text representation and image representation obtained from the CLIP text encoder and image encoder, respectively, based on concatenation. (4) **MMBT** [21] is a multimodal model that uses pre-trained text and image features for text-image pair classification. (5)**BAN**$_{\text{BERT}}$ utilizes bilinear attention for feature fusion between text and image. The original BAN model employs Glove pre-trained word embeddings as the text embedding layer, while this paper replaces the text embedding layer in the BAN [7] model with the BERT pre-trained

Table 2. Experimental results on the CrisisMMD dataset(%)

Model	Informative			Humanitarian		
	P	R	F1	P	R	F1
BERT	87.41	87.55	87.36	82.48	82.51	82.45
$CLIP_{Text}$	88.22	88.33	88.15	83.48	83.25	83.30
$CLIP_{Image}$	89.52	89.50	89.51	85.30	85.13	84.69
$Concat_{CLIP}$	92.98	93.02	92.99	89.63	89.53	89.34
MMBT	90.44	90.48	90.42	86.50	86.70	86.58
BAN_{BERT}	92.07	92.11	92.09	89.34	89.32	89.30
SSE	90.21	90.15	90.29	85.38	85.34	85.18
ME_{CA}	90.81	90.87	90.82	85.51	85.44	85.36
CLIP-TMFM	**94.04**	**94.07**	**94.05**	**91.16**	**91.20**	**91.09**

model, naming it BAN_{BERT}. (6) **SSE** [2] is a cross-attention-based method proposed by Abavisani et al. for multimodal classification of social media disaster events. (7) **ME_{CA}** [1] is a method for multi-modal classification of text-image pairs using attention-based modules to extend pre-trained single-modal models. The ME_{CA} applies cross-modal attention to fine-grained high-level features of another modality after each module in each single-modal model of SSE, with the aim of achieving modality fusion of different levels of features within the same unimodal model.

Table 2 shows the experimental results of the proposed model CLIP-TMFM and other comparative models. In the single-modal model, $CLIP_{Text}$ outperforms the BERT pre-training model in both informative and humanitarian datasets. The F1 score of $CLIP_{Image}$ on both datasets is more than 1% higher than that of $CLIP_{Text}$, which verifies that CLIP can obtain visually strong generalization ability by using natural language supervision signals.

In the comparative multi-modal models, the $Concat_{CLIP}$ model, which uses concatenation to achieve multi-modal fusion, achieved significant performance improvements compared to the single-modal models $CLIP_{Image}$ and $CLIP_{Text}$. The F1 score improvement of ConcatCLIP on the informative task is over 3% compared to the single-modal models, and the F1 score improvement on the humanitarian task is over 4%. Moreover, the performance of $Concat_{CLIP}$ also exceeded that of other relatively comparative multi-modal models. Among them, the experimental results of MMBT, SSE, and ME_{CA} is relatively close, and the text features of these models are obtained by the BERT pre-training model. MMBT uses ResNet to obtain image features, and SSE and ME_{CA} use pre-trained DenseNet for image features. The $Concat_{CLIP}$ model achieved nearly 1% improvement in F1 score on the informative dataset compared to BAN_{BERT}, and the performance of the two models on the humanitarian dataset is similar. The above experimental results prove that for the pre-trained CLIP model, using

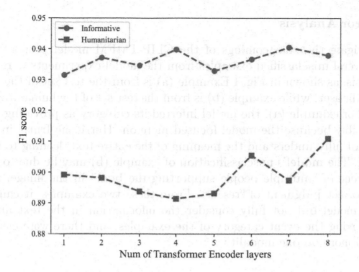

Fig. 3. Results of Transformer encoder with different layers in CLIP-TMFM.

concatenation to fuse text and image representations is an effective multi-modal fusion method.

The proposed CLIP-TMFM model achieves a considerable improvement compared to the best Concat$_{CLIP}$ model in the baseline system, with an increase of approximately 1% in F1 score on the informative dataset and over 1% on the humanitarian dataset. This indicates the effectiveness of using Transformer encoders to implement multimodal fusion methods, which can further enhance the information interaction between text and image modalities and obtain better multimodal joint representations.

4.3   Effect of Different Layers of Transformer

To investigate whether the number of stacked layers in Transformer encoders affects multimodal fusion and subsequently model performance, we conducts ablation experiments on Transformer encoders with different numbers of layers. Figure 3 shows the results of the ablation experiments on the number of layers in the Transformer encoders in the fusion module of the CLIP-TMFM model. The red line graph represents the results of the informative dataset, and the green line graph represents the results of the humanitarian dataset. A total of Transformer encoders with 1 to 8 layers were constructed as fusion modules in this paper. For the informative dataset, the experimental results of Transformer encoders with different numbers of layers fluctuate around an F1 score of 93.5%. However, the fluctuation of F1 values of Transformer encoders with different numbers of layers on the humanitarian dataset is larger than that on the informative dataset, which may be due to the fact that the humanitarian dataset is a multi-classification task and modeling is more complex compared to the binary classification task of the informative dataset.

4.4   Error Analysis

To investigate the shortcomings of the CLIP-TMFM model proposed in this paper, several misclassified examples from the model experiments were selected for analysis, as shown in Fig. 4 Example (a) is from the test set of the informative task dataset, while example (b) is from the test set of the humanitarian task dataset. For example (a), the model inferred its category as providing information, possibly because the model focused more on "Hurricane Irma" in the text but did not fully understand the meaning of the entire text, leading to incorrect reasoning. The model's misclassification of example (b) may be due to its focus on the scene of "multiple people supporting the boat" in the image, resulting in an incorrect judgment of "rescue". From these two examples, it can be seen that the model did not fully consider the information in the text and image when inferring the event category of the examples, and there were cases where it focused more on one modality.

Hurricane Irma could blow out Majesty Building, like all highrises

CLIP-TMFM:informative
golden label: not informative

(a)

Storm Harvey flood victims face displaced alligators

CLIP-TMFM: Rescue volunteering or donation effort
golden label: Affected individuals

(b)

Fig. 4. Examples of misclassified samples of CLIP-TMFM.

5   Conclusion

In this paper, we proposed a multi-modal event classification method, CLIP-TMFM, based on a visual language pre-training model. Our model uses CLIP to obtain text and image features and uses Transformer encoders to fuse the two modalities, promoting information interaction between text and images. Experimental results on the CrisisMMD dataset show that the proposed CLIP-TMFM method is effective and achieves the best performance.

Acknowledgements. The authors would like to thank the three anonymous reviewers for their comments on this paper. This research was supported by the National Natural

Science Foundation of China (Nos. 62276177 and 61836007), and Project Funded by the Priority Academic Program Development of Jiangsu Higher Education Institutions (PAPD).

References

1. Liang, T., Lin, G., Wan, M., Li, T., Ma, G., Lv, F.: Expanding large pre-trained unimodal models with multimodal information injection for image-text multimodal classification. In: CVPR 2022, pp. 15471–15480 (2022)
2. Abavisani, M., Wu, Li., Hu, S., Tetreault, J.R., Jaimes, A.: Multimodal categorization of crisis events in social media. In: CVPR 2020, pp. 14667–14677 (2020)
3. Radford, A., et al.: Learning transferable visual models from natural language supervision. In: ICML 2021, pp. 8748–8763 (2021)
4. Vaswani, A., et al.: Attention is all you need. In: NIPS 2017, pp. 5998–6008 (2017)
5. Alam, F., Ofli, F., Imran, M.: CrisisMMD: multimodal twitter datasets from natural disasters. In: ICWSM 2018, pp. 465–473 (2018)
6. Ofli, F., Alam, F., Imran, M.: Analysis of social media data using multimodal deep learning for disaster response. In: ISCRAM 2020, pp. 802–811 (2020)
7. Kim, J., Jun, J., Zhang, B.: Bilinear attention networks. In: NeurIPS 2018, pp. 1571–1571 (2018)
8. Zhang, T., et al.: Improving event extraction via multimodal integration. In: ACM Multimedia 2017, pp. 270–278 (2017)
9. Li, M., Zareian, A., Zeng, Q., Whitehead S., Lu, D., Ji, H., Chang, S.: Cross-media structured common space for multimedia event extraction. In: ACL 2020, pp. 2557–2568 (2020)
10. Tong, M., Wang, S., Cao, Y., Xu, B., Li, J., Hou, L., Chua, T.: Image enhanced event detection in news articles. In: AAAI 2020, pp. 9040–9047 (2020)
11. Devlin, J., Chang, M., Lee, K., Toutanova, K.: BERT: pre-training of deep bidirectional transformers for language understanding. In: NAACL-HLT(1) 2019, pp. 4171–4186 (2019)
12. Wadden, D., Wennberg, U., Luan, Y., Hajishirzi, H.: Entity, relation, and event extraction with contextualized span representations. In: EMNLP/IJCNLP(1) 2019, pp. 5783–5788 (2019)
13. Lin, Y., Ji, H., Huang, F., Wu, L.: A joint neural model for information extraction with global features. In: ACL 2020, pp. 7999–8009 (2020)
14. Du, X., Cardie, C.: Event extraction by answering (almost) natural questions. In: EMNLP(1)2020, pp. 671–683 (2020)
15. Lu, Y., et al.: Text2Event: controllable sequence-to-structure generation for end-to-end event extraction. In: ACL/IJCNLP(1)2021, pp. 2795–2806 (2021)
16. Raffel, C., et al.: Exploring the limits of transfer learning with a unified text-to-text transformer. J. Mach. Learn. Res. 21(140), 1–67 (2020)
17. Liu, X., Huang, H., Shi, G., Wang, B.: Dynamic prefix-tuning for generative template-based event extraction. In: ACL(1)2022, pp. 5216–5228 (2022)
18. Kim, W., Son, B., Kim, I.: ViLT: vision-and-language transformer without convolution or region supervision. In: ICML 2021, pp. 5583–5594 (2021)
19. Yang, Z., He, X., Gao, J., Deng, L., Smola, A.J.: Stacked attention networks for image question answering. In: CVPR 2016, pp. 21–29 (2016)
20. Lu, J., Yang, J., Batra, D., Parikh, D.: Hierarchical question-image co-attention for visual question answering. In: NIPS 2016, pp. 289–297 (2016)

21. Kiela, D., Bhooshan, S., Firooz, H., Testuggine, D.: Supervised Multimodal Bitransformers for Classifying Images and Text. In:arXiv preprint arXiv:1909.02950 (2019)
22. He, K., Zhang, X., Ren, S., Sun, J.: Deep residual learning for image recognition. In: CVPR 2016, pp. 770–778 (2016)
23. Dosovitskiy, A., et al.: An image is worth 16x16 words: transformers for image recognition at scale. In: ICLR 2021 (2021)

ADV-POST: Physically Realistic Adversarial Poster for Attacking Semantic Segmentation Models in Autonomous Driving

Huan Deng, Minhuan Huang$^{(\boxtimes)}$, Tong Wang, Hu Li, Jianwen Tian, Qian Yan, and Xiaohui Kuang

NKLSTISS, Institute of Systems Engineering, Academy of Military Sciences, Beijing, China
darbean@126.com

Abstract. In recent years, deep neural networks have gained significant popularity in real-time semantic segmentation tasks, particularly in the domain of autonomous driving. However, these networks are susceptible to adversarial examples, which pose a serious threat to the safety of autonomous driving systems. Existing adversarial attacks on semantic segmentation models primarily focus on the digital space and lack validation in real-world scenarios, or they generate meaningless and visually unnatural examples. To address this gap, we propose a method called Adversarial Poster (**ADV-POST**), which generates physically plausible adversarial patches to preserve semantic information and visual naturalness by adding small-scale noise to posters. Specifically, we introduce a dynamic regularization method that balances the effectiveness and intensity of the generated patches. Moreover, we conduct comprehensive evaluations of the attack effectiveness in both digital and physical environments. Our experimental results demonstrate the successful misguidedness of real-time semantic segmentation models in the context of autonomous driving, resulting in inaccurate semantic segmentation results.

Keywords: Real-Time semantic segmentation · Adversarial example · Physical adversarial attack · Autonomous driving

1 Introduction

Deep neural networks (DNNs) have achieved comparable, and even superior, performance to humans in many computer vision tasks, making them widely applicable in the real world. Among these tasks, semantic segmentation [5,17,22,30] involves classifying each pixel in an image, enabling pixel-level semantic understanding and image segmentation. Real-time semantic segmentation networks [12,29] aim to perform semantic segmentation quickly and efficiently in complex real-time scenarios. They are crucial for the perception module

B. Luo et al. (Eds.): ICONIP 2023, CCIS 1967, pp. 351–364, 2024.
https://doi.org/10.1007/978-981-99-8178-6_27

of autonomous driving systems, providing semantic category information from camera images to support the perception module's understanding of the environment. Real-time semantic segmentation is therefore a critical visual task in autonomous driving. However, recent research has shown that DNNs are vulnerable to adversarial examples and prone to adversarial attacks [10,27]. Therefore, ensuring the safety of DNN-based autonomous driving systems and exploring safety scenarios in their real-world applications, along with performing security tests, has become a critically important task.

Most research on adversarial attacks has largely centered on tasks of image classification and object detection [2,3,13,16,24]. It is also acknowledged that semantic segmentation models, like their counterparts, are susceptible to adversarial attacks [1,11,28]. Such attacks on semantic segmentation models can be perpetrated in both digital space and physical space. Regrettably, current research has predominantly been oriented towards the digital space, with very few instances addressing physical validation. In the digital space, the employment of global adversarial perturbations, which impact all pixels in the input images, is quite common. However, the implementation of these global adversarial perturbations in the physical world is often deemed unfeasible. Thus, physical adversarial attacks typically encompass local adversarial perturbations, particularly physical adversarial patches.

In order to validate these attacks in the physical space, one technique is to create nonsensical pixel perturbations to construct adversarial examples. Unfortunately, this often yields unnatural adversarial patches devoid of semantic information. To improve the subterfuge of physical adversarial patches, we draw upon methods utilized in the generation of adversarial patches for image classification and object detection tasks. Consequently, we propose the creation of a poster-like physical adversarial patch (**ADV-POST**). Although visually natural, this patch can successfully fool real-time semantic segmentation models in physical settings. Roadside signs or posters are a common sight in driving scenarios. When noise is introduced to these advertisement posters within a defined range, the poster is transformed into an adversarial patch. This patch not only retains semantic information but also preserves visual naturalness.

We achieve effective adversarial attack performance by iteratively optimizing the noise in the adversarial patch. Prior works [20,21] tend to directly compute the adversarial loss for all pixels in the image. On the contrary, we refer to the work of Nesti et al. [21], which divides adversarial losses into two parts based on whether or not pixels are attacked, resulting in misclassification. In the actual experiment, we observed that pixels that are not successfully attacked in the attack process are worthy of attention, and they are very important to improve the attack effect. Therefore, the countermeasure loss term of correctly classified pixels can encourage the adversarial example generation process to attack the current correctly classified pixels, resulting in a better attack effectiveness. This is the opposite of what Nesti suggests, who believes that the weight of the adversarial loss for correctly classified pixels should be decreased. Based on our observation, we propose a novel, dynamic regularization approach for the adversarial loss, where the weight of the adversarial loss for correctly classified pixels are dynamically increased to achieve a better balance between the effectiveness and strength of the adversarial attack. To enhance the robustness of adversarial attacks in the physical world, we have employed methods such as Expectation over Transformation (EoT) [2]. We validate the effectiveness of the generated physical adversarial patches in the digital space and further validate them by physically printing the adversarial posters and placing them in the real world. Experimental results demonstrate that our approach can successfully mislead real-time semantic segmentation models into producing erroneous semantic segmentation outputs.

In summary, the key contributions of this work are as follows:

- We proposed a novel form of physical adversarial patch that enhances its visual naturalness and semantic information by introducing small-scale noise onto a poster.
- We observed that the contribution of adversarial loss for correctly classified pixels is crucial for the success of adversarial attacks and, on the basis of this observation, we proposed a novel dynamic regularization approach for adversarial loss.
- We conducted comprehensive verification and analysis of the effectiveness of the generated physical adversarial patches in both digital and physical spaces.

2 Related Work

2.1 Real-Time Semantic Segmentation Networks

Real-Time Semantic Segmentation Networks refer to deep learning models that can perform semantic segmentation rapidly and efficiently in real-world scenarios [12,29]. These models are of significant practical significance in domains such as autonomous driving and intelligent surveillance. Early real-time semantic segmentation networks primarily relied on architectures such as Fully Convolutional Networks (FCN) [17] and UNet [22]. However, due to computational complexity and a large number of parameters, these models faced challenges in terms of real-time performance. To address this issue, researchers have proposed a series of network models specifically designed for real-time semantic segmentation. Among them, BiSeNet [29] and DDRNet [12] are two notable innovative models.

Bilateral Segmentation Network (BiSeNet) [29] adopts a bilateral segmentation strategy, dividing the network into global and local branches. The global branch is responsible for capturing global semantic information, while the local branch focuses on preserving details. This bilateral design allows BiSeNet to effectively address the balance between global and local information in semantic segmentation, achieving a trade-off between accuracy and speed. Dilated Dense Residual Network (DDRNet) [12] is a network architecture based on dense residual connections and dilated convolutions. DDRNet introduces dense connections to facilitate better information flow and promote feature reuse within the network. Additionally, DDRNet employs dilated convolutions to enlarge the receptive field, allowing it to capture a broader context while maintaining high resolution. This design enables DDRNet to achieve high accuracy and efficiency in real-time semantic segmentation tasks.

2.2   Adversarial Attacks

Adversarial attacks aim to perturb input images using carefully crafted adversarial perturbations to mislead the predictions of deep neural networks. Early research [4,10,18,19] conducted attacks in the digital space by perturbing all pixels of the image, but global perturbations are not feasible in physical-world attack scenarios. Subsequently, physical adversarial patches [3,8,14,16,24] were introduced as a pattern of adversarial attack, which perturb local pixels of an image, thereby extending adversarial attacks into the physical space.

However, up to now, research on adversarial attacks in computer vision tasks has primarily focused on image classification and object detection, with limited attention given to semantic segmentation attacks. There are some articles [6,9,11,20,23,25] that discuss adversarial attacks in semantic segmentation, but most of them concentrate on adversarial perturbations in the digital space, which perturb all pixels in the input image. Therefore, even small adversarial perturbations can lead to significant errors in semantic segmentation results. Adversarial perturbations in the digital space cannot be replicated in the real world, while physical-world adversarial attacks can pose real security threats to autonomous vehicles.

Nesti et al. [21] was the first to apply adversarial patch attacks to real-time semantic segmentation models, optimizing the adversarial patches in the digital space and validating them in the physical world. However, they did not strictly limit the range of noise. In order to achieve better attack effectiveness, they increased the perturbation intensity, but the generated adversarial patches lacked semantic information and appeared unnatural. In other words, they sacrificed the stealthiness of adversarial patches to enhance the effectiveness of the attack. We believe that for successful physical-world adversarial attacks, it is crucial to maintain the stealthiness of adversarial patches while achieving a certain level of attack effectiveness. Therefore, compared to existing works, we add small perturbations to advertising posters to generate adversarial patches that preserve semantic information, appear visually natural, and maintain good attack effectiveness.

3   Methodology

3.1   Preliminaries

We define our problem and symbols in this section. In semantic segmentation, let the input image dataset be denoted as $X = \{\mathbf{x}_i\}$, where $\mathbf{x}_i$ represents the i-th input image and is expressed as $\mathbf{x}_i \in [0,1]^{H \times W \times C}$, with H, W, and C being the height, width, and number of channels of $\mathbf{x}_i$ respectively. The ground truth set corresponding to the image dataset X is denoted as $Y = \{\mathbf{y}_i^{true}\}$, where $\mathbf{y}^{true}$ represents the ground truth of semantic segmentation corresponding to $\mathbf{x}_i$ and is expressed as $\mathbf{y}^{true} \in \mathbb{N}^{H \times W \times 1}$. The semantic segmentation model learns the mapping $f : X \rightarrow Y$, where the predicted probability score of class j for the k-th pixel in input image $\mathbf{x}_i$ is denoted as $f_k^j(\mathbf{x}_i) \in [0, 1]$, with $j \in \{1, 2, \cdots, N\}$ and N representing the number of classes. Consequently, the final classification result for that pixel is determined by the class with the highest prediction score. Therefore, the output of the semantic segmentation of the input image $\mathbf{x}_i$ is defined as $f(\mathbf{x}_i) = \underset{j \in \{1, 2, \cdots, N\}}{\arg\max} f_k^j(\mathbf{x}_i), \forall k \in \{1, 2, \cdots, H \times W\}$.

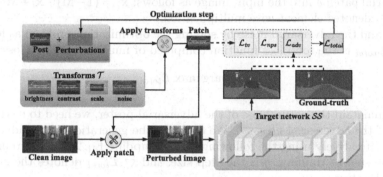

Fig. 1. Overview of the ADV-POST

3.2 Adversarial Poster Attacks Overview

Our objective is to generate natural and semantically meaningful physical adversarial patches that can be used to launch adversarial attacks on real-time semantic segmentation networks in the physical world. To achieve this, we propose the **Adversarial Poster Attack** method. Specifically, 1) we start by selecting a common advertising poster, such as "KEEP A SAFE DISTANCE," as the seed patch. 2) Then we add random noise to the seed patch while constraining the range of the noise. 3) The resulting patch with added noise serves as the adversarial patch, which is then added to different images to create a set of adversarial examples. It should be noted that we randomly transform the patch's position to ensure the robustness of the attack. 4) These adversarial examples are then inputted into the real-time semantic segmentation network to obtain the corresponding predicted outputs. 5) With the predicted outputs, we calculate the loss and optimize adversarial noise, and go back to step 2) for iterative optimization until we reach the iteration counts, as illustrated in Fig. 1.

3.3 Adversarial Poster Generation

Our objective is to find suitable adversarial noise that, when added to a poster, creates a natural and semantically meaningful patch. We start by preparing a poster as the seed patch, denoted as $P \in [0,1]^{h \times w \times C}$, where h and w represent the height and width of the seed patch, respectively, both of which should be smaller than the image size. Similarly to global adversarial perturbations, we initialize a small random noise $\xi \in [-\pi, \pi]^{h \times w \times C}$ on the seed patch, where π is a threshold. The adversarial patch is then obtained by adding the noise ξ to the seed patch, resulting in $\delta = P + \xi$. Furthermore, we generate a perturbation mask $M \in \{0,1\}^{H \times W \times 1}$, which indicates the application scope of the adversarial poster. To enhance the robustness of the adversarial patch in the physical world, we utilize the EoT [2] method, where during each iteration of optimization, we apply a set of physical transformations $\mathcal{T}$, such as randomizing scales, contrast, brightness, and adding noise. We create adversarial samples by incorporating the adversarial patch δ into the input image as follows: $\mathbf{x}_i' = (1-M) \odot \mathbf{x}_i + M \odot \mathcal{T}(\delta)$, where $\odot$ denotes element-wise multiplication.

We find the adversarial noise by solving the optimization problem as follows, where $\mathcal{L}_{total}$ is the total loss function composed of multiple weighted loss terms.

$$\xi^* = \arg\max_{\xi} \mathcal{L}_{total}, \tag{1}$$

To maintain the naturalness of the adversarial poster, we need to restrict the range of the adversarial noise ξ^*. We optimize the generation of the adversarial noise ξ^* iteratively using the following formula update Eq. 2, where t represents the time step of training, η is the step size, and $\nabla_\xi \mathcal{L}_{total}$ denotes the gradient of the objective.

$$\xi^{(0)} = \xi; \xi^{(t+1)} = Clip_{[-\pi,\pi]}(\xi^{(t)} + \eta \cdot \sum_{i=1} \nabla_\xi \mathcal{L}_{total}(f(\mathbf{x}_i'), \mathbf{y}^{true})) \tag{2}$$

3.4 Loss Function

The Adversarial Loss. In this section, we introduce the design of the adversarial loss for semantic segmentation, denoted as $\mathcal{L}_{adv}$. Previous works [6,9,20] on adversarial attacks in semantic segmentation mainly focused on digital adversarial attacks, which typically apply global perturbations to images by adding noise to all pixels. Therefore, the adversarial loss often adopts the pixel-wise cross-entropy (CE) loss, denoted as $\mathcal{J}_{ce}$, which yields good results.

For adversarial patch attack on real-time semantic segmentation models, an approach is used to follow the previous works and continue using the pixel-wise cross-entropy (CE) loss [11,20]. That is, computing the pixel-wise cross-entropy in the image after adding the adversarial patch, which can be expressed as $\mathcal{L}_{adv} = \sum_{i=1} \sum_{k \in \mathcal{P}} \frac{1}{|\mathcal{P}|} \cdot \mathcal{J}_{ce}(f_k(\mathbf{x}_i'), \mathbf{y}^{true})$, where $\mathcal{P} = \{1, 2, \cdots, H \times W\}$ represents the set of all pixels in the adversarial sample $\mathbf{x}_i'$.

However, Nesti et al. [21] used physical adversarial patch attacks on real-time semantic segmentation networks without constraining the noise range of the physical adversarial patch, which improves the pixel-wise cross-entropy loss by computing $\mathcal{J}_{ce}$ outside the region of the adversarial patch δ. It divides $\mathcal{J}_{ce}$ into two parts: the cross-entropy loss of correctly classified pixels $\mathcal{J}_{ce\_cor}$ and the cross-entropy loss of misclassified pixels $\mathcal{J}_{ce\_mis}$, resulting in a stronger attack. Specifically, $p = \{1, 2, \cdots, h \times w\}$ represents the set of all pixels in the adversarial patch, and $\widehat{\mathcal{P}} = \{k \notin p \mid k \in \mathcal{P}\}$ represents the set of pixels outside the region of the adversarial patch δ. Let $\widehat{\mathcal{P}}_{cor} = \{k \in \widehat{\mathcal{P}} \mid f_k(\mathbf{x}_i') = \mathbf{y}_k^{true}\}$ be the set of correctly classified pixels outside the region of the adversarial patch δ, and $\widehat{\mathcal{P}}_{mis} = \{k \in \widehat{\mathcal{P}} \mid f_k(\mathbf{x}_i') \neq \mathbf{y}_k^{true}\}$ be the set of misclassified pixels outside the region of the adversarial patch δ. Thus, Nesti et al. proposes a new adversarial loss as follows:

$$\mathcal{L}_{adv}(\mathbf{x}_i') = \sum_{k \in \mathcal{P}} \frac{1}{|\mathcal{P}|} \cdot (\gamma \cdot \mathcal{L}_{ce\_cor}(\mathbf{x}_i') + (1 - \gamma) \cdot \mathcal{L}_{ce\_mis}(\mathbf{x}_i'))$$

$$\mathcal{L}_{ce\_cor}(\mathbf{x}_i') = \sum_{k \in \widehat{\mathcal{P}}_{cor}} \mathcal{J}_{ce}(f(x_i'), \mathbf{y}^{true}) \tag{3}$$

$$\mathcal{L}_{ce\_mis}(\mathbf{x}_i') = \sum_{k \in \widehat{\mathcal{P}}_{mis}} \mathcal{J}_{ce}(f(x_i'), \mathbf{y}^{true})$$

In its setting, the value of γ decreases gradually with an increase in the number of iterations, that is, the weight of the cross-entropy loss term for correctly classified pixels decreases gradually, while the weight of the cross-entropy loss term for misclassified pixels increases gradually.

In contrast to Nesti's suggestion, we believe that the contribution of the cross-entropy loss term for correctly ones is crucial for adversarial attacks. Therefore, we propose to gradually increased the weight of the cross-entropy loss term for correctly classified pixels and decreased the weight for misclassified ones as the number of iterations increased. In the context of real-time semantic segmentation models for autonomous driving systems, we argue that promoting misclassified pixels in the semantic segmentation prediction image poses a greater threat to the system's safety when facing adversarial attacks, compared to enhancing the adversarial strength for the currently misclassified ones. By gradually increasing the weight of the cross entropy loss term for correctly classified pixels, we can better promote the increase in the number of misclassified pixels during the gradient descent process and achieve better attack effects.

To address this, we propose a novel approach of dynamically regularizing the adversarial loss term, where the weight of the adversarial loss term is dynamically adjusted to achieve a better balance between the effectiveness and strength of adversarial attacks. The new adversarial loss that we propose is defined as Eq. 4, where μ is the dynamically regularized coefficient that we propose.

$$\mathcal{L}_{adv}(\mathbf{x}_i') = \sum_{k \in \mathcal{P}} \frac{1}{|\mathcal{P}|} \cdot (\mu \cdot \mathcal{L}_{ce\_cor}(\mathbf{x}_i') + (1 - \mu) \cdot \mathcal{L}_{ce\_mis}(\mathbf{x}_i')) \tag{4}$$

The definition of μ is as follows,

$$\mu = a \cdot \left(\frac{t}{T}\right)^b + c \tag{5}$$

where a is the linear coefficient, in which its positive or negative value controls the growth or decrease rate of μ, b is the exponent of the iteration quantity, where its value determines the extent of b variation, c is the base, which controls the initial value of μ.

Total Variance Loss. We use total variation loss $\mathcal{L}_{tv}$ [26], which reduces the variation between adjacent pixels to increase the smoothness of the adversarial patch. It is defined as Eq. 6, where $s_{m,n}$ is the pixel value at coordinate (m, n) in δ. The total variation loss helps to improve the naturalness of physical-world adversarial examples.

$$\mathcal{L}_{tv}(\delta) = \sum_{m,n} \sqrt{(s_{m,n} - s_{m+1,n})^2 + (s_{m,n} - s_{m,n+1})^2} \tag{6}$$

Non-printability Score. We use the non-printability score loss $\mathcal{L}_{nps}$ [24], which measures the color distance between the digital adversarial perturbation and the printed adversarial perturbation to ensure their colors are close. It is defined as Eq. 7, where k represents a pixel in the adversarial patch pixel set p, and h represents a vector in the printable color vector set H.

$$\mathcal{L}_{nps}(\delta) = \sum_{k \in p} \prod_{h \in H} |k - h| \tag{7}$$

Overall, the total loss function for our adversarial patch generation process is defined as Eq. 8, where λ_{tv} and λ_{nps} are weight parameters to balance the robustness of physical-world adversarial attacks.

$$\mathcal{L}_{total} = \sum_{i=1} \mathcal{L}_{adv}(\mathbf{x}_i') + \lambda_{tv} \cdot \mathcal{L}_{tv}(\delta) + \lambda_{nps} \cdot \mathcal{L}_{nps}(\delta) \tag{8}$$

We used Adam optimizer [15] with a learning rate of 0.005 when using the EoT transformation, and a learning rate of 0.001 when not using the EoT transformation. The batch size was set to 10.

4   Experiment

In this section, we evaluate the effectiveness of our proposed adversarial poster attack in different models and settings, and validate it in both the digital and physical worlds. Furthermore, we provide several ablation studies with different parameters to obtain physically adversarial patches with better attack performance.

4.1   Experimental Setup

Dataset. In our experiments, we used the Cityscapes dataset [7], which is a commonly used semantic segmentation dataset to study and evaluate urban scene understanding tasks in autonomous driving. Specifically, we randomly sampled 300 images from the training set to optimize the adversarial patches. We evaluated the effectiveness of the adversarial patches in attacking real-time semantic segmentation models on the entire validation set of 500 images.

Models. We selected two real-time semantic segmentation models, BiSeNet and DDRNet, commonly used in autonomous driving tasks, to evaluate the effectiveness of the adversarial patch attack method proposed in this paper. We used the pre-trained models trained on the Cityscapes dataset. BiSeNet uses the Xception39 version provided by the authors [29] as its backbone, while DDRNet uses the DDRNet23Slim version provided by the authors [12].

Criteria. We used semantic segmentation evaluation metrics, mean accuracy (mAcc) and mean Intersection over Union (mIoU), which can comprehensively consider both classification accuracy and segmentation accuracy.

- mAcc: Mean Accuracy represents the average classification accuracy of the model for each pixel, providing an overall measure of the model's classification accuracy.
- mIoU: Mean Intersection over Union measures the overlap between the predicted segmentation results and the ground-truth segmentation results, providing an average measure of the segmentation accuracy.

4.2 Digital Domain Experiment

Adversarial Poster Attack Effectiveness. We validated the effectiveness of our proposed dynamically regularized adversarial loss method. In our experiments, we used adversarial patches in the form of adversarial posters and used the adversarial loss function proposed in [20,21] as the baseline method. We designed the size of the patch is 600 * 300 which occupied 8.6% of the image size. We examine the impact of different noise magnitudes on the effectiveness of the attack and create adversarial patches with thresholds of $\pi = 16/255$ and $\pi = 32/255$. We compared our method with two baseline methods, with and without the EoT method, and evaluated the experimental results using mAcc and mIoU scores against posters augmented with random noise. Table 1 demonstrates the superior attack effectiveness of our proposed method under all experimental conditions. Furthermore, Fig. 2 illustrates that the semantic information and visual naturalness of poster patterns are preserved after small-scale noise is added.

Ablation Study

Setting of the Weight of the Adversarial Loss Components. We validated the effectiveness of our proposed dynamically regularized adversarial loss method. To verify which part of $\mathcal{L}_{ce\_cor}$ and $\mathcal{L}_{ce\_mis}$ is more important for the effectiveness of the attack, we conducted two control experiments, where the initial weights of $\mathcal{L}_{ce\_cor}$ and $\mathcal{L}_{ce\_mis}$ were set at 0.5, that is, $c = 0.5$. In our method, we gradually increased the weight of $\mathcal{L}_{ce\_cor}$ while decreasing the weight of $\mathcal{L}_{ce\_mis}$ which means that $a = 0.5$, $b = 1/3$. The other set of experiments followed the opposite approach which means that $a = -0.5$, $b = 1/3$. The variation of μ is

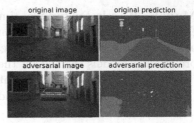

(a) Noise and adversarial poster   (b) Outputs

Fig. 2. In Figure a, the left is the adversarial noise and the right is the visually natural adversarial poster. In Figure b, it is the corresponding semantic segmentation predictions.

Table 1. Adversarial patch attack results in (mAcc/mIoU) from the validation set of the Cityscapes dataset. Bold font indicates better performance.

Method	DDRNet				BiSeNet			
	$\pi = 16$		$\pi = 32$		$\pi = 16$		$\pi = 32$	
	EoT	w/o EoT	EoT	w/o EoT	EoT	w/o EoT	EoT	w/o EoT
Initial Poster	0.844/0.770	0.844/0.770	0.843/0.770	0.843/0.770	0.742/0.648	0.742/0.648	0.740/0.646	0.740/0.646
CE loss [20]	0.714/0.605	0.619/0.517	0.691/0.586	0.602/0.510	0.710/0.599	0.710/0.599	0.713/0.600	0.691/0.574
Nesti [21]	0.669/0.570	0.568/0.464	0.168/0.108	0.075/0.023	0.562/0.471	0.465/0.363	0.491/0.416	0.443/0.347
ADV-POST	**0.219/0.154**	**0.256/0.194**	**0.165/0.106**	**0.064/0.011**	**0.551/0.463**	**0.438/0.352**	**0.478/0.402**	**0.389/0.317**

illustrated in Fig. 3(a). The experimental results, presented in Table 2 and Fig. 3, demonstrate the superior attack performance achieved by dynamically increasing the weight of the loss term for correctly classified pixels.

We also investigated the impact of the increase rate of μ, we set $a = 0.5$, $b = 3$. Table 2 and Fig. 3 reveal that maintaining a higher weight for μ during the optimization process leads to improved attack effectiveness.

The Influence of Location of the Patch. We investigated the robustness of the adversarial patches on BiSeNet generated by our method against variations in attack positions. Specifically, we altered the patch placement within the image while keeping other experimental settings unchanged. The patch was positioned at the center of the image, the center of the top left quarter region, and the corresponding positions in the top right, bottom left, and bottom right regions. The attack effectiveness of the adversarial patches at different positions is presented in Table 3 and Fig. 4. Our method consistently achieved favorable attack performance across various image regions. In particular, placing the patch in the top-left corner yielded enhanced attack effectiveness.

4.3 Real World Experiment

We validated our proposed attack method in the physical world using the Cityscapes dataset on DDRNet. Adversarial posters were generated using the

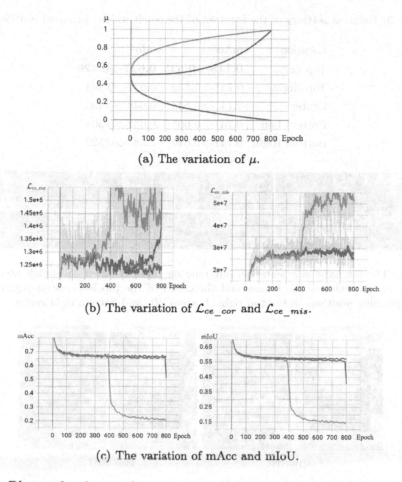

(a) The variation of μ.

(b) The variation of $\mathcal{L}_{ce\_cor}$ and $\mathcal{L}_{ce\_mis}$.

(c) The variation of mAcc and mIoU.

Fig. 3. Blue, red and green line represent μ increasing rapidly, μ decreasing, and μ increasing gradually, respectively. When the value of μ increases rapidly and remains at a high value, $\mathcal{L}_{ce\_mis}$ increases quickly, while mAcc and mIoU decrease rapidly, indicating a stronger attack effect on DDRNet. (Color figure online)

Table 2. Settings of the weight of the adversarial loss components. Experiments 1 and 2 demonstrate that $\mathcal{L}_{ce\_cor}$ is more crucial for the effectiveness of the attack. Experiments 1 and 3 show that maintaining a high weight value μ during the optimization process can lead to better attack performance.

Settings	DDRNet		BiseNet	
	mAcc	mIoU	mAcc	mIoU
μ increases(gradually)	0.8435->0.5121	0.77->0.4064	0.7423->0.586	0.6479->0.49
μ increases(rapidly)	**0.8435->0.2161**	**0.77->0.1513**	**0.7423->0.5686**	**0.6479->0.4737**
μ decreases	0.8435->0.6723	0.77->0.5715	0.7423->0.6749	0.6479->0.5696

Table 3. Different settings of the location of the patch which optimized in BiSeNet.

Location	mAcc	mIoU
Top Left	**0.718->0.523**	**0.617->0.426**
Top Right	0.741->0.530	0.637->0.434
Center	0.742->0.546	0.648->0.459
Bottom Left	0.734->0.605	0.637->0.508
Bottom Right	0.742->0.619	0.643->0.529

Fig. 4. The semantic segmentation outcome on BiSeNet when the patch was positioned at the center of the image, and the center of the top left quarter region, and corresponding positions in the top right, bottom left, and bottom right regions.

Fig. 5. The variation in the physical world. The regions near the adversarial poster are misclassified, indicating the effectiveness of our method in the physical world.

EoT method and printed with dimensions of 90 cm * 45 cm. The attack effectiveness of the generated patches was verified in our urban environment. Our adversarial posters successfully attacked real-time semantic segmentation systems in the physical world, demonstrating notable effectiveness, while the initial poster did not produce significant results, which is presented in Fig. 5.

However, in physical-world validation, the transferability of adversarial patches across data domains and models needs to be considered, highlighting an area for future improvement.

5   Conclusion

We proposed a new form of adversarial attack, Adversarial Poster Attack, which generated physically plausible adversarial patches with semantic information and visual naturalness by constraining the range of noise. We validated the effectiveness of these physical adversarial patches in real-world attacks, demonstrating that autonomous driving systems based on deep neural network models are vulnerable to physical-world adversarial attacks. Our work highlights the existence of visually natural adversarial patches in the form of posters as a potential security threat, as they can deceive human observers and mislead models to produce erroneous results, which can be potentially fatal. Future studies can further explore the robustness and transferability of physical adversarial patches and validate their performance in broader real-world scenarios.

References

1. Arnab, A., Miksik, O., Torr, P.H.: On the robustness of semantic segmentation models to adversarial attacks. In: Proceedings of CVPR (2018)
2. Athalye, A., Engstrom, L., Ilyas, A., Kwok, K.: Synthesizing robust adversarial examples. In: Proceedings of ICML (2018)
3. Brown, T.B., Mané, D., Roy, A., Abadi, M., Gilmer, J.: Adversarial patch. arXiv preprint arXiv:1712.09665 (2017)
4. Carlini, N., Wagner, D.: Towards evaluating the robustness of neural networks. In: Proceedings of SP (2017)
5. Chen, L.C., Papandreou, G., Kokkinos, I., Murphy, K., Yuille, A.L.: Deeplab: semantic image segmentation with deep convolutional nets, atrous convolution, and fully connected crfs. IEEE Trans. Pattern Anal. Mach. Intell. (2017)
6. Chen, Z., Wang, C., Crandall, D.: Semantically stealthy adversarial attacks against segmentation models. In: Proceedings of the IEEE/CVF Winter Conference on Applications of Computer Vision (2022)
7. Cordts, M., et al.: The cityscapes dataset for semantic urban scene understanding. In: Proceedings of CVPR (2016)
8. Eykholt, K., et al.: Robust physical-world attacks on deep learning visual classification. In: Proceedings of CVPR (2018)
9. Fischer, V., Kumar, M.C., Metzen, J.H., Brox, T.: Adversarial examples for semantic image segmentation. arXiv preprint arXiv:1703.01101 (2017)
10. Goodfellow, I.J., Shlens, J., Szegedy, C.: Explaining and harnessing adversarial examples. arXiv preprint arXiv:1412.6572 (2014)
11. Hendrik Metzen, J., Chaithanya Kumar, M., Brox, T., Fischer, V.: Universal adversarial perturbations against semantic image segmentation. In: Proceedings of ICCV (2017)
12. Hong, Y., Pan, H., Sun, W., Jia, Y.: Deep dual-resolution networks for real-time and accurate semantic segmentation of road scenes. arXiv preprint arXiv:2101.06085 (2021)
13. Hu, Y.C.T., Kung, B.H., Tan, D.S., Chen, J.C., Hua, K.L., Cheng, W.H.: Naturalistic physical adversarial patch for object detectors. In: Proceedings of ICCV (2021)
14. Hu, Z., Huang, S., Zhu, X., Sun, F., Zhang, B., Hu, X.: Adversarial texture for fooling person detectors in the physical world. In: Proceedings of CVPR (2022)
15. Kingma, D.P., Ba, J.: Adam: A method for stochastic optimization. arXiv preprint arXiv:1412.6980 (2014)
16. Kong, Z., Guo, J., Li, A., Liu, C.: Physgan: generating physical-world-resilient adversarial examples for autonomous driving. In: Proceedings of CVPR (2020)

17. Long, J., Shelhamer, E., Darrell, T.: Fully convolutional networks for semantic segmentation. In: Proceedings of CVPR (2015)
18. Moosavi-Dezfooli, S.M., Fawzi, A., Fawzi, O., Frossard, P.: Universal adversarial perturbations. In: Proceedings of CVPR (2017)
19. Moosavi-Dezfooli, S.M., Fawzi, A., Frossard, P.: Deepfool: a simple and accurate method to fool deep neural networks. In: Proceedings of CVPR (2016)
20. Nakka, K.K., Salzmann, M.: Indirect local attacks for context-aware semantic segmentation networks. In: Vedaldi, A., Bischof, H., Brox, T., Frahm, J.-M. (eds.) ECCV 2020. LNCS, vol. 12350, pp. 611–628. Springer, Cham (2020). https://doi.org/10.1007/978-3-030-58558-7_36
21. Nesti, F., Rossolini, G., Nair, S., Biondi, A., Buttazzo, G.: Evaluating the robustness of semantic segmentation for autonomous driving against real-world adversarial patch attacks. In: Proceedings of the IEEE/CVF Winter Conference on Applications of Computer Vision (2022)
22. Ronneberger, O., Fischer, P., Brox, T.: U-net: convolutional networks for biomedical image segmentation. In: Navab, N., Hornegger, J., Wells, W.M., Frangi, A.F. (eds.) MICCAI 2015. LNCS, vol. 9351, pp. 234–241. Springer, Cham (2015). https://doi.org/10.1007/978-3-319-24574-4_28
23. Rony, J., Pesquet, J.C., Ben Ayed, I.: Proximal splitting adversarial attack for semantic segmentation. In: Proceedings of CVPR (2023)
24. Sharif, M., Bhagavatula, S., Bauer, L., Reiter, M.K.: Accessorize to a crime: Real and stealthy attacks on state-of-the-art face recognition. In: Proceedings of the 2016 ACM SIGSAC Conference on Computer and Communications Security (2016)
25. Shen, G., Mao, C., Yang, J., Ray, B.: Advspade: realistic unrestricted attacks for semantic segmentation. arXiv preprint arXiv:1910.02354 (2019)
26. Strong, D., Chan, T.: Edge-preserving and scale-dependent properties of total variation regularization. Inverse problems (2003)
27. Szegedy, C., et al.: Intriguing properties of neural networks. arXiv preprint arXiv:1312.6199 (2013)
28. Xie, C., Wang, J., Zhang, Z., Zhou, Y., Xie, L., Yuille, A.: Adversarial examples for semantic segmentation and object detection. In: Proceedings of ICCV (2017)
29. Yu, C., Wang, J., Peng, C., Gao, C., Yu, G., Sang, N.: Bisenet: bilateral segmentation network for real-time semantic segmentation. In: Proceedings of ECCV (2018)
30. Zheng, S., et al.: Conditional random fields as recurrent neural networks. In: Proceedings of ICCV (2015)

Uformer++: Light Uformer for Image Restoration

Honglei Xu, Shaohui Liu$^{(\boxtimes)}$ ⓘ, and Yan Shu

Harbin Institute of Technology, Heilongjiang, China
shliu@hit.edu.cn

Abstract. Based on UNet, numerous outstanding image restoration models have been developed, and Uformer is no exception. The exceptional restoration performance of Uformer is not only attributable to its novel modules but also to the network's greater depth. Increased depth does not always lead to better performance, but it does increase the number of parameters and the training difficulty. In this paper, we propose Uformer++, a reconstructed Uformer based on an efficient ensemble of UNets of varying depths that partially share an encoder and co-learn simultaneously under deep supervision. Our proposed new architecture has significantly fewer parameters than the vanilla Uformer, but still with promising results achieved. Considering that different channel-wise features contain totally different weighted information and so are pixel-wise features, a novel Nonlinear Activation Free Feature Attention (NAFFA) module combining Simplified Channel Attention (SCA) and Simplified Pixel Attention (SPA) is added to the model. The experimental results on various challenging benchmarks demonstrate that Uformer++ has the least computational cost while maintaining performance.

Keywords: Uformer · UNet · Image Denoising · Image Deblurring

1 Introduction

With the rapid development of consumer and industry cameras and smartphones, the requirements for removing undesired degradation in images are constantly growing. Recovering genuine images from their degraded versions (e.g., noise, blur) is a classic task in computer vision. Most of the most recent state-of-the-art methods [4,6,22] are based on CNNs, which achieve impressive results but have trouble capturing long-term dependencies. To address this problem, several recent works [8,9] have begun to incorporate transformer-based architecture [16] into their frameworks, with promising results. However, due to the limitation of self-attention computational complexity, only low-resolution feature maps are applied.

In order to leverage the capability of self-attention in feature maps at multi-scale resolutions to recover more image details, Uformer [17], an effective and efficient Transformer-based structure for image restoration has been presented. Uformer is built upon an elegant architecture, UNet [14], in which the convolution layers are modified to Transformer blocks while keeping the same overall

© The Author(s), under exclusive license to Springer Nature Singapore Pte Ltd. 2024
B. Luo et al. (Eds.): ICONIP 2023, CCIS 1967, pp. 365–376, 2024.
https://doi.org/10.1007/978-981-99-8178-6_28

hierarchical encoder-decoder structure and the skip-connections. Uformer has achieved state-of-the-art performance on multiple image restoration tasks, benefiting from two core designs: Locally-enhanced Window (LeWin) Transformer block and the learnable multi-scale restoration modulator. Extensive experiments conducted on several image restoration tasks have demonstrated these designs are effective and efficient, but UNet is so simple that Uformer needs more depth to perform better, and the optimal depth is unknown.

In this paper, we propose Uformer++, a reconstructed Uformer based on the UNet++ [25] of three depths which consists of UNets of varying depths whose decoders are densely connected at the same resolution via the redesigned skip connections. By training Uformer++ with deep supervision, all the constituent Uformer are trained simultaneously while benefiting from a shared image representation. Compared with Uformer, Uformer++ has significantly fewer parameters while maintaining promising performance. In addition to the new architecture, we also redesign Locally-enhanced Feed-Forward Network (LeFF) of Uformer with Nonlinear Activation Free Feature Attention (NAFFA) module. Since the attention mechanism has been widely used in neural networks, it has played an important role in the performance of networks. Inspired by the FFA-Net [13], we further design a novel Nonlinear Activation Free Feature Attention (NAFFA) module combining Simplified Channel Attention (SCA) and Simplified Pixel Attention (SPA) which pay attention in channel-wise and pixel-wise features respectively. NAFFA treats different features and pixels unequally, which can provide additional flexibility in dealing with different types of information. NAFFA is simpler than the FA of FFA-Net and there is no loss of performance.

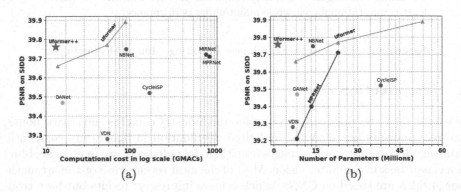

Fig. 1. (a) PSNR vs. computational cost and (b) PSNR vs. parameters amount on SIDD.

To evaluate the performance of different image restoration networks, peak-signal-to-noise-ratio (PSNR) and structure similarity index (SSIM) are commonly used to quantify image restoration quality. We compare Uformer++ with previous state-of-the-art methods on multiple image restoration tasks and mainly conduct experiments on SIDD [1] for image denoising and GoPro [11] for image deblurring. Experimental results demonstrate that Uformer++ has the least computational cost while maintaining performance on all benchmarks. Moreover, we conduct many ablation experiments to prove that our key components have an excellent performance (Fig. 2).

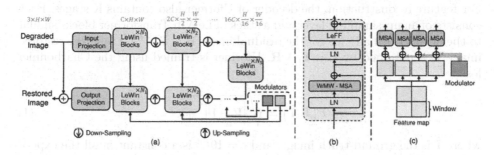

Fig. 2. (a) Overview of the Uformer structure. (b) LeWin Transformer block. (c) Illustration of how the modulators modulate the W-MSAs in each LeWin Transformer block which is named MW-MSA.

Overall, we summarize the contributions of this paper as follows:

1) We reconstructed Uformer based on UNet++, an efficient ensemble of UNets of varying depths that partially share an encoder and co-learn simultaneously under deep supervision. The new architecture has significantly fewer parameters than the vanilla Uformer, but still with promising results achieved.
2) We propose a novel Nonlinear Activation Free Feature Attention (NAFFA) module combining Simplified Channel Attention (SCA) and Simplified Pixel Attention (SPA) which pay attention to channel-wise and pixel-wise features respectively. Compared with previous CA and PA, SCA and SPA have fewer parameters but better performance.

2   Materials and Methods

In this section, we first describe the overall pipeline, especially the new architecture of Uformer++ for image restoration. Then, we provide the details of the Nonlinear Activation Free Feature Attention (NAFFA) module combining Simplified Channel Attention (SCA) and Simplified Pixel Attention (SPA).

2.1   Review Uformer

As described in previous work, the overall structure of the Uformer is a U-shaped hierarchical network [14] with skip-connections between the encoder and the decoder. Uformer firstly applies a 3×3 convolutional layer with LeakyReLU to extract low-level features $\mathbf{X}_0 \in \mathbb{R}^{C \times H \times W}$ from the input image $\mathbf{I}_0 \in \mathbb{R}^{3 \times H \times W}$. Next, the feature maps $\mathbf{X}_0$ are passed through K encoders stages. Each stage contains a down-sampling layer and a stack of the proposed LeWin Transformer blocks. The LeWin Transformer block takes advantage of the self-attention mechanism for capturing long-range dependencies and also cuts the computational cost due to self-attention usage through non-overlapping windows on the feature maps. The l-th stage of encoder produces the feature maps $\mathbf{X}_l \in \mathbb{R}^{2^l C \times \frac{H}{2^l} \times \frac{H}{2^l}}$. For feature reconstruction, the decoder of Uformer also contains K stages. Each consists of an up-sampling layer and a stack of LeWin Transformer blocks similar to the encoder. Finally we got the residual image $\mathbf{R} \in \mathbb{R}^{3 \times H \times W}$ and the restored image $\mathbf{I}'$ is obtained by $\mathbf{I}' = \mathbf{I}_0 + \mathbf{R}$. Uformer is trained using the Charbonnier loss:

$$\ell\left(\mathbf{I}', \hat{\mathbf{I}}\right) = \sqrt{\left\|\mathbf{I}' - \hat{\mathbf{I}}\right\|^2 + \epsilon^2} \tag{1}$$

where $\hat{\mathbf{I}}$ is the ground-truth image, and $\epsilon = 10^{-3}$ is a constant in all the experiments of Uformer.

However, previous works have proved the U-shaped hierarchical network for image restoration comes with two limitations. First, the optimal depth of an encoder-decoder network can vary from one application to another, depending on the task difficulty and the amount of labeled data available for training. A simple approach would be to train models of varying depths separately and then ensemble the resulting models during inference time. Furthermore, being trained independently, these networks do not enjoy the benefits of multi-task learning. Second, the design of skip connections used in an encoder-decoder network is unnecessarily restrictive, demanding the fusion of the same scale encoder and decoder feature maps. While striking as a natural design, the same-scale feature maps from the decoder and encoder networks are semantically dissimilar and no solid theory guarantees that they are the best match for feature fusion.

2.2   Uformer++

Uformer++ is a reconstructed Uformer based on an efficient ensemble of UNets of varying depths that partially share an encoder and co-learn simultaneously under deep supervision that aims at overcoming the above limitations and reducing computational cost while maintaining performance.

Uformer++ embeds Uformer of varying depths in its architecture. They partially share an encoder, while their decoders are intertwined. Let $x^{i,j}$ denote the output of node $X^{i,j}$ where i index the down-sampling layer along the encoder and j indexes the convolution layer of the dense block along the skip connection. The stack of feature maps represented by $x^{i,j}$ is computed as:

$$x^{i,j} = \begin{cases} \mathcal{H}\left(\mathcal{D}\left(x^{i-1,j}\right)\right) & j = 0 \\ \mathcal{H}\left(\left[\left[x^{i,k}\right]_{k=0}^{j-1}, \mathcal{U}\left(x^{i+1,j-1}\right)\right]\right) & j > 0 \end{cases} \tag{2}$$

where function $\mathcal{H}$ is a convolution operation followed by an activation function, $\mathcal{D}$ and $\mathcal{U}$ denote a down-sampling layer and an up-sampling layer respectively, and [] denotes the concatenation layer. It is obvious that nodes at level $j = 0$ receive only one input from the previous layer of the encoder and nodes at level $j > 1$ receive $j + 1$ inputs, of which j inputs are the outputs of the previous j nodes in the same skip connection and the $j + 1^{th}$ input is the up-sampled output from the lower skip connection. All prior feature maps accumulate and arrive at the current node in every layer make every layer works like DenseNet making use of a dense convolution block along each skip connection (Fig. 3).

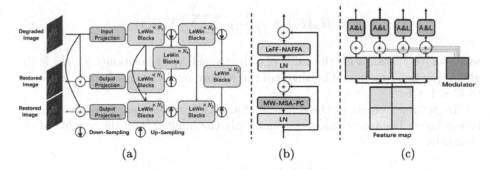

(a)   (b)   (c)

Fig. 3. (a) Overview of the Uformer++ structure (b) LeWin Transformer block. (c) Illustration of how the modulators modulate the MSA-PL in each LeWin Transformer block which is named MW-MSA-PL.

2.3 NAFFA

To treat different features and pixels region unequally and provide additional flexibility in dealing with different types of information, Channel Attention [5] and Pixel Attention [13] are proposed. Channel Attention squeezes the spatial information into channels first and then a multilayer perceptual applies to it to calculate the channel attention, which will be used to weigh the feature map. Pixel Attention is similar to Channel Attention except for the absence of a pooling layer. They could be represented as:

$$CA(\mathbf{X}) = \mathbf{X} * \sigma\left(W_{c2} \max\left(0, W_{c1} \operatorname{pool}(\mathbf{X})\right)\right) \tag{3}$$

$$PA(\mathbf{X}) = \mathbf{X} * \sigma\left(W_{p2} \max\left(0, W_{p1}(\mathbf{X})\right)\right) \tag{4}$$

where $\mathbf{X}$ represents the feature map, and pool indicates the global average pooling operation which aggregates the spatial information into channels. σ is a nonlinear activation function, Sigmoid, W_{*1}, W_{*2} are fully-connected layers and ReLU is adopted between two fully-connected layers. Last, $*$ is a channel-wise product operation.

From Eq. (3) and Eq. (4), we note that the CA and PA themselves contain non-linearity and do not depend on σ and we propose a novel Nonlinear Activation Free Feature Attention (NAFFA) module combining Simplified Channel Attention (SCA) and Simplified Pixel Attention (SPA). SCA is proposed by NAFNet [3] which retains the two most important roles of channel attention: aggregating global information and channel information interaction. Firstly, we take the channel-wise global spatial information into a channel descriptor by using global average pooling:

$$g_c = H_p\left(F_c\right) = \frac{1}{H \times W} \sum_{i=1}^{H} \sum_{j=1}^{W} X_c(i,j) \tag{5}$$

where $X_c(i,j)$ stands for the value of c-th channel X_c at position(i,j), H_P is the global pooling function. The shape of the feature map changes from $C \times H \times W$ to $C \times 1 \times 1$.

To get the weights of the different channels, features pass through a convolution layer. Then we element-wise multiply the input X and the weights of the channels:

$$\text{SCA}(\mathbf{X}) = \mathbf{X} * Conv(g) \tag{6}$$

It has been proved that SCA has fewer parameters but a better performance by NAFNet. Similar to SCA, we redesign the Pixel Attention (PA) to Simple Pixel Attention (SPA). We directly feed the output of SCA into one convolution layer which change the shape of features from $C \times H \times W$ to $1 \times H \times W$:

$$\text{SPA}(\mathbf{X}) = \mathbf{X} * Conv(SCA(\mathbf{X})) \tag{7}$$

Although SPA is simpler than PA, there is no loss of performance (Fig. 4).

2.4 Loss Function

For a fair comparison, we use the same loss function with Uformer: the Charbonnier loss [2]. The differences are that Uformer++ has more than one output and is trained with deep supervision. Therefore, the loss function of Uformer++ can be expressed as:

$$Loss\left(\mathbf{I}', \hat{\mathbf{I}}\right) = \sum_{i=1}^{N} \lambda_i \ell\left(\mathbf{I}'_i, \hat{\mathbf{I}}_i\right) \tag{8}$$

where N is the number of outputs. $\lambda_i = 1$ is the hyper-parameter to balance each Charbonnier loss. ℓ is the Charbonnier loss expressed at Eq.(1).

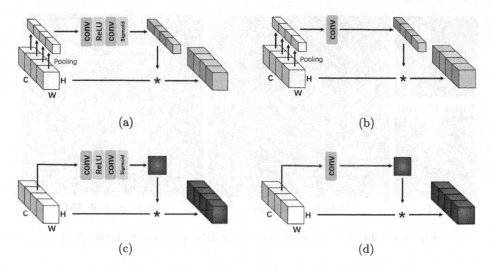

Fig. 4. (a) Channel Attention (CA). (b) Simplified Channel Attention (SCA). (c) Pixel Attention (PA). (d) Simplified Pixel Attention (SPA).

3 Results

3.1 Experimental Setup

Basic Setting. Following the training strategy of Uformer, we train Uformer++ using the AdamW optimizer [10] with the momentum terms of $(0.9, 0.999)$ and the weight decay of 0.02. We randomly augment the training samples using horizontal flipping and rotate the images by $90°$, $180°$ or $270°$. We use the cosine decay strategy to decrease the learning rate to 1e-6 with the initial learning rate 2e-4. The LeWin Transformer blocks of Uformer++ are similar to those of Uformer. The maximum number of encoder/decoder stages K equals 2 by default. And the dimension of each head in Transformer block d_k equals C.

Evaluation Metrics. We adopt the commonly-used PSNR and SSIM [18] metrics to evaluate the restoration performance. These metrics are calculated in the RGB color space following the previous work.

Table 1. Image Denoising Results on SIDD.

Method	VDN [19]	DANet [20]	CycleISP [21]	MPRNet [22]	NBNet [4]	Uformer-T [17]	ours
PSNR	39.28	39.47	39.52	39.71	39.75	39.66	39.76
SSIM	0.909	0.918	0.957	0.958	0.959	0.959	0.959
GMACs(G)	49.8	15.6	118.3	173.5	89.8	12.0	10.1
#Param(M)	4.8	6.2	36	20	13.3	5.23	1.4

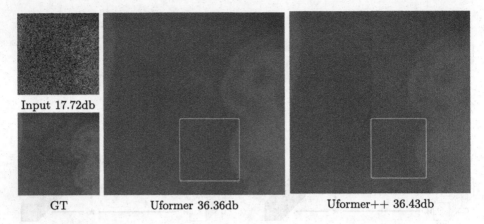

Fig. 5. Qualitative comparison of image deblurring methods on SIDD.

3.2 Image Denoising

Table 1 reports the results of real noise removal on SIDD. We compare Uformer++ with SOTA denoising methods. Our Uformer++ achieves 39.76 dB on PSNR which is comparable to other methods. By presenting the results of PSNR vs. computational cost in Fig. 1, We notice that our Uformer++ have the least computational cost and the least parameters which demonstrates the efficiency and effectiveness of Uformer++. We also show the qualitative results on SIDD in Fig. 5.

3.3 Image Deblurring

We compare the deblurring results of SOTA methods on GoPro dataset, flip and rotate augmentations are adopted. As we shown in Table 2 and Fig. 6, Uformer++ shows a comparable result on GoPro with previous SOTA methods. From the visualization results, we observe that the images recovered by Uformer++ looks almost the same as the images recovered by other SOTA methods.

Table 2. Image Deblurring Results on GoPro.

Method	DBGAN [24]	SPAIRC [12]	SRN [15]	DMPHN [23]	MPRNet [22]	Uformer-T [17]	ours
PSNR	31.10	32.06	30.26	31.20	32.66	29.57	31.60
SSIM	0.942	0.953	0.934	0.940	0.959	0.965	0.965
GMACs(G)	49.8	15.6	118.3	173.5	89.8	12.0	10.1
#Param(M)	4.8	6.2	36	20	13.3	5.23	1.4

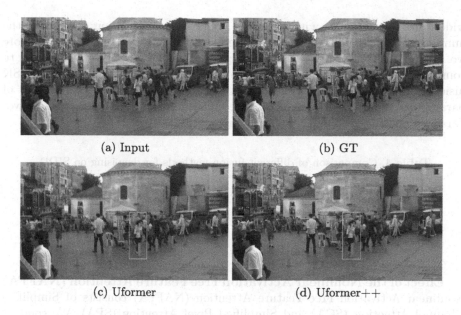

(a) Input   (b) GT

(c) Uformer   (d) Uformer++

Fig. 6. Qualitative comparison of image deblurring methods on GoPro.

4 Ablation Study

In this section, we analyze the effect of each component of Uformer++ in detail. The evaluations are conducted on image denoising (SIDD) and deblurring (GoPro) using different variants.

Unet++ vs. Unet. We rebuild a Uformer-3 whose number of layers is 3 instead of the origin 5 to compare with Uformer++ fairly. Table 3 reports the comparison results. Ufromer-T++ with deep supervision achieves 39.66 dB and outperforms Uformer-T-3 by 0.04 dB. This study indicates the effectiveness of the Unet++ architecture, compared with the traditional Unet architecture but it will introduce additional parameters. More layers of the network will introduce more parameters but the improvement of the result is limited, therefore Unet++ architecture is more efficient in shallow architecture.

Table 3. Comparison of different network architectures for denoising on SIDD.

	GMACs	#Param	PSNR↑
Uformer-3	8.15G	1.30M	38.58
Uformer-T	12.00G	5.23M	39.66
Uformer++ w/o NAFFA	9.47 G	0.73M	39.63

Concatenation vs. SK fusion The SK fusion layer is inspired by SKNet [7], it is designed to fuse multiple branches using channel attention. SK fusion is widely used in networks with Unet-like architecture. SK fusion can reduce the number of channels of feature maps to reduce the computational cost. Therefore we further build Uformer++ with concatenation and SK fusion separately to compare them. The results are presented in Table 4. We observe that the SK fusion has less effect than concatenation but its reduction of the number of parameters is obvious. We need a more efficient fusion method to make fewer parameters while keeping the performance of the network.

Table 4. Comparison of different fusion methods for denoising on SIDD.

	GMACs(G)	#Param(M)	PSNR↑
SK fusion	4.59	1.20	39.35
Concatenation	10.10	1.40	39.76

Effect of the Nonlinear Activation Free Feature Attention (NAFFA) Nonlinear Activation Free Feature Attention (NAFFA) consists of Simplified Channel Attention (SCA) and Simplified Pixel Attention (SPA). We conduct three experiments: 1) only with CA, 2) only with SCA, 3) only with PA, 4) only with SPA, 5) with bath CA and PA (with FA), and 6) with both SCA and SPA (with NAFFA). Table 5 shows the effect of the NAFFA. This study validates that the proposed NAFFA can help the network pay more attention to the most important areas of the features to restore more details with less computational cost and fewer parameters compared with FA.

Table 5. Effect of the NAFFA.

	GMACs(G)	#Param(M)	PSNR↑
Baseline	9.47	0.73	39.63
w CA	10.52	1.11	39.65
w SCA	10.07	1.07	39.70
w PA	9.99	1.11	39.64
w SPA	9.50	1.07	39.68
w FA	11.04	1.48	39.72
w NAFFA	10.10	1.40	39.76

5   Conclusion

In this paper, we have presented a light Uformer called Uformer++ for image restoration. The tiny Uformer has the least computational cost and parameters while performing excellently. Our Uformer++ has less computational cost and parameters than Uformer and has comparable performance with it! Less computation and fewer parameters allow it to be mounted on most devices to quickly process images and even videos. The efficiency of Uformer++ benefits from the new architecture and the novel attention mechanism. Uformer++ embeds Uformer of varying depths in its architecture. They partially share an encoder while their decoders are intertwined. The new architecture addresses the challenge that the depth of the optimal architecture is unknown. Our novel attention mechanism named NAFFA is designed to help the network pay more attention to the most important feature to restore more details. Extensive experiments demonstrate that Uformer++ achieves the least computational cost and parameters while maintaining performance on denoising and deblurring.

There are several limitations in Uformer++. Unet++ is an effective architecture compared with the traditional Unet architecture but it will introduce additional parameters. More layers of the network will introduce more parameters but the improvement of the result is limited, therefore Unet++ architecture is more efficient in shallow architecture. Besides our method only achieves comparable performance instead of state-of-the-art performance.

In the future, we have some work to do. Firstly, we need a more effective and more efficient architecture than Unet++ to solve the problem that it is not suitable for deep architectures. Secondly, concatenation is too simple to introduce many parameters to the network. However, although the commonly used fusion methods can reduce the parameters, they cannot achieve better performance. Therefore a fusion method that can perform better while reducing the parameters is urgently needed. Last but not least, we have not evaluated Uformer++ for more vision tasks. We look forward to investigating Uformer++ for more applications.

References

1. Abdelhamed, A., Lin, S., Brown, M.S.: A high-quality denoising dataset for smartphone cameras. In: CVPR, pp. 1692–1700 (2018)
2. Charbonnier, P., Blanc-Feraud, L., Aubert, G., Barlaud, M.: Two deterministic half-quadratic regularization algorithms for computed imaging. In: Proceedings of 1st International Conference on Image Processing, vol. 2, pp. 168–172. IEEE (1994)
3. Chen, L., Chu, X., Zhang, X., Sun, J.: Simple baselines for image restoration. In: ECCV, pp. 17–33 (2022)
4. Cheng, S., Wang, Y., Huang, H., Liu, D., Fan, H., Liu, S.: NbNet: noise basis learning for image denoising with subspace projection. In: CVPR, pp. 4896–4906 (2021)
5. Hu, J., Shen, L., Sun, G.: Squeeze-and-excitation networks. In: CVPR, pp. 7132–7141 (2018)
6. Jiang, K., et al.: Multi-scale progressive fusion network for single image deraining. In: CVPR, June 2020
7. Li, X., Wang, W., Hu, X., Yang, J.: Selective kernel networks. In: CVPR, pp. 510–519 (2019)
8. Liang, J., Cao, J., Sun, G., Zhang, K., Van Gool, L., Timofte, R.: SwinIR: image restoration using swin transformer. In: ICCV, pp. 1833–1844 (2021)

9. Liu, D., Wen, B., Fan, Y., Loy, C.C., Huang, T.S.: Non-local recurrent network for image restoration. In: NIPS, vol. 31 (2018)
10. Loshchilov, I., Hutter, F.: Decoupled weight decay regularization. arXiv preprint arXiv:1711.05101 (2017)
11. Nah, S., Hyun Kim, T., Mu Lee, K.: Deep multi-scale convolutional neural network for dynamic scene deblurring. In: CVPR, pp. 3883–3891 (2017)
12. Purohit, K., Suin, M., Rajagopalan, A., Boddeti, V.N.: Spatially-adaptive image restoration using distortion-guided networks. In: ICCV, pp. 2309–2319 (2021)
13. Qin, X., Wang, Z., Bai, Y., Xie, X., Jia, H.: FFA-Net: feature fusion attention network for single image dehazing. In: AAAI, vol. 34, pp. 11908–11915 (2020)
14. Ronneberger, O., Fischer, P., Brox, T.: U-net: convolutional networks for biomedical image segmentation. In: MICCAI, pp. 234–241 (2015)
15. Tao, X., Gao, H., Shen, X., Wang, J., Jia, J.: Scale-recurrent network for deep image deblurring. In: CVPR, pp. 8174–8182 (2018)
16. Vaswani, A., et al.: Attention is all you need. In: NIPS, vol. 30 (2017)
17. Wang, Z., Cun, X., Bao, J., Zhou, W., Liu, J., Li, H.: Uformer: a general u-shaped transformer for image restoration. In: CVPR, pp. 17683–17693 (2022)
18. Wang, Z., Bovik, A.C., Sheikh, H.R., Simoncelli, E.P.: Image quality assessment: from error visibility to structural similarity. TIP **13**(4), 600–612 (2004)
19. Yue, Z., Yong, H., Zhao, Q., Meng, D., Zhang, L.: Variational denoising network: toward blind noise modeling and removal. In: NIPS, vol. 32 (2019)
20. Yue, Z., Zhao, Q., Zhang, L., Meng, D.: Dual adversarial network: Toward real-world noise removal and noise generation. In: ECCV, pp. 41–58 (2020)
21. Zamir, S.W., et al.: Cycleisp: real image restoration via improved data synthesis. In: CVPR, pp. 2696–2705 (2020)
22. Zamir, S.W., et al.: Multi-stage progressive image restoration. In: CVPR, pp. 14821–14831 (2021)
23. Zhang, H., Dai, Y., Li, H., Koniusz, P.: Deep stacked hierarchical multi-patch network for image deblurring. In: CVPR, pp. 5978–5986 (2019)
24. Zhang, K., et al.: Deblurring by realistic blurring. In: CVPR, pp. 2737–2746 (2020)
25. Zhou, Z., Siddiquee, M.M.R., Tajbakhsh, N., Liang, J.: Unet++: redesigning skip connections to exploit multiscale features in image segmentation. TMI **39**(6), 1856–1867 (2019)

Can Language Really Understand Depth?

Fangping Chen[✉] and Yuheng Lu

Peking University, 5 Summer Palace Road, Haidian District, Beijing, China
chenfangping@pku.edu.cn

Abstract. Vision-Language Pre-training (e.g., CLIP) bridges image and language. Despite its great success in a wide range of zero-shot downstream tasks, it still underperforms in some abstract or systematic tasks, such as classifying the distance to the nearest car. Recently, however, some researchers found that vision-language pre-trained models are able to estimate monocular depth, and their performance even approaches that of some earlier fully-supervised methods. Given these conflicting findings, in this paper, we focus on the question - Can vision-language pre-trained models really understand depth? If so, how well does it perform? To answer these two questions, we propose MonoCLIP, which attempts to fully exploit the potential of vision-language pre-trained models by introducing three basic depth estimators and global context-guided depth fusion. Results on two mainstream monocular depth estimation datasets demonstrate the ability of vision-language pre-trained model in understanding depth. Moreover, adequate ablation studies further shed light on why and how it works.

Keywords: Vision-Language Pre-training · Monocular Depth Estimation · Multi-modal Learning

1 Introduction

Vision-Language Models (VLM) that pre-trained on vast image-caption pairs have proved to achieve extraordinary success in a wide range of downstream tasks, while as is reported in Contrastive Language-Image Pre-training (CLIP [19]),

> *"CLIP struggles with abstract and novel tasks such as classifying the distance to the nearest car in a photo."*

However, recently, some researchers [1,26] found that it is possible to directly predict or enhance monocular depth with a CLIP pre-trained model. Zhang et al. [26] even find that without further training, the depth predicted by CLIP surpasses existing unsupervised methods and approaches early fully-supervised methods.

Given these conflicting findings, one may be confused and wonder whether the vision-language pre-trained models really understand depth? In this paper, based on CLIP [19], one of the most representative vision-language pre-training

B. Luo et al. (Eds.): ICONIP 2023, CCIS 1967, pp. 377–389, 2024.
https://doi.org/10.1007/978-981-99-8178-6_29

CLIP underperforms in recognizing **Distance** of the nearest car.

But it is able to estimate monocular **Depth**.

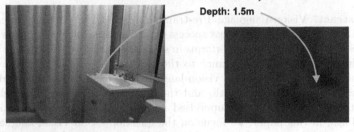

Fig. 1. CLIP underperforms in recognizing distance of nearest car [19], but it is able to estimate monocular depth [26]. One may ask, does CLIP really understand depth?

methods, we propose MonoCLIP which answers the above question by applying CLIP to estimate monocular depth (Fig. 1).

Nevertheless, predicting monocular depth with CLIP along is non-trivial. Specifically, one of the intractable challenges is that CLIP only generates a global feature for each input image, while monocular depth estimation requires estimating the depth for each pixel. Previous works [26,27] address this problem by removing the global pooling layer as shown in Fig. 2a. However, we argue that removing the global pooling layer results in a shift of pre-trained CLIP embedding and eventually leads to a wasted the model capacity. In order to predict a dense depth map without hurting the pre-trained embedding, we propose another two basic CLIP-based depth estimators, named global depth estimator (GDE) and patch depth estimator (PDE).

Moreover, unlike binocular or multi-view depth estimation that rely on stereo matching, there is no reliable cue to perceive depth for a single image. In other words, monocular depth estimation has to understand depth by parsing the semantics of the input image. Therefore, to refine the depth prediction, the proposed MonoCLIP explicitly introduces global semantics in the global context-guided depth fusion step.

To validate the effectiveness of the proposed method and answer the question in the title, we conduct experiments on two mainstream monocular depth datasets NYU-Depth-v2 [21] and SUN RGB-D [22]. Results demonstrate the capability of vision-language models in understanding depth. Moreover, adequate ablation studies further answer why and how vision-language models perform in understanding depth. Our contributions can be summarized as follows.

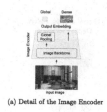   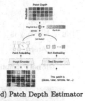

(a) Detail of the Image Encoder (b) Dense Depth Estimator (c) Global Depth Estimator (d) Patch Depth Estimator

Fig. 2. Illustration of basic depth estimators. (a) shows how the image encoder extracts global and dense features. (b), (c), (d) are three basic depth estimators, they are different in how to utilize CLIP pre-trained model.

- We answer an interesting question: Can vision-language pre-trained models really understand depth?
- The proposed MonoCLIP surpasses existing CLIP-based depth estimator by 3.4%, 2.7% in terms of δ_1 accuracy on NYU-Depth-v2 and SUN RGB-D, respectively.
- We conduct sufficient experiments on two mainstream datasets to uncover why and how the vision-language models work in depth understanding.

2 Related Work

2.1 Vision-Language Modeling

Text-image pairs are ubiquitous on the Internet, and despite their noisy nature, paired text is inherently related to the semantics of the image. Vision-Language Modeling is designed to mine and exploit the inherent relationships of text-image pairs. Commonly speaking, according to the pre-training task, vision-language modeling method can be divided into cross-modal masked language modeling [9,16], cross-modal masked region prediction [7,23,24], image-text matching [3, 11,17] and cross-modal contrastive learning [8,12,19]. In this paper, we choose the most representative work in vision-language modeling (i.e. CLIP [19]) to explore whether language can really understand depth.

2.2 Monocular Depth Estimation

Unlike binocular or multi-view depth estimation that rely on the cues from stereo matching, there is no reliable cue to perceive depth from a single image. Therefore, monocular depth estimation is suitable for evaluating whether the language is really able to understand depth, as we do not have to worry about information leakage. Generally speaking, monocular depth estimation methods can be divided into fully-supervised approaches [14,15,20], semi-supervised approaches [4,10,18], unsupervised approaches [6,28], and transfer learning-based approaches [26]. The proposed MonoCLIP does not require depth supervision and uses the knowledge of the CLIP pre-trained model, hence, it belongs to the unsupervised transfer learning-based approaches.

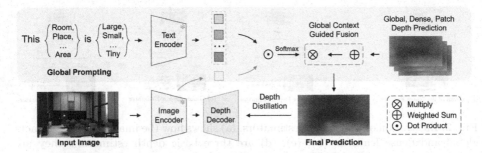

Fig. 3. Framework of MonoCLIP. The basic depth estimators predict depth first (top right of the Figure). Afterwards, the final depth prediction is generated via the Global Context-Guided Fusion module, where the global context is extracted with global prompting and CLIP pre-trained image/text encoders. Finally, in order to reduce the inference latency, the final depth prediction is distilled into a trainable depth decoder.

3   Method

The proposed MonoCLIP aims to verify whether vision-language pre-trained model, e.g., CLIP [19] is really able to understand depth. We first propose three basic depth estimators, which differ in how they utilize CLIP, as shown in Fig. 2b to 2d. After that, a global context-guided depth fusion module is introduced to fuse the outputs of the above depth estimators with the guidance of the global context. Finally, to reduce the inference latency, we further propose depth distillation to distill the depth estimation capability into a depth decoder.

3.1   Basic Depth Estimators

The key for CLIP to understand depth is the inherent relation between position-based words (e.g., close, near, distant, far etc.) and image. However, it is also important how the CLIP pre-trained model is utilized. In this section, as shown in Fig. 2, we introduce three basic depth estimators dubbed Dense Depth Estimator (DDE), Global Depth Estimator (GDE) and Patch Depth Estimator (PDE). They are devised to utilize CLIP pre-trained models in three different ways and in an attempt to maximize CLIP's potential in understanding depth.

Specifically, take the dense depth estimator as an example, as shown in Fig. 2b. It first uses CLIP pre-trained image and text encoder to extract dense image feature (by removing the global pooling layer, as shown in Fig. 2a) and text feature, and then predicts the depth probability of each feature point. Finally, the depth probability is multiplied with a predefined depth bin to generate a depth prediction for each pixel.

However, we argue that removing the global pooling layer shifts the pre-trained CLIP embedding and eventually leads to underutilization of the CLIP pre-trained model. To address this issue, we propose another two basic depth estimators that leave the CLIP pre-trained model unchanged. As shown in Fig. 2c, the global depth estimator generates depth predictions by first explicitly describing the locations in the text prompts, and then aggregating the depth probabilities according to the specific locations. Similarly, patch depth estimators generate depth predictions by rasterizing the image into patches, and then predicting the depth for each patch. Here, we denote the depth prediction of these three basic depth estimators as $\hat{D}_{dense}$, $\hat{D}_{global}$ and $\hat{D}_{patch}$, respectively.

3.2 Global Context-Guided Depth Fusion

Unlike multiple-view geometry that can use stereo matching as a depth prior, monocular depth estimation relies heavily on scene understanding. And the above-mentioned basic depth estimators have their own limitations but may be complementary to each other (e.g., Embedding shift of DDE, low resolution of GDE and global context loss of PDE). Therefore, we propose the Global Context-Guided Depth Fusion (GCGF), which fuses the depth predictions of these base estimators based on the global context semantics.

Specifically, as shown in Fig. 3, the basic depth predictions are first fused via a weighted sum operation, and then the result is further multiplied by a scale factor σ, which is related to the image's global context. More specifically, we prepare another group of global context promptings (e.g., the room is large, the place is small), which is used to extract the global information from the input image. Intuitively, if the room is large, then we upscale the prediction linearly by multiplying by a global scale factor, and vice versa.

The depth prediction for the global context-guided depth fusion is given by:

$$\hat{D} = \sigma \times \sum_{i=1}^{3} w_i \cdot \hat{D}_i, \tag{1}$$

where $\hat{D}_i$ denotes the depth prediction of each basic depth estimator. And σ is computed according to:

$$\sigma = \delta \cdot Softmax(\frac{\mathcal{F}_{img} \cdot \mathcal{F}_{text}}{\|\mathcal{F}_{img}\|\|\mathcal{F}_{text}\|}), \tag{2}$$

where $\mathcal{F}_{img}$ and $\mathcal{F}_{text}$ are features of input image and global text promptings, respectively. $\delta \in \mathcal{R}^n$ is a predefined scale bin for global promptings (e.g., 0.5 for a small room and 2.0 for large.). Intuitively, Eq. (2) first computes the global context probability by soft maxing the similarity between image and global context promptings, and then multiplies it with the predefined scale factor δ.

Table 1. Monocular Depth Estimation results on NYU-Depth-v2

Method	Pre-training	Supervision	$\delta_1 \uparrow$	$\delta_2 \uparrow$	$\delta_3 \uparrow$	$REL \downarrow$	$RMSE \downarrow$	$log_{10} \downarrow$
DepthFormer [14]	–	depth	0.921	0.989	0.998	0.096	0.339	0.041
BinsFormer [15]	–	depth	0.925	0.989	0.997	0.094	0.330	0.040
Lower Bound	–	–	0.299	0.558	0.760	0.645	1.405	0.197
Upper Bound	CLIP	depth	0.628	0.886	0.967	0.237	0.712	0.093
Zhang et al. [25]	KITTI video	0-shot	0.350	0.617	0.799	0.513	1.457	–
DepthCLIP [26]	CLIP	0-shot	0.394	0.683	0.851	0.388	1.167	0.156
Ours	CLIP	0-shot	**0.428**	**0.715**	**0.882**	**0.363**	**1.072**	**0.143**

Table 2. Monocular Depth Estimation results on SUN RGB-D

Method	Pre-training	Supervision	$\delta_1 \uparrow$	$\delta_2 \uparrow$	$\delta_3 \uparrow$	$REL \downarrow$	$RMSE \downarrow$	$log_{10} \downarrow$
DepthFormer [14]	NYU-Dpeth-v2	0-shot	0.815	0.970	0.993	0.137	0.408	0.059
BinsFormer [15]	NYU-Dpeth-v2	0-shot	0.805	0.963	0.990	0.143	0.421	0.061
Lower Bound	–	–	0.259	0.537	0.771	0.653	1.235	0.201
Upper Bound	CLIP	depth	0.572	0.868	0.965	0.250	0.690	0.102
DepthCLIP [26]	CLIP	0-shot	0.381	0.664	0.835	0.342	1.075	0.164
Ours	CLIP	0-shot	**0.408**	**0.689**	**0.862**	**0.331**	**0.987**	**0.152**

3.3 Depth Distillation

So far, we are able to estimate monocular depth. However, since the basic depth estimators are independent between each other, we have to invoke CLIP multiple times. In order to reduce the inference latency, we propose to distill the depth into a depth decoder, as shown in Fig. 3. Specifically, $\hat{\mathbf{D}}$ is regarded as a pseudo depth which is used to supervise the depth decoder. Note that, during distillation, the image encoder uses CLIP pre-trained weights, and only the depth decoder is trainable. The training objective for distillation is given by:

$$\mathcal{L}_{distill} = \alpha \sqrt{\frac{1}{T} \sum_i g_i^2 - \frac{\lambda}{T^2} (\sum_i g_i)^2}, \tag{3}$$

where $\mathcal{L}_{distill}$ is a pixel level scale invariant loss [5]. And $g_i = \log \tilde{d}_i - \log d_i$, d_i and T denote pseudo depth and the number of pixels, respectively. Following [2], α and λ are set to 0.85 and 10, respectively.

4   Experiment

To validate the effectiveness of the proposed method, we conduct experiments on two widely used monocular depth estimation datasets, NYU-Depth-v2 [21] and SUN RGB-D [22]. Besides, we also perform a series of ablations and analysis to reveal why and how our approach works. More results can be found in the supplementary material due to page limitations.

4.1 Datasets and Metrics

NYU-Depth-v2 [21] and **SUN RGB-D** [22] are two mainstream datasets for evaluating monocular depth, in which NYU-Depth-v2 provides images and depth maps at a unify resolution of 640×480, while SUN RGB-D is collected with four different sensors of different resolutions. Different from the supervised methods, MonoCLIP is a zero-shot depth estimator that can be evaluated directly on the test set without seeing the training samples.

For both datasets, followed by [14,26], below metrics are used to evaluate the proposed method, including: accuracy under the threshold ($\delta_i < 1.25^i$, $i = 1, 2, 3$), mean absolute error (REL), root mean squared error (RMSE) and root mean $\log 10$ error (log_{10}), which is following the previous works [14,26].

4.2 Implementation Detail

Our implementation is mainly based on [13], CLIP pre-trained ViT-Base model is used as the image encoder, and the depth decoder employs as one layer of deconvolution operation. Followed by [27], to extract dense feature, although the global pooling layer is removed, we keep the weights of projection matrix. There are three kinds of text prompting used in MonoCLIP, for dense and patch depth estimator, text prompts take the form like "The part is [distance words]". And for the global depth estimator, the text prompts go like "The [location words] is [distance words]". In addition, for the global context promoting, we use "This room is [scale words]" as the text prompts. Specifically, the chosen of the distance words and depth bin follows [26], the location words are the combination between [top, middle, down] and [left, center, right]. And the scale words and factor σ are defined as ["tiny":0.4, "small":0.7, "middle":1.0, "large":1.5, "huge":2.0]. Moreover, to fuse the basic depth estimator, balance weights in Eq. (1) are set to [0.3, 0.6, 0.1]. Additionally, the predicted depth maps are interpolated to the same ground truth resolution for evaluation and loss calculation.

4.3 Main Results

In this section, we compare our method with 1) State-of-the-art supervised depth estimators [14,15], 2) Transfer-learning based depth estimators [25,26] and 3) Upper and lower bounds. Specifically, the upper bound loads CLIP pre-trained weights into the image encoder and only trains the depth decoder with the ground truth depth, while the lower bound simply returns a uniform depth map that is filled with the average depth of the training set. Results are shown in Tables 1 and 2, as we can see, in both NYU-Depth-v2 and SUN RGB-D, the proposed method surpass baselines (transfer-learning based methods) by at least 3.4% and 2.7% in terms of δ_1. which demonstrates the effectiveness of the proposed MonoCLIP. Besides, the large margin between the proposed method and lower bound indicates that MonoCLIP does estimate depth rather than random guessing. Furthermore, compared with the upper bound which only the

Table 3. Ablation on Basic Depth Estimators

Dense Depth	Global Depth	Patch Depth	$\delta_1 \uparrow$	$RMSE \downarrow$
✓			0.391	1.148
	✓		0.407	1.118
		✓	0.385	1.160
✓	✓	✓	**0.412**	**1.111**

Table 4. Ablation on Input Resolution

Dense Depth	Global Depth	Patch Depth	Resolution	$\delta_1 \uparrow$	$RMSE \downarrow$
✓			224×224	0.389	1.159
	✓		224×224	**0.407**	**1.118**
		✓	224×224	0.375	1.182
✓			640×480	**0.391**	**1.148**
	✓		640×480	0.383	1.164
		✓	640×480	**0.385**	**1.160**

Table 5. Ablation on GCGF and Depth Distillation

GCGF	Depth Distillation	$\delta_1 \uparrow$	$RMSE \downarrow$	$Latency \downarrow$
		0.412	1.111	3533 ms
✓		0.428	**1.072**	3657 ms
✓	✓	**0.429**	1.073	**74 ms**

depth decoder is trainable (the depth decoder is one layer of deconvolution that projects the dense CLIP feature into depth), RMSE reduces from 1.072 to 0.712, demonstrating the dense CLIP feature does implicitly contain the depth information. Finally, as we can see, there is still a large margin compared with fully-supervised methods, indicating that even though the CLIP pre-trained model is able to perceive depth, yet, it is not an expert in this area.

4.4 Ablation on Basic Depth Estimators

To further analyze the intrinsic properties of each basic depth estimator, in this section we perform a series of ablation studies, the results of which are presented in Table 3. We can see that, the global depth estimator outperforms the other two, demonstrating that the global information is crucial for understanding depth. Moreover, even though the dense and patch depth estimators generate depth maps with a higher resolution, both of them perform worse than the global depth estimator due to the embedding shift and global context loss. The best results are achieved by fusing these three depth maps together, specifically, in this ablation, fusing is implemented as a weighted sum operation.

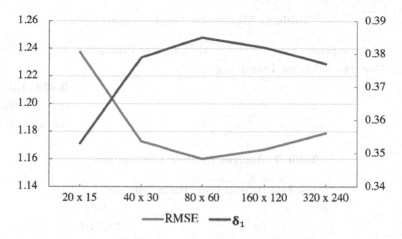

Fig. 4. Performance *vs.* Patch Size

Ablation on Input Resolution . The CLIP pre-trained model accepts input at the resolution of 224×224 by default, however, the image resolution of NYU-Depth-v2 is 640×480. We can either slightly modify the pooling layer of the CLIP pre-trained model to accept the 640×480 input or resize the input image into 224×224. In this section, we investigate the effect of the input resolution. Results are presented in Table 4. As we can see, the dense and patch depth estimators prefer the high resolution setting, while the global depth estimator prefers to the resolution of 224×224. Actually, changing the input resolution will also shift the CLIP embedding, yet the global depth estimator relies heavily on the pre-trained CLIP embedding. In addition, for the patch depth estimator, no matter what the input size is, each patch will be resized to 224×224, therefore, the higher the resolution the better the performance. Finally, for the dense depth estimator, there is no significant difference between input resolutions since the CLIP embedding has already been shifted.

Ablation on Patch Size . The patch depth estimator needs to first rasterize the input image into patches. In this ablation, we would like to investigate the influence of patch size. Results are shown in Fig. 4, as we can see, with the increasing of patch size, the performance increases first and then drops, and achieves the best performance at the path size of 80×60, demonstrating that the patch depth estimator is struggling between the patch size and the depth resolution.

4.5   Ablation on GCGF and Depth Distillation

In this section, we investigate the effect of global context-guided fusion and depth distillation. Results are shown in Table 5. As we can see, compared with the setting that disable both global context-guided fusion and depth distillation modules, enabling global context-guided fusion improves the δ_1 accuracy

Table 6. Ablation on Text Prompting

Qualitative Prompting		Quantitative Prompting	$\delta_1 \uparrow$	$RMSE \downarrow$
Regular Prompting	Inverse Prompting			
✓			**0.428**	**1.072**
	✓		0.283	1.460
		✓	0.316	1.354

Table 7. Analysis by Scene Classification

Dense Feat	Global Feat	Patch Feat	$Acc \uparrow$
✓			31.02%
	✓		**75.54%**
		✓	53.61%

by 1.6% and reduces the RMSE error by 0.039, indicating that global context-guided fusion is beneficial for monocular depth estimation. In addition, if we further enable the depth distillation, δ_1 accuracy and RMSE error are basically unchanged, yet the inference latency dramatically drops.

4.6   Ablation on Text Prompting

To further illustrate whether the language is really able to respond to the depth, we design this ablation study. Specifically, first, we inverse the depth bin that maps text prompting "far away" to a close depth (e.g., 1 m), and text prompting "close" to a distant depth (e.g., 10 m). After that, we further compare qualitative prompting (e.g., the object is far.) with quantitative prompting (e.g., the object is 10 m away). Results are presented in Table 6, as we can see, MonoCLIP degenerates to the lower bound when inverse prompting is enabled, which indicates that language is able to respond to the depth. Besides, we also find that quantitative prompting barley outperforms the lower bound, demonstrating that MonoCLIP fails when depth is directly described in digital numbers. We believe it may be the reason why CLIP is able to predict depth, but fails to predict the distance to the nearest car.

4.7   Analysis by Scene Classification

In this section, we analyze the embedding shift via scene classification. Intuitively, the larger the embedding shift the worse the accuracy. Experiments are conducted in NYU-Depth-V2, which provides scene classification label (e.g., living room, classroom). Results are presented in Table 7, where we can see that global features, which keep the CLIP embedding without any shift, achieve scene classification accuracy of 75.54%. The accuracy of dense features, on the contrary, drops dramatically, indicating that dropping the global pooling layer does result in the embedding shifts. Moreover, even though the patch features preserve the CLIP embedding, the accuracy drops considerably due to the loss of global information.

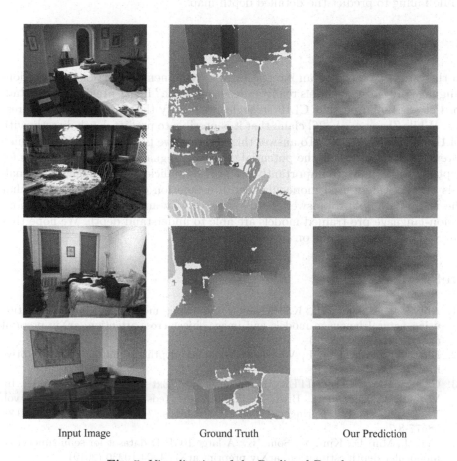

| Input Image | Ground Truth | Our Prediction |

Fig. 5. Visualization of the Predicted Depth

4.8   Visualization

To provide an intuitive understanding, we visualize the predicted depth of MonoCLIP. As shown in Fig. 5, basically, MonoCLIP is able to predict which part of the input image is close or far. However, the prediction is somewhat unclear and noisy. Actually, the feature map of the dense depth estimator is only 13×17, which is far more small than that of the ground truth, that is why the prediction is unclear. Additionally, the depth prediction is inconsistent across patches and estimators due to the embedding shift of dense features and the rasterization of patch features, which is why the prediction looks noisy. Furthermore, comparing the top three samples with the last one, we find that MonoCLIP generates more reasonable results if there is a high depth range in the input image, which further demonstrates that MonoCLIP is able to predict the coarse depth distribution while failing to predict the detailed depth map.

5   Conclusion

In this work, we focus on an interesting and fundamental question - Can vision-language pre-trained models really understand depth? In the previous work, Radford et al. [19] found that CLIP struggles to classify the distance of the nearest car. While Zhang et al. [26] claim that it is possible to predict depth directly with CLIP pre-trained model. To answer this question, we propose MonoCLIP, which attempts to fully exploit the potential of vision-language pre-trained models for depth prediction. More importantly, we provide sufficient experiments and analysis on two widely used monocular depth estimation datasets. We believe that the results are able to answer the question of why and under what conditions vision-language pre-trained models are able to understand depth. We hope that our work can shed a light on future research.

References

1. Auty, D., Mikolajczyk, K.: Objcavit: improving monocular depth estimation using natural language models and image-object cross-attention. arXiv preprint arXiv:2211.17232 (2022)
2. Bhat, S.F., Alhashim, I., Wonka, P.: Adabins: depth estimation using adaptive bins. In: CVPR (2021)
3. Chen, Y.-C., et al.: UNITER: UNiversal image-TExt representation learning. In: Vedaldi, A., Bischof, H., Brox, T., Frahm, J.-M. (eds.) ECCV 2020. LNCS, vol. 12375, pp. 104–120. Springer, Cham (2020). https://doi.org/10.1007/978-3-030-58577-8_7
4. Cho, J., Min, D., Kim, Y., Sohn, K.: A large RGB-D dataset for semi-supervised monocular depth estimation. arXiv preprint arXiv:1904.10230 (2019)
5. Eigen, D., Puhrsch, C., Fergus, R.: Depth map prediction from a single image using a multi-scale deep network. In: NIPS (2014)
6. Garg, R., B.G., V.K., Carneiro, G., Reid, I.: Unsupervised CNN for single view depth estimation: geometry to the rescue. In: Leibe, B., Matas, J., Sebe, N., Welling, M. (eds.) ECCV 2016. LNCS, vol. 9912, pp. 740–756. Springer, Cham (2016). https://doi.org/10.1007/978-3-319-46484-8_45
7. Huang, Z., Zeng, Z., Huang, Y., Liu, B., Fu, D., Fu, J.: Seeing out of the box: end-to-end pre-training for vision-language representation learning. In: CVPR (2021)
8. Jia, C., et al.: Scaling up visual and vision-language representation learning with noisy text supervision. In: ICML (2021)
9. Kim, W., Son, B., Kim, I.: Vilt: vision-and-language transformer without convolution or region supervision. In: ICML (2021)
10. Kuznietsov, Y., Stuckler, J., Leibe, B.: Semi-supervised deep learning for monocular depth map prediction. In: CVPR (2017)

11. Li, G., Duan, N., Fang, Y., Gong, M., Jiang, D.: Unicoder-vl: a universal encoder for vision and language by cross-modal pre-training. In: AAAI (2020)
12. Li, J., Selvaraju, R., Gotmare, A., Joty, S., Xiong, C., Hoi, S.C.H.: Align before fuse: vision and language representation learning with momentum distillation. In: NIPS (2021)
13. Li, Z.: Monocular depth estimation toolbox (2022). https://github.com/zhyever/Monocular-Depth-Estimation-Toolbox
14. Li, Z., Chen, Z., Liu, X., Jiang, J.: Depthformer: exploiting long-range correlation and local information for accurate monocular depth estimation. arXiv preprint arXiv:2203.14211 (2022)
15. Li, Z., Wang, X., Liu, X., Jiang, J.: Binsformer: revisiting adaptive bins for monocular depth estimation. arXiv preprint arXiv:2204.00987 (2022)
16. Lin, J., Yang, A., Zhang, Y., Liu, J., Zhou, J., Yang, H.: Interbert: vision-and-language interaction for multi-modal pretraining. arXiv preprint arXiv:2003.13198 (2020)
17. Lu, J., Batra, D., Parikh, D., Lee, S.: Vilbert: pretraining task-agnostic visiolinguistic representations for vision-and-language tasks. In: NIPS (2019)
18. Luo, Y., et al.: Single view stereo matching. In: CVPR (2018)
19. Radford, A., et al.: Learning transferable visual models from natural language supervision. In: ICML (2021)
20. Ranftl, R., Bochkovskiy, A., Koltun, V.: Vision transformers for dense prediction. In: ICCV (2021)
21. Silberman, N., Hoiem, D., Kohli, P., Fergus, R.: Indoor segmentation and support inference from RGBD images. In: Fitzgibbon, A., Lazebnik, S., Perona, P., Sato, Y., Schmid, C. (eds.) ECCV 2012. LNCS, vol. 7576, pp. 746–760. Springer, Heidelberg (2012). https://doi.org/10.1007/978-3-642-33715-4_54
22. Song, S., Lichtenberg, S.P., Xiao, J.: Sun RGB-D: a RGB-D scene understanding benchmark suite. In: CVPR (2015)
23. Tan, H., Bansal, M.: LXMERT: learning cross-modality encoder representations from transformers. arXiv preprint arXiv:1908.07490 (2019)
24. Xue, H., et al.: Probing inter-modality: visual parsing with self-attention for vision-and-language pre-training. In: NIPS (2021)
25. Zhang, M., Ye, X., Fan, X., Zhong, W.: Unsupervised depth estimation from monocular videos with hybrid geometric-refined loss and contextual attention. In: Neurocomputing (2020)
26. Zhang, R., Zeng, Z., Guo, Z., Li, Y.: Can language understand depth? In: ACM MM (2022)
27. Zhou, C., Loy, C.C., Dai, B.: Extract free dense labels from clip. In: Avidan, S., Brostow, G., Cisse, M., Farinella, G.M., Hassner, T. (eds.) ECCV 2022. LNCS, vol. 13688, pp. 696–712. Springer, Heidelberg (2022). https://doi.org/10.1007/978-3-031-19815-1_40
28. Zhou, T., Brown, M., Snavely, N., Lowe, D.G.: Unsupervised learning of depth and ego-motion from video. In: CVPR (2017)

Remaining Useful Life Prediction of Control Moment Gyro in Orbiting Spacecraft Based on Variational Autoencoder

Tao Xu[1](✉), Dechang Pi[1], and Kuan Zhang[2]

[1] College of Computer Science and Technology, Nanjing University of Aeronautics and Astronautics, Nanjing 211106, China
xutao_swu_edu@163.com
[2] Beijing Aerospace Control Center, Beijing 100094, China

Abstract. For the telemetry data generated by the key components of space-craft during the orbital operation contain a lot of degradation information, and these telemetry data have the characteristics of large data volume and high dimensionality which are difficult to process. In this paper, we present CMG-VAE, a variational autoencoder-based method for predicting the remaining useful life of control moment gyro in orbiting spacecraft. The method improves the structure of the variational autoencoder. In the encoding phase, the temporal convolutional network is used to extract time-dependent information from the telemetry data, while a graph representation learning approach is used to obtain structural information about the data. The final output of the encoding part is obtained by weighted fusion using a feature fusion approach. One part of the newly fused features is fed into the decoder for data reconstruction and the other part is fed into the remaining useful life prediction module for prediction. To evaluate the effectiveness of the proposed method, this paper uses a set of control moment gyro data obtained from a space station and NASA's C-MAPSS simulation dataset for validation. The experimental results show that our proposed method achieves the best results compared to other state-of-the-art benchmarks. In particular, on the control moment gyro dataset, the root mean square error (RMSE) obtained by the method proposed in this paper is reduced by 24% compared to the obtained by the best-performing baseline method.

Keywords: Variational Autoencoder · Remaining Useful Life Prediction · Time Series · Temporal Convolutional Network · Graph Attention Network

1 Introduction

Control moment gyro (CMG) is one of the key components of attitude control actuators on modern large spacecraft. It works by obtaining angular momentum through a rotor rotating at high speed and changing the direction of angular momentum to generate control torque. It has the advantages of big output torque and high energy efficiency ratio, so it has been widely used in the aerospace field. As a core component of the spacecraft, the health status of the CMG is related to the normal work of the entire spacecraft [1]. The remaining useful life study can identify the problems of CMGs as early as possible, determine their health status during the orbital operation, predict the possible failure and occurrence time, and provide a basis for the ground controllers to replace the equipment or switch the operation status in time.

Currently, the remaining useful life prediction for most spacecraft products is mainly obtained by taking the 1:1 full-life experiment. Although this method is very accurate in obtaining results, it takes a long time and is too costly due to the high reliability and long life of spacecraft products. With the development of artificial intelligence technology, the rise of Prognostics Health Management (PHM) technology provides a new solution to this problem. For complex equipment in service, PHM technology performs a dynamic assessment of reliability and real-time prediction of failure based on modeling of repair and replacement data and real-time degradation data, as well as formulating scientific and effective health management strategies based on the information from the assessment and prediction [2]. One of the most important of the many items of PHM technology is the Remaining Useful Life (RUL) prediction of critical equipment, which is the task of estimating the remaining useful life of the equipment based on its historical sensor data [3]. For the control moment gyro, a large amount of telemetry time series data is generated during its on-orbit work, and these data contain rich degradation information, so the PHM technique can be used to model the telemetry data to achieve its RUL prediction.

The main contributions of this paper are as follows:

- A method for predicting the remaining useful life of the control moment gyro based on a variational autoencoder is proposed, which shows excellent results on a real control moment gyro dataset;
- The coding structure of the variational autoencoder is designed so that it can extract the time-dependent information and structural information present in the time series data;
- In order to extract structural information from time series data, a method for converting time series data into a graph based on the maximal information coefficient (MIC) is proposed;
- Explore the roles played by various components in the CMG-VAE model.

The rest of this paper is organized as follows: Sect. 2 reviews the work related to RUL prediction. Section 3 introduces the VAE-based control moment gyro remaining useful life prediction model (CMG-VAE) proposed in this paper. Section 4 describes the experimental content, including the dataset and the preprocessing method, the evaluation criteria, the benchmarking methods, the model hyperparameter settings, and the experimental results and discussion. Finally, Sect. 5 concludes the paper and provides an outlook for future work.

2 Related Work

In general, existing RUL prediction methods can be divided into three main categories, namely, physical model-based methods, data-driven methods, and hybrid methods [4, 5]. Physical model-based approaches usually require the researcher to have a solid knowledge of physics and mathematics and to design the model with the specific domain knowledge of the device under study. This type of approach usually shows good results and can provide a clear explanation when a device fails [6]. But the physical model-based approach is often used for components or simple systems. In the case of complex systems, designing such models first requires a great deal of domain knowledge, and even with that, the design of the model is an impossible task for most engineers due to the complexity of the system itself. Therefore, for complex systems, the physical model-based approach is not very feasible [7]. With the development of data science and technology, data-driven methods are becoming more and more popular and have achieved good results in RUL prediction. These methods model the degradation process based on historical data collected from sensors as a function between sensor monitoring data and the remaining lifetime [8]. Data-driven approaches do not need to rely on domain knowledge for modeling, thus reducing the difficulty of constructing RUL prediction models for complex systems, and have gained the attention of most researchers [9]. The hybrid method is proposed by combining the above two methods in order to make use of their advantages [10]. However, developing an effective hybrid approach is still very challenging, as it is often unrealistic for developers to have knowledge of both the problem domain and the data science domain. Therefore, the dominant approach for RUL prediction is currently a data-driven approach.

The earliest data-driven approaches were based on statistical methods. These methods have played a role in RUL prediction, but they are less commonly used because they are not effective for long-term prediction and are susceptible to noise and changes in equipment operating conditions [11, 12]. The development of machine learning has provided new avenues for RUL prediction, and the use of methods such as neural networks, support vector machine, and random forest has shown good results. But for a good machine learning model, feature engineering is a very important step [13]. It is often a very complex and tedious process, and requires a lot of experience and continuous experimentation by researchers [14]. Moreover, feature engineering methods vary for data acquired in different scenarios, which is a barrier limiting the success of machine learning methods for RUL prediction.

With the improvement of hardware technology, deep learning has been developed rapidly in recent years. Deep learning is a branch of machine learning that uses deep neural networks to extract features from data adaptively, without the need for cumbersome feature engineering. At present, deep learning has been widely used in image processing, speech recognition, and natural language processing, and it also provides a new solution for RUL prediction [15, 16]. Although deep learning methods are currently achieving some results in the field of RUL prediction, there is still a gap in the research on remaining useful life prediction for in-orbit control moment gyros using deep learning methods. And the current deep RUL prediction methods also suffer from some problems such as complicated model training processes and weak interpretation due to using deep networks. Also, most of the models extract only one aspect feature of the data, such as extracting only the time-dependent information of the data for RUL prediction. In practice, however, there are also interaction relationships between different sensors and these relationships change accordingly with the mode of operation or with time. Most previous models have neglected the acquisition of this structural aspect information, i.e., few studies have combined time-dependent information with structural information for RUL prediction.

Based on the above analysis, this paper proposes a model for predicting the remaining useful life of control moment gyro in orbit based on the variational autoencoder (CMG-VAE). The unique encoding structure is used to obtain time-dependent information as well as structural information of telemetry data, which enables multifaceted extraction of data features. Finally, new features with more expressive capability are obtained by weighted fusion of the two, and then the RUL prediction is performed.

3 CMG-VAE Model

3.1 Problem Formulation

Data-driven RUL prediction methods are all carried out by processing multivariate time series data generated by the equipment. Suppose there is multidimensional time series data $T \in \mathbb{R}^{l \times k}$ collected by sensors, where l denotes the length of the data, which corresponds to the time index of the data, and k denotes the feature dimension of the data, i.e., the number of sensors required to collect the data. Our goal is to generate a data $RUL \in \mathbb{R}^{l \times 1}$ of the same length l, representing the remaining useful life value of the device at each moment, according to the model.

3.2 Overview

The control moment gyro RUL prediction method CMG-VAE proposed in this paper is based on a variational autoencoder. The time-dependent and structural information in the multidimensional time series data is extracted by designing the structure of its encoder and decoder. Finally, a new feature with stronger performance ability is obtained by means of feature fusion, and then the remaining useful life of the equipment is predicted by using this feature. The CMG-VAE consists of the following four components:

- Time encoding module: obtain time-dependent information in time series data by using temporal convolutional network (TCN).
- Structure encoding module: transform time series data into graph by maximal information coefficient, and then use graph representation learning method to obtain the structural information in the data.
- Feature fusion module: the features obtained from the above two modules are fused to obtain a new fusion feature as the output of the encoder.
- Decoder and RUL prediction module: this module uses the obtained fused features to perform the final remaining useful life prediction task with the help of a multilayer perceptron while taking care of reconstructing the data during the model training phase.

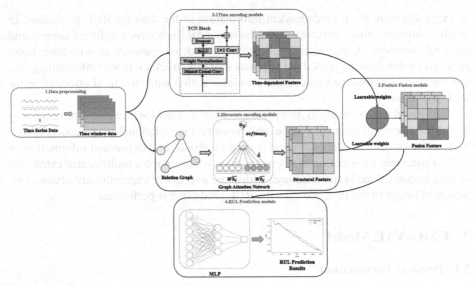

Fig. 1. The proposed CMG-VAE model.

Figure 1 shows the overall workflow of our proposed framework.

Only the encoder part and the RUL prediction module will be kept in the final trained model, and the decoder part will be discarded, so the decoder is not shown in Fig. 1.

3.3 Variational Autoencoder

The CMG-VAE model uses a variational autoencoder (VAE) as the underlying structure, so a brief introduction to the VAE is given before introducing the individual components. VAE is developed from the autoencoder, which also has an encoder and decoder structure. But the difference is that VAE is based on variational inference for data encoding, and the latent space obtained by its encoding is not a simple set of low-dimensional features, but a set of probability distributions approximating the standard Gaussian distribution. Sampling from these distributions yields the low dimensional representation of the original data, which is then input into the decoder for data reconstruction [17]. The above process can be expressed as Eq. (1).

$$p(x) = p(x|z)p(z)dz \tag{1}$$

where z is a continuous domain, for which integration is difficult to handle. For this reason, variational inference is used since this problem can be solved by maximizing the lower bound of the log-likelihood $\mathcal{L}_{vae}$. as shown in Eq. (2).

$$\mathcal{L}_{vae} = E_{q_\phi(z|x)}[\log p_\theta(x|z)] - D_{KL}(q_\phi(z|x)||p_\theta(z)) \tag{2}$$

In this paper, VAE is chosen as the basic structure of the CMG-VAE model mainly for two reasons: 1) the control moment gyro telemetry data collected by sensors are high-dimensional time series data, and the encoder-decoder structure can realize the dimensionality reduction of the data and speed up the processing rate of the model. 2) VAE has certain interpretability, and using VAE as the basic structure of the model can increase the interpretability of the model.

3.4 Time Encoding Module

The time encoding module, as one of the sub-modules of the CMG-VAE model encoder, is designed to extract time-dependent information from the time series data. In this paper, we use the temporal convolutional network (TCN) to form this module.

TCN is a combination of one-dimensional full convolutional network (1DFCN) and causal convolution [18]. By zero-padding the 1DFCN to process the data, it makes the input and output of each layer of TCN have the same time step, and the convolution kernel of 1D convolution only slides in one direction, which is very suitable for processing time series data. Compared with recurrent neural network (RNN) and its variants (LSTM, GRU, etc.), which are commonly used to process sequence data, TCN can perform parallel computation and obtain more historical information through multi-layer convolution, so this paper chooses to use TCN as the network structure of the time encoding module.

3.5 Structure Encoding Module

The structure encoding module is divided into two steps, first converting the time series data into a relational graph, and then extracting the structural information from it by graph representation learning. In this paper, the graph is constructed by calculating the maximal information coefficient (MIC) between each dimension. MIC uncovers linear or nonlinear relationships between two sequences. Its calculation is shown in Eqs. (3) to (4).

$$M(D)_{(a,b)} = \frac{\max_C I(D|C)}{\log \min(a, b)} \tag{3}$$

$$MIC(D)_{(a,b)} = \max_{a \times b < B} \{M(D)_{(a,b)}\} \tag{4}$$

where D denotes a finite set of ordered pairs, and C denotes a grid space divided into two dimensions according to the a, and b parameters, where a denotes the number of grids in the x-value direction and b denotes the number of grids in the y-value direction. And $B=|D|^{0.6}$ is a constraint on the grid division. The value of the calculated MIC is within [0, 1], and the larger the value, the stronger the correlation between the data.

Suppose there exist nodes v_i and v_j in the relationship graph, whose feature vectors are denoted as $H_i = \{h_{i,1}, h_{i,2}, \ldots, h_{i,n}\}$, $H_j = \{h_{j,1}, h_{j,2}, \ldots, h_{j,n}\}$. The features of the two nodes are transformed into a finite set $D = \{(h_{i,1}, h_{j,1}), \ldots, (h_{i,n}, h_{j,n})\}$ of ordered pairs, and the correlation coefficient $r_{i,j}$ between the two nodes is calculated according to Eqs. (3) to (4). Finally, the existence of edges between nodes v_i and v_j is calculated according to Eq. (5).

$$\begin{cases} e_{i,j} = 1, if\ r_{i,j} \geq \delta \\ e_{i,j} = 0, if\ r_{i,j} < \delta \end{cases} \tag{5}$$

where δ denotes the threshold value for the presence or absence of edges between nodes, which takes values between [0,1]. It can be set by the user based on the overall situation of correlation between the dimensions in the dataset.

After converting the time series data into graphs, this paper uses graph attention network (GAT) for representation learning. Of course, other graph neural networks and their variants such as GNN, GCN, etc. can also be used. The final output of the module is obtained [19].

3.6 Feature Fusion Module

The time series data is processed by the time encoding module and the structural encoding module to obtain two sets of depth features, which need to be fused into one set as the output of the encoder. The most direct means of feature fusion are feature concatenation and feature addition, but these operations simply turn two sets of vectors into one set without weighing the magnitude of the role played by the two different sources in the newly fused features. For this reason, in this paper, when taking the feature addition approach for feature fusion, the different weights are given to the features from different sources, and these weights are learned by setting them as learnable parameters. The calculation is shown in Eq. (6).

$$\vec{H} = \alpha \times \vec{H_t} + \beta \times \vec{H_s} \tag{6}$$

where $\vec{H}$ denotes the new features, $\vec{H_t}$, and $\vec{H_s}$ denote the output of the time encoding module and the structural encoding module, respectively, and α, β are the weights for performing feature fusion.

Compared with the direct feature fusion by feature concatenation and feature addition, the method in this paper can balance the role played by the two features in the new features to a greater extent and show better performance in downstream tasks.

3.7 Decoder and RUL Prediction Module

After the encoder obtains the fusion features, the task of the decoder is to restore these new features to the original data. Here we use TCN to achieve decoding operation, and the reconstructed data size of the output of the decoder is the same as that of the original data. The distance between the decoder output result $X\prime$ and the input data X is calculated to obtain the data reconstruction loss of CMG-VAE. In this paper, the mean square error is used as the calculation of the data reconstruction loss, and Eq. (7) shows the specific calculation method.

$$loss_r = \frac{1}{n} \sum_{i=1}^{n} \frac{1}{2} \|x_i - x_{i\prime}\|^2 \tag{7}$$

Since the CMG-VAE method proposed in this paper uses VAE as the underlying structure, it is necessary to calculate the KL divergence between the latent space and the standard normal distribution, in addition to calculating the data reconstruction loss. The final loss of the encoding part is shown in Eq. (8).

$$loss_e = loss_r + KL(N(\mu, \sigma^2)||N(0, 1)) \tag{8}$$

where μ, and σ denote the expectation and variance of the latent space.

In the model proposed in this paper, a multilayer perceptron (MLP) with one hidden layer is used for remaining useful life prediction. We input the new fused features into the MLP for RUL prediction, and finally output the predicted remaining useful life $RUL' \in \mathbb{R}^{T \times 1}$, where T denotes the length of the time series data processed. The prediction loss is calculated based on the prediction results and the RUL label $RUL \in \mathbb{R}^{T \times 1}$ in the dataset, and the root mean square error (RMSE) is used here as the loss function, and Eq. (9) shows the specific calculation process.

$$loss_p = \sqrt{\frac{1}{T} \sum_{i=1}^{T} \frac{1}{2} \|rul_i - rul_i'\|^2} \tag{9}$$

So, the total loss of the CMG-VAE model is calculated as shown in Eq. (10).

$$loss = \theta \times loss_e + (1 - \theta) \times loss_p \tag{10}$$

The loss function consists of two parts, one is the loss generated by the coding structure and the other is the RUL prediction loss. The two losses are weighted and summed, where the weight factor measures the proportional size of both in the total loss function and takes the value in the interval (0, 1).

4 Experiments

4.1 Dataset and Preprocessing

Table 1. C-MAPSS Dataset Details.

Dataset	FD001	FD002	FD003	FD004
Engine units for training	100	260	100	249
Engine units for testing	100	259	100	248
Operating conditions	1	6	1	6
Fault conditions	1	1	2	2

To verify the effectiveness of the proposed method in this paper, we conduct experiments using two datasets, which include a real control moment gyro telemetry dataset (CMGD) from a space station and NASA's Commercial Modular Aero-Propulsion System Simulation (C-MAPSS) dataset. The dataset of the control moment gyro is composed of two sets of data extracted for a single frame CMG on a space station. The data cover all data obtained from October 2011 to March 2015. The data are sampled at a frequency of 1 (one data was obtained by sampling within each second), followed by a secondary sampling at intervals of one thousandth. This results in a set of control moment gyro full life cycle data. Each data in this data set has 8 dimensions and the length is 12234. The details of the C-MAPSS dataset are shown in Table 1.

The range of values for different dimensions of data in the two datasets varies greatly, and in this paper, the data are scaled equally using linear normalization. And in the time series data modeling problem, the time window method is usually used for data preparation, which is also adopted in this paper to transform the original long series into multiple fixed-length windows of data. This not only helps to reduce the parameters of the model, but also allows to obtain richer information from the sequences, which in turn improves the accuracy of the model.

4.2 Comparison Metrics

In order to ensure a reasonable comparison with models built by other methods, root mean square error (RMSE) as well as scoring function are chosen for model evaluation in this paper. In the RUL prediction task, RMSE is often used as an evaluation criterion and is calculated as shown in Eq. (11).

$$RMSE = \sqrt{\frac{1}{N} \sum_{i=1}^{N} h_i^2} \tag{11}$$

where N denotes the number of control moment gyroscopes and h_i denotes the distance between the true RUL and the predicted RUL obtained. The advantage of RMSE is that it gives equal weight to early and late predictions, which is more responsive to the overall performance of the model compared to other evaluation criteria.

The scoring function is derived from the evaluation criteria provided in the PHM 2008 Data Challenge and Eq. (12) shows how the scoring function is calculated.

$$S = \begin{cases} \sum_{i=1}^{N} (e^{-\frac{h_i}{13}} - 1), \, for \, h_i < 0 \\ \sum_{i=1}^{N} (e^{\frac{h_i}{10}} - 1), \, \, for \, h_i \geq 0 \end{cases} \tag{12}$$

where S is the calculated score, N is the number of CMG in the test dataset, and h_i denotes the difference between the predicted RUL and the true RUL. The scoring function can distinguish well between early predictions and late predictions, and it has a greater penalty for late predictions compared to early predictions, which is in line with the practice in the field of equipment maintenance.

4.3 Experimental Settings

All experiments in this paper were conducted on the PC with an Intel Core i5 CPU, 8-GB RAM, and GEFORCE GTX 950M GPU. We used Pytorch version 1.11.0 and PyTorch

Geometric Library version 2.0.4. The model was trained using the Adam optimizer with a learning rate of 0.001 and a total of 20000 epochs. The time window sizes for the CMGD and C-MAPSS dataset divisions are 60 and 30, respectively.

4.4 Evaluation Results

The results of comparing the proposed method in this paper with other benchmarks, including MLP, SVM, LSTM [15, 20], and DCNN [16], will be shown first. These four methods include both traditional machine learning methods and the latest and recent deep learning methods. The comparative experimental results are shown in Table 2.

Table 2. Evaluation results of various algorithms on two datasets.

Model	CMGD		FD001		FD002		FD003		FD004	
	RMSE	Score	RMSE	Score	RMSE	Score	RMSE	Score	RMSE	Score
MLP	42.64	253675	21.29	1542	45.05	362000	23.10	1730	44.18	573700
SVM	39.02	196547	20.73	1375	43.71	271800	21.03	1570	45.29	401000
LSTM	32.53	9352	16.53	356	25.02	5740	16.23	325	27.36	5709
DCNN	26.51	4759	18.45	1286	30.28	15790	19.82	1580	29.16	7890
CMG-VAE	20.17	1793	15.02	251	17.19	2162	14.13	263	18.42	1607

The results in Table 2 show that our proposed CMG-VAE method achieves the best results compared to other benchmarks, which proves the effectiveness of the method. This is because CMG-VAE extracts not only time-dependent information but also structural information from the data, and fuses the two to form new deep fusion features. The fused features contain more information and have more expressive power. And when using deep learning models, extracting features of the data from multiple perspectives can more fully explore the essence of the data and facilitate the implementation of downstream tasks.

We can also observe the results that deep learning (LSTM, DCNN) methods are generally better than machine learning methods (MLP, SVM). This is mainly because deep learning algorithms can extract deeper features of the data and obtain as much information as possible through the stacking of multilayer networks. The advantage of deep learning algorithms is more obvious for complex datasets, as they can uncover the most essential features in complex data, which is difficult to do with general machine learning methods.

4.5 Ablation Experiments

To further validate the role played by each module in the model, we conducted ablation experiments. The role played by the time encoding modules and structural encoding modules, respectively, in the CMG-VAE method was first verified. When verifying the time encoding module, the structural encoding module is removed and only the former is used for encoding operations; the opposite is true when verifying the structural encoding module. When only one encoding module is used for encoding, no feature fusion operation is required. The result is shown in Fig. 2.

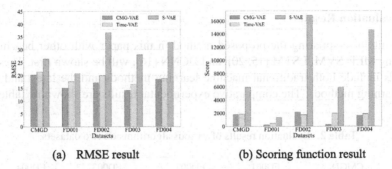

(a) RMSE result   (b) Scoring function result

Fig. 2. Results of each sub-dataset under single module encoding.

Where Time-VAE indicates that only the time encoding module is used for encoding operations and S-VAE indicates that only the structure encoding module is used. From the results, when a single module is used to build the model, the predictions obtained are much worse compared to the CMG-VAE. This finding once again illustrates the importance of obtaining data features from multiple perspectives. And those using only the structural encoding module for feature extraction achieved worse results. This indicates that the time-dependent information plays a primary role and the structural information plays a secondary role in the new fusion features.

5   Conclusion

For the control moment gyro remaining useful life prediction problem, this paper proposes a method CMG-VAE based on variational autoencoder. The method includes the encoder, the feature fusion module, and the decoder and remaining lifetime prediction modules. The encoder is further divided into a time encoding module and a structural encoding module to obtain the time-dependent and structural information of the data, respectively. The two extracted features are weighted and fused to form new features. The decoder and the remaining useful life prediction module perform data reconstruction and RUL prediction based on the fused resulting features, respectively. To verify the effectiveness of our proposed method, we validate it on a real on-orbit control moment gyro dataset and compare it with some of the most used RUL prediction methods. The experimental results show that our proposed method achieves optimal results. In order to demonstrate the generalizability of our proposed method, we applied it to the public dataset C-MAPSS and achieved good results.

References

1. Higashiyama, D., Shoji, Y., Satoh, S., et al.: Attitude control for spacecraft using pyramid-type variable-speed control moment gyros. Acta Astronaut. **173**, 252–265 (2020)

2. Zhang, Z., Si, X., Hu, C., et al.: Degradation data analysis and remaining useful life estimation: a review on wiener-process-based methods. Eur. J. Oper. Res. **271**(3), 775–796 (2018)
3. Peng, Y., Pan, X., Wang, S., et al.: An aero-engine RUL prediction method based on VAE-GAN. In: 2021 IEEE 24th International Conference on Computer Supported Cooperative Work in Design (CSCWD), pp. 953–957. IEEE (2021)
4. Mao, W., He, J., Zuo, M.J.: Predicting remaining useful life of rolling bearings based on deep feature representation and transfer learning. IEEE Trans. Instrum. Meas. **69**(4), 1594–1608 (2019)
5. Youness, G., Aalah, A.: An explainable artificial intelligence approach for remaining useful life prediction. Aerospace **10**(5), 474 (2023)
6. Wang, B., Lei, Y., Li, N., et al.: A hybrid prognostics approach for estimating remaining useful life of rolling element bearings. IEEE Trans. Reliab. **69**(1), 401–412 (2018)
7. Lei, Y., Li, N., Gontarz, S., et al.: A model-based method for remaining useful life prediction of machinery. IEEE Trans. Reliab. **65**(3), 1314–1326 (2016)
8. Li, X., Ding, Q., Sun, J.Q.: Remaining useful life estimation in prognostics using deep convolution neural networks. Reliab. Eng. Syst. Saf. **172**, 1–11 (2018)
9. Zhang, W., Yang, D., Wang, H.: Data-driven methods for predictive maintenance of industrial equipment: a survey. IEEE Syst. J. **13**(3), 2213–2227 (2019)
10. Polenghi, A., Roda, I., Macchi, M., et al.: Ontology-augmented prognostics and health management for shopfloor-synchronised joint maintenance and production management decisions. J. Ind. Inf. Integr. **27**, 100286 (2022)
11. Sharma, A.K., Punj, P., Kumar, N., et al.: Lifetime prediction of a hydraulic pump using ARIMA model. Arab. J. Sci. Eng. 1–13 (2023)
12. Zhai, Q., Ye, Z.S.: RUL prediction of deteriorating products using an adaptive wiener process model. IEEE Trans. Industr. Inf. **13**(6), 2911–2921 (2017)
13. Chen, Z., Cao, S., Mao, Z.: Remaining useful life estimation of aircraft engines using a modified similarity and supporting vector machine (SVM) approach. Energies **11**(1), 28 (2017)
14. Wu, D., Jennings, C., Terpenny, J., et al.: Cloud-based machine learning for predictive analytics: tool wear prediction in milling. In: 2016 IEEE International Conference on Big Data (Big Data), pp. 2062–2069. IEEE (2016)
15. Costa, N., Sánchez, L.: Variational encoding approach for interpretable assessment of remaining useful life estimation. Reliab. Eng. Syst. Saf. **222**, 108353 (2022)
16. Li, H., Zhao, W., Zhang, Y., et al.: Remaining useful life prediction using multi-scale deep convolutional neural network. Appl. Soft Comput. **89**, 106113 (2020)
17. Su, C., Li, L., Wen, Z.: Remaining useful life prediction via a variational autoencoder and a time-window-based sequence neural network. Qual. Reliab. Eng. Int. **36**(5), 1639–1656 (2020)
18. Yan, J., Mu, L., Wang, L., et al.: Temporal convolutional networks for the advance prediction of ENSO. Sci. Rep. **10**(1), 1–15 (2020)
19. Wu, Z., Pan, S., Chen, F., et al.: A comprehensive survey on graph neural networks. IEEE Trans. Neural Networks Learn. Syst. **32**(1), 4–24 (2020)
20. Ordonez, C., Lasheras, F.S., Roca-Pardinas, J., et al.: A hybrid ARIMA–SVM model for the study of the remaining useful life of aircraft engines. J. Comput. Appl. Math. **346**, 184–191 (2019)

Dynamic Feature Distillation

Xinlei Huang[1,3], Ning Jiang[1,3(✉)], Jialiang Tang[2], and Wenqing Wu[1]

[1] School of Computer Science and Technology, Southwest University of Science
and Technology, Mianyang 621000, Sichuan, China
jiangning@swust.edu.cn
[2] School of Computer Science and Engineering, Nanjing University of Science
and Technology, Nanjing 210094, Jiangsu, China
[3] Jiangxi Qiushi Academy for Advanced Studies, Nanchang 330036, Jiangxi, China

Abstract. Feature-based knowledge distillation utilizes features from
superior and complex teacher networks as knowledge to help portable
student networks improve their generalization capability. Recent feature
distillation algorithms focus on various feature processing and transmission methods while ignoring the flexibility of feature selection, resulting in limited distillation effects for students. In this paper, we propose
Dynamic Feature Distillation to increase the flexibility of feature distillation by dynamically managing feature transfer sites. Our method
leverages Online Feature Estimation to monitor the learning status of
the student network in the feature dimension. Adaptive Position Selection then dynamically updates valuable feature transmission locations for
efficient feature transmission. Notably, our approach can be easily integrated as a strategy for feature management into other feature-based
knowledge transfer methods to improve their performance. We conduct
extensive experiments on the CIFAR-100 and Tiny-ImageNet datasets
to validate the effectiveness of Dynamic Feature Distillation.

Keywords: Knowledge distillation · Feature transmission · Image
classification

1 Introduction

As the performance of deep neural networks (DNNs) in various computer vision
applications improves, their computing and storage requirements increase continuously. In recent, many works have been done to compress redundant DNNs to
be suitable for resource-limited equipment, including lightweight networks [20],
parameter quantification [13,19], model pruning [4,5], and knowledge distillation. Among these technologies, this paper focuses on knowledge distillation,
which employs a teacher model with excessive parameters and exceptional performance to train a lightweight student model.

In order to improve the generalization ability of student model, many knowledge distillation methods focus on transferring the informative features from
teacher to student. Early studies [14,23] simply extract features from the middle
layers of teacher networks and promote student networks to learn these features

B. Luo et al. (Eds.): ICONIP 2023, CCIS 1967, pp. 402–413, 2024.
https://doi.org/10.1007/978-981-99-8178-6_31

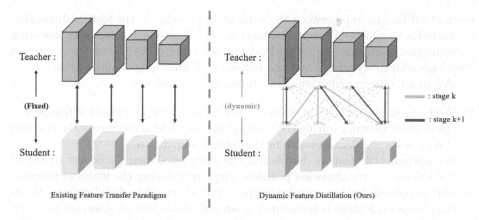

Fig. 1. The comparison of feature transmission strategy between existing feature distillation paradigm and our proposed Dynamic Feature Distillation. **Left**: Existing feature distillation methods transfer feature knowledge at fixed positions throughout training. **Right**: DFD dynamically adjusts feature transmission positions at different training stages based on the learning state of the student network.

by minimizing a ℓ_2 loss. Recently, a series of advanced works explore different feature transmission methods. For example, Chen et al. [2] assign different weights to all teacher features for student layers to learn. Zhang et al. [25] utilize wavelet transform to extract high-frequency information from features for distillation. Gao et al. [12] introduce an assistant network to capture residual features during the distillation process and combine them as supplementary knowledge to the student network. These methods can effectively transfer feature information from teacher to student. However, the feature transmission positions remain constant throughout training as illustrated in Fig. 1 (Left).

Drawing inspiration from [10], which clarified that the knowledge of each sample shows different values to the student network to varying phases of training, we notice that transferring features at manually selected fixed locations may limit feature distillation. Specifically, during the training process, the impact of the features from different teacher locations on the student network will vary as the training steps. In a nutshell, in the beginning, the randomly initialized student network is eager to learn the features from all layers of the teacher network due to the subpar feature processing capabilities. Fixed feature transmission positions throughout the training will result in two problems: 1) After students have successfully simulated some of the teacher's features, continuing to learn the features in repeated positions will reduce the efficiency of knowledge transmission. 2) As the training progresses, the fixed transmission locations cannot meet students' needs for learning other more valuable features.

To address the above two problems, we propose Dynamic Feature Distillation (DFD), a flexible feature management strategy that automatically adjusts the learning position according to the gap between the student's and teacher's feature sets at each training phase. (Fig. 1 right). Figure 2 shows the overall framework of DFD. The gap between the student and teacher in the feature dimension is statistics by Online Feature Evaluation, which provides a basis for selecting transmission locations. Adaptive Position Selection picks feature transmission locations adaptively to increase the efficiency of knowledge transmission.

All in all, our contributions are summarized as follows:

1) We focus on the changes in the value of features at the layer level of teachers to the student network during the training process, which represents the training status of the student network in the feature dimension.
2) We optimize the feature distillation efficiency by allowing students to adjust the knowledge transmission position adaptively during the learning process.
3) Our proposed method can be simply applied to other distillation methods that focus on feature transmission as an automatic feature selection strategy.

2   Related Work

Knowledge distillation aims to transfer knowledge from the burdensome pre-training teacher model to the lightweight student model. According to the definition of knowledge, the existing knowledge distillation work can be divided into response-based [8], relationship-based [12,18], and feature-based methods. Taking features as knowledge, as opposed to the other two distillation methods, focuses on the feature processing procedure within the DNNs, which is directly related to the model's performance. FitNet [14] advances feature distillation methods by developing the selection of knowledge as features in the hint layer. AT [23] allows each student layer to learn the attention features of the teacher network's corresponding position, allowing the student to learn attention from the teacher. SRRL [21] utilizes teacher classifiers to optimize features from the penultimate layer of students. MGD [22] instructs students on how to repair mask features from the backbone network to improve the presentation ability of the student network. These works are dedicated to exploring transmission methods and feature categories. Differently, we consider the flexibility of feature transmission location, which enables the student network to dynamically adjust teacher features for learning instead of manually selecting them.

A recent related study is KCD [10], which summarizes informative knowledge from samples by dynamically evaluating the value of each sample to perform response-based knowledge distillation methods efficiently. Inspired by this work, we adaptively update the location of feature transmission by evaluating the global value of teachers' features, further boosting the performance of existing feature distillation methods.

3   Method

In this section, we first define the formulaic representation of feature-based knowledge distillation framework and related notations in image classification tasks (Sect. 3.1). Then, we describe how to evaluate the relationship between the student's and teacher's features in different positions (Sect. 3.2). Finally, we propose an adaptive position selection strategy to improve traditional feature loss flexibly and effectively (Sect. 3.3).

3.1   Preliminary

When training a student network S for an image classification task, given an input image x and its corresponding label y, the cross-entropy loss $\mathcal{L}_{ce}$ is usually used to reduce the distance between the network prediction $S(x)$ and the truth label y. On this basis, knowledge distillation introduces a pre-trained teacher network T with a powerful classification ability to help the training of S. Traditional knowledge distillation loss function is used to bridge the classification probability of the student's output to that of the teacher:

$$\mathcal{L}_{KD} = \tau^2 \mathcal{H}_{\mathrm{KL}} \left(\sigma \left(\frac{S(x)}{\tau} \right), \sigma \left(\frac{T(x)}{\tau} \right) \right), \tag{1}$$

where $\mathcal{H}_{\mathrm{KL}}(\cdot, \cdot)$ represents Kullback-Leibler divergence to measure the distance between student's and teacher's predicted probability, and $\sigma(\cdot)$ is the softmax function with a temperature hyper-parameter τ to soft the inputted probabilities.

Feature-based knowledge distillation further introduces feature transmission between student and teacher network layers. We let F_S and F_T to represent feature sets from the student and the teacher. Note that the notations with subscripts T and S denote they are related to the teacher and student network, respectively. Feature loss usually describes the distance between processed F_S and F_T:

$$\mathcal{L}_{Feature} = \mathcal{H}_{\mathrm{FT}} \left(\Psi_S(F_S), \Psi_T(F_T) \right), \tag{2}$$

where $\mathcal{H}_{\mathrm{FT}}$ refers to the feature transfer function to transfer features from teacher to student, e.g., ℓ_2 loss. $\Psi_S(\cdot)$ and $\Psi_T(\cdot)$ represent the processing of feature sets. In previous works [7,14,21–23], feature loss effectively increased students' generalization ability. However, during training, feature sets F_S and F_T are fixed due to manually selected feature transmission positions before training begins, limiting the further improvement of student network performance, which is also the focus of this paper.

3.2   Online Feature Estimation

During the distillation process, the contributions of teacher features for each student network layer continue to change as student network performance improves. Therefore, it is necessary to adjust the feature transmission position flexibly based on the student's learning status at different training stages. We propose online feature estimation as a basis for students to choose the locations of feature transmission by counting the impact of features from different locations of teachers on student networks.

We indicate the processed feature set collected by the student network at m locations as $F_S = \left\{ f_S^i \right\}_{i=1}^m$. Similarly, the teacher feature set is marked as

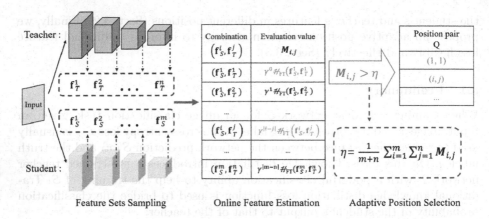

Fig. 2. The general framework of the proposed Dynamic Feature Distillation (DFD). Feature sets from the teacher and student are evaluated pairwise by Online Feature Estimation and further filtered by Adaptive Position Selection for optimal feature positions combination.

$F_{\mathcal{T}} = \left\{ f_{\mathcal{T}}^{j} \right\}_{j=1}^{n}$, where n represents the number of feature extraction locations of teacher network. In order to quantify training states of the student network in the feature dimension, we calculate the global feature distance between $F_{\mathcal{S}}$ and $F_{\mathcal{T}}$ as the evaluation data $M \in \mathbb{R}^{m \times n}$:

$$ M = \left\{ \mathcal{H}_{\mathrm{FT}} \left(f_{\mathcal{S}}^{i}, f_{\mathcal{T}}^{j} \right) \right\}_{i=1, j=1}^{i=m, j=n} . \tag{3} $$

Considering the cross-layer features, i.e., when i ≠ j in Eq. (3), the inherent semantic gap [2] will introduce additional loss value, interfering with the evaluation results. We introduce a discount coefficient γ for the cross-layer feature distance to account for the additional impact of semantic gap. We use a smaller discount coefficient to punish feature pairs that span multiple layers. Inspired by [3], which uses the projector ensemble consisted of multiple MLPs to align feature dimensions, we average multiple parallel projectors, e.g. 1×1 convolution layers, to align students' feature dimensions with teachers. Consequently, the feature evaluation data for teachers and students are rewritten as:

$$ M = \left\{ \gamma^{|i-j|} \mathcal{H}_{\mathrm{FT}} \left(\mathrm{Project} \left(f_{\mathcal{S}}^{i} \right), f_{\mathcal{T}}^{j} \right) \right\}_{i=1, j=1}^{i=m, j=n}, \tag{4} $$

where Project($\cdot$) represents the projection function. We further validated the effect of using different numbers of projector in Sect. 4.3.

3.3   Adaptive Position Selection

By calculating our evaluation data M at the start of a student training epoch, we can intuitively see the impact of features from different teachers' positions on various student network layers. In order to enable students to learn the required teacher features automatically, we propose adaptive position selection to screen the valuable feature position combination (*i.e.*, data coordinates in the evaluation data M).

We calculate the mean value in the evaluation data M as the value threshold η:

$$\eta = \frac{1}{m+n} \sum_{i=1}^{m} \sum_{j=1}^{n} M_{i,j}. \qquad (5)$$

Then, we extract all subscript combinations with values greater than threshold η as the selected position pair set Q:

$$Q = \{(i,j) \mid M_{i,j} > \eta\}. \qquad (6)$$

The position pair set Q is calculated before each training epoch, and it includes the feature positions that students desire to learn at the current learning stage. To mitigate the interference of semantic gaps from cross-layer features on training, γ is still used in feature loss calculation. Based on the position pair Q, we improved the feature loss into a more flexible form:

$$\mathcal{L}_{Feature} = \sum_{(i,j) \in Q} \gamma^{|i-j|} \mathcal{H}_{FT} \left(\text{Project} \left(\boldsymbol{f}_S^i \right), \boldsymbol{f}_T^j \right). \qquad (7)$$

Our method is easily applicable to other distillation methods focusing on feature information transmission because it only optimizes the selection of feature transmission locations without interfering with the feature transmission process. The total loss of the training student network consists of three parts: cross-entropy loss, distillation loss, and feature loss:

$$\mathcal{L}_{\text{Total}} = \alpha \mathcal{L}_{ce} + \beta \mathcal{L}_{KD} + \delta \mathcal{L}_{Feature}, \qquad (8)$$

where α, β, and δ are hyperparameters used to balance loss components. Following most distillation methods, the default values for α and β are set to 1, whereas the value of δ depends on the specific feature transmission method.

3.4   Implementation Procedure

The overall algorithm process of our method is described in Algorithm 1. Given the training phase's length, the training data features are obtained from the student and teacher networks at the beginning of each training phase. *Online Feature Estimation* generate a feature evaluation matrix M for quantifying the learning state of student networks. Further, the optimized feature transmission position is updated based on M using *Adaptive Position Selection*. Finally, the feature distillation method can be performed on the feature transfer sites obtained by DFD. Since our method serves only as a strategy for managing features and does not interfere with the feature distillation process, it can be easily and effectively combined with other feature distillation methods to optimize their efficacy.

Algorithm 1. DFD: Dynamic Feature Distillation

Input: A student network $\mathcal{S}$ with trainable parameters Θ_s, maximum number of epoches E, a pretrained teacher network $\mathcal{T}$ with parameters Θ_t;
  Hyper-parameters: Discount factor γ and number of epochs in a training phase Tp.

1: **for** $e = 1, 2, \ldots, E$ Epoch **do**
2:   **if** e % Tp = 0 **then**
3:     # Extract the features f_S of the student network for training
     data x at m locations to form a feature set F_S
4:     $F_S = \left\{ f_S^i \right\}_{i=1}^m = \mathcal{S}(x, \Theta_s)$;
5:     # Extract the features f_T of the teacher network for training
     data x at n locations to form a feature set F_T
6:     $F_T = \left\{ f_T^j \right\}_{j=1}^n = \mathcal{T}(x, \Theta_t)$;
7:     # Online Feature Estimation
8:     Calculate feature evaluation matrix $M \in \mathbb{R}^{m \times n}$ with Discount factor γ via
     Eq. (4)
9:     # Adaptive Position Selection
10:     Calculate value threshold η on M
11:     Update feature transfer location Q via Eq. (6)
12:   **end if**
13:   # Feature distillation
14:   Perform feature distillation based on feature transmission position Q by minimize Eq. (8)
15: **end for**
Output: Θ_s

---

4 Experiment

4.1 Datasets and Experimental Setting

We conduct experiments on two benchmark datasets. The CIFAR-100 [9] dataset has 100 categories, each containing 500 training and 100 test 32×32 RGB images. Tiny-ImageNet [17] dataset contains 10000 training images and 1000 test 64×64 RGB images from the 200 classes.

Four network architectures [6,15,24] are selected as pre-trained teacher networks, including VGG13, WideResNet40-2, ResNet56, and ResNet-32×4. Correspondingly, the student network consists of five types of networks in our experiment, including VGG8, WideResNet16-2, ResNet20, ShuffleNetV1 [26], and ShuffleNetV2 [11]. The performance of student networks is evaluated by the classification accuracy of Top-1.

We follow the experimental settings of the previous work [16] both on CIFAR-100 and Tiny-ImageNet. The Stochastic Gradient Descent (SGD) [1] optimizer with a momentum of 0.9 and weight decay of 5e–4 is used throughout the training. All student networks are trained for 240 epochs. The learning rate is initially set to 0.05, decreasing by 0.1 at 150, 180, and 210 epochs. The hyperparameter γ is set to the default value of 0.6. We set the default number of projectors used to project student features into the teacher feature dimension to 3. The length of a training phase is one epoch in our experiments. Further exploration results

Table 1. Test Top-1 classification accuracy on the CIFAR-100 dataset. The impact of our method on existing feature distillation methods is marked as (+/−). The highest classification accuracy of the same teacher-student architecture is denoted as **bold**.

TeacherNet	Vgg13	ResNet56	WideResNet40-2	ResNet-32×4	ResNet-32×4
StudentNet	Vgg8	ResNet20	WideResNet16-2	ShuffleNetV1	ShuffleNetV2
Teacher (base)	74.64%	72.34%	75.61%	79.42%	79.42%
Student (base)	70.36%	69.06%	73.26%	70.50%	71.82%
KD [8]	72.98%	70.66%	74.92%	74.07%	74.45%
SP [18]	72.68%	69.67%	73.83%	73.48%	74.56%
FitNet [14]	71.02%	69.21%	73.58%	73.59%	73.54%
AT [23]	71.43%	70.55%	74.08%	71.73%	72.73%
RKD [12]	71.48%	69.61%	73.35%	72.28%	73.21%
CRD [16]	74.29%	71.63%	75.64%	75.12%	76.05%
SRRL [21]	73.44%	69.87%	75.59%	75.18%	76.19%
MGD [22]	73.83%	71.52%	75.68%	74.6%	75.53%
(DFD+AT)	73.26%	71.80%	**75.91%**	75.95%	76.47%
	(+1.83%)	(+1.25%)	(+1.83%)	(+3.77%)	(+3.74%)
(DFD+MGD)	**74.02%**	**72.04%**	75.77%	**75.95%**	**76.65%**
	(+0.19%)	(+0.52%)	(+0.09%)	(+1.35%)	(+1.12%)

for the number of projectors, the length of a training stage, and γ are presented in Sect. 4.3, 4.4, and 4.5.

4.2 Main Results

Classification Results on CIFAR-100. We compare our method to other previous distillation methods [8,12,14,16,18,21–23], and combine a classical attention transfer method AT [23] and an advanced feature distillation MGD [22] as the baseline method with DFD. Table 1 shows the classification results of three teacher-student architectures on CIFAR-100.

Benefiting from the dynamic change of feature transmission locations, our method relieves the limitations of AT and MGD in feature selection and achieves a better training effect. The ShuffleNetV1 trained by DKD+AT achieves 75.95% top-1 accuracy, 3.77% higher than the baseline method AT. For the advanced method MGD, our DFD brings 0.19%, 0.52%, 0.09%, 1.35%, and 1.12% classification performance optimization when using VGG8, ResNet20, WideResNet16-2, ShuffleNetV1, and ShuffleNetV2. The experimental results demonstrate the effectiveness of our method on different network architectures.

Classification Results on Tiny-ImageNet. We conduct further experiments on Tiny-ImageNet. Varied feature distillation methods are selected as comparison methods. As in the CIFAR-100 experiment, we continue to use AT and

Table 2. Test Top-1 classification accuracy on the Tiny-ImageNet dataset. The impact of our method on existing feature distillation methods is marked as $(+/-)$. The highest classification accuracy of the same teacher-student architecture is denoted as **bold**.

TeacherNet	Vgg13	ResNet56	WideResNet40-2
StudentNet	Vgg8	ResNet20	WideResNet16-2
Teacher (base)	62.23%	58.10%	62.45%
Student (base)	56.81%	53.75%	58.84%
KD [8]	61.86%	55.34%	61.53%
FitNet [14]	58.16%	53.67%	58.63%
AT [23]	59.56%	56.04%	60.32%
MGD [22]	61.97%	56.42%	62.68%
(DFD+AT)	62.01%	56.52%	**62.84%**
	(+2.45%)	(+0.48%)	(+2.52%)
(DFD+MGD)	**62.34%**	**56.70%**	62.77%
	(+0.37%)	(+0.28%)	(+0.09%)

Table 3. The impact of the number of projectors on student network performance. Both ShuffleNetV1 and ShuffleNetV1 are distilled by teacher networks ResNet-32×4 on CIFAR-100.

Method	Student	1-Projector	2-Projector	3-Projector	4-Projector
DFD+AT	ShuffleNetV1	75.68%	75.12%	**75.95%**	75.06%
	ShuffleNetV2	76.25%	76.35%	**76.47%**	76.24%
DFD+MGD	ShuffleNetV1	75.55%	**76.11%**	75.95%	75.59%
	ShuffleNetV2	76.63%	76.57%	**76.65%**	76.38%

MGD as baseline methods and DFD for improvement. Table 2 shows the experimental results. Due to the adaptive change of feature transmission position, our method improves AT performance by 0.48%–2.52% and MGD performance by 0.09%–0.37%, respectively. Extensive experimental results demonstrate that our DFD can effectively optimize the feature distillation methods.

4.3 The Numbers of Projectors

Similar to [3], we use a projector ensemble consisting of multiple projectors to align the dimensions of student features to teacher features. The difference is that the projector is a 1×1 convolution layer instead of a MLP in our method. The projected features are obtained by averaging the outputs of multiple projectors. Table 3 shows the effect of different numbers of projectors on our method performance. In most cases, student networks can achieve superior classification performance with three projectors. Therefore, in our experiments, the default value for the number of projections is set to 3.

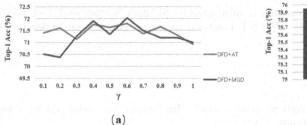
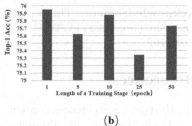

(a)   (b)

Fig. 3. Accuracy with discount coefficient γ and the length of a training phase.

4.4 The Length of a Training Phase

We explored the influence of selecting different training stage lengths on the effectiveness of our DFD. Figure 3(b) shows the performance of student network ShuffleNetV1 trained by method DFD+MGD at different lengths of a training phase. When the length of the training phase is one epoch, the classification performance of the student reaches the highest 75.95%, which means that timely adjusting the appropriate feature transmission position before each epoch is essential for improving the feature distillation effect.

4.5 Parameter Sensitivity Test

In order to counteract the influence of the semantic gap and increase the fairness of evaluation, we include γ in Eq. (4). We use ResNet20 trained by DFD+AT and DFD+MGD to test the sensitivity of γ on CIFAR-100. As shown in Fig. 3(a), when γ is too small ($\gamma < 0.3$) or too large ($\gamma > 0.8$), the balancing effect of γ on cross-layer features is extreme, i.e., excessive punishment or almost no effect, either of which will affect the evaluation results of DFD on features. The student performs better when γ takes the appropriate value ($0.4 \leq \gamma \leq 0.6$) and achieves the highest accuracies of 71.80% and 72.04% when γ equals 0.6.

5   Conclusion

In this work, we examine the limitations imposed by the rigid placement of feature transmission and suggest an innovative solution to enhance feature distillation by developing a new method named Dynamic Feature Distillation (DFD). During the training phase, our proposed DFD automatically selects the optimal locations for feature transmission, increasing the efficiency and efficacy of feature distillation. Our DFD approach, as a feature selection tactic, can be integrated into other feature transmission techniques, allowing for the transfer of features with flexibility and without the need for additional training parameters. Extensive experiments on the CIFAR-100 and Tiny-ImageNet datasets confirm the efficacy of our proposed method.

Acknowledgement. This research is supported by Sichuan Science and Technology Program (No. 2022YFG0324), SWUST Doctoral Research Foundation under Grant 19zx7102.

References

1. Bottou, L.: Stochastic gradient descent tricks. In: Neural Networks: Tricks of the Trade, 2nd edn., pp. 421–436 (2012)
2. Chen, D., et al.: Cross-layer distillation with semantic calibration. In: Proceedings of the AAAI Conference on Artificial Intelligence, vol. 35, pp. 7028–7036 (2021)
3. Chen, Y., Wang, S., Liu, J., Xu, X., de Hoog, F., Huang, Z.: Improved feature distillation via projector ensemble. arXiv preprint arXiv:2210.15274 (2022)
4. Elkerdawy, S., Zhang, H., Ray, N.: Lightweight monocular depth estimation model by joint end-to-end filter pruning. In: 2019 IEEE International Conference on Image Processing (ICIP), pp. 4290–4294. IEEE (2019)
5. Ghosh, S., Srinivasa, S.K., Amon, P., Hutter, A., Kaup, A.: Deep network pruning for object detection. In: 2019 IEEE International Conference on Image Processing (ICIP), pp. 3915–3919. IEEE (2019)
6. He, K., Zhang, X., Ren, S., Sun, J.: Deep residual learning for image recognition. In: Proceedings of the IEEE Conference on Computer Vision and Pattern Recognition, pp. 770–778 (2016)
7. Heo, B., Kim, J., Yun, S., Park, H., Kwak, N., Choi, J.Y.: A comprehensive overhaul of feature distillation. In: Proceedings of the IEEE/CVF International Conference on Computer Vision, pp. 1921–1930 (2019)
8. Hinton, G., Vinyals, O., Dean, J.: Distilling the knowledge in a neural network. arXiv preprint arXiv:1503.02531 (2015)
9. Krizhevsky, A., Hinton, G., et al.: Learning multiple layers of features from tiny images (2009)
10. Li, C., et al.: Knowledge condensation distillation. In: Avidan, S., Brostow, G., Cisse, M., Farinella, G.M., Hassner, T. (eds.) Computer Vision-ECCV 2022: 17th European Conference, Tel Aviv, Israel, 23–27 October 2022, Proceedings, Part XI, pp. 19–35. Springer, Heidelberg (2022). https://doi.org/10.1007/978-3-031-20083-0_2
11. Ma, N., Zhang, X., Zheng, H.T., Sun, J.: Shufflenet v2: practical guidelines for efficient CNN architecture design. In: Proceedings of the European Conference on Computer Vision (ECCV), pp. 116–131 (2018)
12. Park, W., Kim, D., Lu, Y., Cho, M.: Relational knowledge distillation. In: Proceedings of the IEEE/CVF Conference on Computer Vision and Pattern Recognition, pp. 3967–3976 (2019)
13. Polino, A., Pascanu, R., Alistarh, D.: Model compression via distillation and quantization. arXiv preprint arXiv:1802.05668 (2018)
14. Romero, A., Ballas, N., Kahou, S.E., Chassang, A., Gatta, C., Bengio, Y.: Fitnets: hints for thin deep nets. arXiv preprint arXiv:1412.6550 (2014)
15. Simonyan, K., Zisserman, A.: Very deep convolutional networks for large-scale image recognition. arXiv preprint arXiv:1409.1556 (2014)
16. Tian, Y., Krishnan, D., Isola, P.: Contrastive representation distillation. arXiv preprint arXiv:1910.10699 (2019)
17. Torralba, A., Fergus, R., Freeman, W.T.: 80 million tiny images: a large data set for nonparametric object and scene recognition. IEEE Trans. Pattern Anal. Mach. Intell. **30**(11), 1958–1970 (2008)

18. Tung, F., Mori, G.: Similarity-preserving knowledge distillation. In: Proceedings of the IEEE/CVF International Conference on Computer Vision, pp. 1365–1374 (2019)
19. Wang, K., Liu, Z., Lin, Y., Lin, J., Han, S.: HAQ: hardware-aware automated quantization with mixed precision. In: Proceedings of the IEEE/CVF Conference on Computer Vision and Pattern Recognition, pp. 8612–8620 (2019)
20. Wang, Y., et al.: Lednet: a lightweight encoder-decoder network for real-time semantic segmentation. In: 2019 IEEE International Conference on Image Processing (ICIP), pp. 1860–1864. IEEE (2019)
21. Yang, J., Martinez, B., Bulat, A., Tzimiropoulos, G., et al.: Knowledge distillation via softmax regression representation learning. In: International Conference on Learning Representations (ICLR) (2021)
22. Yang, Z., Li, Z., Shao, M., Shi, D., Yuan, Z., Yuan, C.: Masked generative distillation. In: Avidan, S., Brostow, G., Cisse, M., Farinella, G.M., Hassner, T. (eds.) Computer Vision-ECCV 2022: 17th European Conference, Tel Aviv, Israel, 23–27 October 2022, Proceedings, Part XI, pp. 53–69. Springer, Heidelberg (2022). https://doi.org/10.1007/978-3-031-20083-0_4
23. Zagoruyko, S., Komodakis, N.: Paying more attention to attention: improving the performance of convolutional neural networks via attention transfer. arXiv preprint arXiv:1612.03928 (2016)
24. Zagoruyko, S., Komodakis, N.: Wide residual networks. arXiv preprint arXiv:1605.07146 (2016)
25. Zhang, L., Chen, X., Tu, X., Wan, P., Xu, N., Ma, K.: Wavelet knowledge distillation: towards efficient image-to-image translation. In: Proceedings of the IEEE/CVF Conference on Computer Vision and Pattern Recognition, pp. 12464–12474 (2022)
26. Zhang, X., Zhou, X., Lin, M., Sun, J.: Shufflenet: an extremely efficient convolutional neural network for mobile devices. In: Proceedings of the IEEE Conference on Computer Vision and Pattern Recognition, pp. 6848–6856 (2018)

Detection of Anomalies and Explanation in Cybersecurity

Durgesh Samariya[1(✉)], Jiangang Ma[1], Sunil Aryal[2], and Xiaohui Zhao[1]

[1] Institute of Innovation, Science and Sustainability, Federation University,
Ballarat, Australia
{d.samariya,j.ma,x.zhao}@federation.edu.au
[2] School of Information Technology, Deakin University, Geelong, VIC, Australia
sunil.aryal@deakin.edu.au

Abstract. Histogram-based anomaly detectors have gained significant attention and application in the field of intrusion detection because of their high efficiency in identifying anomalous patterns. However, they fail to explain why a given data point is flagged as an anomaly. Outlying Aspect Mining (OAM) aims to detect aspects (a.k.a subspaces) where a given anomaly significantly differs from others. In this paper, we have proposed a simple but effective and efficient histogram-based solution - HMass. In addition to detecting anomalies, HMass provides explanations on why the points are anomalous. The effectiveness and efficiency of HMass are evaluated using comparative analysis on seven cyber security datasets, covering the tasks of anomaly detection and outlying aspect mining.

Keywords: anomaly detection · outlying aspect mining · outlier explanation · cyber security · anomaly explanation

1 Introduction

While the Internet undoubtedly offers numerous advantages, it has some well-known negative aspects. Specifically, cyber security is a critical issue for many businesses and organizations. Cyber attacks can devastate businesses, leading to financial losses, reputation damage, and regulatory penalties. As a result, there has been a growing interest in developing effective techniques for detecting cyber attacks. Cyber security encompasses various technologies and processes to protect computer systems, networks, programs, and data from potential threats, including unauthorized access, alteration, or destruction. Researchers have made significant progress in developing Intrusion Detection Systems (IDS) that can detect attacks in various environments. However, given the diversity of environments and attack types, certain approaches often perform more effectively than others in specific contexts.

Anomaly (or Outlier) detection has emerged as a promising approach for identifying potential cyber attacks, a critical need for many organizations in

B. Luo et al. (Eds.): ICONIP 2023, CCIS 1967, pp. 414–426, 2024.
https://doi.org/10.1007/978-981-99-8178-6_32

today's digital era. Anomaly detection algorithms play a crucial role in identifying abnormal patterns or behaviors within a network or system that could indicate the presence of a cyber threat.

Many anomaly detection methods have been developed to detect anomalies, such as density-based, distance-based, histogram-based, isolation-based, etc. Histogram-based anomaly detection methods are easy to construct and intuitive to understand [1,4]; such methods are computationally faster than distance/density-based methods. In these methods, data is first divided into bins, and the anomaly score of the data instance is computed by summing the log probability data mass of the histogram where the instance falls into. Histogram-based methods outperform existing state-of-the-art anomaly detectors in terms of runtimes and produce competitive results [1,4]. Although all these algorithms are good at detecting anomalies/outliers, they cannot explain why those data points are flagged as anomalies/outliers.

To find such an explanation researchers are interested in *Outlying Aspect Mining* (**OAM**) [3,12–15,17–19], where the goal is to detect aspects (a subset of features), where a given anomaly/outlier is dramatically different than others. In the context of Cyber Security, OAM aims at identifying aspects or subsets of features in which a particular anomaly or outlier exhibits significantly different behavior compared to others. These aspects can be various security-related dimensions, such as network traffic patterns, user behavior, system logs, or application usage. By employing OAM techniques, cyber security professionals can effectively identify and understand the specific features or dimensions that contribute to an anomaly, aiding in the detection and prevention of potential security threats.

The research reported in this paper makes the following contributions to anomaly detection in cyber security domain:

- Propose a new scoring measure called HMass that detects anomalies and explains by giving a set of features showing how a given anomaly is dramatically different than others.
- Compare HMass against six anomaly detectors and four OAM scoring measures using synthetic and real-world datasets.

The rest of this paper is organized as follows. Section 2 reviews related work. HMass is introduced in Sect. 3. Section 4 reports the empirical evaluation, and Sect. 5 concludes the paper.

2 Preliminaries and Related Work

2.1 Preliminaries

Let $D = \{x_1, x_2, \cdots, x_N\}$ be a data set $x_i \in \Re^d$, where $i \in \{1, 2, \cdots, N\}$ represents the position of data point x in D and N is the number of data points in data set. The feature set $\mathbb{F} = \{F_1, F_2, \cdots, F_d\}$ denotes the full feature space, where d is the number of dimensions. Outlier detection problem is identifying

x_i, which remarkably deviates from others in full feature set $\mathbb{F}$. Whereas the problem of OAM is to identify a set of feature(s) (a.k.a subspace) in which a given query object $x_i \in D$ is significantly different from the rest of the data. The query object is referred as **q**.

2.2 Anomaly Detection

In 2000, Breunig et al. [2] introduced a local outlier detection approach called LOF (Local Outlier Factor). LOF calculates an anomaly score for each data object based on its local density and the average local density of its k-nearest neighbors. The LOF score of data object x is defined as follows:

$$LOF(x) = \frac{\sum\limits_{y \in N^k(x)} lrd(y)}{|N^k(x)| \times lrd(x)}$$

where $lrd(x) = \frac{|N^k(x)|}{\sum\limits_{y \in N^k(x)} max(dist^k(y,D), dist(x,y))}$, $N^k(x)$ is a set of k-nearest neighbours of x, $dist(x,y)$ is a distance between x and y and $dist^k(x,D)$ is the distance between x and its k-NN in D.

Liu et al. (2008) proposed a framework called iForest (Isolation Forest) [7], which aims to isolate individual data points through axis-parallel partitioning of the feature space. The iForest framework constructs an ensemble of trees known as isolation trees or iTrees. Each iTree is built using randomly selected subsamples (ψ) from the dataset, without replacement. At each tree node, a random split is performed on a randomly selected data point from the attribute space. The partitioning process continues recursively until either all nodes contain only one data point or the nodes reach the height limit specified for that particular iTree.

The anomaly score of a data point x using iForest is defined as:

$$iForest(x) = \frac{1}{t} \sum_{i=1}^{t} l_i(x)$$

where $l_i(x)$ is the path length of x in tree T_i.

In 2012, Goldstein and Dengel [4] proposed HBOS (Histogram-Based Outlier Score) as an intuitive and straightforward statistical-based outlier detection technique. Firstly, HBOS creates a histogram in each data dimension. Each dimension is divided into multiple bins, and the histogram captures the frequency or density of data points falling within each bin. The outlier score for each data object is calculated by combining the inverse probabilities of the bins in which the data object resides across all dimensions.

In an independent study, Aryal et al. (2016) [1] introduced SPAD (Simple Probabilistic Anomaly Detector), which is a histogram-based approach for anomaly detection. The underlying idea behind SPAD is that an outlier will likely exhibit a significantly low probability in a few specific features. The outlier score of x is computed as:

$$\text{SPAD}(x) = \sum_{i=1}^{d} \log \frac{|B_i(x)| + 1}{N + b}$$

where B_i is the bin where data instance x falls into dimension i, $|\cdot|$ measure the cardinality of a set and b is number of bins.

2.3 Outlying Aspect Mining Algorithms

Duan et al. (2016) [3] introduced a novel approach called OAMiner (Outlying Aspect Miner). OAMiner utilizes Kernel Density Estimation (KDE) as a scoring measure to calculate the outlyingness of a query $\mathbf{q}$ in a subspace S :

$$\tilde{f}_S(\mathbf{q}) = \frac{1}{n(2\pi)^{\frac{m}{2}} \prod_{i \in S} h_i} \sum_{\mathbf{x} \in \mathcal{O}} e^{-\sum_{i \in S} \frac{(\mathbf{q}.i - x.i)^2}{2\,h_i^2}}$$

where $\tilde{f}_S(\mathbf{q})$ is a kernel density estimation of $\mathbf{q}$ in subspace S, m is the dimensionality of subspace S ($|S| = m$), h_i is the kernel bandwidth in dimension i.

When comparing subspaces with different dimensionality, a scoring measure needs to be unbiased w.r.t. dimensionality. An example of a dimensionally bias scoring measure is the density measure, which decreases as the dimension increases. As a result, density is biased towards higher-dimensional subspaces.

Vinh et al. (2016) [18] captured the concept of dimensionality unbiasedness and further investigate dimensionally unbiased scoring functions. They proposed to use the density Z-score as an outlying scoring measure. The density Z-score of a query $\mathbf{q}$ in a subspace S is defined as follows:

$$Z\text{-Score}(\tilde{f}_S(\mathbf{q})) \triangleq \frac{\tilde{f}_S(\mathbf{q}) - \mu_{\tilde{f}_S}}{\sigma_{\tilde{f}_S}}$$

where $\mu_{\tilde{f}_S}$ and $\sigma_{\tilde{f}_S}$ are the mean and standard deviation of the density of all data instances in subspace S, respectively.

Wells and Ting (2019) [19] introduced a density estimator called sGrid, which is an enhanced version of the conventional grid-based estimator, also known as a histogram. To address the issue of dimensionality bias, they applied Z-score normalization to the outlier scores. By employing the sGrid density estimator, the computational speed of the Beam search was significantly improved, enabling it to run much faster compared to using KDE. Samariya et al. (2022) [16] detected shortcomings of sGrid and proposed a simpler version of sGrid, called sGrid++. sGrid++ is not only efficient and effective but also dimensionally unbiased. Thus, it does not require Z-score normalization.

Samariya et al. (2020) [12,14] proposed a **S**imple **I**solation score using **N**earest **N**eighbor **E**nsemble (SiNNE in short) measure. SiNNE constructs t ensemble of models. Each model is constructed from randomly chosen sub-samples ($\mathcal{D}_i$). Each model has ψ hyperspheres, where a radius of hypersphere is the Euclidean distance between x ($x \in \mathcal{D}_i$) to its nearest neighbor in $\mathcal{D}_i$.

The outlying score of $\mathbf{q}$ in a model $\mathcal{M}_i$, $I(\mathbf{q}\|\mathcal{M}_i) = 0$ if $\mathbf{q}$ falls in any of the hyperspheres and 1 otherwise. The final outlying score of $\mathbf{q}$ using t models is :

$$\text{SiNNE}(\mathbf{q}) = \frac{1}{t} \sum_{i=1}^{t} I(\mathbf{q}\|\mathcal{M}_i)$$

3   The Proposed Method: HMass

Our proposed method – HMass, consists of two main stages: (i) the Training stage and (ii) the Evaluation stage. In the training stage, we create b equal-width histograms in each dimension. These histograms capture the distribution of data points in each dimension and serve as the foundation for the HMass.

Definition 1. *A histogram is a set of b equal width bins, $H = \{B_1, B_2, \cdots, B_b\}$.*

The HMass method starts by creating a univariate histogram in each dimension. These histograms are constructed based on the value range of all data instances in each dimension i, denoted as $[min_i, max_i]$. The minimum value min_i and maximum value max_i represent the boundaries of the data range in dimension i.

To create the histogram, HMass divides the value range $[min_i, max_i]$ into a specified number of equally-sized bins. Each bin has a unique identifier ranging from 1 to the total number of bins b.

Once bins are constructed, the histogram will know the mass of each bin.

Definition 2. *A mass is defined as the number of data points that falls into the region(bin).*

By constructing histograms in each dimension, HMass captures the distribution of data values within specific ranges. Each bin in the histogram represents a subinterval of the value range, and the mass of a bin is determined by the number of data samples that falls within that bin. The mass estimation process allows HMass to quantify the density of data points in each bin, providing valuable information for anomaly detection.

In the evaluation stage, the HMass algorithm calculates the HMass score for a given data point as an average of the masses of the bins in each dimension where the data point falls into.

Definition 3. *The HMass score for a data point $x \in \mathbb{R}^d$ can be defined as follows:*

$$HMass(x) = \frac{1}{d} \sum_{i=1}^{d} m\Big(B_i(x)\Big) \tag{1}$$

where $B_i(x)$ is the bin in i^{th} dimension where x falls into, $m\Big(B_i(x)\Big)$ is the mass of a bin.

To identify anomalies, the data instances in a dataset are ranked in ascending order based on their anomaly scores. The top-ranked instances with the lowest anomaly scores are considered anomalies. These anomalies are characterized by having a low probability mass compared to normal instances.

Theorem 1. *The proposed measure HMass(x) is dimensionally unbiased as per dimensionality unbiasedness definition [18, Definition 4].*

Proof. Given a data set D of N data instances drawn from a uniform distribution $\mathcal{U}([0,1]^d)$.

As data is drawn from the uniform distribution, each bin has the same mass,

$$m\Big(B_i(\cdot)\Big) = \frac{N}{b}$$

where b is the total number of bins.

If we substitute mass in Eq. 1, for query $\mathbf{q}$, final outlying score is,

$$\text{HMass}(x) = \frac{1}{d}\sum_{i=1}^{d} m\Big(B_i(x)\Big) = \frac{1}{d} \cdot d \cdot \frac{N}{b} = \frac{N}{b}$$

Thus, the average value of the HMass scoring measure of each data instance is,

$$E[\text{HMass}(\mathbf{q})|\mathbf{q} \in \mathcal{O}] = \frac{1}{N}\sum_{\mathbf{q} \in \mathcal{O}} \text{HMass}(\mathbf{q})$$

$$= \frac{1}{N}\sum_{\mathbf{q} \in \mathcal{O}} \frac{N}{b} = \frac{1}{N} \cdot N \cdot \frac{N}{b} = \frac{N}{b}, constant\ w.r.t\ |S|$$

In the training stage, HMass calculates the mass count for each bin. This process involves iterating over the dataset, which consists of N data instances, and computing the bin assignments in each of the d dimensions. Therefore, the time complexity of the training stage is $O(Nd)$, as each data point needs to be processed in each dimension to update the corresponding bin mass count.

In the evaluation stage, HMass relies on the precomputed mass table. Given a new data point, HMass performs a lookup operation to determine the corresponding bin masses in each dimension. This process has a time complexity of $O(d)$ since it involves accessing the mass values for each dimension. Overall, the time complexity of the HMass method is dominated by the training stage, which contributes $O(Nd)$.

Comparing the time complexity of HMass with other competing methods, it can be observed that HMass offers competitive performance. Table 1 compares the proposed method's time complexity with its competitors.

Table 1. Time complexities. N is total number of instances in data set; d is number of dimensions; ψ is number of sub-samples; t is number of trees in forest.

Methods	Time complexity
LOF	$O(N^2 d)$
iForest	$O(Nt \log_2(\psi))$
HMass	$O(Nd)$

3.1 Relation to HBOS and SPAD

The HMass method is influenced by the concepts of HBOS and SPAD, both of which are histogram-based outlier detection approaches. However, HMass distinguishes itself from these methods through its histogram generation method and outlier scoring mechanism.

In terms of histogram generation, HMass creates a fixed number of bins, denoted as b, with equal width in a given feature range $[min_i, max_i]$ for each dimension i. In contrast, SPAD employs a different strategy for histogram generation. It constructs b bins in a range of $[\mu_i - 3\sigma_i, \mu_i + 3\sigma_i]$ for each dimension i, where μ_i and σ_i represent the mean and standard deviation of the data values in dimension i.

Furthermore, it is essential to note that SPAD and HBOS are not dimensionally unbiased. It means these methods may not be suitable for applications in the outlying aspect mining domain, where the objective is to detect outliers in specific dimensions or subspaces.

4 Empirical Evaluation

In this section, we compare HMass with six anomaly detectors in terms of performance on real-world cyber security datasets. After that, we also compare the proposed method to three outlying aspect mining methods.

4.1 Contenders and Their Parameter Settings

LOF [2], iForest [7], HBOS [4] and SPAD [1] are selected as competitors of HMass. We select LOF and iForest as contenders because they are state-of-the-art methods, whereas SPAD and HBOS are closely related to HMass.

We used default parameters as suggested in respective papers unless specified otherwise. For LOF, we set a number of nearest neighbors $(k) = 10$. iForest employed 100 trees (t) and a subsample size (ψ) of 256. For SPAD, we set a number of bins in each dimension as $[\log_2(N) + 1]$, where N is the number of data instances. The parameter w block size for a bit set operation in sGrid and sGrid++ was 64. SiNNE employed a sub-sample size (ψ) 8 and an ensemble size (t) 100. Parameters beam width (W) and maximum dimensionality of subspace (ℓ) in Beam search procedure were set to 100 and 3, respectively, as done in [18].

Table 2. Characteristics of data sets used.

Data set	Data size ($\#N$)	Dimensions ($\#d$)	Reference
TON-IoT-Linux-Memory	1000000	9	[9]
TON-IoT-Linux-Disk-Data	1000000	8	[9]
TON-IoT-Linux-Process-Data	1000000	13	[9]
TON-IoT-Windows7	28367	132	[10]
TON-IoT-Windows10	35975	124	[10]
Network	1000000	42	[8]
UNSW-NB15	700001	48	[11]

We use PyOD [20] Python library to implement LOF, iForest, and HBOS anomaly detection algorithms. We used SPAD, sGrid, SiNNE and sGrid++ Java implementations made available by [1,12,19] and [16], respectively.

All experiments were conducted in a macOS machine with a 2.3 GHz 8-core Intel Core i9 processor and 16 GB memory running on macOS Monterey 12.4. We ran all jobs for 1hr and killed all uncompleted jobs.

4.2   Datasets and Performance Measure

We evaluated the performance of the outlier detector on 7 real-world cybersecurity datasets. A summary of the data set used in this study is provided in Table 2.

We used the area under the ROC curve (AUC) [5] as a measure of effectiveness for outlier ranking produced by an anomaly detector. Anomaly detector with a high AUC indicates better detection accuracy, whereas a low AUC indicates low detection accuracy.

4.3   Outlier Detector Performance

The AUC comparison of HMass and its contenders is presented in Table 3. HMass is the best-performing method with 0.62 average AUC. While iForest is its closest competitor with 0.60 average AUC. SiNNE and HBOS perform similarly, with an average AUC of 0.58. sGrid++ perform better than SPAD and LOF with an average AUC of 0.55. LOF is the worst-performing measure, with an average AUC of 0.51.

4.4   Outlying Aspect Mining Performance

Mining Outlying Aspects on Synthetic Datasets. We evaluate the performance of HMass and four contending scoring measures on three synthetic datasets[1], where number of instances (N) are 1,000 and number of dimensions (d): 10 to 50.

---

[1] The synthetic datasets are from Keller et al. (2012) [6]. Available at https://www.ipd.kit.edu/~muellere/HiCS/.

Table 3. AUC scores of HMass, iForest(IF), SPAD, LOF , HBOS, SiNNE and sGrid++ outlier detection methods on 7 real-world cyber security datasets.

Data set	HMass	IF	SPAD	LOF	HBOS	SiNNE	sGrid++
TON-IoT-Linux-Memory	**0.58**	0.57	0.32	0.52	**0.58**	0.49	0.44
TON-IoT-Linux-Disk-Data	**0.51**	0.49	0.49	0.50	**0.51**	0.50	0.50
TON-IoT-Linux-Process-Data	0.52	0.65	0.55	0.51	**0.76**	0.41	0.53
TON-IoT-Windows7	0.57	**0.61**	0.53	0.48	0.51	0.43	0.31
TON-IoT-Windows10	0.59	0.43	0.40	0.50	0.19	**0.62**	0.29
TON-IoT-Network	**0.60**	0.47	0.41	0.55	0.58	0.46	0.53
UNSW-NB15	0.94	**0.95**	**0.95**	0.49	0.94	0.93	0.49
Avg. AUC	0.62	0.60	0.52	0.51	0.58	0.58	0.55

Table 4. Comparison of the proposed measure and its three contenders on five synthetic datasets. q-id represents the query index, and GT represents the ground truth. The numbers in a bracket are feature indices (i.e. subspaces).

	q-id	GT	HMass	Z (sGrid)	SiNNE	sGrid++	iForest
10D	172	{8, 9}	**{8, 9}**	**{8, 9}**	**{8, 9}**	**{8, 9}**	{0, 1, 2}
	245	{2, 3, 4, 5}	**{2, 3, 4, 5}**	**{2, 3, 4, 5}**	**{2, 3, 4, 5}**	**{2, 3, 4, 5}**	{0}
	577	{2, 3, 4, 5}	**{2, 3, 4, 5}**	{6, 7}	**{2, 3, 4, 5}**	**{2, 3, 4, 5}**	{0}
20D	43	{0, 1, 2}	**{0, 1, 2}**	**{0, 1, 2}**	**{0, 1, 2}**	**{0, 1, 2}**	{4, 13, 17}
	86	{18, 19}	**{18, 19}**	**{18, 19}**	**{18, 19}**	**{18, 19}**	{0, 1}
	665	{0, 1, 2}	**{0, 1, 2}**	**{0, 1, 2}**	**{0, 1, 2}**	**{0, 1, 2}**	{0, 1, 7}
50D	121	{21, 22, 23}	**{21, 22, 23}**	**{21, 22, 23}**	**{21, 22, 23}**	**{21, 22, 23}**	{0}
	248	{13, 14, 15}	**{13, 14, 15}**	**{13, 14, 15}**	**{13, 14, 15}**	**{13, 14, 15}**	{0, 1}
	427	{5, 6, 7, 8}	**{5, 6, 7, 8}**	{8, 9, 48}	**{5, 6, 7, 8}**	**{5, 6, 7, 8}**	{0, 1, 2, 3}

Table 4 presents a summary of the discovered subspaces for three specific queries[2], as evaluated by different measures, on three synthetic datasets.

Three measures stand out as the best performers when considering exact matches: HMass, sGrid++, and SiNNE. These measures successfully detected the ground truth of all 9 queries across all datasets. In comparison, the measure Z(sGrid) achieved exact matches for 7 out of the 9 queries. On the other hand, the iForest measure failed to produce any exact matches for the queries.

Mining Outlying Aspects on Cyber Security Datasets. In this section, we conducted experiments using seven real-world cyber security datasets to find outlying aspects. Due to the computational time required, we only compared HMass with SiNNE and sGrid++, excluding Z(sGrid) and iForest.

In real-world datasets, the outliers and their outlying aspects are not available. Thus, we used HMass, SiNNE, and sGrid++ outlier detectors to find the

---

[2] Due to space constraints, we only reported results for three queries.

top 3 common outliers detected by each method. These identified outliers were then used as queries.

Table 5 summarizes the discovered subspaces for three specific queries[3]. Since we do not have ground truth for the real-world datasets and there are no quality assessment measures for discovered subspaces, we compare the results of contending measures visually where the dimensionality of subspaces are up to 3.

Table 5. Comparison of the proposed measure and its two contenders on seven cyber security data set. q-id represents query index. The numbers in the brackets are feature indices (i.e., subspaces).

	q-id	HMass	SiNNE	sGrid++
Linux Memory	640811	{3, 8}	{6, 8}	{8}
	642935	{4, 8}	{2, 7}	{8}
	650021	{3, 8}	{1, 4}	{8}
Linux Disk	10900	{3, 4}	{2, 4}	{4}
	13260	{3, 4}	{4}	{4}
	15393	{3, 4}	{1, 4}	{4}
Linux Process	641093	{4, 11}	{2, 5}	{4}
	647649	{5}	{0, 5}	{4, 12}
	713529	{7, 12}	{1, 11}	{12}
Windows 7	13482	{20, 119}	{25}	{106, 117}
	14200	{3, 9}	{95, 109}	{9, 17}
	16201	{3, 102}	{8, 129}	{3, 11}
Windows 10	10355	{81, 118}	{102, 114}	{102, 114}
	12716	{02, 100}	{3, 102}	{3, 102}
	15060	{81, 97}	{56, 97}	{92, 97}
Network	200907	{1, 3}	{2, 32}	{1}
	203520	{1, 3}	{2, 11}	{1}
	211228	{2, 5}	{0, 5}	{1, 5}
UNSW-NB15	22	{10, 36}	{31, 38}	{9, 13}
	111	{2, 9}	{33}	{0, 13}
	3018	{2, 36}	{14, 46}	{9}

Table 6 shows the subspace discovered by each scoring measure on real-world datasets. We used one query (presented as a red square) from each data set to examine the quality of discovered subspace by different scoring measures. Note that we plotted all one dimension subspaces using the histogram, where the number of bins is set to 10.

---

[3] Due to space constraints, we only reported results for three queries.

Table 6. Visualization of discovered subspaces by HMass, SiNNE, and sGrid++ on real-world datasets.

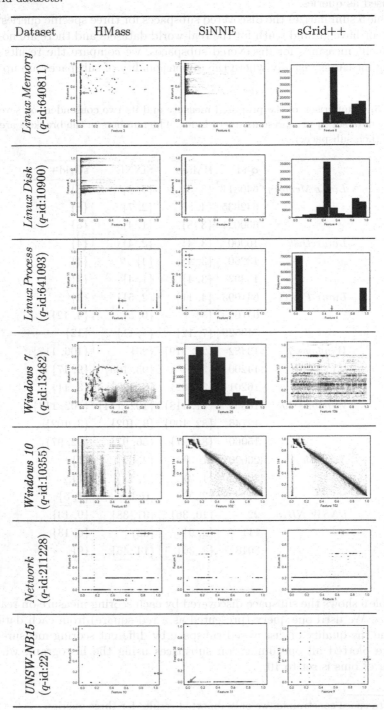

Overall, subspace discovered by HMass, SiNNE and sGrid++ are similar in most of the cases. For the *Linux Memory* and *Linux Disk* datasets, the subspace identified by sGrid++ is not as interesting or noteworthy when compared to the subspaces discovered by the other methods. Similarly, in the case of the *Windows 7* dataset, the subspace found by SiNNE is not as interesting as the subspaces discovered by the other methods.

5   Conclusion

Existing anomaly detector techniques heavily rely on the distance or density of data to identify outlier data instances. These methods fail to run when datasets are huge. This paper discussed a new histogram-based technique using mass to detect anomalies from the data set. HMass is a simpler and more intuitive than other scoring measures based on density/distance. This makes it the most efficient anomaly detector, which can easily scale up to a data set with millions of data instances and thousands of dimensions. In anomaly detection domain, HMass have better performance compare to others, in addition to that, HMass is fastest measure. In outlying aspect mining domain, HMass perform similar or better compare to it's competitors.

Acknowledgments. This work is supported by Federation University Research Priority Area (RPA) scholarship, awarded to Durgesh Samariya.

References

1. Aryal, S., Ting, K.M., Haffari, G.: Revisiting attribute independence assumption in probabilistic unsupervised anomaly detection (2016)
2. Breunig, M.M., Kriegel, H.P., Ng, R.T., Sander, J.: LoF: identifying density-based local outliers. SIGMOD Rec. **29**(2), 93–104 (2000). https://doi.org/10.1145/335191.335388
3. Duan, L., Tang, G., Pei, J., Bailey, J., Campbell, A., Tang, C.: Mining outlying aspects on numeric data. Data Min. Knowl. Disc. **29**(5), 1116–1151 (2015). https://doi.org/10.1007/s10618-014-0398-2
4. Goldstein, M., Dengel, A.: Histogram-based outlier score (hbos): a fast unsupervised anomaly detection algorithm. In: KI-2012: Poster and Demo Track, pp. 59–63 (2012)
5. Hand, D.J., Till, R.J.: A simple generalisation of the area under the roc curve for multiple class classification problems. Mach. Learn. **45**(2), 171–186 (2001)
6. Keller, F., Muller, E., Bohm, K.: HICS: high contrast subspaces for density-based outlier ranking. In: 2012 IEEE 28th International Conference on Data Engineering, pp. 1037–1048 (2012). https://doi.org/10.1109/ICDE.2012.88
7. Liu, F.T., Ting, K.M., Zhou, Z.H.: Isolation forest. In: 2008 Eighth IEEE International Conference on Data Mining, pp. 413–422 (2008). https://doi.org/10.1109/ICDM.2008.17
8. Moustafa, N.: A new distributed architecture for evaluating AI-based security systems at the edge: Network ton_iot datasets. Sustain. Cities Soc. **72**, 102994 (2021)

9. Moustafa, N., Ahmed, M., Ahmed, S.: Data analytics-enabled intrusion detection: evaluations of ton_iot linux datasets. In: 2020 IEEE 19th International Conference on Trust, Security and Privacy in Computing and Communications (TrustCom), pp. 727–735 (2020). https://doi.org/10.1109/TrustCom50675.2020.00100

10. Moustafa, N., Keshky, M., Debiez, E., Janicke, H.: Federated ton_iot windows datasets for evaluating AI-based security applications. In: 2020 IEEE 19th International Conference on Trust, Security and Privacy in Computing and Communications (TrustCom), pp. 848–855 (2020). https://doi.org/10.1109/TrustCom50675.2020.00114

11. Moustafa, N., Slay, J.: Unsw-nb15: a comprehensive data set for network intrusion detection systems (unsw-nb15 network data set). In: 2015 Military Communications and Information Systems Conference (MilCIS), pp. 1–6 (2015). https://doi.org/10.1109/MilCIS.2015.7348942

12. Samariya, D., Aryal, S., Ting, K.M., Ma, J.: A new effective and efficient measure for outlying aspect mining. In: Huang, Z., Beek, W., Wang, H., Zhou, R., Zhang, Y. (eds.) WISE 2020. LNCS, vol. 12343, pp. 463–474. Springer, Cham (2020). https://doi.org/10.1007/978-3-030-62008-0_32

13. Samariya, D., Ma, J.: Mining outlying aspects on healthcare data. In: Siuly, S., Wang, H., Chen, L., Guo, Y., Xing, C. (eds.) HIS 2021. LNCS, vol. 13079, pp. 160–170. Springer, Cham (2021). https://doi.org/10.1007/978-3-030-90885-0_15

14. Samariya, D., Ma, J.: A new dimensionality-unbiased score for efficient and effective outlying aspect mining. In: Data Science and Engineering, pp. 1–16 (2022)

15. Samariya, D., Ma, J., Aryal, S.: A comprehensive survey on outlying aspect mining methods. arXiv preprint arXiv:2005.02637 (2020)

16. Samariya, D., Ma, J., Aryal, S.: sGrid++: revising simple grid based density estimator for mining outlying aspect. In: Chbeir, R., Huang, H., Silvestri, F., Manolopoulos, Y., Zhang, Y. (eds.) WISE 2022. LNCS, vol. 13724, pp. 194–208. Springer, Cham (2022). https://doi.org/10.1007/978-3-031-20891-1_15

17. Samariya, D., Ma, J., Aryal, S., Zhao, X.: Detection and explanation of anomalies in healthcare data. Health Inf. Sci. Syst. **11**(1), 20 (2023)

18. Vinh, N.X., Chan, J., Romano, S., Bailey, J., Leckie, C., Ramamohanarao, K., Pei, J.: Discovering outlying aspects in large datasets. Data Min. Knowl. Disc. **30**(6), 1520–1555 (2016). https://doi.org/10.1007/s10618-016-0453-2

19. Wells, J.R., Ting, K.M.: A new simple and efficient density estimator that enables fast systematic search. Pattern Recogn. Lett. **122**, 92–98 (2019)

20. Zhao, Y., Nasrullah, Z., Li, Z.: PYOD: a python toolbox for scalable outlier detection. J. Mach. Learn. Res. **20**(96), 1–7 (2019). https://jmlr.org/papers/v20/19-011.html

Document-Level Relation Extraction with Relation Correlation Enhancement

Yusheng Huang and Zhouhan Lin[✉]

Shanghai Jiao Tong University, Shanghai, China
huangyusheng@sjtu.edu.cn,lin.zhouhan@gmail.com

Abstract. Document-level relation extraction (DocRE) is a task that focuses on identifying relations between entities within a document. However, existing DocRE models often overlook the correlation between relations and lack a quantitative analysis of relation correlations. To address this limitation and effectively capture relation correlations in DocRE, we propose a relation graph method, which aims to explicitly exploit the interdependency among relations. Firstly, we construct a relation graph that models relation correlations using statistical co-occurrence information derived from prior relation knowledge. Secondly, we employ a re-weighting scheme to create an effective relation correlation matrix to guide the propagation of relation information. Furthermore, we leverage graph attention networks to aggregate relation embeddings. Importantly, our method can be seamlessly integrated as a plug-and-play module into existing models. Experimental results demonstrate that our approach can enhance the performance of multi-relation extraction, highlighting the effectiveness of considering relation correlations in DocRE.

Keywords: Document-level relation extraction · Relation correlation · Relation graph construction

1 Introduction

Relation extraction (RE) plays a vital role in information extraction by identifying semantic relations between target entities in a given text. Previous research has primarily focused on sentence-level relation extraction, aiming to predict relations within a single sentence [7]. However, in real-world scenarios, valuable relational facts are often expressed through multiple mentions scattered across sentences, such as in Wikipedia articles [17]. Consequently, the extraction of relations from multiple sentences, known as document-level relation extraction, has attracted significant research attention in recent years.

Compared to sentence-level RE, document-level RE presents unique challenges in designing model structures. In sentence-level RE, a single relation type is associated with each entity pair, as observed in SemEval 2010 Task 8 [10] and TACRED [33]. However, in document-level RE, an entity pair can be associated with multiple relations, making it more challenging than sentence-level

© The Author(s), under exclusive license to Springer Nature Singapore Pte Ltd. 2024
B. Luo et al. (Eds.): ICONIP 2023, CCIS 1967, pp. 427–440, 2024.
https://doi.org/10.1007/978-981-99-8178-6_33

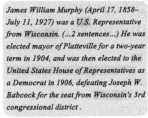

James William Murphy (April 17, 1858–July 11, 1927) was a U.S. Representative from Wisconsin. (...2 sentences...) He was elected mayor of Platteville for a two-year term in 1904, and was then elected to the United States House of Representatives as a Democrat in 1906, defeating Joseph W. Babcock for the seat from Wisconsin's 3rd congressional district.

Subject: Wisconsin
Object: U.S.
Relation: country;
    located in the administrative territorial entity

Subject: the United States House of Representatives
Object: U.S.
Relation: country;
    applies to jurisdiction

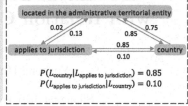

$P(L_{country}|L_{applies\ to\ jurisdiction}) = 0.85$
$P(L_{applies\ to\ jurisdiction}|L_{country}) = 0.10$

(a) A document with entities.   (b) Multi-label entity pairs.   (c) Conditional probability between relations.

Fig. 1. Examples of relation correlation for multi-relation extraction. (a) presents a document containing multiple entities. (b) illustrates the multi-relation entity pairs. For instance, the subject entity *Wisconsin* and the object entity *U.S.* express the *country* and *located in the administrative territorial entity* relations. (c) demonstrates the conditional probabilities between three relations, which are derived from the DocRED dataset.

RE. Figure 1(b) illustrates multi-relation examples extracted from the DocRED dataset [27], where each entity pair is associated with two distinct relations. Moreover, in document-level RE, the number of relation types to be classified can be large (e.g., 97 in the DocRED dataset), further increasing the difficulty of extracting multiple relations.

To address this challenge, previous studies have commonly approached it as a multi-label classification problem, where each relation is treated as a label. Binary cross-entropy loss is typically employed to handle this multi-label scenario [16,31]. During inference, a global threshold is applied to determine the relations. More recently, [14] utilize the asymmetric loss (ASL) [1] to mitigate the imbalance between positive and negative classes. Additionally, [32] propose to employ a balanced softmax method to mitigate the imbalanced relation distribution, where many entity pairs have no relation. [38] introduce the adaptive thresholding technique, which replaces the global threshold with a learnable threshold class. However, previous studies have rarely quantitatively analyzed the co-occurred relations and have not explicitly utilized this feature.

According to the statistics in DocRED dataset, we find that relations co-occur with priors. As illustrated in Fig. 1(c), for entity pairs with multiple relations, the conditional probability of relation *country* appears given that relation *applies to jurisdiction* appears is 0.85, while the conditional probability of relation *applies to jurisdiction* appears given that relation *country* appears is 0.10. Besides, with great chance, relation *country* and relation *located in the administrative territorial entity* appear together. Considering the relations exhibit combinatorial characteristics, it is desirable to employ the relation correlations to ameliorate the model structure and boost the multi-relation extraction.

In this paper, we aim to tackle the challenge of multi-relation extraction in document-level RE by leveraging the correlation characteristics among relations. Specifically, we propose a relation graph method that leverages the prior knowledge of interdependency between relations to effectively guide the extraction of multiple relations. To model the relation correlations, we estimate it by calculating the frequency of relation co-occurrences in the training set [3,24]. To avoid overfitting, we filter out noisy edges below a certain threshold and create a conditional probability matrix by dividing each co-occurrence element by the occurrence numbers of each relation. This matrix is then binarized to enhance the model's generalization capability, and the relation graph is constructed as a binary directed graph. Additionally, we employ a re-weighting scheme to construct an effective relation correlation matrix, which guides the propagation of relation information [5]. We employ Graph Attention Networks (GAT) [21] with the multi-head graph attention to aggregate relation embeddings. Based on the adaptive thresholding technique [38], the loss function in our method is also amended by emphasizing the multi-relation logits. Our method is easy for adoption as it could work as a plug-in for existing models. We conduct extensive experiments on the widely-used document-level RE dataset DocRED, which contains around 7% multi-relation entity pairs. Experimental results demonstrate the effectiveness of our method, achieving superior performance compared to baseline models. In summary, our contributions are as follows:

- We conduct comprehensive quantitative studies on relation correlations in document-level RE, providing insights for addressing the challenge of multi-relation extraction.
- We propose a relation graph method that explicitly leverages relation correlations, offering a plug-in solution for other effective models.
- We evaluate our method on a large-scale DocRE dataset, demonstrating its superior performance compared to baselines.

2 Related Work

Relation extraction, a crucial task in natural language processing, aims to predict the relations between two entities. It has widespread applications, including dialogue generation [9] and question answering [11]. Previous researches largely focus on sentence-level RE, where two entities are within a sentence. Many models have been proposed to tackle the sentence-level RE task, encompassing various blocks such as CNN [19,30], LSTM [2,37], attention mechanism [23,29], GNN [8,39], and transformer [4,26].

Recent researches work on document-level relation extraction since many real-world relations can only be extracted from multiple sentences [27]. From the perspective of techniques, various related approaches could be divided into the graph-based category and the transformer-based category. For the graph-based models that are advantageous to relational reasoning, [16] propose LSR that empowers the relational reasoning across multiple sentences through automatically inducing the latent document-level graph. [31] propose GAIN with two constructed graphs that captures complex interaction among mentions and entities. [35] propose GCGCN to model the complicated semantic interactions among multiple entities. [12] propose to characterize the complex interaction between multiple sentences and the possible relation instances via GEDA networks. [34]

introduce DHG for document-level RE to promote the multi-hop reasoning. For the transformer-based models that are capable of implicitly model long-distance dependencies, [22] discover that using pre-trained language models can improve the performance of this task. [20] propose HIN to make full use of the abundant information from entity level, sentence level and document level. [28] present CorefBERT to capture the coreferential relations in context. Recent works normally directly leverage the pre-trained language models such as BERT [6] or RoBERTa [15] as word embeddings.

3   Methodology

In this section, we provide a detailed explanation of our LAbel Correlation Enhanced (LACE) method, for document-level relation extraction. We begin by formulating the task in Sect. 3.1 and then introduce the overall architecture in Sect. 3.2. In Sect. 3.3, we discuss the encoder module for obtaining the feature vectors of entity pairs. The relation correlation module, outlined in Sect. 3.4, is designed to capture relation correlations. Finally, we present the classification module with multi-relation adaptive thresholding loss for model optimization in Sect. 3.5.

3.1   Task Formulation

Given an input document that consists of N entities $\mathcal{E} = \{e_i\}_{i=1}^{N}$, this task aims to identity a subset of relations from $\mathcal{R} \cup \{NA\}$ for each entity pair (e_s, e_o), where $s, o = 1, ..., N; s \neq o$. The first entity e_s is identified as the *subject* entity and the second entity e_o is identified as the *object* entity. $\mathcal{R}$ is a pre-defined relation type set, and NA denotes no relation expressed for the entity pair. Specifically, an entity e_i can contain multiple mentions with different surface names $e_i^k, k = 1, ..., m$. During testing, the trained model is supposed to predict labels of all the entity pairs $(e_s, e_o)_{s,o=1,...,N;s \neq o}$ within documents.

3.2   Overall Architecture

As illustrated in Fig. 2, the overall architecture consists of three modules. The encoder module first yields the contextual embeddings of all the entity mentions, and then each entity embedding is obtained by integrating information from the corresponding entity mentions, i.e. surface name merging. Afterward, entity pair features are calculated to enhance the entity pair embedding. The relation correlation module generates relation feature vectors. The correlation matrix is built in a data-driven manner, which is based on the statistics of the provided training set. We employ the edge re-weighting scheme to create a weighted adjacency matrix, which is beneficial for deploying graph neural networks. GAT is applied to the correlation matrix and relation features to generate more informative relation feature vectors. In the classification module, a bi-linear layer is utilized for prediction. Besides, based on the adaptive thresholding technique, we resort to a refined loss function for better multi-label classification.

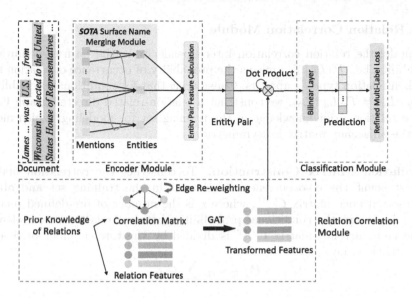

Fig. 2. Architecture of our LACE method, consisting of three modules: the encoder module, relation correlation module, and the classification module.

3.3 Encoder Module

For a document consisting of l words $[x_t]_{t=1}^l$ and N entities (each entity containing several mentions), we obtain the word embedding x_i^w and entity type embedding x_i^t. Then we concatenate them and feed them to BiLSTM layers to generate the contextualized input representations h_i:

$$h_i = \text{BiLSTM}([x_i^w; x_i^t]). \tag{1}$$

Mention representations m_i are obtained by conducting a max-pooling operation on the words, and entity representations E_i are generated by the log-sum-exp pooling over all the entity mention representations m_i:

$$E_i = \log \sum_{i=1}^j \exp(m_i), \tag{2}$$

where j is the number of entity mentions.

In this way, we can generate the embeddings of the head entity and tail entity, denoted as $E_s \in \mathbb{R}^{d_B}$ and $E_o \in \mathbb{R}^{d_B}$, respectively. The entity pair features are obtained by concatenating these embeddings.

Many studies fall in this module, which focus on generating more contextual representations by leveraging Transformer [38] or document graphs [1]. These studies can be seamlessly integrated into our LACE method, or conversely, LACE can be incorporated as a plug-in for other models.

3.4   Relation Correlation Module

We model the relation correlation interdependency in the form of conditional probability, i.e., $P(L_b \mid L_a)$ means the probability of occurrence of relation type L_b when relation type L_a appears. Considering that for conditional probabilities $P(L_b \mid L_a) \neq P(L_a \mid L_b)$, we construct a relation-related directed graph based on the relation prior knowledge of the training set for modeling, which means that the adjacency matrix is asymmetric.

Correlation Matrix Construction. To construct the correlation matrix, we first count the co-occurrence of relations in the training set and obtain the co-occurrence matrix $C^{r \times r}$, where r is the number of pre-defined relation types. To obtain the conditional probabilities between relations, each element in the co-occurrence matrix $C^{r \times r}$ is divided by the total number of relation co-occurrences, i.e.,

$$P_{ij} = C_{ij} / \sum_j C_{ij}, \tag{3}$$

where $P_{ij} = P(L_j \mid L_i)$ denotes the probability of relation type L_j when relation type L_i appears.

However, the above method for correlation matrix construction may suffer two drawbacks. Firstly, some relations rarely appear together with others. This will lead to a large probability value for the co-occurred relation, which is unreasonable. Secondly, there may be a deviation between the statistics of the training dataset and the statistics of the test dataset. Using the exact numbers tend to overfit the training dataset, which might hurt the generalization capacity. Therefore, to alleviate these issues, we set a threshold τ to filter these rare co-occurred relations. Then we binarize the conditional probability matrix P by

$$B_{ij} = \begin{cases} 0, & \text{if } P_{ij} < \delta \\ 1, & \text{if } P_{ij} \geq \delta \end{cases}, \tag{4}$$

where B is the binarized correlation matrix. δ is the conditional probability threshold. Besides, We add the self-loop by setting $B_{ii} = 1, i = 1, ..., r$.

Edge Re-weighting Scheme. One concern for utilizing the binary correlation matrix B for graph neural networks is the over-smoothing issue [36] that the node attribute vectors tend to converge to similar values. There is no natural weight difference between the relation features and its neighbor nodes'. To mitigate this issue, we employ the following re-weighting scheme,

$$R_{ij} = \begin{cases} p / \sum_{\substack{j=1 \\ i \neq j}}^{r} B_{ij}, & \text{if } i \neq j \\ 1 - p, & \text{if } i = j \end{cases}, \tag{5}$$

where R is the re-weighted relation correlation matrix and p is a hyperparameter. In this way, the fixed weights for the relation feature and its neighbors will be applied during training, which is beneficial for alleviating this issue.

Relation features are the embedding vectors obtained in the same way as word embeddings. We then exploit GAT networks with a K-head attention mechanism to aggregate relation features for two reasons. First, GAT is suitable for directed graphs. Second, GAT maintains a stronger representation ability since the weights of each node can be different. The transformed features by GAT are denoted as $R \in \mathbb{R}^{r \times d_B}$.

3.5 Classification Module

Given feature vectors $E_s, E_o \in \mathbb{R}^{d_B}$ of the entity pair (e_s, e_o) and the transformed relation features $R \in \mathbb{R}^{r \times d_B}$, we map them to hidden representations $I_s, I_o \in \mathbb{R}^r$ followed by the layer normalization operation,

$$I_s = \text{LayerNorm} \left(R \cdot E_s \right), \tag{6}$$

$$I_o = \text{LayerNorm} \left(R \cdot E_o \right). \tag{7}$$

Then, we obtain the prediction probability of the relation r' via a bilinear layer,

$$P_{r'} = \sigma \left((E_s \oplus I_s)^\top W_r (E_o \oplus I_o) + b_r \right), \tag{8}$$

where σ is the sigmoid activation function. $W_r \in \mathbb{R}^{(d_B+r) \times (d_B+r)}$, $b_r \in \mathbb{R}$ are model parameters, and $\oplus$ denotes the concatenation operation.

Previous study [38] has shown the effectiveness of the Adaptive Thresholding loss (AT loss), where a threshold class is set such that logits of the positive classes are greater than the threshold class while the logits of the negative classes are less than the threshold class. However, their designed loss function does not quite match the multi-label problem, since they implicitly use the softmax function in the calculation of the positive-class loss function. Therefore, during each loss calculation, the AT loss is unable to extract multiple relations. The superposition of multiple calculations would result in a significant increase in time overhead. To mitigate this issue, we propose a novel loss function called Multi-relation Adaptive Thresholding loss (MAT loss), which is defined as follows,

$$\mathcal{L}_+ = -\log(1 - P(\text{TH})) - \sum_{r' \in L_p} \left(y^{r'} \log P(r') + \left(1 - y^{r'}\right) \log(1 - P(r')) \right), \tag{9}$$

$$\mathcal{L}_- = -\log \left(\frac{\exp\left(\text{logit}_{\text{TH}}\right)}{\sum_{r' \in L_o \cup \{\text{TH}\}} \exp\left(\text{logit}_{r'}\right)} \right), \tag{10}$$

where the threshold class TH is the NA class. L_p and L_o denote the relations the exist and do not exist between the entity pair, respectively. logit means the number without σ in Eq. 8. The final loss function is $\mathcal{L} = \alpha \mathcal{L}_+ + (1 - \alpha)\mathcal{L}_-$, where α is a hyper-parameter. In this way, our MAT loss enables the extraction of multiple relations.

During inference, we assign labels to entity pairs whose prediction probabilities meet the following criteria,

$$P\left(r' \mid e_s, e_o\right) \geq (1 + \theta) \, P(\text{TH}), \tag{11}$$

where θ is a hyper-parameter that maximizes evaluation metrics.

4 Experiments

4.1 Dataset

We evaluate our proposed approach on a large-scale human-annotated dataset for document-level relation extraction DocRED [27], which is constructed from Wikipedia articles. DocRED is larger than other existing counterpart datasets in aspects of the number of documents, relation types, and relation facts. Specifically, DocRED contains 3053 documents for the training set, 1000 documents for the development set, and 1000 documents for the test set, with 96 relation types and 56354 relational facts. For entity pairs with relations, around 7% of them express more than one relation type, and an entity pair can express up to 4 relations. [1]

4.2 Implementation Details

We employ GloVe [18] and BERT-based-cased [6] word embeddings in the encoder module, respectively. When employing GloVe word embeddings, we use Adam optimizer with learning rate being e^{-3}. When employing BERT-based-cased, we use AdamW with a linear warmup for the first 6% steps. The learning rate for BERT parameters is $5e^{-5}$ and e^{-4} for other layers. In the relation correlation module, we set the threshold τ to be 10 for filtering noisy co-occurred relations, and δ is set to be 0.05 in Eq. 4. We set p to be 0.3 in Eq. 5 and θ to be 0.85 in Eq. 11. We employ 2-layer GAT networks with $k = 2$ attention heads computing 500 hidden features per head. We utilize the exponential linear unit (ELU) as the activation function between GAT layers. α in the classification module is 0.4. All hyper-parameters are tuned on the development set.

4.3 Baseline Systems

We compare our approach with the following models, including three categories.

GloVe-based Models. These models report results using GloVe word embeddings and utilize various neural network architectures including CNN, BiLSTM, and Context-Aware [27], to encode the entire document, and then obtain the embeddings of entity pairs for relation classification. The recent mention-based-reasoning model MRN [1] also present the results with GloVe word embedding.

Transformer-based Models. These models directly exploit the pre-trained language model BERT for document encoding without document graph construction, including HIN-BERT [20], CorefBERT [28], and ATLOP-BERT [38]. We mainly compare our method LACE with ATLOP model that aims to mitigate the multi-relation problem.

---

[1] We conduct no experiments on the CDR [13] and GDA [25] datasets in the biomedical domain, because they do not suffer the multi-relation issue. Therefore, they do not match our scenario.

Graph-based Models. Homogeneous or heterogeneous graphs are constructed based on the document features for reasoning. Then, various graph-based models are leveraged to perform inference on entity pairs, including BiLSTM-AGGCN [8], LSR-BERT [16], GAIN-BERT [31]. The MRN-BERT [1] aims to capture the local and global interactions via multi-hop mention-level reasoning.

When compared to GloVe-based models and graph-based models, we integrate the MRL layer from MRN into the encoder module. When compared to Transformer-based models, we incorporate the localized context pooling technique from ATLOP into the encoder module.

Table 1. Results on the development set and test set of DocRED.

Model	Dev		Test	
	Ign F_1	F_1	Ign F_1	F_1
With GloVe				
CNN [27]	41.58	43.45	40.33	42.26
BiLSTM [27]	48.87	50.94	48.78	51.06
Context-Aware [27]	48.94	51.09	48.40	50.70
MRN [14]	56.62	58.59	56.19	58.46
LACE	**57.01**	**58.92**	**56.61**	**58.64**
With BERT+Transformer				
HIN-BERT [20]	54.29	56.31	53.70	55.60
CorefBERT [28]	55.32	57.51	54.54	56.96
ATLOP-BERT [38]	59.22	61.09	59.31	61.30
LACE-BERT	**59.58**	**61.43**	**59.40**	**61.50**
With BERT+Graph				
BiLSTM-AGGCN [8]	46.29	52.47	48.89	51.45
LSR-BERT [16]	52.43	59.00	56.97	59.05
GAIN-BERT [31]	59.14	61.22	59.00	61.24
MRN-BERT [14]	59.74	61.61	59.52	61.74
LACE-MRL-BERT	**59.98**	**61.75**	**59.85**	**61.90**

4.4   Quantitative Results

Table 1 shows the experimental results on the DocRED dataset. Following previous studies [27,38], we adopt the Ign F_1 and F_1 as the evaluation metrics, where Ign F1 is calculated by excluding the shared relation facts between the training set and development/test set.

For the GloVe-based models, our method LACE achieves 56.61% Ign F_1 and 58.64% F_1-score on the test set, outperforming all other methods. For the transformer-based models using BERT, our method LACE-BERT achieves 61.50% F1-score on the test set, which outperforms the ATLOP-BERT model. These experimental results also show that the pre-trained language model can cooperate well with the LACE method. For the graph-based models, we achieve 61.90% F_1-score on the test set. The result demonstrates that capturing the mention-level contextual information is helpful and our proposed method could work well with the mention-based reasoning method. Overall, results demonstrate the effectiveness of leveraging the relation information.

4.5   Analysis of Relation Correlation Module

We investigate the effect of key components in the relation correlation module.

Matrix Construction Threshold. As shown in Table 2, we analyze the effect of probability filtering threshold δ in Eq. 4. We obtain the highest F_1 score when δ equals 0.05 for all experiments. Besides, results indicates that $\delta = 0.03$ will lead to more performance degradation compared with $\delta = 0.07$. We believe that this is due to the smaller threshold value resulting in more noise edges.

Table 2. F1-score on the development set when tuning the probability filtering threshold δ.

Model	3%	5%	7%
LACE	58.74	**58.92**	58.82
LACE-BERT	61.38	**61.43**	61.40
LACE-MRL-BERT	61.67	**61.75**	61.71

Table 3. F1-score on the development set with different GAT layers. L denotes layer.

Model	1-L	2-L	3-L
LACE	58.80	**58.92**	58.64
LACE-BERT	61.40	**61.43**	61.23
LACE-MRL-BERT	61.69	**61.75**	61.63

GAT layer. We report the results of different GAT layers with two heads in Table 3. Results demonstrates that 1-layer and 2-layer GAT networks achieves relatively similar results, while 3-layer GAT networks leads to greater performance degradation. The probable reason for the performance degradation might be the over-smoothing issue, that is, the node feature vectors are inclined to converge to comparable values.

Table 4. F1-score for multi-relation extraction on the development set. *Rel* denotes relations.

Model	2-*Rel*	3-*Rel*	Overall
ATLOP-BERT	40.13	29.59	39.62
LACE-BERT	**42.03**	**32.58**	**41.55**

4.6 Performance on Multi-Label Extraction

In order to evaluate the performance of multi-label extraction, we re-implement ATLOP-BERT model and report the experimental results of multi-relation extraction as shown in Table 4. As seen, our approach LACE-BERT gains 1.9% and 2.99% F1-score improvement on the 2-relation and 3-relation extraction, respectively, which demonstrates the effectiveness of leveraging the relation correlations. Overall, our approach achieves 1.93% F1-score improvements on multi-label extraction compared with ATLOP-BERT.

Table 5. F1-score on the development set for ablation study. RCM denotes the relation correlation module.

Model	2-*Relation*	3-*Relation*	Overall
LACE-BERT	**42.03**	**32.58**	**41.55**
- RCM	40.74	30.62	40.25
- $\mathcal{L}_{MAT}$	41.43	31.78	40.94

4.7 Ablation Study

We conduct ablation studies to verify the necessity of two critical modules in LACE-BERT for multi-relation extraction as depicted in Table 5. Results show that two modules contribute to the final improvements. Firstly, removing the relation correlation module causes more performance degradation, and we thus believe that leveraging the prior knowledge of relation interdependency is helpful for the multi-relation extraction. Secondly, we replace our multi-relation adaptive thresholding loss $\mathcal{L}_{MAT}$ with the adaptive thresholding loss [38] for comparison. We believe that the reason for the improvement is that the MAT loss enlarges the margin values between all the positive classes and the threshold class.

4.8 Case Study

Figure 3 shows a case study of our proposed approach LACE-BERT, in comparison with ATLOP-BERT baseline. We can observe that ATLOP-BERT can only identify the *P17* and *P131* relations for the entity pair *(Ontario, Canada)*, where

[S1] *Alfred and Plantagenet is a township in eastern Ontario, Canada, in the United Counties of Prescott and Russell.*
[S2] *Located approximately from downtown Ottawa at the confluence of the Ottawa River and the South Nation River.*

ATLOP:
Subject: *Ontario* Object: *Canada*
Relation: P17: country
           P131: located in the admini-
           strative territorial entity

LACE:
Subject: *Ontario* Object: *Canada*
Relation: P17: country
           P131: located in the admini-
           strative territorial entity
           P205: basin country

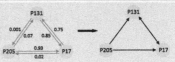

Fig. 3. Case study of a triple-relation entity pair from the development set of DocRED. We visualize the conditional probabilities among these relations and exhibit the constructed directed sub-graph.

the two relations frequently appear together. However, ATLOP-BERT fails to identity the *P205* relation, while LACE-BERT deduces this relation. By introducing the label correlation matrix, this relation *P205* establishes connections with other relations with high conditional probabilities, which is advantageous for multi-relation extraction.

5   Conclusion

In this work, we propose our method LACE for document-level relation extraction. LACE includes a relation graph construction approach which explicitly leverages the statistical co-occurrence information of relations. Our method effectively captures the interdependency among relations, resulting in improved performance on multi-relation extraction. Experimental results demonstrate the superior performance of our proposed approach on a large-scale document-level relation extraction dataset.

Acknowledgements. The authors would like to thank the support from the National Natural Science Foundation of China (NSFC) grant (No. 62106143), and Shanghai Pujiang Program (No. 21PJ1405700).

References

1. Baruch, E.B., et al.: Asymmetric loss for multi-label classification. CoRR (2020)
2. Cai, R., Zhang, X., Wang, H.: Bidirectional recurrent convolutional neural network for relation classification. In: Proceedings of ACL (2016)
3. Che, X., Chen, D., Mi, J.: Label correlation in multi-label classification using local attribute reductions with fuzzy rough sets. In: FSS (2022)
4. Chen, M., Lan, G., Du, F., Lobanov, V.S.: Joint learning with pre-trained transformer on named entity recognition and relation extraction tasks for clinical analytics. In: ClinicalNLP@EMNLP 2020, Online, November 19, 2020 (2020)
5. Chen, Z., Wei, X., Wang, P., Guo, Y.: Multi-label image recognition with graph convolutional networks. In: CVPR (2019)
6. Devlin, J., Chang, M., Lee, K., Toutanova, K.: BERT: pre-training of deep bidirectional transformers for language understanding. In: Proceedings of ACL (2019)

7. Feng, J., Huang, M., Zhao, L., Yang, Y., Zhu, X.: Reinforcement learning for relation classification from noisy data. In: Proceedings of AAAI (2018)
8. Guo, Z., Zhang, Y., Lu, W.: Attention guided graph convolutional networks for relation extraction. In: Proceedings of ACL (2019)
9. He, H., Balakrishnan, A., Eric, M., Liang, P.: Learning symmetric collaborative dialogue agents with dynamic knowledge graph embeddings. In: Proceedings of ACL (2017)
10. Hendrickx, I., et al.: Semeval-2010 task 8. In: SEW@NAACL-HLT 2009, Boulder, CO, USA, June 4, 2009 (2009)
11. Hixon, B., Clark, P., Hajishirzi, H.: Learning knowledge graphs for question answering through conversational dialog. In: ACL (2015)
12. Li, B., Ye, W., Sheng, Z., Xie, R., Xi, X., Zhang, S.: Graph enhanced dual attention network for document-level relation extraction. In: Proceedings of COLING (2020)
13. Li, J., et al.: Biocreative V CDR task corpus: a resource for chemical disease relation extraction. Database J. Biol. Databases Curation 2016 (2016)
14. Li, J., Xu, K., Li, F., Fei, H., Ren, Y., Ji, D.: MRN: a locally and globally mention-based reasoning network for document-level relation extraction. In: Proceedings of ACL (2021)
15. Liu, Y., et al.: Roberta: a robustly optimized BERT pretraining approach. CoRR (2019)
16. Nan, G., Guo, Z., Sekulic, I., Lu, W.: Reasoning with latent structure refinement for document-level relation extraction. In: ACL (2020)
17. Peng, N., Poon, H., Quirk, C., Toutanova, K., Yih, W.: Cross-sentence N-ary relation extraction with graph LSTMs. In: TACL (2017)
18. Pennington, J., Socher, R., Manning, C.D.: Glove: Global vectors for word representation. In: EMNLP, pp. 1532–1543. ACL (2014)
19. dos Santos, C.N., Xiang, B., Zhou, B.: Classifying relations by ranking with convolutional neural networks. In: ACL (2015)
20. Tang, H., et al.: HIN: hierarchical inference network for document-level relation extraction. In: Proceedings of KDD (2020)
21. Velickovic, P., Cucurull, G., Casanova, A., Romero, A., Liò, P., Bengio, Y.: Graph attention networks. In: ICLR (2018)
22. Wang, H., Focke, C., Sylvester, R., Mishra, N., Wang, W.Y.: Fine-tune BERT for docred with two-step process. CoRR (2019)
23. Wang, L., Cao, Z., de Melo, G., Liu, Z.: Relation classification via multi-level attention CNNs. In: Proceedings of ACL (2016)
24. Wang, Y., et al.: Multi-label classification with label graph superimposing. In: Proceedings of AAAI (2020)
25. Wu, Y., Luo, R., Leung, H.C.M., Ting, H., Lam, T.W.: RENET: a deep learning approach for extracting gene-disease associations from literature. In: RECOMB 2019, Washington, DC, USA, May 5–8, 2019, Proceedings (2019)
26. Xiao, Y., Tan, C., Fan, Z., Xu, Q., Zhu, W.: Joint entity and relation extraction with a hybrid transformer and reinforcement learning based model. In: Proceedings of AAAI (2020)
27. Yao, Y., et al.: Docred: a large-scale document-level relation extraction dataset. In: ACL (2019)
28. Ye, D., et al.: Coreferential reasoning learning for language representation. In: Proceedings of EMNLP (2020)
29. Ye, Z., Ling, Z.: Distant supervision relation extraction with intra-bag and inter-bag attentions. In: Proceedings of ACL (2019)

30. Zeng, D., Liu, K., Lai, S., Zhou, G., Zhao, J.: Relation classification via convolutional deep neural network. In: Proceedings of COLING (2014)
31. Zeng, S., Xu, R., Chang, B., Li, L.: Double graph based reasoning for document-level relation extraction. In: EMNLP (2020)
32. Zhang, N., et al.: Document-level relation extraction as semantic segmentation. In: Proceedings of IJCAI (2021)
33. Zhang, Y., Zhong, V., Chen, D., Angeli, G., Manning, C.D.: Position-aware attention and supervised data improve slot filling. In: Proceedings of EMNLP (2017)
34. Zhang, Z., et al.: Document-level relation extraction with dual-tier heterogeneous graph. In: Proceedings of COLING (2020)
35. Zhou, H., Xu, Y., Yao, W., Liu, Z., Lang, C., Jiang, H.: Global context-enhanced graph convolutional networks for document-level relation extraction. In: COLING (2020)
36. Zhou, J., et al.: Graph neural networks: a review of methods and applications. AI Open (2020)
37. Zhou, P., et al.: Attention-based bidirectional long short-term memory networks for relation classification. In: Proceedings of ACL (2016)
38. Zhou, W., Huang, K., Ma, T., Huang, J.: Document-level relation extraction with adaptive thresholding and localized context pooling. In: Proceedings of AAAI (2021)
39. Zhu, H., Lin, Y., Liu, Z., Fu, J., Chua, T., Sun, M.: Graph neural networks with generated parameters for relation extraction. In: Proceedings of ACL (2019)

Multi-scale Directed Graph Convolution Neural Network for Node Classification Task

Fengming Li[1], Dong Xu[1,2], Fangwei Liu[1], Yulong Meng[1,2(✉)], and Xinyu Liu[3]

[1] College of Computer Science and Technology, Harbin Engineering University,
Harbin, China
mengyulong@hrbeu.edu.cn
[2] Modeling and Emulation in E-Government National Engineering Laboratory,
Harbin Engineering University, Harbin, China
[3] Jiangsu JARI Technology Group Co. Ltd, Lianyungang, China

Abstract. The existence of problems and objects in the real world which can be naturally modeled by complex graph structure has motivated researchers to combine deep learning techniques with graph theory. Despite the proposal of various spectral-based graph neural networks (GNNs), they still have shortcomings in dealing with directed graph-structured data and aggregating neighborhood information of nodes at larger scales. In this paper, we first improve the Lanczos algorithm by orthogonality checking method and Modified Gram-Schmidt orthogonalization technique. Then, we build a long-scale convolution filter based on the improved Lanczos algorithm and combine it with a short-scale filter based on Chebyshev polynomial truncation to construct a multi-scale directed graph convolution neural network (MSDGCNN) which can aggregate multi-scale neighborhood information of directed graph nodes in larger scales. We validate our improved Lanczos algorithm on the atom classification task of the QM8 quantum chemistry dataset. We also apply the MSDGCNN on various real-world directed graph datasets (including WebKB, Citeseer, Telegram and Cora-ML) for node classification task. The result shows that our improved Lanczos algorithm has much better stability, and the MSDGCNN outperforms other state-of-the-art GNNs on such task of real-world datasets.

Keywords: Graph neural network · Lanczos algorithm · Directed graph · Node classification

1 Introduction

Traditional deep learning techniques based on artificial neurons such as RNN [36] and CNN [22] have achieved great success in classification tasks [15] and recognition tasks [17] in euclidean space data such as images and texts. The success of the deep learning techniques is that they effectively leverage the statistical properties of Euclidean space data such as permutation invariance [46].

B. Luo et al. (Eds.): ICONIP 2023, CCIS 1967, pp. 441–456, 2024.
https://doi.org/10.1007/978-981-99-8178-6_34

However, in the real world, graph-structured data is ubiquitous, and a large number of problems and objects need to be modeled based on complex graph structures. Unfortunately, graph-structured data do not belong to the Euclidean space, which means the data do not possess the aforementioned statistical properties. Thus, applying deep learning techniques on graph-structured data faces great challenges.

In recent years, many researchers have applied deep learning techniques to process graph-structured data in non-Euclidean space and have achieved success in various applications such as recommendation systems [18], program analysis [27], software mining [24], drug discovery [3] and anomaly detection [48]. As mentioned in [45], there are four kinds of graph neural networks currently available, including recurrent graph neural networks, graph convolution neural networks, graph autoencoders and spatio-temporal graph neural networks. Gated graph neural networks [27] aims to learn node embeddings through the construction of a recurrent neural network with gated units. Graph autoencoders [20] encodes nodes or entire graph into a latent vector through an encoder and then decodes the latent vector by a decoder for node-level or graph-level learning tasks. Spatio-temporal graph neural networks [26] aims to learn the hidden patterns of a graph from spatio-temporal graphs. Graph convolution neural networks generalizes the convolution operation from grid data to graph data, aggregating features of nodes and their neighbors through the convolution operation to generate node embeddings. More details about graph neural networks can be found in [45,51].

In spectral-based graph neural networks, there exist two main issues. Firstly, spatial-based graph neural networks such as GAT [42] can naturally be implemented for directed graph data by symmetrizing the adjacency matrix and treating it as an undirected graph. However, spectral-based graph neural networks that involves Fourier transformation require the adjacency matrix of a graph to be symmetric to ensure the matrix's eigenvalues are real. Since the adjacency matrix of a directed graph often does not have symmetry, spectral-based graph neural networks can not be directly applied to directed graphs. Secondly, how to obtain the multi-scale information of graphs by convolution operations has not been effectively explored. Due to the involvement of large sparse matrix eigen-decomposition and multiplication in spectral-based graph convolution theory, performing multi-scale convolution operations is itself a computational challenge.

Recently, many researchers ([6,12,16,21], etc.) have used the Magnetic Laplacian Matrix [29] from the field of particle physics to model directed graph data, which involves introducing a phase parameter q to control the strength of the directional flow in a graph and converting the adjacency matrix of the graph to a Magnetic Laplacian Matrix. This approach transforms the real-valued asymmetric adjacency matrix of a directed graph into a complex Hermitian matrix, which partially overcomes the difficulty of processing directed graphs directly with spectral-based graph neural networks. Compared with DGCN [41] and Diag-Graph [40], which are specifically used for directed graphs, the method reduces the complexity of neural networks and makes spectral-based directed graph convolution neural networks can be easily trained. However, in the research aforementioned, the filter used for the convolution operation faces extremely high computational overhead in the problem of expanding the convolution scale.

Most of the current spectral-based graph neural networks use the K_{th} order truncation of Chebyshev polynomials to obtain filtering operators, which avoids directly performing eigendecomposition on the Laplacian matrix of a graph and reduces computational complexity to some extent. However, there is still a high computational cost when using the method to aggregate larger-scale information about neighbors of a node. The Lanczos algorithm [23], which is commonly used in quantum systems, has been applied to compute the eigendecomposition of large-scale sparse graphs [39]. In their experiments, the algorithm exhibits lower error and time cost than the Chebyshev polynomial truncation method. Subsequently, AdaLanczosNet [28] was proposed based on the Lanczos algorithm. Experimental results showed that the spectral-based graph neural network with the algorithm has lower training and testing errors as well as better computational efficiency than other spectral-based GNNs with the Chebyshev polynomial truncation method. However, although the aforementioned studies considered the problem of loss of orthogonality of *Krylov* subspace vectors due to rounding errors during the iterative process of the Lanczos algorithm, their methods did not effectively correct the orthogonal vectors. Even worse, they lowered the efficiency of the Lanczos algorithm since executing re-orthogonalization at each iteration. In this paper, we aims to addressed these problems and the main contributions are listed here:

* We improve Lanczos algorithm using Modify Gram-Schmdit orthogonalization technique and the orthogonality checking method.
* We model directed graphs using the Magnetic Laplacian matrix and build a multi-scale graph convolution neural network(MSDGCNN) for node classification tasks combining ChebyShev polynomials and our improved Lanczos algorithm.
* We evaluate the performance of our improved Lanczos algorithm on the QM8 quantum chemistry dataset and validate the effectiveness of MSDGCNN on the WebKB, Telegram, CiteSeer and Cora-ML datasets.

2   Related Work

2.1   Spectral Graph Convolution Theory

Given an undirected graph $G = (V, E, X)$, where V denotes the set of nodes of G, $|V| = N$; E denotes the set of edges, $E = \{e_{ij}|\ if\ \exists\ v_i \leftrightarrow v_j;\ i, j = 1, ..., N\} \subseteq V \times V$; W is the set of weights of the edges, $W = \{w_{ij}|\ if\ e_{ij} \in E\}$ and $W \in \mathcal{R}^{N \times N}$, if G is an unweighted graph then $w_{ij} \in \{0, 1\}$; X denotes the set of node features, $X = \{x_i|\ i = 1, ..., N\}$; A denotes the adjacency matrix of G and we have:

$$A_{ij} = \begin{cases} w_{ij} & \text{if } \exists\ v_i \leftrightarrow v_j \in E \\ 0 & \text{other cases} \end{cases}$$

According to the spectral graph theory [4], define the unnormalized Laplacian matrix of G: $L_U = D - A \in \mathcal{R}^{N \times N}$, where $D \in \mathcal{R}^{N \times N}$ is the degree diagonal matrix of G and $D_{ii} = \sum_{j=1}^{N} A_{ij}$. Then the normalized Laplacian matrix of G is defined as $L_N = I_N - D^{-\frac{1}{2}}AD^{-\frac{1}{2}} \in \mathcal{R}^{N \times N}$ where I_N is the identity matrix of order N. Clearly, $L_N = L_N^T$ (L_N^T denotes the transpose of L_N), thus, L_N has non-negative real eigenvalues $\lambda_1, ..., \lambda_N$ and the orthogonal eigenvectors $u_1, ..., u_N$ corresponding to the eigenvalues. Let $U = [u_1|...|u_N] \in \mathcal{R}^{N \times N}, \Lambda = diag(\lambda_1, ..., \lambda_N)$, then we have $L_N = U\Lambda U^T$, where $U = [u_1|...|u_N]$ is called the Fourier orthogonal basis. According to [37], the graph Fourier transform of the feature signal $x_i \in \mathcal{R}^N$ of node $v_i(i = 1, ..., N)$ is $\hat{x}_i = U^T x_i \in \mathcal{R}^N$ and the inverse transform of the graph Fourier transform is $x_i = U\hat{x}_i$. Thus, the convolution operation on the feature signal of node v_i of the graph G in spectral domain is defined as

$$y * x_i = U\left((U^T y) \odot (U^T x_i)\right)$$
$$= U(\hat{y} \odot (U^T x_i)) \tag{1}$$
$$= U diag(\hat{y}) U^T x_i$$

where $y \in \mathcal{R}^N$, $\hat{y}$ is called the Fourier coefficient and $\odot$ denotes Hadamard product. Let $g_\theta(\Lambda) = diag(\hat{y})$, according to Eq. (1) the feature signal x_i of node v_i is filtered by $g_\theta(\Lambda) = diag(\hat{y})$. Also, in practice, $g_\theta(\Lambda)$ can be regarded as a function of the diagonal matrix Λ of eigenvalues of the Laplacian matrix L_N of G [43].

2.2   Chebyshev Polynomials Approximate

Due to the involvement of large sparse matrix multiplication and eigendecomposition, computing $g_\theta(\Lambda)$ is computationally expensive. [14] approximated the Fourier filter $g_\theta(\Lambda)$ by truncating the Chebyshev polynomials at order K. Then, [5] used this method to construct ChebNet, which avoids the huge cost of eigendecomposition of the Laplacian matrix of graph G and improves stability in the face of perturbations [25].

Let $\hat{\Lambda} = \frac{1}{\lambda_{max}}\Lambda - I_N$ be the normalized Laplacian eigen matrix, then we have the filter which is based on truncating the Chebyshev polynomials at order K:

$$g_\theta(\hat{\Lambda}) = \sum_{m=0}^{K} \theta_m T_m(\hat{\Lambda}) \tag{2}$$

where the K_{th} order Chebyshev polynomial T_K is recursively defined as $T_0(\hat{\Lambda}) = I_N$, $T_1(\hat{\Lambda}) = \hat{\Lambda}$, $T_K(\hat{\Lambda}) = 2\hat{\Lambda}T_{K-1}(\hat{\Lambda}) + T_{K-2}(\hat{\Lambda})$. Then the feature h_i obtained by the convolution operation on the feature x_i of the node v_i of graph G is

$$h_i = \sum_{m=0}^{K} U\theta_m T_m(\hat{\Lambda})U^T x_i$$
$$= \sum_{m=0}^{K} \theta_m T_m(\hat{L})x_i \tag{3}$$

3   Method

In this section, we will provide a detailed description of the construction details of our proposed MSDGCNN which can aggregate larger scale information of a node of G with lower computational overhead by using our improved Lanczos algorithm. Firstly, we magnetize the Laplacian Matrix of the graph G to obtain the Magnetic Laplacian Matrix. Then, we utilize our improved Lanczos algorithm to obtain the low-rank approximation of the eigenvalues and eigenvectors of the Magnetic Laplacian Matrix of graph G. Finally, through the MSDGCNN, we perform convolution operations using short Scale Filters and song Scale Filters, and output the node classification results.

3.1   Problem Formulation

Given a directed graph $G = (V, E, X)$, where V denotes the set of nodes of G, $|V| = N$; E denotes the set of edges, $E = \{e_{ij}| \ if \ \exists \ v_i \to v_j; \ i, j = 1, ..., N\} \subseteq V \times V$; W denotes the set of weights of edges, $W = \{w_{ij}| \ if \ e_{ij} \in E\} \subseteq \mathcal{R}^{N \times N}$, and if G is unweighted then $w_{ij} \in \{0, 1\}$. Let A be the adjacency matrix of G, if there exists $v_i \to v_j$, i.e. $\exists \ e_{ij} \in E$ then $A_{ij} = w_{ij}$, otherwise $A_{ij} = 0$. Let X denotes the set of features of nodes, $X = \{x_i| \ i = 1, ..., N\}$.

3.2   Constructing the Magnetic Laplacian Matrix of G

To improve the stability of the training process, let $\tilde{A} = \frac{1}{2}(A^T + A) + I_N$; $\tilde{D}$ is the degree diagonal matrix of $\tilde{A}$, where $\tilde{D}_{ii} = \sum_{j=1}^N \tilde{A}_{ij}$ and $\tilde{D}_{ij} = 0 \ (i \neq j)$. We have the magnetization process of graph G by follows:

$$\Theta^{(q)} = 2\pi q(A - A^T), \ q \geq 0 \tag{4}$$

$$\begin{aligned} H^{(q)} &= \tilde{A} \odot exp\left(i\Theta^{(q)}\right) \\ &= \tilde{A} \odot \left(\cos(\Theta^{(q)}) + i\sin(\Theta^{(q)})\right) \end{aligned} \tag{5}$$

where $\Theta^{(q)}$ is a skew-symmetric matrix; $H^{(q)}$ is an Hermitian matrix with complex elements; $sin(.)$ and $cos(.)$ are element-wise functions; The phase matrix $\Theta^{(q)}$ is used to capture the directional information of the directed edges of G. For example, $\Theta^{(0)} = 0$ when $q = 0$, $H^{(0)} = \tilde{A}$ degenerates to the adjacency matrix of the undirected graph. When $q \neq 0$, $\Theta^{(q)}$ will be able to capture the directional information of G. If $q = 0.25$ and $\exists \ e_{ij} \in E$ then we have:

$$H_{(i,j)}^{(0.25)} = \pm \frac{iw_{ij}}{2} = -H_{(i,j)}^{(0.25)}$$

In this case, an edge $v_i \to v_j$ will be regarded as an edge opposite to $v_j \to v_i$. Thus the Magnetic Laplacian Matrix of G can be defined as

$$L_U^{(q)} = \tilde{D} - H^{(q)} = \tilde{D} - \tilde{A} \odot exp\left(i\Theta^{(q)}\right) \tag{6}$$

$$L_N^{(q)} = I - (\tilde{D}^{-\frac{1}{2}} \tilde{A} \tilde{D}^{-\frac{1}{2}}) \odot exp(i\Theta^{(q)}) \qquad (7)$$

where $L_U^{(q)}$ denotes the unnormalized Magnetic Laplacian Matrix of G and $L_N^{(q)}$ denotes the normalized Magnetic Laplacian Matrix of G. From Eqs. (6) and (7), we have two theorems as follows:

Theorem 1. *Both $L_U^{(q)}$ and $L_N^{(q)}$ have non-negative real eigenvalues.*

Proof. The proof of the Theorem 1 can be found in [7].

Theorem 2. *For $\forall q \geqslant 0$, the eigenvalues of the normalized Magnetic Laplacian Matrix $L_N^{(q)}$ taking the interval $[0, 2]$.*

Proof. The proof of the Theorem 2 can be found in [50].

The two theorems above, ensure that we can use the K_{th} order Chebyshev polynomial to build short-scale filter.

We follow a similar setup to [50]: let $K = 1$, set $\lambda_{max} \approx 2$ and $\theta_1 = -\theta_0$. Let $\tilde{L} = \frac{2}{\lambda_{max}} L_N^{(q)} - I_N$, then the output of the node feature x_i after filtering is:

$$
\begin{aligned}
h_i &= \sum_{m=0}^{1} \theta_m T_m(\tilde{L}) x_i \\
&= \theta_0 (I_N + (\tilde{D}^{-\frac{1}{2}} \tilde{A} \tilde{D}^{-\frac{1}{2}}) \odot exp(i\Theta^{(q)})) x_i
\end{aligned}
\qquad (8)
$$

3.3   Imporved Lanczos Algorithm

Lanczos algorithm is a kind of *Krylov* subspace iteration method, which aims to find a set of standard orthogonal bases $Q_k = [q_1|...|q_k] \in C^{N \times k}$ (where Q_k^H denotes the conjugate transpose of Q_k and $Q_k^H Q_k = I_k$) and a tridiagonal matrix $T_k \in R^{k \times k}$, thus approximating the eigenvalues and eigenvectors of the Hermitian matrix A. Given the Hermitian matrix $A \in C^{N \times N}$, if $Q_k^H A Q_k = T_k$ is a tridiagonal matrix and $Q_k^H Q_k = I_k$ then:

$$
\begin{aligned}
\mathcal{K}_k(A, q_1, k) &= Q_k Q_k^H K_k(A, q_1, k) \\
&= Q_k [e_1 | T_k e_1 | ... | T_k^{k-1} e_1]
\end{aligned}
$$

is the QR decomposition of $\mathcal{K}_k(A, q_1, k)$. Where e_1 and q_1 are the first columns of the identity matrix I_N and Q_k, respectively. Using the orthogonal matrix whose first column is q_1 to tridiagonalize A can effectively generate the columns of Q_k. Define the tridiagonal matrix $T_k \in \mathcal{R}^{k \times k}$:

$$
T_k = \begin{pmatrix}
\alpha_1 & \beta_1 & & & \\
\beta_1 & \alpha_2 & \beta_2 & & \\
& \ddots & \ddots & \ddots & \\
& & \ddots & \ddots & \beta_{k-1} \\
& & & \beta_{k-1} & \alpha_k
\end{pmatrix}
$$

By transforming $Q_k^H A Q_k = T_k$ to $A Q_k = Q_k T_k$, for $i = 1, ..., k-1$, we have:

$$Aq_i = \beta_{i-1}q_{i-1} + \alpha_i q_i + \beta_i q_{i+1}$$

where $\beta_0 q_0 \equiv 0$, according to the orthogonality of vector q there is: $\alpha_i = q_i^H A q_i$. By shifting the term, the vector r_i is defined as

$$r_i = (A - \alpha_i I_N)q_i - \beta_{i-1}q_{i-1}$$

Then we have $q_{i+1} = \frac{r_i}{\beta_i}$, where $\beta_i = \pm\|r_i\|_2$. If $r_i = 0$ then the iteration stops, at which point useful information about the $Krylov$ invariant subspace has been obtained.

Although the Lanczos algorithm is more computationally efficient than the power method, rounding errors during its iteration process can cause the orthonormal basis $q_1, ..., q_k$ of the $Krylov$ subspace to lose orthogonality [31]. To overcome the problem above, it is necessary to perform reorthogonalization of the vectors $q_1, ..., q_i(i = 1, ..., k)$ at each iteration of the Lanczos algorithm. However, this approach significantly reduces the computational efficiency of the algorithm. Fortunately, numerical experiments and theoretical analyses indicate that the results of the Lanczos algorithm still have high accuracy when the loss of orthogonality of the basis vectors is within an acceptable scope ([11,32,38], etc.). Therefore, we turn to check the orthogonality of the orthonormal basis generated by the Lanczos algorithm at each iteration, and use the Modify Gram-Schmdit orthogonalization technique ([30] suggest that Modify Gram-Schmdit is a more effective method.) to reorthogonalize the basis vectors when the loss of orthogonality exceeds a certain threshold.

Our improved Lanczos algorithm is shown in Algorithm 1. In Step 6, we perform an orthogonality check on the $Krylov$ subspace basis generated by the iteration, where ϵ_M is machine precision and the $max()$ is an element-wise function which finds the maximum value of matrix. In Step 13, we perform a Schur decomposition on the three-term recurrence relation matrix T_k, i.e. $V_k^H T_k V_k = diag(r_1, ..., r_k)$, to obtain the diagonal matrix $R_k \in \mathcal{R}^{k \times k}$ which is consisted of Ritz values. In Step 14, we construct the matrix $Y_k = [y_1|...|y_k]$, where y_i and r_i form a Ritz pair. Finally, the matrix Y_k formed by Ritz vectors and diagonal matrix R_k obtained from k steps of the algorithm can be used to approximate the Hermitian matrix A, i.e. $A \approx Y_k R_k Y_k^H$. The upper bound on the approximation error of the Lanczos algorithm is given by Theorem 3:

Theorem 3. *Let $U \Lambda U^H$ be the Schur decomposition of a Hermitian matrix $S \in \mathcal{C}^{N \times N}$, where $\Lambda = diag(\lambda_1, ..., \lambda_N)$, $\lambda_1 \geq ... \geq \lambda_N$; $U = [u_1, ..., u_N]$. Let $\mathcal{U}_j = span\{u_1, ..., u_j\}$, the initial vector for K-step Lanczos algorithm is q, and it outputs an orthogonal matrix $Q \in \mathcal{C}^{N \times K}$ and a tridiagonal matrix $T \in \mathcal{R}^{K \times K}$. For any $j(1 < j < K < N)$, we have:*

$$\left\| S - QTQ^H \right\|_F^2 \leq \sum_{i=1}^{j} \lambda_i^2 \left(\frac{\sin(q, \mathcal{U}_i) \prod_{k=1}^{j-1} (\lambda_k - \lambda_N)/(\lambda_k - \lambda_j)}{\cos(q, u_i) T_{K-i}(1 + 2\gamma_i)} \right)^2 + \sum_{i=j+1}^{N} \lambda_i^2$$

Algorithm 1: Improved Lanczos Algorithm

Input: $A \in C^{N \times N}, A^H = A; q_1$ is a unit $- 2$ norm vector $\in$
$\quad\quad C^N;$ LanczosStep $\in N_+,$ LanczosStep $\ll N.$
Result: Ritz vector $y_1, .. y_k$ and Ritz value $r_1, .., r_k.$
Data: set $k = LanczosStep, i = 0, \beta_0 = 1$

1 **while** $i \leq k$ and $\beta_i \neq 0$ **do**
2 $\quad$ $q_{i+1} \leftarrow \frac{r_i}{\beta_i};$
3 $\quad$ $i \leftarrow i + 1;$
4 $\quad$ $\alpha_i \leftarrow q_i^H A q_i;$
5 $\quad$ $r_i \leftarrow (A - \alpha_i I)q_i - \beta_{i-1}q_{i-1};$
6 $\quad$ **if** $\max(|I_i - Q_i^H Q_i|) > \sqrt{\epsilon_M}$ **then**
7 $\quad\quad$ | $\quad$ do Modify Gram-Schmdit Process
8 $\quad$ **end**
9 $\quad$ $\beta_i \leftarrow \|r_i\|_2;$
10 **end**
11 $Q_k = [q_1, ..., q_k];$
12 $T_k \leftarrow Tridiagonal[\alpha_1, ..., \alpha_k]$ and $[\beta_1, ..., \beta_k - 1];$
13 $R_k, V_k \leftarrow decompose$ $T_k;$
14 $Y_k \leftarrow Q_k V_k;$
15 **return** $Y_k = [y_1 | ... | y_k], R_k = diag(r_1, ..., r_k);$

---

where $T_{K-i}(x)$ is a Chebyshcv polynomial of order $K - i$, $\gamma_i = \frac{\lambda_i - \lambda_{i+1}}{\lambda_{i+1} - \lambda_N}$. The proof details can be found in [10].

3.4   MSDGCNN Architecture

In this section, we propose a multi-scale directed graph convolution neural network (MSDGCNN) that combines Chebyshev polynomials and the improved Lanczos algorithm. The MSDGCNN uses K-order truncation of Chebyshev polynomials to approximate low-order filters of G during short-scale convolution operations and uses our improved Lanczos algorithm to obtain high-order filters of G during long-scale convolution operations. The network architecture is shown in Fig. 1.

Short-Scale Filter. Let $X^{(0)}_{N \times F_0} = [x_1^{(0)} | ... | x_N^{(0)}]$, $x_i^{(0)} \in \mathcal{R}^{F_0}$ is the matrix of initial node features, where the feature dimensions of each node is F_0. Let $Z^{(0)} = X^{(0)}$, where $Z^{(l)}$ is the output features of the l_{th} convolution layer. The output of the short-scale convolution filter in the $l_{th}(l = 1, ..., L)$ layer can be represented as $h_{short}^{(l)}$. According to Eq. (8), we have:

$$h_{short}^{(l)} = \theta_0(I_N + (\tilde{D}^{-\frac{1}{2}}\tilde{A}\tilde{D}^{-\frac{1}{2}}) \odot exp(i\Theta^{(q)}))Z^{(l-1)} \tag{9}$$

Long-Scale Filter. To obtain the structure information of G in larger scale, we use the low-rank approximation of the Magnetic Laplacian Matrix of G obtained

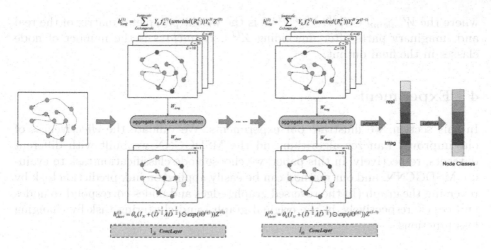

Fig. 1. Network architecture: for example we set $longscale = [10, 20, 30, 40]$ and use short scale filter given by Eq. (8) with K=1.

by the Algorithm 1, i.e. $L_N^{(q)} \approx Y_k R_k Y_k^H$, then the long-scale convolution filter output $h_{long}^{(l)}$ of the l_{th} layer is:

$$h_{long}^{(l)} = \sum_{\mathcal{L} \in longscale} Y_k f_{\mathcal{L}}^{(l)}(unwind(R_k^{\mathcal{L}})) Y_k^H Z^{(l-1)} \qquad (10)$$

where the function $f_{\mathcal{L}}$ is a shallow neural network formed by MLP, the function $unwind()$ unwinds the diagonal matrix R_k into a one-dimensional vector, and $longscale$ is a long-scale list. For example, we set $longscale = [10, 20, 30, 40]$, and the maximum value of the long-scale list $longscale$ should not exceed the number of nodes i.e. N. We set up a shallow neural network for each scale $\mathcal{L}$ in our long-scale list.

Aggregate Scale. According to Eqs. (9) and (10), we can finally construct the l_{th} convolution layer.

$$Z^{(l)} = \sigma \left(W_{short}^{(l)} h_{short}^{(l)} + W_{long}^{(l)} h_{long}^{(l)} + B^{(l)} \right) \qquad (11)$$

where $W_{short}^{(l)}$ and $W_{long}^{(l)}$ are the weight parameters of the short-scale and the long-scale convolution filters, respectively. $B^{(l)}$ is the bias parameter matrix and $\sigma()$ is the complex ReLU activation function [50]. After passing through L convolution layers, the output $Z_{N \times F_L}^{(L)}$ is obtained, then we unwind $Z^{(L)}$ into real and imaginary parts, and use the softmax function to output the result of node classification:

$$class = softmax \left(unwind(Z^{(L)}) W_{unwind} \right)$$

where the $W_{unwind} \in \mathcal{R}^{2F_L \times Numclass}$ is the weight parameter matrix of the real and imaginary parts after unwinding $Z^{(L)}$. $Numclass$ is the number of node classes in the final output.

4   Experiment

In this section, we illustrate our experiments. We validate the effectiveness of our improved Lanczos algorithm and the MSDGCNN we built with different datasets, respectively. In this paper, we choose node classification task to evaluate MSDGCNN, and our method can be easily applied on link prediction task by reversing the graph (In the reversed graph, edges and nodes correspond to nodes and edges, respectively, in the original graph.) and clustering task by changing loss functions.

4.1   Datasets

The QM8 dataset is derived from summarizing chemical structures and quantum chemistry computation output values of molecules [34]. The dataset contains 21,786 molecules with 6 different types of chemical bonds and 70 different types of atoms. In our experiment, each molecule is treated as a graph, where each type of chemical bond represents a type of connection (or edge). Each atom is treated as a node in the graph. Since there may be multiple types of chemical bonds between two different atoms in a molecule, the graph for each molecule is a multi-graph. We evaluated our improved Lanczos algorithm for the node classification task on the QM8 quantum chemistry dataset.

Table 1. Real world dataset details.

Dataset	Nodes	Edges	Class	Features
Telegram	245	8912	4	1
Citeseer	3312	4715	6	3703
Cora-ML	2995	8416	7	2879
Wiscosin	251	499	5	1703
Cornell	183	295	5	1703
Texas	183	309	5	1703

The Telegram dataset [2] is a network of pairwise interactions between 245 telegram channels, comprising 8,912 relationships between channels. The nodes in the telegram channel network are divided into four classes according to the information provided in [2]. The Wiscosin, Cornell and Texas datasets are taken from the WebKB dataset which model the relationships between different university websites [33]. In our experiment, the nodes in these datasets are labeled to cover identity information for five categories of personnel. The Cora-ML and CiteSeer datasets, provided by [1], are citation networks for scientific articles. The articles in these datasets belong to seven and five different scientific fields, respectively. We perform node classification tasks on these datasets to evaluate the performance of our multi-scale directed graph convolution neural network. The detailed information about the datasets is presented in Table 1.

4.2   Implementation Details

We implement our improved Lanczos algorithm using the Python language and build our multi-scale directed graph convolution neural network using the Pytorch1.13 deep learning framework. We use the Ubuntu20.04 operating system as our system environment. We are conducting experiments using a computer equipped with an Intel i5-12600K with 64GB DDR4 memory and Nvidia RTX 3060 with 12GB memory.

Evaluation of Improved Lanczos Algorithm. We use the method provided by DeepChem [35] to split the QM8 quantum chemistry dataset. Also, we use the approach provided by [9,44], which means that we use MSE as loss function to update the parameters in training and use MAE function to evaluate the training performance. We apply the improved Lanczos algorithm to AdaLanczosNet [28] by replacing the original Lanczos algorithm and compare it with the original AdaLanczosNet which is the baseline in the experiment. To ensure the reliability of the experiment, our hyperparameter settings follow the same long-scale list $longscale = [5, 7, 10, 20, 30]$ and short-scale list $shortscale = [1, 2, 3]$ as in [28]. We set the threshold for rounding error of the improved Lanczos algorithm to $\epsilon_M = 1.192 \times 10^{-7}$ which is the machine precision of 32-bit floating-point numbers. In the experiment, since the number of atoms contained in each molecular graph is different and the average number of atoms per molecule is around 16, we set $LanczosStep$ to 10 and train both the AdaLanczosNet using the original Lanczos algorithm and the AdaLanczosNet using the improved Lanczos algorithm for 200 epochs.

Evaluation of MSDGCNN. We use the splliting method, as in many literatures ([33,49], etc.) mentioned, to split the WebKB and Telegram datasets, i.e., 60%, 20%, 20% for training, validation and testing sets. During training process, we also use the data of testing set without the node labels. The spliting method of Cora-ML and CiteSeer we used is similar to [40]. We randomly split each dataset into 10 subsets. When applying our improved Lanczos algorithm, we set the size of $LanczosStep$ to half of the number of nodes in each dataset (i.e. $LanczosStep \leq \frac{N}{2}$), therefore, the number of iterations of our improved Lanczos algorithm will not exceed $\frac{N}{2}$. We set different long-scale lists for different datasets with different node numbers (e.g. for the node classification task on the Telegram dataset, we set $longscale = [15, 25, 35, 45]$.), and adjust the dimensions of the hidden layers of the function $f_{\mathcal{L}}$ from Eq. (10) according to the node feature dimensions and number of nodes in the dataset. To speed up the training process of our network, we pre-train the short-scale convolution filters and use the trained parameters as the initialization parameters for the model for all node classification tasks. We use phase parameters q as hyperparameters in experiment to set the appropriate direction strength for each dataset. We choose several popular graph neural networks as our baselines, including MagNet [50], APPNP [8], GraphSAGE [13], GIN [47], GAT [42], ChebNet [5], GCN [19], as well as DGCN [41] and DiGraph [40]. According to [45], These graph neural networks, can be divided into spatial-based as well as spectral-based GNNs and specialized directed graph GNNs. For all baseline models, we choosed best hyperparameters from literature and used the best result as baselines. We set the number of convolution layers to 2 for all models and set the order of the Chebyshev polynomial to 1. In our experiments, we train all models for 4000 epochs using the Adam optimizer.

4.3 Experimental Results and Analysis

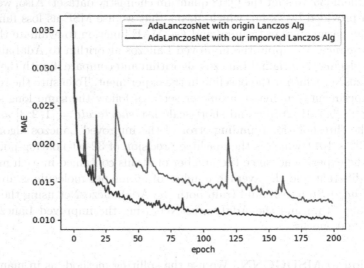

Fig. 2. Validation MAE on QM8 dataset during the training process

Results and Analysis on Improved Lanczos Algorithm. The validation MAE during the experiment on the QM8 dataset is shown in Fig. 2. It is evident that AdaLanczosNet with our improved Lanczos algorithm has better stability during the training process compared to AdaLanczosNet with the original Lanczos algorithm. The test MAE is shown in Table 2, obliviously, our improved algorithm achieved lower classification error by correctly modifying the orthogonal basis. Through experiments, we can clearly observe that the loss of orthogonality

Table 2. Test MAE on QM8 dataset.

Method	Test MAE($\times 10^{-3}$)
Origin AdaLanczosNet	11.77($\pm$0.8)
AdaLanczosNet with our improvement Lanczos Alg	9.89($\pm$0.04)

Table 3. Node classification accuracy on real world datasets.

Method	DataSet					
	Telegram	CiteSeer	Cora-ML	Wisconsin	Cornell	Texas
GCN	73.4($\pm$5.8)%	66.0($\pm$1.5)%	82.0($\pm$1.1)%	55.9($\pm$5.4)%	59.0($\pm$6.4)%	58.7($\pm$3.8)%
ChebNet	70.2($\pm$6.8)%	66.7($\pm$1.6)%	80.0($\pm$1.8)%	81.6($\pm$6.3)%	79.8($\pm$5.0)%	79.2($\pm$7.5)%
APPNP	67.3($+3.0$)%	66.9($\pm$1.8)%	**82.6**($\pm$1.4)%	51.8($\pm$7.4)%	58.7($\pm$4.0)%	57.0($\pm$4.8)%
SAGE	56.6($\pm$6.0)%	66.0($\pm$1.5)%	82.3($\pm$1.2)%	83.1($\pm$4.8)%	80.0($\pm$6.1)%	84.3($\pm$5.5)%
GIN	74.4($\pm$8.1)%	63.3($\pm$2.5)%	78.1($\pm$2.0)%	58.2($\pm$5.1)%	57.9($\pm$5.7)%	65.2($\pm$6.5)%
GAT	72.6($\pm$7.5)%	67.3($\pm$1.3)%	81.9($\pm$1.0)%	54.1($+4.2$)%	57.6($\pm$4.9)%	61.2($\pm$5.0)%
DGCN	90.4($\pm$5.6)%	66.3($\pm$2.0)%	81.3($\pm$1.4)%	65.5($\pm$4.7)%	66.3($\pm$2.0)%	71.7($\pm$7.4)%
Digraph	82.0($\pm$3.1)%	62.6($\pm$2.2)%	79.4($\pm$1.8)%	59.6($\pm$3.8)%	66.8($\pm$6.2)%	64.9($\pm$8.1)%
DigraphIB	64.1($\pm$7.0)%	61.1($\pm$1.7)%	79.3(1.2)%	64.1($\pm$7.0)%	64.4($\pm$9.0)%	64.9($\pm$13.7)%
MagNet	87.6($\pm$2.9)%	67.5($\pm$1.8)%	79.8($\pm$2.5)%	85.7($\pm$3.2)%	84.3($\pm$7.0)%	83.3($\pm$6.1)%
Our Method	**92.5**($\pm$4.7)%	**69.1**($\pm$1.7)%	81.7($\pm$1.5)%	**87.4**($\pm$5.8)%	**86.7**($\pm$4.3)%	**89.4**($\pm$8.2)%
q	0.15	0.0	0.0	0.05	0.25	0.15
$longscale$	$[15, 25, 35, 45]$	$[10, 20, 30, 40]$	$[10, 25, 30, 45]$	$[10, 20, 25, 35]$	$[5, 10, 15, 25]$	$[5, 10, 15, 20]$
$LanczosStep$	125	1500	1400	125	90	90

caused by rounding errors during the iteration process of the Lanczos algorithm has a significant impact on both training stability and final classification accuracy.

Results and Analysis on MSDGCNN. Our multi-scale directed graph convolution neural network achieved outstanding performance in node classification tasks on multiple datasets as shown in Table 3. We achieved the best classification accuracy on the large-scale CiteSeer dataset. Although we did not achieve the highest classification accuracy on the Cora-ML dataset, our method outperformed MagNet which only uses short-scale convolution filters by nearly 2%. This indicates that our long-scale convolution filter constructed by improved Lanczos algorithm is effective. In experiment on four small-scale datasets (including Wisconsin, Cornell, Texas and Telegram), our method achieved the best node classification accuracy. Experimental results on the real-world directed graph datasets demonstrate that the MSDGCNN has better overall performance compared to most other state-of-the-art GNNs.

5   Conclusion

We improved the Lanczos algorithm by using the modified Gram-Schmidt orthogonalization technique combining orthogonalization check method, and the result on QM8 dataset demonstrated that our improved algorithm has better stability. Similarly, our designed Multi-scale Directed Graph Convolution Neural Network (MSDGCNN) which aggregates larger scale information of a node with lower computational overhead showed outstanding performance on numerous real-world directed graph datasets. However, from the results of experiments on MSDGCNN, it can be observed that increasing the convolution scale of the convolution layer has limited contribution to improving the accuracy of node classification tasks on large-scale sparse graphs, and increasing the convolution scale will further increase the training time and difficulty at same time. Furthermore, there is still a need for further discussion on how to utilize spectral-based graph convolution neural networks to handle multi-graph structures with multiple types of edges between nodes.

Acknowledgements. This work is financially supported by: The National Key R&D Program of China (No. 2020YFB1712600); The Fundamental Research Funds for Central University (No. 3072022QBZ0601); and The State Key Laboratory of Underwater Robotics Technology at Harbin Engineering University (No. KY70100200052).

References

1. Bojchevski, A., Günnemann, S.: Deep gaussian embedding of graphs: Unsupervised inductive learning via ranking. arXiv preprint arXiv:1707.03815 (2017)
2. Bovet, A., Grindrod, P.: The activity of the far right on telegram (2020)
3. Chen, B.: Molecular Graph Representation Learning and Generation for Drug Discovery. Ph.D. thesis, Massachusetts Institute of Technology (2022)
4. Chung, F.R.: Spectral graph theory, vol. 92. American Mathematical Soc. (1997)
5. Defferrard, M., Bresson, X., Vandergheynst, P.: Convolutional neural networks on graphs with fast localized spectral filtering. Advances in neural information processing systems 29 (2016)
6. Fanuel, M., Alaiz, C.M., Suykens, J.A.: Magnetic eigenmaps for community detection in directed networks. Phys. Rev. E **95**(2), 022302 (2017)
7. Furutani, S., Shibahara, T., Akiyama, M., Hato, K., Aida, M.: Graph signal processing for directed graphs based on the hermitian laplacian. In: Brefeld, U., Fromont, E., Hotho, A., Knobbe, A., Maathuis, M., Robardet, C. (eds.) ECML PKDD 2019. LNCS (LNAI), vol. 11906, pp. 447–463. Springer, Cham (2020). https://doi.org/10.1007/978-3-030-46150-8_27
8. Gasteiger, J., Bojchevski, A., Günnemann, S.: Predict then propagate: Graph neural networks meet personalized pagerank. arXiv preprint arXiv:1810.05997 (2018)
9. Gilmer, J., Schoenholz, S.S., Riley, P.F., Vinyals, O., Dahl, G.E.: Neural message passing for quantum chemistry. In: International Conference on Machine Learning, pp. 1263–1272. PMLR (2017)
10. Golub, G.H., Van Loan, C.F.: Matrix computations. JHU press (2013)
11. Grcar, J.F.: Analyses of the Lanczos Algorithm and of the Approximation Problem in Richardson's Method. University of Illinois at Urbana-Champaign (1981)

12. Guo, K., Mohar, B.: Hermitian adjacency matrix of digraphs and mixed graphs. J. Graph Theory **85**(1), 217–248 (2017)

13. Hamilton, W., Ying, Z., Leskovec, J.: Inductive representation learning on large graphs. Advances in neural information processing systems 30 (2017)

14. Hammond, D.K., Vandergheynst, P., Gribonval, R.: Wavelets on graphs via spectral graph theory. Appl. Comput. Harmon. Anal. **30**(2), 129–150 (2011)

15. He, K., Zhang, X., Ren, S., Sun, J.: Deep residual learning for image recognition. In: Proceedings of the IEEE Conference on Computer Vision and Pattern Recognition, pp. 770–778 (2016)

16. He, Y., Perlmutter, M., Reinert, G., Cucuringu, M.: Msgnn: a spectral graph neural network based on a novel magnetic signed laplacian. In: Learning on Graphs Conference, pp. 40–1. PMLR (2022)

17. Hinton, G.E., Srivastava, N., Krizhevsky, A., Sutskever, I., Salakhutdinov, R.R.: Improving neural networks by preventing co-adaptation of feature detectors. arXiv preprint arXiv:1207.0580 (2012)

18. Huang, T., Dong, Y., Ding, M., Yang, Z., Feng, W., Wang, X., Tang, J.: Mixgcf: An improved training method for graph neural network-based recommender systems. In: Proceedings of the 27th ACM SIGKDD Conference on Knowledge Discovery & Data Mining. pp. 665–674 (2021)

19. Kipf, T.N., Welling, M.: Semi-supervised classification with graph convolutional networks. arXiv preprint arXiv:1609.02907 (2016)

20. Kipf, T.N., Welling, M.: Variational graph auto-encoders. arXiv preprint arXiv:1611.07308 (2016)

21. Ko, T., Choi, Y., Kim, C.K.: A spectral graph convolution for signed directed graphs via magnetic laplacian. Neural Networks (2023)

22. Krizhevsky, A., Sutskever, I., Hinton, G.E.: Imagenet classification with deep convolutional neural networks. Commun. ACM **60**(6), 84–90 (2017)

23. Lanczos, C.: An iteration method for the solution of the eigenvalue problem of linear differential and integral operators (1950)

24. LeClair, A., Haque, S., Wu, L., McMillan, C.: Improved code summarization via a graph neural network. In: Proceedings of the 28th International Conference on Program Comprehension, pp. 184–195 (2020)

25. Levie, R., Huang, W., Bucci, L., Bronstein, M., Kutyniok, G.: Transferability of spectral graph convolutional neural networks. J. Mach. Learn. Res. **22**(1), 12462–12520 (2021)

26. Li, Y., Yu, R., Shahabi, C., Liu, Y.: Diffusion convolutional recurrent neural network: data-driven traffic forecasting. arXiv preprint arXiv:1707.01926 (2017)

27. Li, Y., Tarlow, D., Brockschmidt, M., Zemel, R.: Gated graph sequence neural networks. arXiv preprint arXiv:1511.05493 (2015)

28. Liao, R., Zhao, Z., Urtasun, R., Zemel, R.S.: Lanczosnet: multi-scale deep graph convolutional networks. arXiv preprint arXiv:1901.01484 (2019)

29. Lieb, E.H., Loss, M.: Fluxes, laplacians, and kasteleyn's theorem. Statistical Mechanics: Selecta of Elliott H. Lieb, pp. 457–483 (2004)

30. Paige, C.C.: Computational variants of the lanczos method for the eigenproblem. IMA J. Appl. Math. **10**(3), 373–381 (1972)

31. Paige, C.C.: Error analysis of the lanczos algorithm for tridiagonalizing a symmetric matrix. IMA J. Appl. Math. **18**(3), 341–349 (1976)

32. Parlett, B.N., Scott, D.S.: The lanczos algorithm with selective orthogonalization. Math. Comput. **33**(145), 217–238 (1979)

33. Pei, H., Wei, B., Chang, K.C.C., Lei, Y., Yang, B.: Geom-gcn: geometric graph convolutional networks. arXiv preprint arXiv:2002.05287 (2020)

34. Ramakrishnan, R., Hartmann, M., Tapavicza, E., Von Lilienfeld, O.A.: Electronic spectra from tddft and machine learning in chemical space. J. Chem. Phys. **143**(8), 084111 (2015)
35. Ramsundar, B., Eastman, P., Walters, P., Pande, V., Leswing, K., Wu, Z.: Deep Learning for the Life Sciences. O'Reilly Media (2019)
36. Schuster, M., Paliwal, K.K.: Bidirectional recurrent neural networks. IEEE Trans. Signal Process. **45**(11), 2673–2681 (1997)
37. Shuman, D.I., Narang, S.K., Frossard, P., Ortega, A., Vandergheynst, P.: The emerging field of signal processing on graphs: extending high-dimensional data analysis to networks and other irregular domains. IEEE Signal Process. Mag. **30**(3), 83–98 (2013)
38. Simon, H.D.: The lanczos algorithm with partial reorthogonalization. Math. Comput. **42**(165), 115–142 (1984)
39. Susnjara, A., Perraudin, N., Kressner, D., Vandergheynst, P.: Accelerated filtering on graphs using lanczos method. arXiv preprint arXiv:1509.04537 (2015)
40. Tong, Z., Liang, Y., Sun, C., Li, X., Rosenblum, D., Lim, A.: Digraph inception convolutional networks. Adv. Neural. Inf. Process. Syst. **33**, 17907–17918 (2020)
41. Tong, Z., Liang, Y., Sun, C., Rosenblum, D.S., Lim, A.: Directed graph convolutional network. arXiv preprint arXiv:2004.13970 (2020)
42. Veličković, P., Cucurull, G., Casanova, A., Romero, A., Lio, P., Bengio, Y.: Graph attention networks. arXiv preprint arXiv:1710.10903 (2017)
43. Wu, L., Cui, P., Pei, J., Zhao, L., Guo, X.: Graph neural networks: foundation, frontiers and applications. In: Proceedings of the 28th ACM SIGKDD Conference on Knowledge Discovery and Data Mining, pp. 4840–4841 (2022)
44. Wu, Z., et al.: Moleculenet: a benchmark for molecular machine learning. Chem. Sci. **9**(2), 513–530 (2018)
45. Wu, Z., Pan, S., Chen, F., Long, G., Zhang, C., Philip, S.Y.: A comprehensive survey on graph neural networks. IEEE Trans. Neural Networks Learn. Syst. **32**(1), 4–24 (2020)
46. Xu, B., Shen, H., Cao, Q., Qiu, Y., Cheng, X.: Graph wavelet neural network. arXiv preprint arXiv:1904.07785 (2019)
47. Xu, K., Hu, W., Leskovec, J., Jegelka, S.: How powerful are graph neural networks? arXiv preprint arXiv:1810.00826 (2018)
48. Zhang, C., et al.: Deeptralog: trace-log combined microservice anomaly detection through graph-based deep learning (2022)
49. Zhang, J., Hui, B., Harn, P.W., Sun, M.T., Ku, W.S.: Mgc: a complex-valued graph convolutional network for directed graphs. arXiv e-prints pp. arXiv-2110 (2021)
50. Zhang, X., He, Y., Brugnone, N., Perlmutter, M., Hirn, M.: Magnet: a neural network for directed graphs. Adv. Neural. Inf. Process. Syst. **34**, 27003–27015 (2021)
51. Zhou, J., et al.: Graph neural networks: a review of methods and applications. AI Open **1**, 57–81 (2020)

Probabilistic AutoRegressive Neural Networks for Accurate Long-Range Forecasting

Madhurima Panja[1]($\boxtimes$), Tanujit Chakraborty[2], Uttam Kumar[1],
and Abdenour Hadid[2]

[1] International Institute of Information Technology Bangalore, Bengaluru, India
{madhurima.panja,uttam}@iiitb.ac.in
[2] Sorbonne University Abu Dhabi, Abu Dhabi, UAE
{tanujit.chakraborty,abdenour.hadid}@sorbonne.ae

Abstract. Forecasting time series data is a critical area of research with applications spanning from stock prices to early epidemic prediction. While numerous statistical and machine learning methods have been proposed, real-life prediction problems often require hybrid solutions that bridge classical forecasting approaches and modern neural network models. In this study, we introduce a Probabilistic AutoRegressive Neural Network (PARNN), capable of handling complex time series data exhibiting non-stationarity, nonlinearity, non-seasonality, long-range dependence, and chaotic patterns. PARNN is constructed by improving autoregressive neural networks (ARNN) using autoregressive integrated moving average (ARIMA) feedback error. Notably, the PARNN model provides uncertainty quantification through prediction intervals and conformal predictions setting it apart from advanced deep learning tools. Through comprehensive computational experiments, we evaluate the performance of PARNN against standard statistical, machine learning, and deep learning models. Diverse real-world datasets from macroeconomics, tourism, epidemiology, and other domains are employed for short-term, medium-term, and long-term forecasting evaluations. Our results demonstrate the superiority of PARNN across various forecast horizons, surpassing the state-of-the-art forecasters. The proposed PARNN model offers a valuable hybrid solution for accurate long-range forecasting. The ability to quantify uncertainty through prediction intervals further enhances the model's usefulness in various decision-making processes.

Keywords: Forecasting · ARIMA · Neural networks · Hybrid model

1 Introduction

Time series forecasting has been a potential arena of research for the last several decades. Predicting the future is fundamental in various applied domains like economics, healthcare, demography, and energy, amongst many others, as it aids in better decision-making and formulating data-driven business strategies [7, 24, 34]. Forecasts for any time series arising from different domains are usually obtained

B. Luo et al. (Eds.): ICONIP 2023, CCIS 1967, pp. 457–477, 2024.
https://doi.org/10.1007/978-981-99-8178-6_35

by modeling their corresponding data-generating process based on the analysis of the historical data. With the abundance of past data, there has been an increasing demand for forecasts in the industrial sectors as well [25,46]. To serve the need to generate accurate and reliable forecasts, numerous statistical and machine learning methods have been proposed in the literature [8,21,22,43]. One of the most popular statistical forecasting models is the autoregressive integrated moving average (ARIMA) [5] model that tracks the linearity of a stationary data-generating process. The ARIMA model describes the historical patterns in a time series as a linear combination of autoregressive (lagged inputs) and moving average (lagged errors) components. Despite its vast applicability, ARIMA is not suitable for modeling nonlinear datasets, as the model assumes the future values of the series to be linearly dependent on the past and current values. Thus, the modeling capability of linear models like ARIMA shrinks while dealing with real-world datasets that exhibit complex nonlinear and chaotic patterns [47].

Modern tools of deep learning such as multilayer perceptron (MLP) [37], auto-regressive neural network (ARNN) [17], and ensemble deep learning [36] methods leverage the ground truth data (with nonlinear trends) to learn the temporal patterns in an automated manner. However, these methods fail to model long-term dependencies in time series. Current progress in computationally intelligent frameworks has brought us deep autoregressive (DeepAR) [38] model, neural basis expansion analysis (NBeats) [32], temporal convolutional network (TCN) [11], Transformers [45], and many others for modeling nonlinear and non-stationary datasets [15,39]. The innovation in these advanced deep learning frameworks has demonstrated tremendous success in modeling and extrapolating the long-range dependence and interactions in temporal data; hence they are pertinent in the current literature on time series forecasting. Albeit these models are applied in several forecasting applications, their accuracy depends largely on the appropriate choice of hyperparameters, which are commonly data-dependent. Any misspecification of these parameters may lead to a significant reduction of the forecast accuracy for out-of-sample predictions. Moreover, these complex models often suffer from the problem of overfitting the given time series data, i.e., it learns the pattern and the noise in the underlying data to such an extent that it negatively impacts the prediction capability in the unseen data [19]. Another major drawback of some of the approaches mentioned above is the lack of explainability and the "black-box-like" behavior of these models.

To overcome the limitations arising from the stand-alone forecasting methods and to simplify the model selection procedure, several hybrid forecasting techniques that decompose a given time series into linear and nonlinear components have been proposed in the literature [1,8,47]. The hybrid architectures comprise three steps: firstly, the linear patterns of the series are forecasted using linear models which are followed by an error re-modeling step using nonlinear models, and finally, the forecasts from both the steps are combined to produce the final output. These hybrid systems have been extended for forecasting applications in various domains such as finance [6,7], earth science [41], transportation

[46], energy [35], agriculture [40], epidemiology [3, 9] and their forecasting performance has surpassed their counterpart models. Although the hybrid forecasting models have outperformed their counterpart models along with other individual baseline forecasters in various applications, they are constructed based on certain assumptions. For instance, it is assumed that a series's linear and nonlinear patterns can be modeled separately or that the residuals comprise only the nonlinear trends. Alternatively, an additive or multiplicative relationship exists between the linear and nonlinear segments of the datasets [9]. However, if these assumptions do not hold, i.e., if the existing linear and nonlinear components cannot be modeled separately or if the residuals of the linear model do not contain valid nonlinear patterns, or if there is a lack of linear relationship between the components of a series, then the forecast accuracy of these hybrid models might substantially degrade.

Motivated by the above observations, we propose a hybrid probabilistic autoregressive neural network (PARNN) that is designed to overcome the limitations of hybrid time series forecasting models while improving their predictive accuracies. The first phase of our proposal encompasses a linear ARIMA model fitting and generating in-sample residuals. During the second phase, these unexplained residuals and the original input-lagged series are remodeled using a nonlinear ARNN model. In the proposed hybridization, neural network autoregression is used to model not only the time series data but also the feedback errors of the linear ARIMA model to improve the learning ability of the network. This combined architecture is easier to train and can accurately make long-range predictions for a wide variety of irregularities of real-world complex time series. The proposed framework is extensively evaluated on twelve publicly available benchmark time series datasets. The experimental results show that our proposed model provides accurate and robust predictions for nonlinear, non-stationary, chaotic, and non-Gaussian data, outperforming benchmark models in most real-data problems.

The main contributions of the paper can be summarized as follows:

1. We introduce the mathematical formulation of our PARNN model, tailored for complex real-world datasets with nonlinear, non-stationary, chaotic, and non-Gaussian characteristics. We assess PARNN's uncertainty using various methods.
2. We apply our model to twelve diverse time series datasets, exploring their global features. Using a rolling window approach with short, medium, and long-term test horizons, we gauge forecasting performance through three widely used performance measures.
3. Comparing against thirteen advanced forecasters, including traditional models and recent deep learning algorithms, we confirm our model's superiority, particularly in long-range predictions. This is substantiated by a non-parametric statistical test, demonstrating the robustness of our proposed PARNN approach.

2   Background

Most time series models tend to suffer in modeling the complexities of real-world datasets. The performance of hybrid models in such situations is encouraging. We aim to improve the forecasting performance of hybrid models by proposing a hybrid PARNN model. Our proposed model comprises the linear ARIMA model and the nonlinear ARNN model. In this section, we briefly describe the constituent models along with popularly used hybrid and ensemble models before describing our proposed framework.

2.1   ARIMA Model

The classical autoregressive integrated moving average (ARIMA) model, often termed as Box-Jenkins method [5], is one of the most widely used statistical models in the forecasting literature. The $\text{ARIMA}(p, d, q)$ model comprises three parameters where p and q denote the order of the AR and MA terms, and d denotes the order of differencing. The mathematical formulation of the ARIMA model is given by

$$y_t = \beta_0 + \alpha_1 y_{t-1} + \alpha_2 y_{t-2} + \ldots + \alpha_p y_{t-p} + \varepsilon_t - \beta_1 \varepsilon_{t-1} - \beta_2 \varepsilon_{t-2} - \ldots - \beta_q \varepsilon_{t-q}$$

where, y_t is the actual time series, ε_t is the random error at time t, and α_i and β_j are the model parameters. The ARIMA model is constructed using three iterative steps. Firstly, we convert a non-stationary series into a stationary one by applying the difference of order d. Then, once a stationary series is obtained, we select the model parameters p and q from the ACF plot and PACF plot, respectively. Finally, we obtain the "best fitted" model by analyzing the residuals.

2.2   ARNN Model

Autoregressive neural networks (ARNN), derived from the artificial neural network (ANN), is specifically designed for modeling nonlinear time series data [17]. The ARNN model comprises a single hidden layer embedded within its input and output layers. The $\text{ARNN}(u, v)$ model passes u lagged input values from its input layer to the hidden layer comprising of v hidden neurons. The value of v is determined using the formula $v = \lceil (u + 1)/2) \rceil$, where $\lceil \cdot \rceil$ is the ceiling function (also known as least integer function) as proposed in [22,34]. After being trained by a gradient descent back-propagation approach [37], the final forecast is obtained as a linear combination of its predictors. The mathematical formulation of the ARNN model is given by:

$$g(\underline{y}) = \alpha_0^* + \sum_{j=1}^{v} \alpha_j^* \phi \left(\beta_j^* + \theta_j^{*'} \underline{y} \right),$$

where g denotes a neural network, $\underline{y}$ is a u-lagged inputs, $\alpha_0^*, \beta_j^*, \alpha_j^*$ are connecting weights, θ_j^* is a weight vector of dimension u and ϕ is a bounded nonlinear activation function.

2.3   Ensemble and Hybrid Models

In the context of forecasting literature, although ARIMA and ARNN models have individually achieved significant successes in their respective domains [27,31], their difficulties in modeling the complex autocorrelation structures within real-world datasets have led to the development of ensemble and hybrid forecasting approaches. The idea of the ensemble forecasting model was first introduced in 1969 [2]. The final output of these models was a weighted combination of the forecasts generated by its component models i.e.,

$$\hat{y} = \sum_{i \in \mathcal{M}} \gamma_i \hat{y}_i,$$

where γ_i denotes the weights and $\hat{y}_i$ denotes the forecast generated from the ith, $i = 1, 2, \ldots, \mathcal{M}$ individual model. The ensemble models could significantly improve forecast accuracy for various applications. However, the appropriate selection of weights, often termed as "forecast combination puzzle" posed a significant challenge to its universal success.

The concept of hybrid forecasting was led by Zhang in 2003 [47] where the author assumed that a time series could be decomposed into its linear and nonlinear components as follows:

$$y_t = y_{Lin_t} + y_{Nlin_t},$$

where y_{Lin_t} and y_{Nlin_t} denote the linear and nonlinear elements, respectively. Using a linear model (e.g., ARIMA), y_{Lin_t} was estimated from the available dataset. Let the forecast value of the linear model be denoted by $\hat{y}_{Lin_t}$. Then, the model residuals, y_{Res_t} were computed as:

$$y_{Res_t} = y_t - \hat{y}_{Lin_t}.$$

The left-out autocorrelations in the residuals were further re-modelled using a nonlinear model (e.g., ANN or ARNN) which generates the nonlinear forecasts as $\hat{y}_{Nlin_t}$. The final forecast $\hat{y}_t$ is computed by adding the out-of-sample forecasts generated by the linear and the nonlinear models i.e.,

$$\hat{y}_t = \hat{y}_{Lin_t} + \hat{y}_{Nlin_t}.$$

Various combinations of hybrid models have been proposed in the literature, showing forecasting performance improvement [4,13,33]. Among these models, ARIMA-ARNN models, for instance, have shown higher accuracies for epidemic [8] and econometric modeling [7].

3   Proposed Model

In this section, we discuss the formulation of the proposed PARNN model and the computation of its prediction intervals. Practical usage of the proposal for real-world time series datasets is presented in Sect. 4.

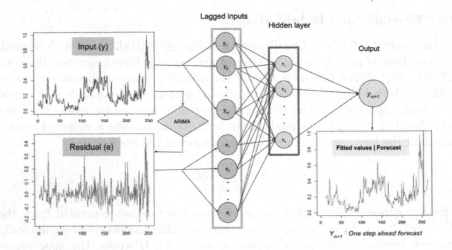

Fig. 1. Schematic diagram of proposed PARNN model, where y_t and e_t denote the original time series and residual series obtained using ARIMA, respectively. The value of k in the hidden layer is set to $k = \lceil (m + l + 1)/2 \rceil$.

3.1   PARNN Model

We propose a hybrid probabilistic autoregressive neural networks (PARNN) which is a modification to the artificial neural network ($\text{ANN}(p, d, q)$) [26] and recurrent neural network (RNN) [12] models used for nonlinear time series forecasting. The current hybrid models in time series [47], such as the hybrid ARIMA-ARNN model [8], made a significant improvement in the predictive accuracy of individual forecasters like ARIMA and ARNN models. These hybrid models assume an additive relationship between linear and nonlinear components of the time series which is not always obvious. This paper builds a probabilistic ARNN framework that blends classical time series with scalable neural network architecture. Our proposed $\text{PARNN}(m, k, l)$ model considers the future values of the time series to be a nonlinear function of m-lagged values of past observations and l-lagged values of ARIMA residuals (feedback errors). Therefore, we have,

$$y_t = f(y_{t-1}, \ldots, y_{t-m}, e_{t-1}, \ldots, e_{t-l}),$$

where the nonlinear function f is a single hidden-layered autoregressive neural network having k hidden neurons, y_t and e_t denote the actual time series and the residual series at time t. Using the connection weights α and a bounded nonlinear sigmoidal activation function $G(\cdot)$ of the neural network, y_t can be denoted as

$$y_t = \alpha_0 + \sum_{j=1}^{k} \alpha_j G\left(\alpha_{0,j} + \sum_{i=1}^{m} \alpha_{i,j} y_{t-i} + \sum_{i=m+1}^{m+l} \alpha_{i,j} e_{t+m-i}\right) + \epsilon_t. \quad (1)$$

The learning phase of the proposed PARNN model comprises two stages. In the first stage, the original series is standardized and the ARIMA residuals (e_t)

are generated from the standardized series (y_t) using the best-fitted ARIMA model with minimum Akaike Information Criterion (AIC). In the second stage, a single hidden layered feed-forward neural network is built to model the non-linear and linear relationships using ARIMA residuals and original observations. This neural network receives m lagged values of y_t and l lagged values of e_t as inputs. We denote the proposed model as PARNN(m, k, l), where k is the number of neurons arranged in a single hidden layer and set to $k = \lceil \frac{m+l+1}{2} \rceil$ (where $\lceil \cdot \rceil$ is the ceiling function) as in [22,34]. The proposed network architecture generates a one-step-ahead forecast for the series (y_t) at each iteration. The model employs historical observations and recent forecasts as inputs for generating multi-step ahead forecasts iteratively. The proposed PARNN is useful for modeling the underlying data-generating process with non-stationary, non-linear, and non-Gaussian structures in the time series data. Nonlinear ARMA and simple ARNN models can be seen as special cases of our proposed model. It can be seen that in the proposed PARNN(m, k, l) model, in contrast to the traditional hybrid models, no prior assumption is considered for the relationship between the linear and nonlinear components. Additionally, it can be generally guaranteed that the performance of the probabilistic ARNN model will not be worse than either of the components – ARIMA and ARNN – used separately or hybridized. We determine the optimal values of the hyperparameters m and l by performing a grid search algorithm on the validation dataset. The maximum values of both the hyperparameters are usually set to 10, and the model is iterated over a grid of m and l. At each iteration, the model is fitted for different pairs of values of m and l, and the forecasts $\hat{y}_t$ are generated for the validation set (a part of the training set is kept separately for validation to avoid data leakage). The performance of these fitted models is evaluated for all possible pairs of m and l on the validation set, and eventually, the pair of values exhibiting the best performance w.r.t. the mean absolute scaled error (MASE) metric is selected. Thus, our proposed PARNN can be considered as a hybrid framework in which ARIMA is embedded into an ARNN model that converges very fast unlike the ANN(p, d, q) and RNN models for nonlinear forecasting. The proposed PARNN model is preferable for long-range forecasting which is one of the most attractive features of autoregressive neural networks [30]. A workflow diagram of the proposed PARNN model is presented in Fig. 1. An algorithmic version of our proposal is given in Algorithm 1.

3.2   Prediction Interval of the PARNN Model

Owing to the hybrid nature of the PARNN model, we can't calculate prediction intervals analytically. Therefore, we use two methods to quantify the model's uncertainty namely, conformal prediction [42] and simulating future paths with bootstrapped residuals. Conformal prediction transforms point estimates into prediction regions, ensuring convergence in a distribution-free and model-agnostic manner. For its computation, we employ the "caretForecast" package in R software and determine the intervals by analyzing residual distributions. This approach addresses data and modeling uncertainties, resulting in trustworthy prediction intervals. In the case of bootstrapped residuals, the forecast of our proposed model at time t can be expressed as

$$y_t = f(\mathbf{y_{t-1}}, \mathbf{e_{t-1}}) + \epsilon_t,$$

where, $\mathbf{y_{t-1}} = (y_{t-1}, \ldots, y_{t-m})'$ are the lagged inputs and $\mathbf{e_{t-1}} = (e_{t-1}, \ldots, e_{t-l})'$ are the lagged ARIMA feedback, and f is the neural network with k hidden nodes. We assume that the error, ϵ_t follows a Gaussian distribution and ensure that it is homoscedastic in nature (by applying Box-Cox transformation with parameter 0.5). We can then iteratively draw random samples from this distribution and simulate future sample paths of the model. Thus, if ϵ_{t+1} is the sample drawn at time $t + 1$ then,

$$y_{t+1} = f(\mathbf{y_t}, \mathbf{e_t}) + \epsilon_{t+1},$$

where, $\mathbf{y_t} = (y_t, \ldots, y_{t-m+1})'$ and $\mathbf{e_t} = (e_t, \ldots, e_{t-l+1})'$. We perform this simulation 1000 times at each forecast time step, accumulating insights into the value distribution. Using these distributions, we calculate 80% prediction intervals by determining percentiles for the PARNN model's future values. Wider intervals for higher prediction levels are excluded due to their limited practicality.

4   Experimental Analysis

In this section, we analyze the performance of our proposed PARNN model by conducting extensive experiments on twelve open-access time series datasets with varying frequencies and compare the results with other state-of-the-art forecasters from statistical, machine learning, and deep learning paradigms.

4.1   Data

The datasets used for the experimental evaluation exhibit distinct global characteristics and are collected from various applied domains as listed below:

Finance and Economics: In our study we consider the daily adjusted close price of three NASDAQ stock exchange datasets, namely Alphabet Inc Class C (*GOOG Stock*), Microsoft Corporation (*MSFT Stock*), and Amazon Inc. (*AMZN Stock*) collected during the financial years 2020–2022 from Yahoo finance. Additionally, we use three macroeconomic datasets comprising of the economic policy uncertainty index for the United States during 2000–2021 (*US-EPU Index*), the unemployment rate in the United Kingdom during the period 1971–2021 (*UK Unemployment*), and the national exchange rate of Russian currency units per U.S. dollars (*Russia Exchange*) during 2000–2021. These datasets provide monthly averages for the relevant series and are collected from the Fred Economic Data repository. Analyzing the forecasting performance of the proposed framework for these financial and economic datasets is crucial because accurate forecasts in these domains aid in decision-making, enable risk management, guide policy formulation, and promote market efficiency.

Algorithm 1. PARNN (m, k, l) model

Input: The standardized time series y_t
Output: Forecasts of y_t $(\hat{y}_{t+1}, \hat{y}_{t+2}, \ldots)$ for a desired test horizon.
Steps:

1. Split the series into in-sample (training and validation) and test data (out-of-sample).
2. Fit an ARIMA(p, d, q) model using the training data.
   - Obtain the predicted values by using the ARIMA model.
   - Generate the residuals e_t by subtracting the ARIMA predicted values from the observed time series y_t.
3. Fit a single hidden layered feed-forward neural network using the actual time series y_t and ARIMA feedback residuals e_t.
   - Initiate the network with a random starting weight and provide m lagged values of the original series and l lagged values of the error as the input to the network with hidden neurons $k = \lceil (m + l + 1)/2 \rceil$ (where $\lceil \cdot \rceil$ is the ceiling function).
   - We determine the optimal values of m and l by performing a grid search algorithm on the validation set (twice the size of the test set). Several PARNN models with performance measures are recorded. Eventually, the model with the least MASE score is selected.
4. Final forecast is acquired by considering the average output of 500 neural networks with optimized hyperparameters selected in the previous step and applying inverse standardization.
5. The above procedure generates a one-step-ahead forecast for the given series y_t. To generate the multi-step ahead forecasts we iteratively repeat the above procedure with the latest forecast as lagged input.

Epidemics: Alongside, economic datasets, we utilize four epidemiological data to assess the efficacy of our proposed PARNN model for predicting future infectious disease dynamics and guide health practitioners in devising appropriate interventions that mitigate the impacts of epidemics. These open-access datasets, collected from the CDC data repository, indicate the weekly crude incidence rate of dengue and malaria diseases in Colombia and Venezuela regions from 2005–2016 and 2002–2014, respectively.

Demography and Business: Furthermore, we consider two benchmark data, namely *Births* and *Tourism* in our experimentation. The Births dataset, collected from the "mosaicData" package of R software, denotes the daily number of births occurring in the US from 1968 to 1988. On the contrary, the Tourism dataset indicates the number of domestic holiday trips happening in Melbourne, Australia reported quarterly from 1998 to 2016, publicly available at [22]. Generating accurate forecasts for these datasets is essential for practitioners to adapt, seize opportunities, and effectively address challenges in their respective domains.

4.2 Analysing Global Characteristics in the Datasets

Twelve open-access time series datasets considered in this study are free from missing observations and are of different lengths. We have examined several global characteristics such as long-term dependency (LTD), non-stationarity, nonlinearity, seasonality, trend, normality, and chaotic behavior of these datasets using Hurst exponent, KPSS test, Teräsvirta test (nonlinearity test), Ollech and Webel's combined seasonality test, Mann-Kendall test, Anderson-Darling test, and Lyapunov exponent, respectively, to understand their structural patterns [10,22]. We performed all the statistical tests using R statistical software. The global characteristics of these datasets evaluated based on the above-stated statistical measures are summarized in Table 1.

Table 1. Global characteristics of the real-world datasets (green circle indicates the presence of feature and the red circle indicates the absence of feature).

Datasets	Frequency	Time Span	Length	LTD	Stationary	Linear	Seasonal	Trend	Gaussian	Chaotic
GOOG Stock	Daily	2020–2022	504	○	●	●	●	○	●	○
MSFT Stock		2020–2022	504	○	●	○	●	○	●	●
AMZN Stock		2020–2022	504	○	●	○	●	○	●	○
Births		1968–1988	7305	○	●	●	○	○	●	●
Colombia Dengue	Weekly	2005–2016	626	○	●	●	○	○	●	○
Colombia Malaria		2005–2016	626	○	●	○	○	○	●	○
Venezuela Dengue		2002–2014	660	○	●	○	○	●	●	○
Venezuela Malaria		2002–2014	669	○	●	○	○	●	●	○
US EPU Index	Monthly	2000–2021	264	○	●	○	●	○	●	●
UK unemployment		1971-2016	552	○	●	○	●	●	●	●
Russia Exchange		2000-2021	264	○	●	●	●	○	●	●
Tourism	Quarterly	1998-2017	80	○	●	○	●	○	●	○

4.3 Performance Measures

We adopt the Mean Absolute Scaled Error (MASE), Root Mean Square Error (RMSE), and Symmetric Mean Absolute Percent Error (SMAPE) as the key performance indicators in our study [22]. The mathematical formulae of these metrics are given below:

$$MASE = \frac{\sum_{i=M+1}^{M+n} |y_i - \hat{y}_i|}{\frac{n}{M-S} \sum_{i=S+1}^{M} |y_i - y_{i-s}|}; \quad RMSE = \left(\frac{1}{n} \sum_{i=1}^{n} (y_i - \hat{y}_i)^2\right)^{\frac{1}{2}};$$

$$SMAPE = \frac{1}{n} \sum_{i=1}^{n} \frac{|y_i - \hat{y}_i|}{(|y_i| + |\hat{y}_i|)/2} * 100\%;$$

where y_i is the actual value, $\hat{y}_i$ is the predicted output, n is the forecast horizon, M denotes the training data size, and S is the seasonality of the dataset. The model with the least error measure is considered the 'best' model.

4.4 Baselines

In this study, we have compared the performance of our proposed PARNN framework with several forecasters from different paradigms. These benchmarks include statistical methods, namely random walk with drift (RWD) [16], exponential smoothing (ETS) [44], autoregressive integrated moving average (ARIMA) [5], and TBAS (T: Trigonometric. B: Box-Cox transformation. A: ARIMA errors. T: Trend. S: Seasonal components) [14]; machine learning models – multilayer perceptron (MLP) [37], autoregressive neural networks (ARNN) [17]; deep learning frameworks namely NBeats [32], DeepAR [38], Temporal convolutional networks (TCN) [11], and Transformers [45], and hybrid models specifically hybrid ARIMA-ANN model (named as Hybrid-1) [47], hybrid ARIMA-ARNN model (named as Hybrid-2) [8], and hybrid ARIMA-LSTM model (named as Hybrid-3) [13].

4.5 Model Implementations

In this section, we outline the implementation of the proposed PARNN framework and other advanced models. To assess the scalability of our proposal, we conduct the experimentation with three different forecast horizons (test data) as short-term, medium-term, and long-term spanning over (5%, 10%, 20%) observations for daily datasets, (13, 26, 52) weeks for weekly datasets, (6, 12, 24) months for monthly datasets, and (4, 8, 12) quarters for quarterly dataset, respectively. Each dataset is divided chronologically into train, validation (twice the test size), and test segments. For PARNN, we initially standardize training data and fit an ARIMA model using 'auto.arima' function of R software. The best ARIMA model is selected via the Akaike information criterion (AIC) and used to estimate residuals. Further, an error correction approach models the original time series and ARIMA residuals with PARNN(m, k, l). The model receives input of m lagged original time series values and l lagged ARIMA residuals and implements the proposed framework as described in Sect. 3. Optimal (m, k, l) are chosen by minimizing the MASE score and are reported in Table 2. We iterate the training of the PARNN model over 500 repetitions with the chosen (m, k, l) values and obtain the final forecast as the average output of these networks. For instance, in the Colombia Malaria dataset for medium-term forecast, ARIMA$(1, 1, 2)$ model was fitted and $AIC = 8168.85$ was obtained. Further, the model residuals and the training data were remodeled using the PARNN model, where each parameter m and l were tuned over a grid of values between $(1, 10)$. This grid search yields a minimum MASE score at PARNN$(1, 2, 2)$ model. Following this, the fitted PARNN$(1, 2, 2)$ model was utilized to generate the final forecast as the average output of 500 networks. The predicted values are then used to evaluate the model performance. In a similar manner, we fit the PARNN model with optimal hyperparameters to the other datasets and generate the final forecast for desired horizons. Additionally, we have implemented the other individual statistical and machine learning models like RWD, ETS, ARIMA,

TBATS, and ARNN models using the "forecast" package in R statistical software [23]. The MLP framework is fitted using the "nnfor" package in R [29]. For the deep learning frameworks, we have utilized the "darts" library in Python [20], and the hybrid models are fitted using the implementation provided in [10]. Owing to the variance of machine learning and deep learning models we repeat the experiments ten times and report the mean error measures and their respective standard deviation.

Table 2. Estimated parameter values of the proposed PARNN(m, k, l) model for forecasting the pre-defined short (ST), medium (MT), and long (LT) term forecast horizons of the chosen time series data sets.

Data	GOOG Stock	MSFT Stock	AMZN Stock	Births	Colombia Dengue	Colombia Malaria	Venezuela Dengue	Venezuela Malaria	US-EPU Index	UK Unemp	Russia Exchange	Tourism
ST	2, 2, 1	10, 6, 1	1, 2, 1	6, 6, 4	2, 2, 1	3, 3, 1	5, 6, 6	3, 6, 7	4, 3, 1	10, 7, 3	3, 3, 1	5, 4, 1
MT	10, 8, 5	7, 7, 5	5, 4, 1	7, 5, 2	3, 3, 1	1, 2, 2	3, 3, 2	3, 6, 7	6, 4, 1	8, 6, 2	3, 3, 1	1, 3, 3
LT	1, 3, 3	8, 7, 4	1, 3, 3	10, 8, 5	2, 2, 1	3, 3, 1	3, 3, 2	1, 5, 7	3, 3, 1	7, 6, 3	2, 3, 2	5, 4, 3

4.6 Benchmark Comparsions

We evaluated the performance of the proposed PARNN model and thirteen benchmark forecasters for different forecast horizons. The experimental results for the short, medium, and long-term horizons are summarized in Tables 3, 4 and 5, respectively. It is observable from the given tables that there is an overall drop in the performance of all the baseline forecasters with the increasing forecast horizon since the more elongated the horizon more difficult the forecast problem. However, in the case of the proposed PARNN model, the long-range forecasting performance has significantly improved as evident from the experimental results. Moreover, owing to the distinct scale and variance of the different datasets, there is a significant difference between their corresponding accuracy metrics. For example, the RMSE values for all the models are quite high due to the scale of the data. MASE [22] is often preferred more by forecasters for which the error metrics can be better compared. The proposal generates the best short-term forecast for daily datasets except for AMZN Stock data, where the MLP model demonstrates significant improvement compared to the other forecasters. However, for the medium-term forecasts of these datasets, the statistical forecasting technique, namely ETS, shows competitive performance alongside the proposal. For the long-term counterparts, the proposed PARNN framework can successfully diminish the MASE score by 20.05%, 60.34%, 25.96%, and 1.37% for AMZN Stock, Births, MSFT Stock, and GOOG Stock prediction task, respectively in comparison with ARNN model. This improvement in the forecast accuracy of the proposed model is primarily attributed to the augmentation of the input series with ARIMA residuals, which enables the model to generate accurate long-range forecasts. Furthermore, in the case of the epidemiological forecasting

Table 3. Short-term forecast performance comparison in terms of MASE, RMSE, and SMAPE of proposed PARNN model with statistical, machine learning, and deep learning forecasters. Mean values and (standard deviations) of 10 repetitions are reported in the table and the best-performing models are **highlighted**.

Dataset	Metric	RWD [16]	ETS [44]	ARIMA [5]	TBATS [14]	MLP [37]	ARNN [17]	Hybrid-1 [47]	Hybrid-2 [8]	NBeats [32]	DeepAR [38]	TCN [11]	Transformers [45]	Hybrid-3 [13]	PARNN (Proposed)
AMZN Stock	MASE	3.352 (0)	3.648 (0)	3.679 (0)	3.087 (0)	**2.527** (0.04)	2.566 (0.01)	3.639 (0.001)	3.631 (0.02)	2.943 (0.43)	8.428 (11.3)	8.296 (15.3)	3.889 (2.07)	3.705 (0.11)	3.568 (0.01)
	RMSE	12.00 (0)	13.17 (0)	13.26 (0)	10.76 (0)	**8.444** (0.14)	8.502 (0.03)	13.17 (0.003)	13.14 (0.07)	10.75 (1.41)	26.52 (33.1)	26.94 (44.6)	13.26 (6.13)	13.56 (0.40)	10.82 (0.03)
	SMAPE	6.514 (0)	7.113 (0)	7.177 (0)	5.985 (0)	**4.896** (0.08)	4.973 (0.02)	7.098 (0.002)	7.080 (0.05)	5.710 (0.82)	21.08 (33.1)	25.56 (59.5)	7.676 (4.56)	7.236 (0.23)	6.308 (0.02)
GOOG Stock	MASE	3.034 (0)	2.671 (0)	3.074 (0)	4.017 (0)	1.869 (0.07)	2.149 (0.01)	3.067 (0.001)	3.078 (0.004)	2.931 (1.59)	65.95 (1.22)	11.81 (20.8)	27.79 (19.4)	3.038 (0.06)	**1.437** (0.02)
	RMSE	137.5 (0)	119.3 (0)	139.3 (0)	186.4 (0)	84.09 (2.83)	94.75 (0.46)	139.1 (0.03)	139.4 (0.15)	136.4 (74.2)	2620 (48.5)	507.1 (830)	1109 (768)	137.6 (2.76)	**68.85** (0.34)
	SMAPE	4.506 (0)	3.957 (0)	4.567 (0)	6.019 (0)	2.768 (0.11)	3.178 (0.02)	4.557 (0.001)	4.573 (0.006)	4.391 (2.50)	188.3 (6.79)	27.92 (59.0)	60.31 (56.2)	4.513 (0.09)	**2.135** (0.03)
MSFT Stock	MASE	2.759 (0)	3.213 (0)	2.083 (0)	3.198 (0)	2.159 (0.08)	2.594 (1.03)	2.684 (0.001)	2.682 (0.001)	2.266 (0.64)	20.26 (17.9)	8.649 (18.5)	11.77 (6.73)	2.629 (0.06)	**1.385** (0.07)
	RMSE	15.02 (0)	17.84 (0)	14.61 (0)	17.76 (0)	11.64 (0.45)	14.12 (5.49)	14.61 (0.004)	14.61 (0.003)	12.66 (3.57)	98.65 (87.9)	44.55 (32.0)	57.54 (32.0)	14.54 (0.32)	**8.302** (0.22)
	SMAPE	4.536 (0)	5.305 (0)	4.408 (0)	5.279 (0)	3.536 (0.13)	4.284 (1.78)	4.409 (0.002)	4.407 (0.001)	3.719 (1.07)	47.19 (49.7)	23.52 (59.9)	21.98 (15.9)	4.319 (0.10)	**2.282** (0.11)
Births	MASE	2.416 (0)	1.370 (0)	1.210 (0)	1.222 (0)	1.487 (0.01)	1.552 (0.02)	1.893 (0.97)	1.173 (0.01)	2.117 (0.05)	11.35 (2.83)	3.520 (3.16)	1.625 (0.003)	1.205 (0.004)	**0.940** (0.29)
	RMSE	2538 (0)	1376 (0)	1293 (0)	1286 (0)	1686 (18.7)	1700 (24.6)	1374 (4.98)	1283 (5.55)	2247 (17.8)	3648 (78.5)	3873 (2963)	1671 (3.40)	1302 (2.93)	**1087** (309)
	SMAPE	22.41 (0)	12.05 (0)	10.59 (0)	10.72 (0)	14.98 (0.19)	14.05 (0.18)	10.67 (0.07)	10.24 (0.04)	20.16 (7.83)	19.1 (23.9)	37.09 (55.6)	14.42 (0.03)	10.55 (0.03)	**8.255** (2.48)
Colombia Dengue	MASE	2.129 (0)	2.096 (0)	1.754 (0)	1.388 (0)	2.214 (0.06)	1.868 (0.34)	1.751 (0.003)	1.683 (0.08)	**1.349** (0.49)	7.467 (1.68)	4.113 (2.12)	6.851 (5.23)	1.737 (0.08)	2.030 ($3E^{-15}$)
	RMSE	254.3 (0)	250.4 (0)	227.3 (0)	196.3 (0)	260.0 (5.13)	199.5 (41.9)	227.1 (0.16)	227.5 (6.72)	**188.2** (36.3)	699.0 (150)	454.6 (180)	658.4 (471)	227.4 (5.46)	245.9 ($2E^{-13}$)
	SMAPE	19.76 (0)	19.50 (0)	16.84 (0)	**13.90** (0)	20.39 (0.44)	20.01 (4.03)	16.81 (0.02)	16.24 (0.61)	14.13 (5.54)	110.2 (39.4)	44.11 (43.1)	51.95 (33.7)	16.69 (0.60)	19.01 ($2E^{-14}$)
Colombia Malaria	MASE	1.475 (0)	1.481 (0)	1.472 (0)	1.617 (0)	1.424 (0.04)	1.546 (0.09)	1.488 (0.001)	1.463 (0.03)	1.670 (0.19)	6.470 (0.86)	3.945 (1.69)	3.099 (1.04)	1.478 (0.03)	**1.093** (0.02)
	RMSE	258.8 (0)	276.4 (0)	266.2 (0)	306.3 (0)	259.0 (14.9)	288.0 (22.3)	265.9 (0.10)	263.0 (3.82)	237.8 (14.7)	822.9 (102)	588.4 (246)	446.3 (121)	259.4 (4.01)	**181.8** (2.01)
	SMAPE	21.22 (0)	21.12 (0)	21.11 (0)	22.37 (0)	20.59 (0.39)	21.74 (0.86)	21.31 (0.004)	21.05 (0.29)	24.45 (2.81)	132.4 (31.6)	45.94 (10.7)	41.70 (23.5)	21.24 (0.34)	**16.59** (0.36)
Venezuela Dengue	MASE	4.248 (0)	3.877 (0)	4.129 (0)	4.198 (0)	4.058 (0.03)	4.059 (0.04)	4.114 (0.002)	4.128 (0.001)	4.839 (0.45)	6.437 (1.03)	4.528 (1.22)	4.061 (1.32)	4.089 (0.05)	**3.875** (0.04)
	RMSE	813.2 (0)	**742.3** (0)	794.8 (0)	804.6 (0)	778.3 (5.46)	777.3 (7.49)	792.8 (0.20)	794.7 (0.14)	926.4 (72.9)	1135 (156)	853.8 (209)	784.5 (209)	789.8 (6.62)	753.9 (6.73)
	SMAPE	62.99 (0)	55.66 (0)	60.48 (0)	61.95 (0)	59.14 (0.60)	59.21 (0.86)	60.18 (0.03)	60.47 (0.02)	76.32 (10.7)	125.3 (35.1)	75.19 (26.1)	62.19 (29.0)	59.65 (0.99)	**55.61** (0.67)
Venezuela Malaria	MASE	0.798 (0)	**0.797** (0)	0.824 (0)	0.801 (0)	0.921 (0.05)	1.071 (0.02)	0.802 (0.001)	0.815 (0.001)	1.292 (0.42)	10.17 (1.22)	3.139 (1.49)	6.271 (1.41)	0.832 (0.04)	0.825 (0.02)
	RMSE	121.0 (0)	**120.6** (0)	121.2 (0)	120.7 (0)	129.4 (5.34)	152.1 (5.13)	120.8 (0.01)	120.9 (0.04)	193.9 (63.6)	1337 (159)	501.9 (212)	831.5 (183)	122.9 (2.33)	138.8 (3.76)
	SMAPE	6.519 (0)	**6.510** (0)	6.731 (0)	6.536 (0)	7.528 (0.40)	8.783 (0.20)	6.552 (0.01)	6.653 (0.01)	10.86 (3.99)	140.9 (28.8)	27.96 (20.8)	69.13 (24.7)	6.791 (0.29)	6.725 (0.16)
US-EPU Index	MASE	0.799 (0)	0.753 (0)	0.849 (0)	0.823 (0)	0.615 (0.03)	0.820 (0.04)	0.871 (0.01)	0.868 (0.003)	2.128 (0.52)	2.139 (1.97)	2.745 (1.88)	1.332 (1.37)	0.970 (0.11)	**0.534** (0.06)
	RMSE	17.24 (0)	15.88 (0)	19.75 (0)	17.90 (0)	13.73 (0.55)	13.84 (0.45)	20.40 (0.09)	20.41 (0.04)	50.29 (10.7)	41.42 (34.6)	55.65 (34.6)	27.17 (24.1)	22.49 (2.19)	**12.91** (0.48)
	SMAPE	10.14 (0)	9.523 (0)	10.49 (0)	10.45 (0)	7.773 (0.35)	8.040 (0.47)	10.74 (0.05)	10.70 (0.03)	24.32 (5.71)	35.46 (40.8)	43.46 (47.8)	19.96 (26.3)	11.84 (1.27)	**6.671** (0.67)
UK unem-ployment	MASE	2.674 (0)	0.957 (0)	0.720 (0)	0.951 (0)	0.763 (0.06)	0.716 (0.01)	0.720 (0.002)	0.719 (0.001)	1.427 (0.52)	2.760 (1.18)	5.187 (3.33)	2.426 (1.61)	0.737 (0.03)	**0.493** (0.003)
	RMSE	0.123 (0)	0.048 (0)	0.031 (0)	0.047 (0)	0.036 (0.004)	0.036 (0.003)	0.031 (0.001)	0.031 (0.001)	0.067 (0.03)	0.128 (0.05)	0.251 (0.16)	0.106 (0.07)	0.033 (0.003)	**0.023** (0.001)
	SMAPE	2.211 (0)	0.798 (0)	0.598 (0)	0.793 (0)	0.636 (0.05)	0.598 (0.01)	0.601 (0.002)	0.600 (0.001)	1.187 (0.43)	2.284 (0.97)	4.265 (2.56)	1.998 (1.31)	0.614 (0.02)	**0.411** (0.002)
Russia Exchange	MASE	1.865 (0)	0.928 (0)	0.878 (0)	**0.645** (0)	0.748 (0.07)	1.078 (0.17)	0.856 (0.004)	0.818 (0.01)	2.141 (0.93)	8.808 (14.1)	10.37 (11.6)	10.46 (9.68)	1.392 (0.24)	0.875 (0.02)
	RMSE	2.078 (0)	0.981 (0)	1.129 (0)	**0.792** (0)	0.860 (0.08)	1.266 (0.18)	1.113 (0.003)	1.084 (0.002)	2.431 (1.015)	8.811 (13.8)	11.74 (12.5)	10.35 (9.54)	1.741 (0.27)	0.973 (0.03)
	SMAPE	2.497 (0)	1.258 (0)	1.191 (0)	**0.874** (0)	1.012 (0.09)	1.465 (0.23)	1.161 (0.01)	1.109 (0.01)	2.849 (1.22)	15.14 (27.2)	17.15 (26.1)	16.41 (18.5)	1.886 (0.32)	1.184 (0.02)
Tourism	MASE	**0.647** (0)	0.751 (0)	0.716 (0)	0.842 (0)	0.956 (0.07)	0.751 (0.01)	0.701 (0.003)	0.708 (0.001)	1.118 (0.19)	7.022 (0.09)	1.672 (1.89)	6.328 (0.64)	0.710 (0.003)	0.917 (0.24)
	RMSE	86.71 (0)	**80.46** (0)	83.62 (0)	83.59 (0)	94.63 (5.63)	104.4 (0.59)	84.04 (0.06)	84.06 (0.01)	109.8 (19.6)	652.5 (8.51)	172.1 (71.5)	589.4 (58.3)	83.82 (0.31)	98.19 (29.4)
	MASE	**8.674** (0)	10.19 (0)	9.679 (0)	11.44 (0)	12.94 (0.86)	10.22 (0.08)	9.460 (0.04)	9.563 (0.01)	15.06 (2.40)	191.5 (5.04)	31.05 (51.2)	159.3 (28.3)	9.590 (0.03)	12.56 (3.23)

Table 4. Medium-term forecast performance comparison in terms of MASE, RMSE, and SMAPE of proposed PARNN model with statistical, machine learning, and deep learning forecasters. Mean values and (standard deviations) of 10 repetitions are reported in the table and the best-performing models are **highlighted**.

Dataset	Metric	RWD [16]	ETS [44]	ARIMA [5]	TBATS [14]	MLP [37]	ARNN [17]	Hybrid-1 [47]	Hybrid-2 [8]	NBeats [32]	DeepAR [38]	TCN [11]	Transformers [45]	Hybrid-3 [13]	PARNN (Proposed)
AMZN Stock	MASE	2.987 (0)	2.552 (0)	2.550 (0)	3.420 (0)	3.472 (0.13)	3.313 (0.01)	2.577 (0.001)	2.588 (0.01)	4.098 (0.37)	8.278 (9.37)	9.096 (13.8)	4.071 (3.09)	2.610 (0.10)	**2.440** ($3E^{-4}$)
	RMSE	12.98 (0)	10.93 (0)	10.93 (0)	14.35 (0)	14.53 (0.41)	14.01 (0.02)	11.03 (0.004)	11.07 (0.03)	16.53 (1.19)	30.74 (32.3)	38.41 (56.9)	16.44 (10.4)	11.19 (0.37)	**10.40** ($1E^{-3}$)
	SMAPE	6.816 (0)	5.867 (0)	5.864 (0)	7.749 (0)	7.860 (0.26)	7.520 (0.01)	5.924 (0.002)	5.947 (0.02)	9.186 (0.77)	24.29 (33.8)	23.30 (40.9)	9.820 (8.74)	5.993 (0.22)	**5.620** ($8E^{-4}$)
GOOG Stock	MASE	2.793 (0)	2.843 (0)	2.870 (0)	1.804 (0)	2.697 (0.09)	2.410 (0.004)	2.869 (0.004)	2.867 (0.002)	5.364 (0.61)	55.12 (0.96)	12.05 (17.9)	25.43 (17.0)	2.874 (0.03)	**1.758** (0.01)
	RMSE	156.8 (0)	159.1 (0)	160.3 (0)	103.5 (0)	152.5 (4.24)	139.5 (0.19)	160.2 (0.18)	160.2 (0.06)	275.4 (29.2)	2626 (45.4)	673.9 (996)	1217 (806)	160.3 (1.48)	**100.9** (0.31)
	SMAPE	4.856 (0)	4.940 (0)	4.985 (0)	3.175 (0)	4.699 (0.16)	4.216 (0.01)	4.984 (0.01)	4.981 (0.01)	9.065 (0.99)	189.1 (6.38)	23.67 (41.5)	68.02 (59.8)	4.992 (0.06)	**3.093** (0.03)
MSFT Stock	MASE	2.828 (0)	**1.684** (0)	2.968 (0)	1.713 (0)	3.774 (0.14)	3.084 (0.01)	2.968 (0.001)	2.968 (0.001)	7.013 (0.58)	20.49 (17.6)	11.63 (18.9)	12.06 (8.11)	2.952 (0.11)	1.845 (0.03)
	RMSE	17.38 (0)	**10.94** (0)	18.08 (0)	11.11 (0)	22.37 (0.73)	18.84 (0.04)	18.08 (0.01)	18.08 (0.004)	37.99 (2.67)	104.4 (87.9)	70.89 (112)	61.68 (40.7)	17.84 (0.59)	11.77 (0.16)
	SMAPE	4.740 (0)	**2.867** (0)	4.966 (0)	2.916 (0)	6.257 (0.23)	5.155 (0.01)	4.966 (0.002)	4.966 (0.001)	11.27 (0.87)	50.36 (51.3)	22.13 (41.5)	24.29 (21.4)	4.942 (0.18)	3.136 (0.05)
Births	MASE	2.447 (0)	1.362 (0)	1.352 (0)	1.344 (0)	2.169 (0.01)	2.180 (0.001)	1.254 (0.01)	1.276 (0.01)	6.439 (0.56)	11.25 (13.8)	9.102 (11.7)	1.571 (0.002)	1.313 (0.03)	**0.636** (0.05)
	RMSE	2553 (0)	1354 (0)	1376 (0)	1369 (0)	2229 (0.001)	2227 (0.01)	1365 (4.87)	1349 (5.07)	7417 (83.9)	$10E^3$ (97.4)	$10E^3$ ($14E^3$)	1597 (1.98)	1352 (20.6)	**760.8** (94.1)
	SMAPE	23.00 (0)	12.09 (0)	12.01 (0)	11.93 (0)	20.53 (0.1)	20.59 (0.01)	11.25 (0.57)	11.30 (0.06)	39.54 (1.95)	191.2 (11.7)	64.20 (68.2)	14.05 (0.02)	11.64 (0.24)	**5.610** (0.42)
Colombia Dengue	MASE	8.720 (0)	8.718 (0)	9.069 (0)	8.613 (0)	10.33 (0.13)	10.69 (0.59)	9.075 (0.003)	9.098 (0.08)	10.17 (0.99)	10.09 (1.48)	56.87 (68.5)	6.197 (2.76)	9.118 (0.12)	**1.762** (0.23)
	RMSE	922.4 (0)	917.2 (0)	949.8 (0)	906.0 (0)	1083 (14.7)	1122 (63.6)	953.0 (0.39)	955.2 (6.16)	1089 (136)	1043 (136)	7421 (9200)	680.4 (259)	957.5 (13.3)	**222.6** (20.1)
	SMAPE	53.29 (0)	53.33 (0)	54.79 (0)	52.90 (0)	59.58 (0.47)	60.87 (2.11)	54.79 (0.01)	54.86 (0.32)	58.36 (3.25)	126.1 (32.7)	109.6 (36.9)	49.47 (32.4)	54.96 (0.49)	**14.76** (1.76)
Colombia Malaria	MASE	4.436 (0)	4.997 (0)	4.661 (0)	4.916 (0)	3.330 (0.23)	2.733 (0.24)	4.652 (0.01)	4.631 (0.04)	4.599 (1.29)	8.834 (0.91)	10.87 (4.24)	3.911 (1.94)	4.481 (0.04)	**2.543** (0.002)
	RMSE	591.1 (0)	655.3 (0)	617.0 (0)	646.8 (0)	444.8 (26.4)	404.8 (27.7)	615.3 (0.56)	610.9 (5.64)	637.1 (156)	1059 (99.1)	1507 (526)	519.7 (211)	595.6 (4.24)	**361.7** (0.15)
	SMAPE	38.59 (0)	42.03 (0)	39.99 (0)	41.54 (0)	31.28 (1.63)	26.55 (1.89)	39.94 (0.03)	39.81 (0.26)	38.53 (8.75)	143.8 (27.5)	71.52 (27.4)	42.79 (31.9)	38.87 (0.25)	**25.26** (0.02)
Venezuela Dengue	MASE	3.381 (0)	3.881 (0)	3.179 (0)	3.347 (0)	3.223 (0.01)	3.167 (0.01)	3.183 (0.001)	3.180 (0.001)	3.833 (0.78)	5.536 (1.19)	7.674 (3.35)	3.157 (0.98)	3.149 (0.02)	**2.628** (0.001)
	RMSE	513.5 (0)	588.0 (0)	486.6 (0)	508.6 (0)	492.6 (1.95)	485.2 (1.06)	487.3 (0.07)	486.8 (0.05)	604.5 (133)	875.7 (138)	1292 (564)	518.5 (144)	482.9 (2.76)	**447.7** (0.01)
	SMAPE	41.89 (0)	46.25 (0)	39.95 (0)	41.58 (0)	40.42 (0.13)	39.87 (0.09)	39.99 (0.01)	39.97 (0.01)	48.40 (10.9)	107.8 (40.7)	76.20 (27.9)	44.28 (24.2)	39.65 (0.23)	**33.79** (0.003)
Venezuela Malaria	MASE	0.815 (0)	0.992 (0)	1.217 (0)	1.426 (0)	0.894 (0.05)	0.865 (0.08)	1.179 (0.003)	1.164 (0.04)	1.518 (0.63)	7.783 (0.92)	3.976 (2.63)	4.587 (1.31)	1.144 (0.05)	**0.768** (0.01)
	RMSE	**162.6** (0)	204.6 (0)	244.6 (0)	278.4 (0)	182.2 (12.6)	186.4 (13.4)	238.5 (0.36)	237.1 (5.56)	303.8 (100)	1332 (154)	803.2 (525)	794.8 (219)	232.8 (7.84)	171.2 (2.44)
	SMAPE	8.677 (0)	10.68 (0)	13.32 (0)	15.86 (0)	9.562 (0.60)	9.236 (0.79)	12.87 (0.03)	12.69 (0.48)	17.39 (8.13)	141.6 (28.5)	51.77 (34.3)	65.97 (29.7)	12.46 (0.53)	**8.249** (0.11)
US EPU Index	MASE	4.939 (0)	4.716 (0)	4.590 (0)	2.264 (0)	3.927 (0.36)	5.076 (1.00)	4.729 (0.004)	4.712 (0.003)	6.132 (1.04)	2.147 (1.74)	6.125 (3.81)	**1.567** (1.33)	4.772 (0.22)	1.727 (0.27)
	RMSE	104.9 (0)	100.3 (0)	97.62 (0)	51.93 (0)	84.43 (7.18)	110.5 (20.8)	100.5 (0.08)	100.1 (0.05)	136.4 (26.0)	48.42 (35.2)	145.6 (82.7)	**39.42** (25.8)	101.3 (4.48)	40.84 (0.51)
	SMAPE	52.22 (0)	50.49 (0)	49.49 (0)	27.82 (0)	43.90 (3.13)	52.56 (8.05)	50.59 (0.03)	50.45 (0.02)	59.06 (6.41)	39.58 (41.6)	65.66 (42.5)	26.63 (30.7)	50.91 (1.71)	**22.36** (2.98)
UK unem-ployment	MASE	6.154 (0)	2.167 (0)	1.365 (0)	2.128 (0)	1.634 (0.33)	2.159 (0.20)	1.357 (0.01)	1.361 (0.002)	9.866 (2.82)	3.483 (3.59)	17.30 (3.72)	10.14 (4.33)	1.344 (0.09)	**0.906** (0.01)
	RMSE	0.258 (0)	0.089 (0)	0.060 (0)	0.087 (0)	0.072 (0.01)	0.091 (0.01)	0.059 (0.001)	0.059 (0.001)	0.396 (0.11)	0.160 (0.16)	0.858 (0.21)	0.382 (0.15)	0.059 (0.004)	**0.038** (0.001)
	SMAPE	4.501 (0)	1.612 (0)	1.024 (0)	1.584 (0)	1.215 (0.25)	1.607 (0.15)	1.018 (0.01)	1.021 (0.002)	7.072 (1.94)	2.534 (2.54)	12.15 (2.96)	7.245 (2.98)	1.008 (0.07)	**0.671** (0.01)
Russia Exchange	MASE	4.816 (0)	4.172 (0)	3.632 (0)	2.926 (0)	2.527 (0.19)	5.567 (0.47)	3.631 (0.01)'	3.642 (0.004)	12.07 (8.85)	13.06 (13.7)	23.79 (36.1)	13.89 (10.7)	3.399 (0.09)	**1.250** (0.08)
	RMSE	4.912 (0)	4.197 (0)	3.874 (0)	3.028 (0)	2.646 (0.17)	5.750 (0.48)	3.873 (0.01)	3.883 (0.004)	12.31 (8.78)	12.81 (13.1)	25.89 (36.4)	13.47 (10.2)	3.766 (0.10)	**1.492** (0.11)
	SMAPE	6.088 (0)	5.301 (0)	4.627 (0)	3.750 (0)	3.248 (0.24)	7.573 (0.66)	4.626 (0.01)	4.640 (0.01)	14.01 (9.47)	20.97 (26.7)	32.26 (49.9)	21.32 (20.8)	4.335 (0.12)	**1.626** (0.12)
Tourism	MASE	0.773 (0)	0.752 (0)	**0.727** (0)	0.762 (0)	0.984 (0.07)	1.057 (0.02)	0.737 (0.01)	0.759 (0.02)	1.024 (0.09)	9.141 (0.12)	5.407 (0.79)	8.257 (0.82)	0.729 (0.003)	1.073 (0.003)
	RMSE	83.21 (0)	79.75 (0)	**76.45** (0)	81.84 (0)	94.74 (5.25)	101.9 (1.17)	77.93 (0.07)	80.99 (0.12)	91.51 (8.42)	652.1 (8.71)	415.4 (533)	590.0 (57.9)	76.80 (0.16)	102.1 (0.13)
	SMAPE	8.104 (0)	7.863 (0)	**7.588** (0)	7.982 (0)	10.55 (0.84)	11.42 (0.20)	7.693 (0.01)	7.947 (0.01)	10.59 (0.98)	191.9 (5.13)	62.26 (83.4)	160.2 (27.9)	7.602 (0.03)	11.62 (0.03)

Table 5. Long-term forecast performance comparison in terms of MASE, RMSE, and SMAPE of proposed PARNN model with statistical, machine learning, and deep learning forecasters. Mean values and (standard deviations) of 10 repetitions are reported in the table and the best-performing models are **highlighted**.

Dataset	Metric	RWD [16]	ETS [44]	ARIMA [5]	TBATS [14]	MLP [37]	ARNN [17]	Hybrid-1 [47]	Hybrid-2 [8]	NBeats [32]	DeepAR [38]	TCN [11]	Transformers [45]	Hybrid-3 [13]	PARNN (Proposed)
AMZN Stock	MASE	7.926	4.674	4.674	3.929	3.384	3.304	4.722	4.724	5.284	12.49	13.00	5.630	4.667	**2.641**
		(0)	(0)	(0)	(0)	(0.11)	(0.004)	(0.002)	(0.01)	(0.71)	(13.8)	(16.6)	(6.12)	(0.19)	(0.74)
	RMSE	27.37	17.01	17.01	14.79	12.55	12.24	17.15	17.16	18.76	37.79	44.09	17.67	17.03	**9.974**
		(0)	(0)	(0)	(0)	(0.39)	(0.02)	(0.004)	(0.03)	(2.38)	(38.2)	(51.8)	(17.2)	(0.49)	(2.66)
	SMAPE	13.22	8.141	8.141	6.917	5.998	5.864	8.219	8.222	9.117	28.94	27.57	10.91	8.125	**4.714**
		(0)	(0)	(0)	(0)	(0.18)	(0.01)	(0.002)	(0.02)	(1.14)	(38.4)	(46.5)	(13.8)	(0.32)	(1.25)
GOOG Stock	MASE	10.49	10.39	10.32	5.413	8.960	2.869	10.30	10.34	11.25	70.52	18.24	36.81	10.31	**2.830**
		(0)	(0)	(0)	(0)	(0.23)	(0.01)	(0.01)		(0.56)	(1.05)	(20.8)	(20.1)	(0.07)	(0.02)
	RMSE	476.5	472.1	469.9	253.8	410.2	132.9	469.3	470.5	521.5	2737	843.5	1435	469.9	**128.4**
		(0)	(0)	(0)	(0)	(10.1)	(0.79)	(0.17)	(0.54)	(24.1)	(40.6)	(897)	(775)	(3.29)	(0.02)
	SMAPE	13.55	13.43	13.34	7.331	11.73	**3.722**	13.32	13.36	14.38	190.7	29.02	77.98	13.32	3.938
		(0)	(0)	(0)	(0)	(0.27)	(0.02)	(0.01)	(0.04)	(0.65)	(5.55)	(45.3)	(58.4)	(0.09)	(0.02)
MSFT Stock	MASE	10.02	9.865	9.947	5.549	10.98	4.741	9.950	9.950	7.792	29.46	15.31	18.78	9.849	**3.240**
		(0)	(0)	(0)	(0)	(0.54)	(0.05)	(0.002)	(0.001)	(0.68)	(19.7)	(20.4)	(10.1)	(0.08)	(0.02)
	RMSE	54.51	53.76	54.20	31.09	59.36	26.73	54.21	54.21	43.20	134.1	81.25	86.97	53.74	**17.59**
		(0)	(0)	(0)	(0)	(2.86)	(0.28)	(0.01)	(0.003)	(4.41)	(87.2)	(100)	(44.7)	(0.38)	(0.33)
	SMAPE	13.52	13.32	13.43	7.841	14.67	6.761	13.43	13.43	10.75	61.63	26.93	32.81	13.31	**4.660**
		(0)	(0)	(0)	(0)	(0.65)	(0.07)	(0.002)	(0.001)	(0.83)	(50.9)	(47.1)	(23.9)	(0.09)	(0.03)
Births	MASE	2.383	1.492	1.462	1.422	2.301	2.289	1.426	1.397	3.862	10.94	$62E^3$	1.539	1.424	**0.554**
		(0)	(0)	(0)	(0)	(0.001)	(0.01)	(1.87)	(0.04)	(2.51)	(1.67)	$(19E^3)$	(0.01)	(0.03)	(0.003)
	RMSE	2432	1480	1442	1404	2187	2199	1407	1427	4165	9877	$15E^6$	1540	1426	**640.5**
		(0)	(0)	(0)	(0)	(0.03)	(0.15)	(19.2)	(25.2)	(2859)	(2176)	$(49E^5)$	(5.35)	(19.4)	(2.42)
	SMAPE	22.03	13.13	12.86	12.49	22.01	21.41	13.98	12.25	28.56	176.8	90.49	13.57	12.51	**4.832**
		(0)	(0)	(0)	(0)	(0.12)	(0.35)	(0.67)	(0.35)	(13.4)	(42.8)	(69.2)	(0.04)	(0.22)	(0.03)
Colombia Dengue	MASE	5.337	4.862	4.865	4.860	4.324	4.096	4.864	4.865	18.535	10.483	7.645	5.928	4.935	**3.228**
		(0)	(0)	(0)	(0)	(0.09)	(0.17)	(0.01)	(0.003)	(4.84)	(0.85)	(3.47)	(2.16)	(0.03)	(0.04)
	RMSE	1111	997.2	998.3	997.5	845.3	840.1	997.8	998.5	3610	1914	1727	1125	1015	**611.0**
		(0)	(0)	(0)	(0)	(22.8)	(39.1)	(1.68)	(0.74)	(1094)	(127)	(788)	(405)	(7.17)	(7.98)
	SMAPE	43.40	40.98	40.99	40.96	37.99	36.42	40.99	40.99	83.74	146.6	59.81	61.39	41.37	**28.94**
		(0)	(0)	(0)	(0)	(0.51)	(1.03)	(0.03)	(0.01)	(8.62)	(24.9)	(29.1)	(37.5)	(0.16)	(0.36)
Colombia Malaria	MASE	3.765	3.501	3.547	3.519	3.837	3.661	3.559	3.575	3.565	8.310	5.533	4.101	3.617	**3.272**
		(0)	(0)	(0)	(0)	(0.23)	(0.19)	(0.001)	(0.01)	(0.49)	(0.54)	(1.83)	(1.65)	(0.04)	(0.01)
	RMSE	850.5	798.6	810.3	803.0	849.4	811.7	813.2	816.4	759.9	1623	1231	880.4	826.3	**724.5**
		(0)	(0)	(0)	(0)	(33.8)	(32.0)	(0.24)	(3.29)	(86.9)	(91.9)	(378)	(294)	(7.63)	(5.09)
	SMAPE	45.08	41.05	41.71	41.31	46.57	43.68	41.89	42.14	41.23	158.1	66.09	55.03	42.74	**37.93**
		(0)	(0)	(0)	(0)	(3.76)	(3.27)	(0.01)	(0.20)	(5.58)	(21.2)	(28.7)	(36.3)	(0.55)	(0.15)
Venezuela Dengue	MASE	3.276	3.387	3.270	3.273	3.512	3.572	3.267	3.269	4.557	6.990	6.445	4.912	3.27	**3.096**
		(0)	(0)	(0)	(0)	(0.07)	(0.07)	(0.001)	(0.001)	(1.19)	(0.98)	(1.49)	(0.96)	(0.01)	(0.04)
	RMSE	624.9	614.9	622.3	622.9	691.7	705.3	621.4	622.0	854.3	1240	1351	927.3	621.8	**592.7**
		(0)	(0)	(0)	(0)	(16.7)	(13.6)	(0.16)	(0.09)	(199)	(141)	(283)	(148)	(1.67)	(26.8)
	SMAPE	39.93	40.61	39.85	39.89	43.34	44.24	39.81	39.84	62.84	126.8	74.68	70.38	39.86	**37.01**
		(0)	(0)	(0)	(0)	(0.98)	(1.01)	(0.01)	(0.01)	(21.7)	(33.7)	(24.6)	(23.9)	(0.09)	(2.06)
Venezuela Malaria	MASE	1.541	1.388	1.279	1.278	1.380	3.577	1.263	1.207	2.381	7.128	19.03	4.256	1.275	**1.067**
		(0)	(0)	(0)	(0)	(0.45)	(0.17)	(0.001)	(0.03)	(1.04)	(0.89)	(47.9)	(1.10)	(0.02)	(0.004)
	RMSE	299.7	271.1	250.2	249.9	273.0	658.3	247.2	237.5	464.7	1208	4941	745.9	249.3	**218.4**
		(0)	(0)	(0)	(0)	(87.0)	(33.2)	(0.16)	(4.65)	(189)	(145)	$(13E^3)$	(177)	(3.26)	(0.72)
	SMAPE	18.25	16.30	14.95	14.94	16.48	50.00	14.76	14.08	24.05	138.5	57.77	64.75	14.90	**12.46**
		(0)	(0)	(0)	(0)	(5.89)	(3.09)	(0.01)	(0.37)	(8.55)	(29.8)	(46.6)	(27.2)	(0.19)	(0.04)
US-EPU Index	MASE	1.595	1.685	1.671	1.948	1.681	1.617	1.658	1.655	1.837	2.511	2.758	2.421	1.591	**1.483**
		(0)	(0)	(0)	(0)	(0.02)	(0.03)	(0.002)	(0.001)	(0.15)	(1.06)	(1.28)	(0.67)	(0.06)	(0.003)
	RMSE	81.41	85.73	85.03	93.35	85.53	82.19	84.50	84.34	89.77	110.2	124.9	108.9	80.93	**68.41**
		(0)	(0)	(0)	(0)	(0.70)	(1.99)	(0.05)	(0.02)	(4.64)	(34.5)	(57.5)	(20.9)	(2.75)	(0.15)
	SMAPE	29.24	31.36	31.01	38.27	31.27	29.77	30.72	30.64	35.53	59.35	59.99	53.67	29.19	**26.95**
		(0)	(0)	(0)	(0)	(0.46)	(0.76)	(0.03)	(0.01)	(4.01)	(38.7)	(27.2)	(25.6)	(1.27)	(0.07)
UK unemployment	MASE	12.69	1.671	9.233	1.802	3.381	5.650	9.225	9.235	12.28	5.191	15.33	10.16	9.227	**1.465**
		(0)	(0)	(0)	(0)	(1.17)	(0.02)	(0.01)	(0.002)	(4.08)	(2.89)	(5.21)	(2.23)	(0.05)	(0.13)
	RMSE	0.687	0.096	0.518	0.103	0.196	0.320	0.518	0.518	0.780	0.274	1.002	0.503	0.518	**0.082**
		(0)	(0)	(0)	(0)	(0.07)	(0.002)	(0.001)	(0.001)	(0.28)	(0.15)	(0.33)	(0.10)	(0.002)	(0.005)
	SMAPE	11.31	1.564	9.497	1.691	3.175	5.259	9.490	9.499	10.73	4.822	13.80	9.100	9.492	**1.362**
		(0)	(0)	(0)	(0)	(1.09)	(0.02)	(0.01)	(0.002)	(3.27)	(2.59)	(4.51)	(1.89)	(0.05)	(0.12)
Russia Exchange	MASE	3.749	5.376	4.676	4.833	4.754	5.120	4.625	4.648	4.199	8.792	10.97	7.442	4.622	**3.162**
		(0)	(0)	(0)	(0)	(0.20)	(0.05)	(0.003)	(0.001)	(0.83)	(5.96)	(11.9)	(5.60)	(0.11)	(0.69)
	RMSE	7.616	10.78	9.402	9.717	9.561	10.28	9.297	9.340	8.550	17.34	24.45	14.67	9.320	**6.555**
		(0)	(0)	(0)	(0)	(0.39)	(0.10)	(0.01)	(0.001)	(1.58)	(10.8)	(27.3)	(10.4)	(0.17)	(1.32)
	SMAPE	10.04	14.75	12.69	13.15	12.92	13.99	12.55	12.61	11.36	27.37	31.26	22.78	12.55	**8.405**
		(0)	(0)	(0)	(0)	(0.58)	(0.16)	(0.01)	(0.001)	(2.37)	(23.9)	(34.7)	(21.5)	(0.31)	(1.92)
Tourism	MASE	0.726	0.801	**0.692**	0.907	0.744	0.935	0.703	0.792	1.050	8.353	6.565	8.217	0.694	0.771
		(0)	(0)	(0)	(0)	(0.05)	(0.02)	(0.001)	(0.001)	(0.05)	(0.12)	(5.86)	(0.23)	(0.003)	(0.003)
	RMSE	**72.53**	87.48	75.21	95.90	83.25	99.46	78.11	80.28	94.55	639.1	587.4	628.9	75.99	85.66
		(0)	(0)	(0)	(0)	(5.04)	(1.29)	(0.05)	(0.01)	(7.28)	(8.90)	(510)	(16.9)	(0.12)	(0.34)
	SMAPE	8.327	9.262	**7.928**	10.60	8.556	10.97	8.050	9.143	11.86	191.8	77.19	185.8	7.941	8.896
		(0)	(0)	(0)	(0)	(0.66)	(0.21)	(0.01)	(0.004)	(0.52)	(5.35)	(71.5)	(9.79)	(0.01)	(0.04)

in the Venezuela region, we notice that statistical ETS and persistence-based RWD models outperform the benchmarks in the case of short-term forecasting, however, for the 26-weeks and 52-weeks ahead counterparts, the results generated by the proposed framework lie closer to the actual incidence cases. In the case of the dengue incidence in the Colombia region, the data-driven NBeats model forecasts the 13-weeks ahead disease dynamics more accurately as measured by the MASE and RMSE metrics, whereas the forecast generated by the conventional TBATS model is more accurate in terms of the relative error measure. On the contrary, for medium-term and long-term forecasting, the proposed PARNN framework outperforms the baseline models in terms of all the key performance indicators. A similar pattern is also prominent for the malaria cases of Colombia region, where the proposal generates the most reliable forecast for all the horizons. Furthermore, in the case of the macroeconomic datasets, namely the US EPU Index series we observe that the deep neural architecture-based Transformers model generates competitive forecasts with the proposed framework for the medium-term forecasting analog. However, the baseline performance significantly deteriorates for the short-term and long-term horizons in comparison to the PARNN model. In the case of the Tourism dataset, the performance of the conventional RWD, ETS, and ARIMA models is significantly better than the other forecasters. However, the forecasts generated by the proposed PARNN model for the UK unemployment dataset outperform all the state-of-the-art forecasters as observed from the experimental evaluations. Moreover, for the Russia Exchange dataset, although the statistical TBATS model generates more accurate 6-month ahead forecasts, but for the 12-month and 24-month horizons the performance of the proposal is significantly better than the competitive models. In general, from the overall experimental results, we can infer that the proposed PARNN model can efficiently model and forecast complex time series datasets, especially for long-range predictions. Furthermore, the approximate run time of the proposed model is significantly less than other deep learning-based methods, and it also controls the model size to prevent the problem of overfitting. Finally, we display the medium-term forecast, the 80% prediction interval, and 80% conformal prediction generated by our proposed PARNN architecture for selected datasets in Fig. 2. From a visualization viewpoint, we can infer that the proposed PARNN overall shows accurate and consistent forecasts compared to other state-of-the-art.

4.7 Significance Test

In this section, we discuss the significance of improvements in accuracy metrics by performing different statistical tests. We initially conducted the non-parametric Friedman test to determine the robustness of different benchmarks and the proposed forecaster [18]. This statistical methodology tests the null hypothesis that the performance of different models is equivalent based on their average ranks for various datasets and rejects the hypothesis if the value of the test statistic is greater than the critical value. In this study, we assign rank $\tilde{r}_{i,j}$ to the ith model (out of κ models) for its prediction task of the jth dataset (out of ζ datasets). The

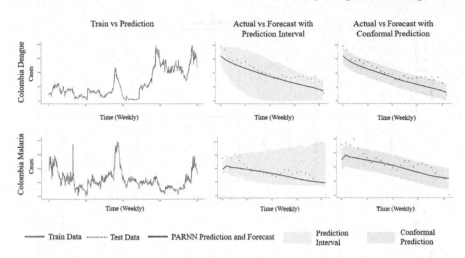

Fig. 2. The plot shows input series (red line), ground truth (red points), 80% prediction interval (yellow shaded region), 80% conformal prediction (green shaded region), predictions (blue), and point forecasts (blue) generated by the PARNN model for the selected datasets. (Color figure online)

Friedman test computes the average rank of each forecaster as $\tilde{R}_i = \frac{1}{\zeta}\sum_{j=1}^{\zeta}\tilde{r}_{i,j}$. Under the null hypothesis, i.e., $\tilde{R}_i$ are equal for all $i = 1, 2, \ldots, \kappa$ the modified version of Friedman test statistic given by:

$$F_{\mathcal{F}} = \frac{(\zeta - 1)\chi_{\mathcal{F}}^2}{\zeta(\kappa - 1) - \chi_{\mathcal{F}}^2} \quad \text{where} \quad \chi_{\mathcal{F}}^2 = \frac{12\zeta}{\kappa(\kappa + 1)}\left[\sum_{i=1}^{\kappa}\tilde{R}_i^2 - \frac{\kappa(\kappa + 1)^2}{4}\right],$$

follows a F distribution with $(\kappa - 1)$ and $(\kappa - 1)(\zeta - 1)$ degrees of freedom [18]. Following the Friedman test procedure, we compute the value of the test statistics for the 14 forecasting models across different test horizons of the 12 datasets and summarize them in Table 6.

Table 6. Observed values of Friedman Test statistic for different accuracy metrics

Test Statistic	MASE	RMSE	SMAPE
$\chi_{\mathcal{F}}^2$	237.2	236.2	250.9
$F_{\mathcal{F}}$	35.96	35.67	40.47

Since the observed value of the test statistic $F_{\mathcal{F}}$ is greater than the critical value $F_{13,455} = 1.742$, so we reject the null hypothesis of model equivalence at a 5% level of significance and conclude that the performance of the forecasters evaluated in this study is significantly different from each other. Furthermore, we utilize post hoc non-parametric multiple comparisons with the best (MCB) test

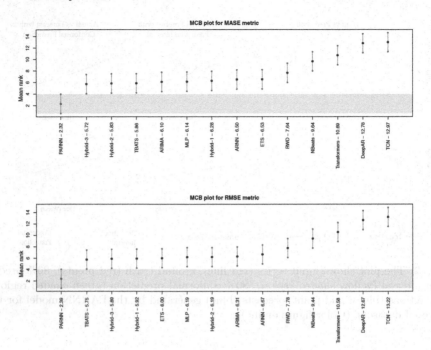

Fig. 3. Plots show the results of the MCB test based on MASE metric (top) and RMSE metric (down). In this figure, PARNN-2.32 means that the average rank of the proposed PARNN model based on the MASE score is 2.32 when tested on all the datasets, similar to others.

[28] to determine the relative performance of different models compared to the 'best' forecaster. To conduct this test, we consider the MASE and RMSE metrics as the key performance indicators and compute the average rank and the critical distance for each model based on their respective scores. The results of the MCB test (presented in Fig. 3) show that the PARNN model achieves the lowest rank, and hence it is the 'best' performing model, followed by hybrid ARIMA-LSTM (Hybrid-3) and hybrid ARIMA-ARNN (Hybrid-2) models in case of the MASE metric. However, in terms of the RMSE scores, the proposed PARNN framework has the minimum rank followed by the TBATS and hybrid ARIMA-LSTM (Hybrid-3) models. Moreover, the upper boundary of the critical interval of the PARNN model (marked by a shaded region) is the reference value for this test. As evident from Fig. 3, the lower boundary of the critical interval for all other forecasters lies above the reference value, meaning that the performance of the baseline forecasters is significantly worse than that of the proposed PARNN model. Hence, based on the statistical tests, we can conclude that the performance of the proposed framework is considerably better in comparison to the baseline methodologies considered in this study.

5   Conclusion and Discussion

This paper presents a hybrid forecasting model that combines the linear ARIMA model with a nonlinear ARNN framework. This integration allows our proposed PARNN model to effectively handle various data irregularities such as nonlinearity, non-stationarity, and non-Gaussian time series. Through extensive experimentation on benchmark time series datasets from diverse domains, we have demonstrated that the PARNN model produces highly competitive forecasts across short, medium, and long-range horizons. Overall, we conclude that our proposed PARNN model introduces a valuable addition to the hybrid forecasting framework, without imposing implicit assumptions regarding the linear and nonlinear components of the dataset. This makes it highly suitable for handling real-world challenges faced by forecasting practitioners. Moreover, we presented a method for uncertainty quantification through confidence intervals, enhancing the model's richness. Our experimental results consistently show the superiority of PARNN across various datasets with different frequencies and forecast horizons, especially for long-range forecasts. An immediate extension of this work would involve extending the PARNN model for multivariate time series forecasting and exploring the asymptotic behavior of the PARNN(m, k, l) model.

References

1. Babu, C.N., Reddy, B.E.: A moving-average filter based hybrid Arima-Ann model for forecasting time series data. Appl. Soft Comput. **23**, 27–38 (2014)
2. Bates, J.M., Granger, C.W.: The combination of forecasts. J. Oper. Res. Soc. **20**(4), 451–468 (1969)
3. Bhattacharyya, A., Chakraborty, T., Rai, S.N.: Stochastic forecasting of COVID-19 daily new cases across countries with a novel hybrid time series model. Nonlinear Dyn. **107**, 1–16 (2022)
4. Bhattacharyya, A., Chattopadhyay, S., Pattnaik, M., Chakraborty, T.: Theta autoregressive neural network: a hybrid time series model for pandemic forecasting. In: 2021 International Joint Conference on Neural Networks (IJCNN), pp. 1–8. IEEE (2021)
5. Box, G.E., Pierce, D.A.: Distribution of residual autocorrelations in autoregressive-integrated moving average time series models. J. Am. Stat. Assoc. **65**(332), 1509–1526 (1970)
6. Cao, J., Li, Z., Li, J.: Financial time series forecasting model based on Ceemdan and LSTM. Phys. A **519**, 127–139 (2019)
7. Chakraborty, T., Chakraborty, A.K., Biswas, M., Banerjee, S., Bhattacharya, S.: Unemployment rate forecasting: a hybrid approach. Comput. Econ. **57**, 1–19 (2020)
8. Chakraborty, T., Chattopadhyay, S., Ghosh, I.: Forecasting dengue epidemics using a hybrid methodology. Phys. A Statist. Mech. Appl. **527**, 121266 (2019)
9. Chakraborty, T., Ghosh, I.: Real-time forecasts and risk assessment of novel coronavirus (COVID-19) cases: a data-driven analysis. Chaos, Solitons Fractals **135**, 109850 (2020)
10. Chakraborty, T., Ghosh, I., Mahajan, T., Arora, T.: Nowcasting of COVID-19 confirmed cases: foundations, trends, and challenges. In: Modeling, Control and Drug Development for COVID-19 Outbreak Prevention, pp. 1023–1064 (2022)

11. Chen, Y., Kang, Y., Chen, Y., Wang, Z.: Probabilistic forecasting with temporal convolutional neural network. Neurocomputing **399**, 491–501 (2020)
12. Connor, J.T., Martin, R.D., Atlas, L.E.: Recurrent neural networks and robust time series prediction. IEEE Trans. Neural Netw. **5**(2), 240–254 (1994)
13. Dave, E., Leonardo, A., Jeanice, M., Hanafiah, N.: Forecasting Indonesia exports using a hybrid model Arima-LSTM. Proc. Comput. Sci. **179**, 480–487 (2021)
14. De Livera, A.M., Hyndman, R.J., Snyder, R.D.: Forecasting time series with complex seasonal patterns using exponential smoothing. J. Am. Stat. Assoc. **106**(496), 1513–1527 (2011)
15. Egrioglu, E., Yolcu, U., Aladag, C.H., Bas, E.: Recurrent multiplicative neuron model artificial neural network for non-linear time series forecasting. Neural Process. Lett. **41**(2), 249–258 (2015)
16. Entorf, H.: Random walks with drifts: nonsense regression and spurious fixed-effect estimation. J. Economet. **80**(2), 287–296 (1997)
17. Faraway, J., Chatfield, C.: Time series forecasting with neural networks: a comparative study using the air line data. J. Roy. Stat. Soc.: Ser. C (Appl. Stat.) **47**(2), 231–250 (1998)
18. Friedman, M.: The use of ranks to avoid the assumption of normality implicit in the analysis of variance. J. Am. Stat. Assoc. **32**(200), 675–701 (1937)
19. Geman, S., Bienenstock, E., Doursat, R.: Neural networks and the bias/variance dilemma. Neural Comput. **4**(1), 1–58 (1992)
20. Herzen, J., et al.: Darts: User-friendly modern machine learning for time series. J. Mach. Learn. Res. **23**(124), 1–6 (2022)
21. Hyndman, R., Koehler, A.B., Ord, J.K., Snyder, R.D.: Forecasting with Exponential Smoothing: the State Space Approach, 1st edn. Springer Science & Business Media, Heidelberg (2008). https://doi.org/10.1007/978-3-540-71918-2
22. Hyndman, R.J., Athanasopoulos, G.: Forecasting: Principles and Practice. OTexts (2018)
23. Hyndman, R.J., et al.: Package 'forecast' (2020). https://cran.r-project.org/web/packages/forecast/forecast
24. Hyndman, R.J., Ullah, M.S.: Robust forecasting of mortality and fertility rates: a functional data approach. Comput. Statist. Data Anal. **51**(10), 4942–4956 (2007)
25. Karmy, J.P., Maldonado, S.: Hierarchical time series forecasting via support vector regression in the European travel retail industry. Expert Syst. Appl. **137**, 59–73 (2019)
26. Khashei, M., Bijari, M.: An artificial neural network (p, d, q) model for time series forecasting. Expert Syst. Appl. **37**(1), 479–489 (2010)
27. Kodogiannis, V., Lolis, A.: Forecasting financial time series using neural network and fuzzy system-based techniques. Neural Comput. App. **11**(2), 90–102 (2002)
28. Koning, A.J., Franses, P.H., Hibon, M., Stekler, H.O.: The m3 competition: statistical tests of the results. Int. J. Forecast. **21**(3), 397–409 (2005)
29. Kourentzes, N.: nnfor: Time series forecasting with neural networks. R package version 0.9. 6 (2017)
30. Leoni, P.: Long-range out-of-sample properties of autoregressive neural networks. Neural Comput. **21**(1), 1–8 (2009)
31. Nochai, R., Nochai, T.: Arima model for forecasting oil palm price. In: Proceedings of the 2nd IMT-GT Regional Conference on Mathematics, Statistics and Applications, pp. 13–15 (2006)
32. Oreshkin, B.N., Carpov, D., Chapados, N., Bengio, Y.: N-beats: neural basis expansion analysis for interpretable time series forecasting. arXiv preprint arXiv:1905.10437 (2019)

33. Panigrahi, S., Behera, H.S.: A hybrid ETS-ANN model for time series forecasting. Eng. Appl. Artif. Intell. **66**, 49–59 (2017)

34. Panja, M., Chakraborty, T., Kumar, U., Liu, N.: Epicasting: an ensemble wavelet neural network for forecasting epidemics. Neural Networks (2023)

35. Qin, Y., et al.: Hybrid forecasting model based on long short term memory network and deep learning neural network for wind signal. Appl. Energy **236**, 262–272 (2019)

36. Ray, A., Chakraborty, T., Ghosh, D.: Optimized ensemble deep learning framework for scalable forecasting of dynamics containing extreme events. Chaos Interdiscip. J. Nonlinear Sci. **31**(11), 111105 (2021)

37. Rumelhart, D.E., Hinton, G.E., Williams, R.J.: Learning representations by backpropagating errors. Nature **323**(6088), 533 (1986)

38. Salinas, D., Flunkert, V., Gasthaus, J., Januschowski, T.: Deepar: probabilistic forecasting with autoregressive recurrent networks. Int. J. Forecast. **36**(3), 1181–1191 (2020)

39. Selvin, S., Vinayakumar, R., Gopalakrishnan, E., Menon, V.K., Soman, K.: Stock price prediction using LSTM, RNN and CNN-sliding window model. In: 2017 International Conference on Advances in Computing, Communications and Informatics (ICACCI), pp. 1643–1647. IEEE (2017)

40. Shahwan, T., Odening, M.: Forecasting agricultural commodity prices using hybrid neural networks. In: Chen, S.H., Wang, P.P., Kuo, T.W. (eds.) Computational Intelligence in Economics and Finance, pp. 63–74. Springer, Heidelberg (2007). https://doi.org/10.1007/978-3-540-72821-4_3

41. Vautard, R., Beekmann, M., Roux, J., Gombert, D.: Validation of a hybrid forecasting system for the ozone concentrations over the Paris area. Atmos. Environ. **35**(14), 2449–2461 (2001)

42. Vovk, V., Gammerman, A., Shafer, G.: Conformal prediction. In: Vovk, V., Gammerman, A., Shafer, G. (eds.) Algorithmic Learning in a Random World, pp. 17–51. Springer, Boston (2005). https://doi.org/10.1007/0-387-25061-1_2

43. Wang, X., Hyndman, R.J., Li, F., Kang, Y.: Forecast combinations: an over 50-year review. arXiv preprint arXiv:2205.04216 (2022)

44. Winters, P.R.: Forecasting sales by exponentially weighted moving averages. Manage. Sci. **6**(3), 324–342 (1960)

45. Wu, N., Green, B., Ben, X., O'Banion, S.: Deep transformer models for time series forecasting: The influenza prevalence case. arXiv preprint arXiv:2001.08317 (2020)

46. Xu, S., Chan, H.K., Zhang, T.: Forecasting the demand of the aviation industry using hybrid time series Sarima-SVR approach. Transp. Res. Part E Logist. Transp. Rev. **122**, 169–180 (2019)

47. Zhang, G.P.: Time series forecasting using a hybrid Arima and neural network model. Neurocomputing **50**, 159–175 (2003)

Stereoential Net: Deep Network for Learning Building Height Using Stereo Imagery

Sana Jabbar[✉] and Murtaza Taj

Computer Vision and Graphics Lab, Lahore University of Management Sciences,
Lahore, Pakistan
{sana.jabbar,murtaza.taj}@lums.edu.pk

Abstract. Height estimation plays a crucial role in the planning and assessment of urban development, enabling effective decision-making and evaluation of urban built areas. Accurate estimation of building heights from remote sensing optical imagery poses significant challenges in preserving both the overall structure of complex scenes and the elevation details of the buildings. This paper proposes a novel end-to-end deep learning-based network (Stereoential Net) comprising a multi-scale differential shortcut connection module (MSDSCM) at the decoding end and a modified stereo U-Net (mSUNet). The proposed Stereoential network performs a multi-scale differential decoding features fusion to preserve fine details for improved height estimation using stereo optical imagery. Unlike existing methods, our approach does not use any multi-spectral satellite imagery, instead, it only employs freely available optical imagery, yet it achieves superior performance. We evaluate our proposed network on two benchmark datasets, the IEEE Data Fusion Contest 2018 (DFC2018) dataset and the 42-cities dataset. The 42-cities dataset is comprised of 42 different densely populated cities of China having diverse sets of buildings with varying shapes and sizes. The quantitative and qualitative results reveal that our proposed network outperforms the SOTA algorithms for DFC2018. Our method reduces the root-mean-square error (RMSE) by 0.31 m as compared to state-of-the-art multi-spectral approaches on the 42-cities dataset. The code will be made publically available via the GitHub repository.

Keywords: Stereo Optical Satellite Imagery · Multi-view Imagery · Multi-scale Features · Stereo U-Net · Building height

1 Introduction

By the year 2050, more than 68% of the world population is expected to live in urban areas causing humongous pressure over urban land and resources [1]. Compound augmented, and compact vertical development of cities is inevitable to shelter the expanding population which has already been practiced in big

B. Luo et al. (Eds.): ICONIP 2023, CCIS 1967, pp. 478–489, 2024.
https://doi.org/10.1007/978-981-99-8178-6_36

cities. This results in an increase in high-rise buildings [11] and thus building height estimation is essential for a plethora of applications. Such as mitigating damage to utility management [6,19], optimising infrastructure [13], urban inventory [20], and autonomous driving [17].

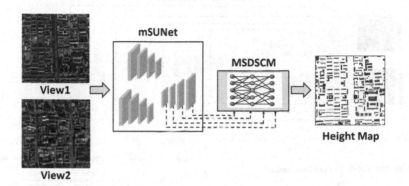

Fig. 1. Overview of the proposed Stereoential network with multi-scale differential shortcut connection modules (MSDSCM) for building height estimation.

Recently, building height estimation has become an active area of research, and many Deep Neural Network (DNN) based methods have been proposed [7, 8,15,18,25]. Many methods incorporate complementary 3D information from an existing tool like Digital Surface models (DSMs), which is useful to sense the relative elevation of the corresponding area [25], [7]. DSM captures both natural and artificial resources in terms of elevation, which can either be obtained from Light Detection and Ranging (LiDAR) or Synthetic Aperture Radar (SAR). However, these methods are expensive and constrained by the availability of strong expertise, high computational resources, and data acquisition cost [10].

The above-mentioned barrier can be overcome using high-resolution single-view satellite imagery [9,10,15,22]. These methods map an RGB image or a multi-spectral image of a specified area into a single-channel elevation map. But single-view satellite imagery is primarily two-dimensional (2D) and is a poor source of vertical information [8]. Moreover, utilization of single-view imagery for building height estimation can be challenging for densely populated areas where buildings are not sparsely located. Concurrently, many stereo-based techniques are developed for building height estimation using the triangulation method from the consecutive views of the same densely populated area on a local scale [21, 24,25]. These methods also make use of multi-spectral imagery for improved estimation of height maps [25].

However, the triangulation method from stereo imagery suffers degradation in performance due to different illumination conditions in multi-view satellite images taken at different times and vantage points. One of the limitations in both single-view [9,10,15,22] and multi-view [21,24,25] based height estimation methods is, despite using both optical and multi-spectral imagery they fail to correctly estimate height for pixels around the building boundary which corresponding to small structures. In this work, we propose a novel encoder-decoder-based architecture that includes a multi-scale novel differential   shortcut connection module

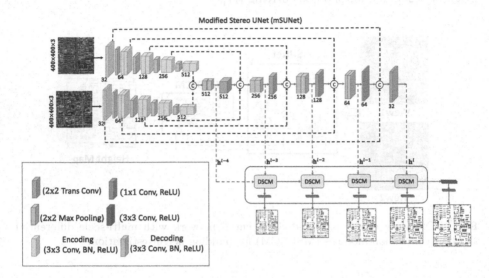

Fig. 2. Detailed schematics of the proposed Stereoential network based on decoding differential connections and stereo imagery. The two symmetrical encoding features are combined at each level and passed to the decoder with shortcut connections. The final height map is obtained after decoding via a multi-scale-differential shortcut connections modules (MSDSCM).

(MSDSCM) for the extraction of these fine details. The key contributions of our work are as follows:

- Unlike [3], we do not use multi-spectral imagery, instead, our method uses only optical imagery.
- Shortcut connections have been used for super-resolution image reconstruction [2]. We also leverage shortcut connections and proposed a novel differential shortcut connection module (DSCM) to introduce fine details in our height estimations.
- Encoder-decoder architectures are commonly used for height map generation [12]. We also propose a novel modified stereo UNet [16] architecture where we utilized shortcut connections at each sub-level from the encoder to the decoder. Our Stereoential network also uses the proposed multi-scale differential shortcut connections (MSDSCM) at the decoding end to get the refined multi-scale height maps (see Fig. 1).
- When evaluated on DFC2018 [14] and 42-cities datasets [3], our proposed Stereoential network showed superior performance as compared to other state-of-the-art methods.

The remainder is organized as follows: Sect. 2 covers the methodology for the proposed framework, results are discussed in Sect. 3 and the conclusions are drawn in Sect. 4.

2 Methodology

In this section, we will provide an overview of the baseline architecture i.e. mSUNet, of the proposed Stereoential network. Subsequently, we will delve into the details of the novel DSSCM module and its various variants, highlighting their unique contributions to the building height estimation task.

2.1 Baseline Architecture

Building height estimation is commonly solved via generative models and many recent approaches [3,8,23,23] used UNet [16] as a baseline architecture. The baseline architecture of our proposed Stereoential network is modified stereo UNet (mSUNet) utilizes stereo optical imagery as input. The mSUNet is comprised of two symmetrically large encoders and a shared decoder. The encoders encompass five encoding levels each with convolution layers of kernel size of 3 (written as 3×3 Conv), with batch normalization (written as BN), rectified linear unit (written as ReLU), followed by a max-pooling (written as 2×2 max-pool). At the encoding level, input images are mapped into feature maps with the combination of Conv, BN, and ReLU twice then followed by a max pooling to reduce the spatial dimension by a factor of 2, and features maps increase with the same factor. The encoding features at each scale from both encoders are combined and passed to the decoder for further processing. At the decoding level, we utilized transpose convolution (written as 2×2 Trans Conv) in order to enlarge the spatial resolution of decoding features by a factor of 2 until it is consistent with the input resolution. Each decoding level applied a combination of concatenation (written as C) of features from two encoders, a pair of 3×3 Conv layers, followed by a trans conv layer. Figure 2 shows the detailed schematics of the proposed Stereoential network along with the MSDSCM module.

In this work, we have designed a multi-scale differential shortcut connection module (MSDSCM) at the decoding end to enhance building height estimation in a progressive manner. The Stereoential network is employed with a multi-scale differential shortcut module (MSDSCM) to improve the fine building elevation details.

2.2 Differential Shortcut Connection Module (DSCM)

We performed experimentation with the proposed Stereoential network along with two variants of DSCM, (1) Single-scale DSCM (SSDSCM), (2) Multi-scale DSCM (MSDSCM). The SSDSCM module is comprised of a single DSCM module and outputs a single height map. Whereas, MSDSCM is employed with five DSCM modules and outputs progressively refined multi-scale height maps.

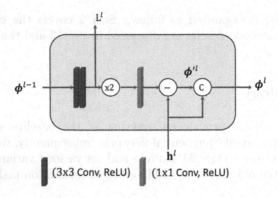

Fig. 3. The structure of the proposed differential shortcut module (DSCM).

DSCM Each DSCM is employed to allow the flow of information from coarse-to-fine multi-scale decoding features. The inputs of DSCM are refined differential shortcut decoding features (ϕ^{l-1}) of the previous scale and decoding features of the current scale ($\mathbf{h}^l$) as described in Fig. 3. The output of DSCM is refined differential shortcut decoding features of the current scale (ϕ^l) and a height map of the previous scale. A pair of 3×3 Conv applied on ϕ^{l-1} and then spatial resolution increased using bilinear upsampling as written in Eq. 1. The intermediate shortcut decoding features ϕ'^l obtained as high frequency component, taking difference between $f(\phi^{l-1})$ and decoding features of current scale ($\mathbf{h}^l$) as evident in Eq. 2. Finally, refined differential shortcut decoding features ϕ^l of the current scale obtained by applying a concatenation operation between ϕ'^l and $\mathbf{h}^l$ as written in Eq. 3.

$$f(\phi^{l-1}) = Up(Conv(Conv(\phi^{l-1}))) \tag{1}$$

$$\phi'^l = Conv(f(\phi^{l-1})) - \mathbf{h}^l \tag{2}$$

$$\phi^l = \phi'^l \parallel \mathbf{h}^l \tag{3}$$

SSDSCM For SSDSCM, we utilized only one DSCM with two decoding features of the last scales. The inputs of SSDSCM are decoding features $\mathbf{h}^{l-1}$ and $\mathbf{h}^l$. The output of SSDSCM is finally refined differential decoding features. As only one height map of the same scale as of input image is obtained at the end, therefore, we utilized a single-scale mean square error loss to update the network parameters. The loss function is described in Eq. 4, where MN is the total number of pixels in the ground truth height map (y_{ij}) and predicted height map y'_{ij}.

$$L = \frac{1}{MN} \sum_{i=1}^{M} \sum_{j=1}^{N} (y_{ij} - y'_{ij})^2 \tag{4}$$

Table 1. Comparative performance analysis of Stereoential network with state-of-the-art methods for DFC2018 dataset. The number of trainable parameters is shown in millions (mn). RMSE and MAE are reported in meters (m).

Method	Semantic Seg.	Parameters	Image Size	RMSE (m)	MAE (m)
Carvalho et al. [4]	✗	≈ 22mn	320 × 320	3.05	1.47
Carvalho et al. [4]	✓	≈ 22mn	320 × 320	2.60	1.26
Liu et al. [9]	✗	≈ 357mn	220 × 220	2.88	1.19
Min et al. [10]	✓	NA	256 × 256	1.96	0.76
Karatsiolis et al. [8]	✗	≈ 104mn	520 × 520	1.63	0.78
Stereoential Net (ours)	✗	≈ 24mn	256 × 256	1.79	0.97

MSDSCM For MSDSCM, we utilized four DSCMs, one for each scale of the decoder.

As at the lowest level, there were no refined differential decoding features, we utilized the lowest resolution decoding features as input as described in Fig. 2. The inputs of MSDSCM are refined decoding features of previous scale ϕ^{l-1}, and decoding features of current scale $\mathbf{h}^l$. The output of MSDSCM is refined differential decoding features and height maps obtained at multi-level scales. For MSDSCM, as multi-scale height maps are obtained at five different scales, therefore, we employed a multi-scale mean square error loss to update network parameters as described in Eq. 5. The inputs to the loss function at l_{th} scale are the predicted height map (y'^l) and ground truth map (y^l). MN is the total number of pixels and ij is the valid pixel in both images. At intermediate scales, ground truth maps are downsampled according to the scale of predicted height maps.

$$L = \sum_{l=1}^{5} \frac{1}{MN} \sum_{i=1}^{M} \sum_{j=1}^{N} (y_{ij}^l - y_{ij}'^l)^2 \qquad (5)$$

3 Results

3.1 Dataset

DFC2018 Dataset: We employed the DFC2018 dataset [14] for the performance evaluation of the proposed Stereoential network. The DFC2018 dataset is comprised of high-resolution (0.05m/pixel) imagery over the University of Houston campus and its neighborhood acquired on February 16, 2017. As our Stereoential network requires multi-view imagery, therefore, we generated a second view with Google Earth Imagery with the time stamp the same as the reference year i.e. 2017. For a fair comparison, we used image patches of resolution 256 × 256 corresponding to 0.5 m/pixel. **42-cities Dataset:** We performed experimentation using the 42-cities dataset of China which is quite diverse in building shapes and heights. Since multi-view optical satellite imagery was not available, only ground truth was provided [3], thus we produced our own dataset of optical stereo pairs.

The acquisition time of Google Earth imagery was close to the reference year (2015 – 2017) of ground truth labels to avoid any inconsistency. We used image patches of resolution 400×400 corresponding to 2.5 m/pixel. For a fair comparison, the same number of training samples were selected after visual assessment and the data split was also the same as described in [3].

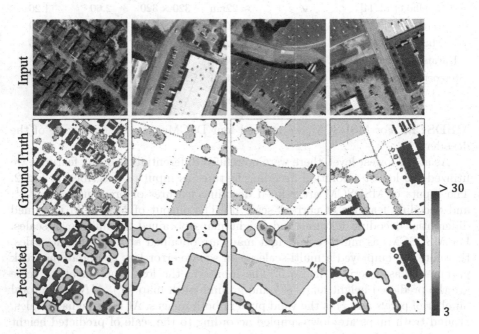

Fig. 4. Qualitative results from DFC2018 dataset collected over the area of Houston University. The spatial extent of 256m×256m is shown for each sample.

3.2   Experimental Detail

All the experiments are conducted using deep learning with four NVIDIA Geforce RTX 2080 Ti GPUs. We implemented our proposed model in Keras [5] framework and adopted the Adam optimizer with a learning rate of 10^{-3} to train the model. The model was initialized with He-Normal weights with a batch size of 16 and trained for 150 epochs. As the network predicts multi-scale height maps we took leverage with multi-scale mean square error loss to access the network performance.

3.3   Quantitative and Qualitative Comparison

Table 1 shows the performance comparison of the proposed Stereoential network on DFC2018 dataset. Our proposed network achieved 1.79m RMSE with less

Table 2. Abaltive and comparative performance analysis of the proposed Stereoential network with state-of-the-art methods for 42-cities dataset. RMSE and MAE are reported in meters (m).

Method	Multi-view	Multi-spectral	RMSE (m)	MAE (m)
RF [3]	✓	✓	7.64	–
MM³ Net [3]	✓	✓	6.26	–
Stereoential Net (MSDSCM)	✓	✗	5.95	2.67
Stereoential Net (mUNet)	✓	✗	6.12	2.78
Steroential Net (SSDSCM)	✓	✗	6.09	2.74
Stereoential Net (MSDSCM)	✗	✗	6.19	2.83
Stereoential Net (MSDSCM)	✓	✗	5.95	2.67

Table 3. Comparative performance analysis of the proposed network with MM³ network on five densely populated cities of China (Bejing, Shanghai, Shenzhen, Wuhan, and X'ian). RMSE and MAE are reported in meters (m).

Cities	MM³		Ours	
	RMSE (m)	MAE (m)	RMSE (m)	MAE (m)
Bejing	5.30	-	5.25	2.00
Shanghai	6.80	-	6.89	3.45
Shenzhen	6.91	-	6.81	2.93
Wuhan	5.52	-	5.84	2.97
X'ian	7.05	-	6.69	3.15

number of trainable parameters ≈ 24mn. Table 2 shows a comparative analysis between our proposed model and state-of-the-art methods for the 42-cities dataset. It can be seen that the RMSE of our proposed network is 0.31 m less than MM³ and 1.70 m less than RF. Despite the fact that MM³ trained on both RGB and multi-spectral imagery our proposed network trained with RGB images only still achieved better performance. Table 2 also comprises the ablative analysis between variants of our proposed architecture. To establish the efficacy of the proposed MSDSCM module we performed an ablative analysis Table 2 without the DSCM module, with a single DSCM, and with the MSDSCM module with a single-view input image. We establish the supremacy of the proposed Stereoential network experimentally.

Table 3 comprised comparative results for five cities. The proposed model achieved RMSE ≈ 0.05 m less than MM³ for Bejing, ≈ 0.1 m less for Shenzhen, and 0.45 m less for X'ian. For the purpose of visual inspection, Figs. 4 and 5, display the predicted results of the proposed model for four cities, Beijing, Shanghai, X'ian, and Wuhan. All these cities are quite diverse in building shape and size due to different urbanization and development planning. In Fig. 4 our predicted results showed better agreement with ground truth images for low-rise buildings in Beijing and Yinchun. Our proposed model also performed well over areas covering a wide range of building heights. Figure 5 shows that the proposed model, along with differential shortcut connections, captured the variation in height of densely populated areas of Shanghai and Wuhan and the other four cities as well.

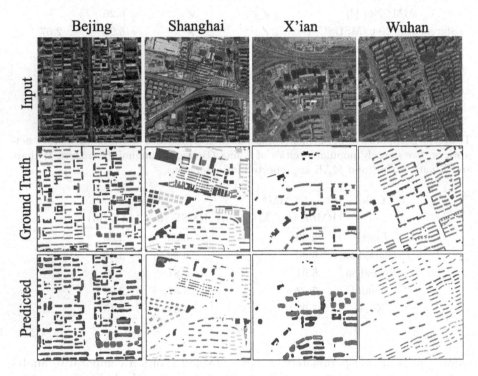

Fig. 5. Qualitative results from 42-cities dataset for Bejing, Shanghai, Xian, and Wuhan. The spatial extent of 1km×1km is shown for each city.

3.4   Error Analysis

Although the training samples were selected after intense visual assessment, the dataset still contained noisy labels as shown in Fig. 6. Where a testing sample was taken from Shenyang city. The network struggled to segment those buildings that were imaged in a single view, such as the area encompassed with a red rectangular box and pointed with a red arrow in both input images, ground truth, and in the predicted height map. The buildings in the yellow boxes are quite visible in both input views, but not labeled in ground truth. Our proposed model predicted labels for such buildings as well which cause degradation in performance.

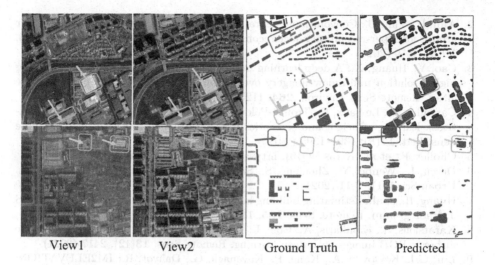

| View1 | View2 | Ground Truth | Predicted |

Fig. 6. Sample result showing error in height estimation. The red box and arrow pointed to those buildings which are present only in one input image. A yellow box and arrow highlighted a building present in both views but was not labeled in the ground truth. (Color figure online)

4   Conclusions

In this work, we proposed two contributions for the estimation of building heights from stereo-pair satellite imagery. First, we proposed a variant of UNet architecture which is modified for stereo imagery. Secondly, we proposed a novel multi-scale differential shortcut connection that introduced fine details in the estimated heights of the building. We have evaluated our method on DFC2018 dataset, where the proposed novel Stereoential network was the best-performing network with less number of training parameters. We also took advantage of 42 different cities and compared it with the best-known method on this dataset i.e. MM^3 [3]. Despite the fact that we only used freely available optical imagery, our algorithm outperformed approaches that used both optical as well as multi-spectral imagery. Our proposed multi-scale differential shortcut connections when combined with our modified stereo UNet give consistently low RMSE on individual cities as well as over the entire dataset. In the future, we also plan to extend our analysis using other benchmark datasets.

Acknowledgements. We thank Dr.Usman Nazir for the assistance with proofreading and comments that greatly improved the manuscript.

References

1. World urban population. https://statisticstimes.com/demographics/world-urban-population.php/. Accessed 21 June 2023

2. Ahn, H., Yim, C.: Convolutional neural networks using skip connections with layer groups for super-resolution image reconstruction based on deep learning. Appl. Sci. **10**(6), 1959 (2020)
3. Cao, Y., Huang, X.: A deep learning method for building height estimation using high-resolution multi-view imagery over urban areas: a case study of 42 Chinese cities. Remote Sens. Environ. **264**, 112590 (2021)
4. Carvalho, M., Le Saux, B., Trouvé-Peloux, P., Champagnat, F., Almansa, A.: Multitask learning of height and semantics from aerial images. IEEE Geosci. Remote Sens. Lett. **17**(8), 1391–1395 (2019)
5. Chollet, F., et al.: Keras (2015). https://github.com/fchollet/keras
6. Deren, L., Wenbo, Y., Zhenfeng, S.: Smart city based on digital twins. Comput. Urban Sci. **1**(1), 1–11 (2021)
7. Huang, H., et al.: Estimating building height in China from ALOS AW3D30. ISPRS J. Photogramm. Remote. Sens. **185**, 146–157 (2022)
8. Karatsiolis, S., Kamilaris, A., Cole, I.: IMG2nDSM: height estimation from single airborne RGB images with deep learning. Remote Sens. **13**(12), 2417 (2021)
9. Liu, C.J., Krylov, V.A., Kane, P., Kavanagh, G., Dahyot, R.: IM2ELEVATION: building height estimation from single-view aerial imagery. Remote Sens. **12**(17), 2719 (2020)
10. Lu, M., Liu, J., Wang, F., Xiang, Y.: Multi-task learning of relative height estimation and semantic segmentation from single airborne RGB images. Remote Sens. **14**(14), 3450 (2022)
11. Mahtta, R., Mahendra, A., Seto, K.C.: Building up or spreading out? Typologies of urban growth across 478 cities of 1 million+. Environ. Res. Lett. **14**(12), 124077 (2019)
12. Mou, L., Zhu, X.X.: IM2HEIGHT: height estimation from single monocular imagery via fully residual convolutional-deconvolutional network. arXiv preprint arXiv:1802.10249 (2018)
13. Perera, A., Javanroodi, K., Nik, V.M.: Climate resilient interconnected infrastructure: co-optimization of energy systems and urban morphology. Appl. Energy **285**, 116430 (2021)
14. Prasad, S., Le Saux, B., Yokoya, N., Hansch, R.: IEEE Data Fusion Challenge - Fusion of Multispectral LiDAR and Hyperspectral data (2020). https://doi.org/10.21227/jnh9-nz89
15. Qi, F., Zhai, J.Z., Dang, G.: Building height estimation using Google Earth. Energy Build. **118**, 123–132 (2016)
16. Ronneberger, O., Fischer, P., Brox, T.: U-net: convolutional networks for biomedical image segmentation. In: Navab, N., Hornegger, J., Wells, W.M., Frangi, A.F. (eds.) MICCAI 2015. LNCS, vol. 9351, pp. 234–241. Springer, Cham (2015). https://doi.org/10.1007/978-3-319-24574-4_28
17. Sautier, C., Puy, G., Gidaris, S., Boulch, A., Bursuc, A., Marlet, R.: Image-to-lidar self-supervised distillation for autonomous driving data. In: Proceedings of CVPR, June 2022
18. Shao, Y., Taff, G.N., Walsh, S.J.: Shadow detection and building-height estimation using IKONOS data. Int. J. Remote Sens. **32**(22), 6929–6944 (2011)
19. Stouffs, R.: Virtual 3D city models. ISPRS Int. J. Geo-Inf. **11**(4), 1–7 (2022)
20. Suwardhi, D., Trisyanti, S.W., Virtriana, R., Syamsu, A.A., Jannati, S., Halim, R.S.: Heritage smart city mapping, planning and land administration (Hestya). ISPRS Int. J. Geo-Inf. **11**(2), 1–10 (2022)

21. Xie, Y., Feng, D., Xiong, S., Zhu, J., Liu, Y.: Multi-scene building height estimation method based on shadow in high resolution imagery. Remote Sens. **13**(15), 2862 (2021)
22. Xing, S., Dong, Q., Hu, Z.: Gated feature aggregation for height estimation from single aerial images. IEEE Geosci. Remote Sens. Lett. **19**, 1–5 (2021)
23. Xue, M., Li, J., Zhao, Z., Luo, Q.: SAR2HEIGHT: height estimation from a single SAR image in mountain areas via sparse height and proxyless depth-aware penalty neural architecture search for Unet. Remote Sens. **14**(21), 5392 (2022)
24. Yu, D., Ji, S., Liu, J., Wei, S.: Automatic 3D building reconstruction from multi-view aerial images with deep learning. ISPRS J. Photogramm. Remote. Sens. **171**, 155–170 (2021)
25. Zhang, C., Cui, Y., Zhu, Z., Jiang, S., Jiang, W.: Building height extraction from GF-7 satellite images based on roof contour constrained stereo matching. Remote sensing **14**(7), 1566 (2022)

FEGI: A Fusion Extractive-Generative Model for Dialogue Ellipsis and Coreference Integrated Resolution

Qingqing Li[1,2] and Fang Kong[1,2(✉)]

[1] Laboratory for Natural Language Processing, Soochow University, Suzhou, China
20214227029@stu.suda.edu.cn
[2] School of Computer Science and Technology, Soochow University, Suzhou, China
kongfang@suda.edu.cn

Abstract. Dialogue systems in open domain have achieved great success due to the easily obtained single-turn corpus and the development of deep learning, but the multi-turn scenario is still a challenge because of the frequent coreference and information omission. In this paper, we aim to quickly retrieve the omitted or coreferred expressions contained in history dialogue and restore them into the incomplete utterance. Jointly inspired by the generative method for text generation and extractive method for span extraction, we propose a fusion extractive-generative dialogue ellipsis and coreference integrated resolution model(FEGI). In detail, we introduce two training tasks OMIT and SPAN to extract missing semantic expressions, then integrate the expressions obtained into the decoding initial and copy stages of the generative model respectively. To support the training tasks, we introduce an algorithm for secondary reconstruction annotation based on existing publicly available corpora via unsupervised technique, which can work in cases of no annotation of the missing semantic expressions. Moreover, We conduct dozens of joint learning experiments on the CamRest676 and RiSAWOZ datasets. Experimental results show that our proposed model significantly outperforms the state-of-the-art models in terms of quality.

Keywords: Dialogue Systems · Ellipsis Recovery · Coreference Resolution

1 Introduction

Dialogue systems have been receiving increasing amounts of attention ([7,8]), and been widely utilized in real-world applications. However, speakers usually omit some self-explanatory content in their responses for reasons of trying to keep the conversation simple and clear in multi-turn dialogue scenario. Table 1 presents a typical example on ellipsis and coreference phenomena in multi-turn human-to-machine dialogue. For the incomplete utterance $x_3(usr)$, the expression '*Hainan*'

Supported by organization x.

is omitted to avoid repetition (Ellipsis), and the pronoun *'this'* in $x_3(usr)$ refers to *'sunny'* in $x_2(sys)$ (Coreference). Lots of works on traditional coreference resolution ([9]) achieved impressive gains. There still exists a major challenge: it is hard to understand the real intention from the original utterance without the context when facing multi-turn dialogue scenario. However, the human brain has a very powerful memory ability, which can record and reconstruct rich history dialogue, and quickly retrieve the omitted or coreferred expressions from history dialogue.

Table 1. An example of multi-turn human-to-machine dialogue, including the dialogue history utterances (x_1, x_2), the incomplete utterance (x_3) and the complete utterance(x_3^*). Purple means omission and orange means coreference.

Turn	Utterance (*Translation*)
$x_1(usr)$	今天海南天气怎么样
	How is the weather in Hainan today
$x_2(sys)$	海南今天是晴天
	Hainan is sunny today
$x_3(usr)$	为什么总是这样
	Why is always this
x_3^*	为什么海南总是晴天
	Why is Hainan always sunny

A series of models of generative-based and extractive-based have been studied for multi-turn dialogue systems. For generative-based models, Researchers heavily employ pointer networks ([11,16]) or sequence to sequence models with copy mechanisms ([3,5,14]) to generate the final semantically complete utterances. As the omitted or co-referred expressions are nearly from either the dialogue history utterances or the incomplete utterances. However, these methods have a strong reliance on the size of the multi-turn corpus, In case of smaller corpus, these methods may lack great generalizations to pay attention to core missing semantic information.

Unlike the generative task, some extractive-based methods have come to appear. Hao et al. [6] propose a sequence tagging methods, which treats utterance rewriting as a text editing task to restore the incomplete utterances. The extractive model pays more attention to the core missing semantic information contained in history dialogue, and improves the semantic integrity of the results. But the extractive methods may suffer from low coverage when phrase that must be added to a source utterance cannot be covered by a single context span. This can occur in languages like English that introduce tokens such as prepositions(e.g., "besides") into the rewrite for grammatically.

As the extractive method can pay more attention to the core missing semantic information, to address the problem of missing important tokens in generative model's rewriting, by adding this information to the original utterance can make the generative model pay more attention to the important missing information. We propose a fusion extractive-generative dialogue ellipsis and coreference integrated resolution model. In particular, our approach enhances the incomplete utterances restoration from three perspectives:

- Inspired by reading comprehension, we introduce a dialogue ellipsis and coreference integration detection task SPAN and build a benchmark platform. The core of the task is how to detect the relevant omitted or coreferred expressions in history dialogue. We regard this problem as a span extractive problem using encoder.
- Not all utterances have information omission and coreference, and the semantically complete utterances do not actually need to review history dialogue. Therefore, we add an auxiliary task OMIT to judge whether the current utterance has ellipsis or coreference phenomena.
- FEGI is a creative fusion of span extraction and auto-regression generation, which integrate the missing semantic expressions obtained with SPAN task into the decoding initial and copy stages of the generative model respectively.

In this way, using the framework of multi-task joint learning, FEGI can effectively extract and aggregate the omitted or coreferred expressions in history dialogue and help dialogue ellipsis and coreference resolution both in copying distribution and vocab generation distribution. Experiments show the effectiveness of our approach.

2　Data Reconstruction

Statistics of the CamRest676 [14] and RiSAWOZ [15] corpus show that 95.41% and 94.68% of incomplete dialogue utterances have missing semantic information appeared in history dialogue accounting. It verifies the authenticity of the argument that "the omitted or coreferred expressions in incomplete utterance always exists in history dialogue", and the necessity of performing missing semantic extraction tasks. We introduce an unsupervised automatic label extraction method, which can accurately and quickly extract labels with missing semantic information and judge whether the current utterance is semantically complete

Table 2. Comparison of the average length between the missing phrase(MP), the incomplete utterance(IU) and the restored utterance(RU).

	Type	MP	IU	RU
CamRest676	All	2.51	8.10	12.14
RiSAWOZ	Train	2.84	9.51	12.31
	Dev	2.90	9.41	12.34
	Test	2.79	10.18	12.83

Algorithm 1: Unsupervised label extraction

Input : S: Current Utterance
         T: Complete Utterance
         C: History Dialogue
Output: A: Missing Semantic Span Labels
         L: Omission Label

---

1 Insert dummy tokens in C
2 $A = [\,]; L = 0; patterns = [\,];$
3 Compute all the *common subsequence* K between S and T
4 **if** $S == T$ **then**
5 $\quad\lfloor\ L = 1$

6 **for** *word in T* **do**
7 $\quad$ **if** *word not in K* **then**
8 $\quad\quad\lfloor$ patterns.append(*word*)

9 Merge contiguous words in patterns
10 **for** *pattern in patterns* **do**
11 $\quad find\_ids = [\,]$
12 $\quad$ **for** $i \in [1, len(C)]$ **do**
13 $\quad\quad$ **if** $C[i] == pattern[0]$ *and* $C[i : i + len(pattern)] == pattern$ **then**
14 $\quad\quad\quad\lfloor\ find\_ids.append((i, i + len(pattern)))$

15 $\quad$ **if** $len(find\_ids) > 0$ **then**
16 $\quad\quad\lfloor\ start, end = find\_ids[-1]$
17 $\quad A.extend([(start, end)])$
18 **return** $A, L;$

---

or not. Given the current utterance, the complete utterance and history dialogue, we then employ Algorithm 1 to extract missing semantic span labels and omission label.

It is worth noting that the current utterance may have multi-position semantic missing. According to statistics of the CamRest676 and RiSAWOZ corpus, the phenomenon of single-position missing is very common, accounting for 90.60% and 86.69% respectively. Therefore, we consider reconstructing multi-position missing instances into single-position missing instances.

Moreover, the comparison of average length between the missing phrase(MP), the incomplete utterance(IU) and the restored utterance(RU) is listed in Table 2, which indicates unsupervised label extraction method can extract the missing semantic information accurately. It lays a good foundation for model training.

3   Methodology

As show in Fig. 1, the model adopts two encoders, which serve in extractive module and generative module respectively. Then in the decoding stage, the extracted missing semantic information is used to guide the generation of the complete utterance. The model will be described in detail as follows.

Following [14], we formulate dialogue ellipsis and coreference resolution as a traditional sequence-to-sequence generative problem. Given the n-th user utterance $U_n = (u_1, u_2, ..., u_s)$ and its dialogue history $H = \{(U_1, R_1), (U_2, R_2), ..., (U_{n-1}, R_{n-1})\}$ corresponding to all the previous dialogue turns, where R_i represents the system response of the i-th turn, the goal is to recover the ellipsis or coreference for the original user utterance U_n. More formally, the dialogue ellipsis and coreference resolution task can be formulated as $((H, U_n) \rightarrow U_c)$ where each token of U_c is generated from current user utterance or its dialogue history.

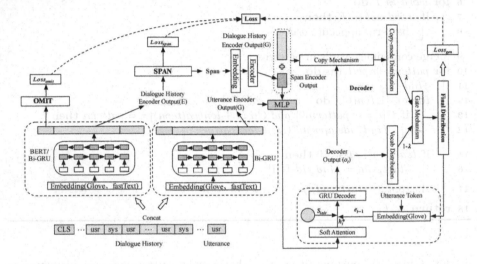

Fig. 1. The architecture of fusion extractive-generative dialogue ellipsis and coreference integrated resolution model.

3.1   Extractive Module

Inspired by the reading comprehension task, the extractive dialogue ellipsis and coreference detection task is reconstructed into a span extraction task called SPAN. First, while constructing the dialogue history sequence $C = \{c_1, c_2, ..., c_n\}$, we insert a tag of $[sys]$ symbol in front of each system utterance, and a tag of $[usr]$ symbol in front of each user utterance to explicitly enhance the model's ability to distinguish between system and user utterances in the history dialogue. We concatenate the dialogue history sequence C and the original user utterance U_n. At the same time, the $[CLS]$ tag and $[SEP]$ tag are added to separate C and U_n. The input sequence $\{[CLS], c_1, c_2, ..., c_n, [SEP], u_1, u_2, ..., u_s\}$ can be acquired, then obtain $S = \{e_1, e_2, ..., e_k\}$ by looking up the embedding tables. Recording the position of the $[SEP]$ tag for segmenting the encoded representation.

Once the input embedding is acquired, we feed such representation into a bi-directional GRU or finetune the pre-trained deep bidirectional transformers for language understanding model(BERT) [2] to capture the contextualized features of the sequence, Which can be further divided into three parts according to the partition of history and original user utterance:

$$[h_{cls}; o_c; o_u] = \text{BiGRU/BERT}(S) \tag{1}$$

h_{cls} is the encodings of $[CLS]$ tag, o_c is the encodings of history, o_u is the encodings of original utterance.Compared with other words already existing in the text, the $[CLS]$ tag itself has no semantic information, so this tag without obvious semantic information will more fairly integrate the rich semantic information contained in each word in utterance, resulting in better characterizing the semantics of entire utterance. Therefore, we regard the $[CLS]$ tag encodings as the global semantic representation of dialogue in extractive module, which is used as the initial hidden state S_{init} for the decoding phase of the generative model in Generative module. We use superscripts $*^e$ and $*^d$ to distinguish the encodings of extractive encoder and generative encoder.

SPAN Task. It is important to detect the omitted tokens from the dialogue context by introducing the SPAN task. Which aims to identify head-tail boundaries where phrases are omitted or referred to in context. We employ a fully connected layer to update the history dialogue encodings o_c^e, then acquire the update encodings x_c. After feeding x_c into a softmax layer, we obtain the probability that each word in the dialogue history is the head and tail boundary of the span, noted as P_{head} and P_{tail}.

$$x_c = w_1 o_c^e + b_1 \tag{2}$$
$$P_{head} = \text{Softmax}(w_2 \cdot x_c + b_2) \in \mathbb{R}^{n \times 2} \tag{3}$$
$$P_{tail} = \text{Softmax}(w_3 \cdot x_c + b_3) \in \mathbb{R}^{n \times 2} \tag{4}$$

Then, apply argmax to the probabilities P_{head} and P_{tail} to obtain the start or end index of missing semantic information in the dialogue history utterances.

$$\hat{I}_{head} = \{i \mid \text{argmax}(P_{head}^{(i)}) = 1\} \tag{5}$$
$$\hat{I}_{tail} = \{j \mid \text{argmax}(P_{tail}^{(j)}) = 1\} \tag{6}$$

Screen all possible span combinations to ensure that the filtered spans satisfy $\{i_{start} < i_{end}, i_{start} \in \hat{I}_{start}, i_{end} \in \hat{I}_{end}\}$. We extract all missing semantic phrase expressions from the dialogue history utterances using spans that satisfy the condition.

OMIT Task. In multi-turn dialogues, user utterances may be incomplete or complete. A robust resolution model needs to be able to accurately identify whether the original user utterance is complete or not, so we introduce an auxiliary task OMIT which judges whether the current utterance has coreference or ellipsis phenomena. Intuitively, we regard that when non-missing dialogue instances are ubiquitous, it will interfere with the model to extract the correct missing semantic information to a certain extent. Given the global semantic representation of dialogue h_{cls}^e, we feed it into a fully connected layer for a more compact representation. After that, the binary classification prediction is performed by a softmax layer, the formula is described as follows:

$$x_{cls} = \tanh(w_4 h_{cls}^e + b_4) \tag{7}$$
$$P_{omit} = \text{Softmax}(w_5 \cdot x_{cls} + b_5) \tag{8}$$

We introduce the auxiliary task not only to achieve a better discriminative effect on the binary classification task, but also hope that it can improve the performance of dialogue ellipsis and coreference missing semantic information detection. The results provide explicit guidance for the span parsing phase. If the predicted label is zero, the boundary recognition of missing semantic information will not be performed, and the span will be directly regarded as empty.

3.2   Generative Module

In generative module, we aim to train the capability of the model to completely extract missing semantic information by extractive module into the original user utterance. We get representation h_{cls}^d, o_c^d, o_u^d through a bi-directional GRU encoder.

Moreover, the missing semantic phrase expressions are forwarded into an embedding layer and an encoding layer to get the missing semantic phrase representation o_{span}. We introduce an MLP layer to incorporate o_{span} into the original user utterance encodings o_u^d as the input of the decoder of generative module.

$$O_u^* = [O_{span}; O_u] = \text{MLP}([o_{span}; o_u^d]) \tag{9}$$

For the consideration of efficiency, we employ a single-layer unidirectional GRU as the backbone of our decoder. Following [1], we use the previous hidden state s_{t-1} and the representation $O_{u_i}^*$ of token u_i from the update user utterance encodings O_u^* to calculate the attention weights at the time t, noted by a^t. Then the attention distribution a^t is used to calculate a weighted sum of the representation $O_{u_i}^*$, which is known as context vector h_t^*.

$$attn_i^t = v^T \tanh(w_6 O_{u_i}^* + w_7 s_{t-1} + b_6) \tag{10}$$
$$a^t = \text{Softmax}(attn_{i-1}^t) \tag{11}$$
$$h_t^* = \sum_i a^t O_{u_i}^* \tag{12}$$

Then the auto-regressive generation is described as follows:

$$o_t, s_t = \text{GRU}([e_{t-1}; h_t^*], s_{t-1}) \tag{13}$$
$$P_{vocab}(y_t | y_{1:t-1}) = w_8 h_t^* + w_9 o_t + b_8 \tag{14}$$

It is worth noting that this method concatenates the history dialogue encodings o_c^d obtained by the generative encoder with the missing semantic phrase representations o_{span} predicted by the extractive module. Through the above operations, an explicit emphasis is given to the relevant omitted or coreferred expressions in history dialogue, so that the fusion model can pay more attention to the missing semantic information in history dialogue in order to generate more relevent utterance. At the same time, the copy mechanism is applied to calculate the probability of copying words from history dialogue information enhanced by the explicit information. The detailed calculations are as follows:

$$P_{copy}(y_t|y_{1:t-1}) = \sum_{i:c_i=y_t}^{|C|} \frac{1}{Z} e^{\psi c^{(c_i)}}, y_t \in C \tag{15}$$

$$\psi_c(y_t = c_i) = \sigma(w_{10}[o_c^d; o_{span}] + b_{10})s_t \tag{16}$$

At the last, We use a gate mechanism to learn whether to copy one token from context dialogue or just generate the token from vocab at different steps. The final probability distribution is as follows:

$$\lambda = \sigma(w_{11}[h_t^*; e_{t-1}; s_t] + b_{11}) \tag{17}$$

$$P(y_t|y_{1:t-1}) = \lambda P_{vocab} + (1 - \lambda)P_{copy} \tag{18}$$

Training. A two-stage cascade training method is adopted. In the first stage, the model is trained with the auxiliary task OMIT. On parameter initializing basis, the second stage of training is carried out, which aims to train the fusion extractive-generative model.

4   Experimentation

In this section, we systematically evaluate our proposed fusion extractive-generative approach in dialogue ellipsis and coreference resolution task.

4.1   Experimental Settings

All experiments are conducted in the data from the English corpus CamRest676 [14] and the Chinese corpus RiSAWOZ [15]. We employed the 50-dimension word embeddings provided by Glove [13] for CamRest676 and 300 dimensional fastText [4] word vectors to initialize word embeddings for RiSAWOZ. We set the vocabulary size V to 800 and 12000 respectively. The size of hidden states was set to 128 and 256 respectively. We use GRU and Bert-base-uncased [2] to initialize encoder of extractive module with the learning rate of 3e-3 and 3e-5 respectively. In order to prevent model from over-fitting, we adopted early stopping strategy and the patience value was 12. It is worth mentioning that we used the standard cross-entropy loss as the loss function to train the entire model. We used exact match rate(EM), F1 score(F1), OMIT-F1(OF1) which means the

F1 for auxiliary task OMIT and Accuracy(Acc) as evaluation metric for extractive module, and adopted the automatic metrics BLEU, EM, EM1, EM2, and F1 score as the main evaluation methods for generation module. Exact match rate(EM) measures whether the generated utterances exactly match the gold utterances. EM1, EM2 respectively indicate the situation that the input utterance is complete or incomplete where EM is the comprehensive evaluation of the two situation. BLEU [12], F1 score(a balance between word-level precision and recall) are also used for the resolution task to evaluate the quality of generated utterances at the n-gram and word level.

4.2   Compared Methods

In this paper, we compare our proposed system with the following competitive models on the CamRest676 and RiSAWOZ corpus:

- Baseline: we build two benchmark models for the extractive module. The encoding layer uses a bidirectional GRU model. Among them, single-task training for missing semantic extraction denoted as **SP**. Individual training for OMIT task denoted as **OM**, which judges whether the current utterance has coreference or ellipsis phenomena.
- **GECOR**: an end-to-end generative model proposed by [14] that uses the copy mechanism [5] to recover omission. They use a LSTM-based Seq2Seq model with attention as their baseline which simply connects all dialogue history utterances in series as the input of history encoder, known as Seq2Seq. Notably, we also reproduce their results over different system settings where GECOR1 means the model with the copy mechanism and GECOR2 means the model with the gated copy mechanism.
- **ELD** [14]: an end-to-end generative model which employs a speaker highlight dialogue history encoder and a top-down hierarchical copy mechanism to generate complete utterances.

4.3   Main Results

We separately analyze the performance of the extractive module and the generative module. FEGI takes two-stage training for extractive module. Which first takes the OMIT task training the model to get the basis of model, then the model is trained on the both SPAN and OMIT tasks. The main results of extraction module on CamRest676 and RiSAWOZ are as shown in Table 3. The main results of generative module on CamRest676 and RiSAWOZ are as shown in Table 4 and Table 5. The performance comparison for the ellipsis and coreference dataset on CamRest676 respectively is shown in Table 6. Comparing the performance of SPAN on the two corpora can be found: the performance of our proposed model has been improved on both corpora. Among them, the improvement on the English corpus CamRest676 is the most significant, improving the exact match rate (EM) by 47 points. For the Chinese corpus RiSAWOZ, improves the EM metric by more than 32 points. The comparative experimental results on the two corpora provide strong support and verify the effectiveness of the improved method proposed. This FEGI results in Table 3 also proves that SPAN task and OMIT task can benefit from each other. The omitted or co-referred expressions in context can bring some implicit guidance to the judgment of whether there is a missing semantic phenomenon in the current utterance. And one possible reason could be that: SPAN always predicts an answer for each instance when parsing the span by the extractive module. The enhanced information-guided joint generative model makes the model have a preliminary judgment on whether there is a semantic missing phenomenon in the current utterance by adding an auxiliary task. The joint learning method provides effective explicit guidance for the task of missing semantic information extraction, which greatly reduces the impact of negative examples on the model.

Table 3. The main results on CamRest676 and RiSAWOZ of our extractive module.

Systems		SPAN		OMIT			
		F1	EM	OF1	Acc	Prec.	Rec.
CamRest676	SP	18.07	9.65	–	–	–	
	OM	–	–	87.36	87.98	83.82	91.20
	FEGI	62.02	57.19	89.51	90.53	90.24	88.80
RiSAWOZ	SP	45.67	35.39	–	–	–	
	OM	–	–	90.57	88.76	87.63	93.72
	FEGI	78.93	68.15	90.93	89.17	87.84	94.24

Table 4. The main results on CamRest676 of our generative method and other methods. † denotes the duplicated systems.

Model	EM	EM1	EM2	BLEU	F1
Seq2Seq †	50.29	71.38	27.78	73.78	90.60
GECOR1 †	65.26	91.82	38.90	83.09	95.59
GECOR2 †	64.09	89.43	37.30	82.83	96.00
ELD †	68.91	94.57	40.00	84.34	95.81
FEGI	68.09	92.96	40.39	84.37	96.24
FEGI-BERT	72.91	94.72	48.62	87.17	96.87
FEGI-G	73.47	**96.48**	47.84	87.02	96.92
FEGI-G-spans	**75.88**	94.72	**54.90**	**89.83**	**97.72**

'FEGI-BERT' means we use BERT as the extractive encoder in extractive module, 'FEGI-G' means only one gold span answer are copied as 'FEGI-G-spans' means extract all the missing phrases.[1] "FEGI" can greatly improve the performance on EM2. This indicates that the FEGI can accurately capture the missing semantic information in history dialogue, and complete them into the incomplete utterance to obtain utterances with complete semantics and coherent word order.

Table 5. The main results on RiSAWOZ of our generative method and other methods.† denotes the duplicated systems.

Model	EM	EM1	EM2	BLEU	F1
GECOR1 †	59.26	86.07	42.36	87.75	97.52
GECOR2 †	58.33	86.25	40.57	87.67	97.02
ELD †	64.89	92.35	47.22	88.68	98.02
FEGI	64.50	91.42	46.90	89.04	97.55
FEGI-BERT	68.05	94.98	38.00	85.04	96.43
FEGI-G	68.92	**95.59**	52.77	**89.95**	97.62
FEGI-G-spans	**70.68**	95.22	**54.78**	87.07	**98.07**

Table 6. Performance comparison on the ellipsis dataset and coreference dataset.

Model	Ellipsis					Coreference				
	EM	EM1	EM2	BLEU	F1	EM	EM1	EM2	BLEU	F1
Seq2Seq †	51.72	71.33	28.15	73.81	91.45	52.72	71.16	32.79	75.98	91.45
GECOR1 †	67.32	92.70	37.82	83.21	96.37	71.15	90.98	47.80	85.41	96.32
GECOR2 †	65.20	90.53	34.87	83.11	96.40	70.31	92.45	44.56	85.30	96.80
ELD †	67.86	92.83	36.86	83.97	95.94	73.58	94.53	49.37	87.66	96.54
FEGI	71.43	92.95	44.81	85.76	96.69	71.99	94.46	46.00	87.18	97.04
FEGI-BERT	72.35	93.29	46.47	86.55	96.82	73.84	93.08	51.60	86.87	96.42
FEGI-G	72.54	94.97	44.81	86.99	96.95	77.18	96.19	55.20	89.16	97.14
FEGI-G-spans	**77.74**	**96.98**	**53.94**	**89.44**	**97.71**	**78.11**	93.43	**60.40**	**90.88**	**97.99**

5   Ablation Study

5.1   Effect of the Auxiliary Task OMIT

Through corpus statistics, we can observe that the dialogue instances without coreference and omission on CamRest676 and RISAWOZ accounted for as high as 51.53% and 38.51% respectively. Which will definitely fuse the model to generate

---

[1] Because of the FEGI-G actually do not finetune on the SPAN task in the second stage, the FEGI-G-BERT equals to the FEGI-G.

right restored utterances. OMIT on both corpora is already at a high level, and the performance indicators are all around 90% in Table 3. The EM performance of FEGI-BERT without doing OMIT task in the first stage is about 2.8 lower than the original, More specific details can be seen from the Fig. 2.

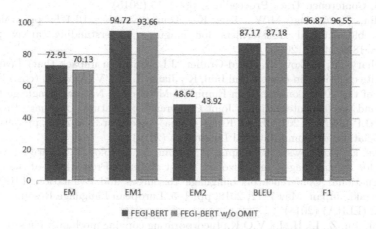

Fig. 2. Different metrics comparision between FEGI-BERT and FEGI-BERT w/o OMIT.

5.2 The Utility of Ellipsis Recovery and Coreference Resolution Respectively

Compared with the benchmark models as shown in Table 6, the performance of "FEGI" on the two version dataset shows that the model is effective for both ellipsis resolution and coreference resolution tasks.

6 Conclusion

We proposed a fusion extractive-generative dialogue ellipsis and coreference integrated resolution model. In the future, we will investigate on extending approach to more areas.

Acknowledgments. This work was supported by Projects 62276178 under the National Natural Science Foundation of China, the National Key RD Program of China under Grant No.2020AAA0108600 and Priority Academic Program Development of Jiangsu Higher Education Institutions.

References

1. Bahdanau, D., Cho, K., Bengio, Y.: Neural machine translation by jointly learning to align and translate. In: Bengio, Y., LeCun, Y. (eds.) 3rd International Conference on Learning Representations, ICLR 2015, San Diego, CA, USA, May 7–9, 2015, Conference Track Proceedings, pp. 1–15 (2015)
2. Devlin, J., Chang, M.W., Lee, K., Toutanova, K.: BERT: pre-training of deep bidirectional transformers for language understanding. arXiv preprint arXiv:1810.04805 (2018)
3. Elgohary, A., Peskov, D., Boyd-Graber, J.L.: Can you unpack that? Learning to rewrite questions-in-context. In: Inui, K., Jiang, J., Ng, V., Wan, X. (eds.) Proceedings of the 2019 Conference on Empirical Methods in Natural Language Processing and the 9th International Joint Conference on Natural Language Processing, EMNLP-IJCNLP 2019, Hong Kong, China, November 3–7, 2019, pp. 5917–5923. Association for Computational Linguistics (2019)
4. Grave, E., Bojanowski, P., Gupta, P., Joulin, A., Mikolov, T.: Learning word vectors for 157 languages. In: Calzolari, N., et al. (eds.) Proceedings of the Eleventh International Conference on Language Resources and Evaluation, LREC 2018, Miyazaki, Japan, May 7–12, 2018, pp. 1–5. European Language Resources Association (ELRA) (2018)
5. Gu, J., Lu, Z., Li, H., Li, V.O.K.: Incorporating copying mechanism in sequence-to-sequence learning. In: Proceedings of the 54th Annual Meeting of the Association for Computational Linguistics, ACL 2016, August 7–12, 2016, Berlin, Germany, Volume 1: Long Papers, pp. 1631–1640. The Association for Computer Linguistics (2016)
6. Hao, J., Song, L., Wang, L., Xu, K., Tu, Z., Yu, D.: Robust dialogue utterance rewriting as sequence tagging. CoRR abs/2012.14535, 1–11 (2020)
7. Huang, M., Zhu, X., Gao, J.: Challenges in building intelligent open-domain dialog systems. ACM Trans. Inf. Syst. (TOIS) **38**, 1–32 (2019)
8. Li, P.: An empirical investigation of pre-trained transformer language models for open-domain dialogue generation. ArXiv abs/2003.04195 (2020)
9. Li, Q., Kong, F.: Transition-based mention representation for neural coreference resolution. In: Huang, D.S., Premaratne, P., Jin, B., Qu, B., Jo, K.H., Hussain, A. (eds.) Advanced Intelligent Computing Technology and Applications, pp. 563–574. Springer, Singapore (2023). https://doi.org/10.1007/978-981-99-4752-2_46
10. Ni, Z., Kong, F.: Enhancing long-distance dialogue history modeling for better dialogue ellipsis and coreference resolution. In: Wang, L., Feng, Y., Hong, Yu., He, R. (eds.) NLPCC 2021. LNCS (LNAI), vol. 13028, pp. 480–492. Springer, Cham (2021). https://doi.org/10.1007/978-3-030-88480-2_38
11. Pan, Z.F., Bai, K., Wang, Y., Zhou, L., Liu, X.: Improving open-domain dialogue systems via multi-turn incomplete utterance restoration. In: Inui, K., Jiang, J., Ng, V., Wan, X. (eds.) Proceedings of the 2019 Conference on Empirical Methods in Natural Language Processing and the 9th International Joint Conference on Natural Language Processing, EMNLP-IJCNLP 2019, Hong Kong, China, November 3–7, 2019, pp. 1824–1833. Association for Computational Linguistics (2019)
12. Papineni, K., Roukos, S., Ward, T., Zhu, W.: BLEU: a method for automatic evaluation of machine translation. In: Proceedings of the 40th Annual Meeting of the Association for Computational Linguistics, July 6–12, 2002, Philadelphia, PA, USA, pp. 311–318. ACL (2002)

13. Pennington, J., Socher, R., Manning, C.D.: GloVe: global vectors for word representation. In: Moschitti, A., Pang, B., Daelemans, W. (eds.) Proceedings of the 2014 Conference on Empirical Methods in Natural Language Processing, EMNLP 2014, October 25–29, 2014, Doha, Qatar, A meeting of SIGDAT, a Special Interest Group of the ACL, pp. 1532–1543. ACL (2014)
14. Quan, J., Xiong, D., Webber, B., Hu, C.: GECOR: an end-to-end generative ellipsis and co-reference resolution model for task-oriented dialogue. In: Inui, K., Jiang, J., Ng, V., Wan, X. (eds.) Proceedings of the 2019 Conference on Empirical Methods in Natural Language Processing and the 9th International Joint Conference on Natural Language Processing, EMNLP-IJCNLP 2019, Hong Kong, China, November 3–7, 2019, pp. 4546–4556. Association for Computational Linguistics (2019)
15. Quan, J., Zhang, S., Cao, Q., Li, Z., Xiong, D.: RiSAWOZ: a large-scale multi-domain Wizard-of-Oz dataset with rich semantic annotations for task-oriented dialogue modeling. In: Webber, B., Cohn, T., He, Y., Liu, Y. (eds.) Proceedings of the 2020 Conference on Empirical Methods in Natural Language Processing, EMNLP 2020, Online, November 16–20, 2020, pp. 930–940. Association for Computational Linguistics (2020)
16. Su, H., et al.: Improving multi-turn dialogue modelling with utterance rewriter. In: Korhonen, A., Traum, D.R., Màrquez, L. (eds.) Proceedings of the 57th Conference of the Association for Computational Linguistics, ACL 2019, Florence, Italy, July 28- August 2, 2019, Volume 1: Long Papers, pp. 22–31. Association for Computational Linguistics (2019)

Assessing and Enhancing LLMs: A Physics and History Dataset and One-More-Check Pipeline Method

Chaofan He, Chunhui Li, Tianyuan Han, and Liping Shen[✉]

Shanghai Jiao Tong University, Shanghai, China
{nafoabehumble,lch2016,hty19980927,lpshen}@sjtu.edu.cn

Abstract. Large language models(LLMs) demonstrate significant capabilities in traditional natural language processing(NLP) tasks and many examinations. However, there are few evaluations in regard to specific subjects in the Chinese educational context. This study, focusing on secondary physics and history, explores the potential and limitations of LLMs in Chinese education. Our contributions are as follows: a PH dataset is established, which concludes secondary school physics and history in Chinese, comprising thousands of multiple-choice questions; an evaluation on three prevalent LLMs: ChatGPT, GPT-3, ChatGLM on our PH dataset is made; a new prompt method called One-More-Check(OMC) is proposed to enhance the logical reasoning capacity of LLMs; finally, three LLMs are set to attend an actual secondary history exam. Our findings suggest that our OMC method improves the performance of LLMs on logical reasoning and LLMs underperform average level of age-appropriate students on the exam of history. All datasets, code and evaluation results are available at https://github.com/hcffffff/PH-dataset-OMC.

Keywords: NLP Application · Large Language Model · Chain-of-Thought · Intelligent Education

1 Introduction

With the advent of ChatGPT, Large Language Models (LLMs) have rapidly become mainstream in solving numerous NLP tasks. As demonstrated in OpenAI's GPT-4 technical report [17], LLMs exhibit excellent proficiency in addressing many English-language tasks. Considerable research has evaluated LLMs in traditional NLP tasks and assorted exam capabilities. But disappointingly, the assessment of LLMs on Chinese datasets, particularly in education, remains scant. Inspired by C-EVAl [8] and GAOKAO-Bench [31], we collect and organize a multiple-choice dataset for Chinese secondary school physics and history, called PH dataset, and evaluate the answer accuracy of ChatGPT, GPT-3 [2], and ChatGLM [6] on this dataset.

Some research has shown that outstanding performance of LLMs occurs when the model's parameter size reaches a certain scale [24]. However, in certain

B. Luo et al. (Eds.): ICONIP 2023, CCIS 1967, pp. 504–517, 2024.
https://doi.org/10.1007/978-981-99-8178-6_38

tasks, particularly logical and mathematical reasoning ones, merely increasing the parameters of LLMs does not significantly improve their performance [19]. To address this challenge, several methods have been proposed. Notably, the Chain-of-Thought (CoT) [25] method prompts LLMs to generate complete reasoning paths and provide answers based on them. Self-Consistency [23] samples multiple reasoning paths and outputs the majority one. In this paper, we propose an improved method based on CoT and Self-Consistency, called One-More-Check (OMC), to enhance the performance of LLMs in more challenging problems. OMC samples multiple reasoning paths from the *generator* model, and applies a *discriminator* model to synthesize all reasoning paths and output a consolidated answer. We test our method on the PH dataset and find that our approach outperformed in various scenarios.

LLMs possess various capabilities such as comprehension, expression, inference and judgment. In our research, we specify two capabilities of LLMs, problem-solving and answer-checking abilities indicating the capability of inference and judgment of LLMs, respectively. We experiment with these two capacities in ChatGPT and believe such abilities have implications for their application in the education field. Lastly, LLMs have participated in numerous examinations [10,16]. We also collaborate with two prominent Chinese secondary schools to carry out actual history exams with junior and senior high school students, utilizing three LLMs and our OMC method.

Our contributions are as follows:

1. We introduce a dataset tailored for Chinese secondary school physics and history, the PH dataset, which has been elaborately reviewed and verified. Three mainstream LLMs are tested on the PH dataset.
2. A new prompting method, OMC, is proposed and validated on the PH dataset, achieving optimal or suboptimal results in multiple tests.
3. In collaboration with two Chinese secondary schools, we test the proficiency of three LLMs and our OMC method on an actual exam compared to age-appropriate students.

2   Related Works

2.1   LLMs Evaluations

LLMs demonstrate outstanding, and in some cases, best performance solving various NLP tasks on publicly available datasets [32]. In the domain of cognitive psychology, GPT-4 shows a high level of accuracy on cognitive psychology tasks compared to previous models [4]. Regarding the medical subject, GPT-4 exceeds the passing score of the United States Medical Licensing Examination (USMLE) by over 20 points and outperforms earlier models [15]. Furthermore, GPT-4 exhibits human-level performance on various professional and academic benchmarks [17], including multi-language and multimodal tasks.

However, in some other fields, LLMs still have limitations. In terms of mathematical abilities, LLMs abilities are significantly below those of an average mathematics graduate student [7]. When faced with frequent challenges involving absurd arguments, LLMs struggle to maintain accurate reasoning [22]. The errors made by LLMs encompass various categories such as reasoning, factual errors, math, coding, etc. [1]. On the C-EVAL dataset, GPT-4 achieved an average score of 68.7 [8]. On the GAOKAO-Bench multiple-choice questions dataset, ChatGPT obtained scores of only 398 and 364 (total score 750) respectively for liberal arts and science subjects [31]. These results indicate significant room for improvement in the application of LLMs in the field of education.

2.2 Chain-of-Thought Improvements

Chain-of-Thought(CoT) [25] is a prompt method that enhances the logical reasoning capability of language models. With CoT, LLMs are able to generate comprehensive intermediate steps, thereby improving their performance in solving complex problems compared to conventional prompt methods.

However, some studies show that the reasoning paths generated by LLMs may be unfaithful and unreliable [21,28]. Subsequently, further research explores more effective methods based on CoT. Chain-of-Questions [33] trains a language model to generate intermediate questions for better solving problems that require multiple steps. Tree-of-Thought [26] considers multiple reasoning paths and employs self-assessment mechanisms to determine the next action while allowing backtracking when necessary for global decision-making. Chain-of-Knowledge [12] enhances LLMs with structured knowledge bases to improve factual accuracy and reduce hallucinations. MultiTool-CoT incorporates various external tools, such as calculators and knowledge retrievers, into the reasoning process to enhance the correctness of LLMs in domains such as mathematics and chemistry [9]. N. Mündler [14] designed a framework to evaluate, detect, and mitigate self-contradictions in sentence pairs generated by LLMs, ensuring logical and informative outputs. Graph-of-Thought models human thought as a graph and demonstrates significant improvements in pure text reasoning tasks and multimodal reasoning tasks when applied to CoT [27].

Self-Consistency [23], improves CoT by sampling a set of different reasoning paths instead of relying solely on greedy inference. Self-Consistency leverages the intuition that a complex reasoning problem often has multiple distinct ways of thinking that lead to its unique correct answer, selects the most consistent answer by marginalizing out the sampled reasoning paths. It achieves significant improvements over CoT in a series of popular arithmetic and commonsense reasoning tasks. Based on Self-Consistency, our approach enhances the model's ability to generate and examine reasoning paths through a Generator-Discriminator framework.

3 Dataset

3.1 Overview

Our PH dataset collects questions from physics and history, pertinent to the junior and senior high school levels within Mainland China's educational frame-

```
{                                          {
  "id": "10496658",                          "id": "10496658",
  "type": "CQ",                              "type": "CQ",
  "question": "若地球表面处的重力加速度为g,      "question": "If the gravitational acceleration at the earth's surface is g,
而物体在距离地面3R (R为地球半径) 处，由于地       and the object is at a distance of 3R from the ground (R is the radius of the
球作用而产生的加速度为g', 则g'/g为（ ）",         earth), the acceleration due to the action of the earth is g', then g'/g is ( )",
  "options": [                               "options": [
    "A . 1",                                    "A . 1",
    "B . 1/9",                                  "B . 1/9",
    "C . 1/4",                                  "C . 1/4",
    "D . 1/16"                                  "D . 1/16"
  ],                                          ],
  "answer": [                                "answer": [
    "D"                                         "D"
  ],                                          ],
  "analysis": "根据万有引力定律，列出公式：        "analysis": "According to the law of universal gravitation and the formula:
g=GM/r^2, 其中M是地球的质量，r是物体在某         g=GM/r^2, where M is the mass of the earth, and r is the distance from the
位置到球心的距离。                             object at a certain position to the center of the sphere.
g'/g=(GM/(3R+R)^2)/(GM/(R)^2)= 1/16。故选：D。  g'/g=(GM/(3R+R)^2)/(GM/(R)^2)= 1/16. Therefore choose: D.",
",                                           "review": "An overview of other application knowledge test points for
  "review": "万有引力定律的其他应用知识考点      the law of universal gravitation..."
概述..."                                     }
}
```

Fig. 1. An example of a physics problem in our PH dataset. Left column shows an original problem in the dataset. Right column shows the translated one for demonstration. Some information is omitted for presentation.

work. Such consideration is taken because these two subjects are fundamental in the sciences and liberal arts parts of the Chinese University Entrance Examination (GAOKAO), and other fields such as mathematics, medicine, and programming already have abundant datasets [15,20,29]. Our dataset solely includes multiple-choice questions, which have definitive answers that expedite the evaluation process.

Table 1. Number of problems of different splits in our PH dataset. Letters in parentheses are abbreviations of the corresponding names used in the following sections.

Category	# Questions
Physics(PS)	2193
Physics Calculation(PC)	210
Junior History(JH)	1000
Senior History(SH)	1000

Table 1 reveals the dataset contains 2193 physics problems and 1000 problems each for junior and senior high history. A subset containing 210 physical calculation problems is listed separately. Every problem carries a unique ID, the question text, four or more options, the correct answer, an explanation, and related knowledge. All problems in the dataset are represented in Chinese. An exemplar physics problem is depicted in Figure 1.

3.2 Data Collection

The dataset, sourced and cleaned from publicly accessible web pages, has undergone extensive review to verify the correctness of format and answers. Since most current public LLMs do not support multimodal inputs such as images, the dataset excludes questions with such elements. A subset of 210 physics problems(PC), all involving calculations, is separated for evaluating the arithmetic capabilities of LLMs more directly. The history dataset includes 1000 questions each on junior high and senior high. All problems have been carefully reviewed for potentially controversial, sensitive, or toxic information to meet security and risk requirements.

In comparison to recent datasets such as GAOKAO-Bench [31] and C-EVAL [8], our dataset prioritizes physics and history, two fundamental subjects in Chinese secondary school education. The larger data volume enables a better evaluation of LLMs' capabilities in knowledge inference, calculation, and formula derivation within the secondary education context.

3.3 Data Verification

In order to verify the correctness and validity of our dataset, we conduct experiments on three different datasets with the ChatGPT (GPT-3.5-turbo) model, including physics and history multiple-choice questions in the C-EVAL val split and the GAOKAO-benchmark. The C-EVAL results are reproduced from official code[1], and the GAOKAO-Bench results are obtained from existing results[2]. All data verification results are presented in Table 2.

Table 2. Comparison of ChatGPT accuracy on three datasets. CoT prompting method is used for all three datasets.

	C-EVAL	GAOKAO-Bench	PH dataset(Ours)
Physics	47.37%(18/38)	35.94%(23/64)	43.41%(952/2193)
Junior history	59.09%(13/22)	–	59.20%(592/1000)
Senior history	60.00%(12/20)	54.36%(156/287)	49.90%(499/1000)

The results indicate that ChatGPT demonstrates consistent accuracy across all three datasets, affirming the effectiveness and validity of our datasets. The result is comparatively lower on the GAOKAO-benchmark physics dataset due to receiving valid model outputs from only 47 out of 64 problems, nearly half of which are indefinite multiple-choice questions where it underperforms.

Additionally, our dataset has been checked by four senior undergraduate students to ensure correctness and validity.

---

[1] C-EVAL official code: https://github.com/SJTU-LIT/ceval.

[2] GAOKAO-Bench existing results: https://github.com/OpenLMLab/GAOKAO-Bench/tree/main/result.

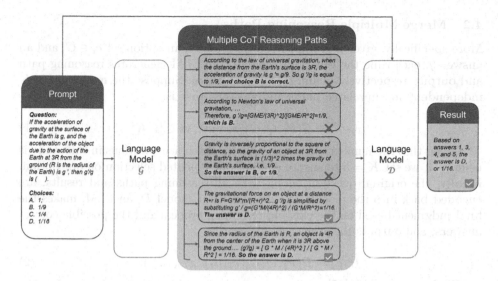

Fig. 2. An example of our OMC method. OMC samples multiple CoT reasoning paths from model $\mathcal{G}$ and combine questions and candidate responses to model $\mathcal{D}$ to generate the final answer.

4 Method

4.1 One-More-Check(OMC)

LLMs display remarkable success in numerous NLP tasks, despite their comparatively modest performance in logical reasoning and mathematical computations. A multitude of methodologies are introduced with the intent to enhance LLMs' proficiency in these areas, principally through improvements in prompting techniques. Strategies such as the Chain-of-Thought [25] and In-context Learning [5] evince certain advancements. Our One-More-Check (OMC) method, based on Chain-of-Thought [25] and Self-Consistency [23], combining the concepts of the generator and discriminator in Generative Adversarial Networks (GANs) [3], further enhances the logical reasoning capabilities of LLMs.

An example of a physics problem illustrating the steps of OMC is shown in Fig. 2. Initially, we pose a Chain-of-Thought prompt to the language model, symbolized as $\mathcal{G}$. However, due to the randomness of the model's generation, errors are likely to occur in the intermediate steps and final outcomes produced by the decoder, particularly when the problem is complex. Regardless of whether the results are correct, we retain any output from the model for subsequent use. Secondly, we repeat the above process five times, yielding five unrelated reasoning paths and results. In the worst-case scenario, all five random results could be different. Finally, we combine the original question with these five output results and encode this into the discriminator model $\mathcal{D}$, allowing it to judge or reconstruct a reasoning path and outcome that has a higher probability of being correct.

4.2   Merge Multiple Reasoning Paths

More specifically, given a prompt P_1, a question Q, an option set $c_i \in C$, and an answer y, each time the generator model $\mathcal{G}$, an LLM, generates reasoning path and output, respectively r_i and a_i, where $r_i \rightarrow a_i$. Suppose the model yields K independent and unrelated paths and answers, that is,

$$(r_i, a_i) = \mathcal{G}(P_1, Q, C), r_i \text{ i.i.d.}, \forall i \in [1, K] \tag{1}$$

Given that the number of options in multiple-choice questions in this study is usually 4, we set $K = 5$ to ensure at least one repeated selection on one option. Finally, the original question, along with all reasoning paths and results, are encoded back into the model. The discriminator model $\mathcal{D}$, an LLM, makes the final judgment based on the existing reasoning process and the possible correct answers, and outputs the answer (r, a), that is,

$$(r, a) = \mathcal{D}(P_2, Q, C | (r_1, a_1)...(r_K, a_K)) \tag{2}$$

Chain-of-Thought [25] utilizes a greedy search method, encouraging the model to output the reasoning path as completely as possible and use it as the final answer, i.e., $(r, a) = \mathcal{G}(P_1, Q, C)$. Self-consistency [23] samples multiple reasoning paths but ultimately selects the output with the majority vote as the most consistent answer, i.e., $\arg\max_a \sum_{i=1}^{K} \mathbb{1}(a_i = a)$, which could potentially lead to errors when the sample size is rather small. Our method, OMC, is not confined to generating a single reasoning path, but instead outputs as many possible reasoning paths and answers as it can, and combines these candidates to further strengthen the certainty of the answer, thus enhancing the model's error tolerance.

An example of a multiple-choice physics question about gravitational acceleration is provided in Fig. 2. For demonstration purposes, we translate it from Chinese into English. For secondary school students with a certain level of physics knowledge, the problem is not difficult. However, a detail to note is that the distance from the point $3R$ above the ground to the center of the Earth is $3R + R = 4R$, where R is the Earth's radius. Among the five independent outputs of the model, the first three times, the language model does not pay attention to this detail and answers incorrectly. The last two times, it notices this detail and responses correctly. If self-consistency were used, the method would have selected the answer with the most votes, B, in this case. Clearly, this would be an incorrect answer. However, after implementing the OMC method, the model selected the more reasonable option D after evaluating the five reasoning paths and answers, thus providing the correct answer.

5   Experiment

5.1   Setup

We perform accuracy tests on various models and prompt methods on our created PH dataset to evaluate the application capabilities and limitations of LLMs in

Chinese educational contexts. We use three transformer-based models, including ChatGPT, GPT-3 [2], and ChatGLM [30].

- **ChatGPT** fine-tuned from GPT-3.5, optimized by Reinforcement Learning with Human Feedback(RLHF) [34]. We evaluate ChatGPT via OpenAI API[3].
- **GPT-3** an autoregressive language model with 175 billion parameters. We use text-davinci-003, built on InstructGPT [18], via OpenAI API.
- **ChatGLM** an open-source, bilingual (Chinese and English) conversational language model based on the General Language Model [6] architecture, with 6.2 billion parameters. We deployed the ChatGLM-6b-int4[4] version for evaluation.

In addition to our OMC method discussed in this paper, we compare the following prompt methods as well.

- **Answer-only** provides the answer directly after presenting the question and options.
- **Chain-of-Thought** [25] provides a detailed reasoning path after presenting the problem and options, then outputs the correct answer based on this reasoning.
- **Few-shot** [13] provides a few input-label pairs (demonstrations) and optimizes LLMs for new tasks. In contrast, zero-shot prompt provides only instructions describing the task [11].
- **CoT with Self-Consistency** [23] samples 5 CoT reasoning paths and selects the option with the most votes. We only use the default zero-shot, 5-candidate-choice setting in Self-Consistency, thus no few-shot experiments are conducted in this method.

Furthermore, a more challenging subset containing 210 physics questions is selected from the dataset(shown in Table 1), which includes numerical or formula calculations and necessitates more complex logical reasoning. We assess whether models exhibit improved reasoning capabilities after using the new prompting method.

5.2   Main Results

Table 3 displays the performance of various models and prompt methods under the physics dataset in our PH dataset. In zero-shot setting, our OMC method outperforms Self-Consistency across all models of varying model parameter sizes, achieving the optimal results. In few-shot setting, our OMC method yields optimal or suboptimal outcomes under both ChatGPT and GPT-3 models, but performs somewhat poorer on ChatGLM. Since models with fewer parameters impose limitations on the length of input text, such as GPT-3 only supports 2049 tokens in each query, and OMC involves significantly longer inputs than other methods, the performance of such models is compromised.

---

[3] Use OpenAI API at: https://openai.com/api/..
[4] THUDM/ChatGLM-6B: https://github.com/THUDM/ChatGLM-6B.

Table 3. Accuracy results(%) of ChatGPT, GPT-3, ChatGLM on PS subset in PH dataset. The best result for each task is shown in bold.

		ChatGPT	GPT-3	ChatGLM
Zero Shot	Answer-only	48.70	38.94	14.68
	CoT	45.60	39.17	11.72
	Self-Consistency	51.39	39.40	11.99
	OMC(Ours)	**52.26**	**39.76**	**15.05**
Few Shot	Answer-only	48.75	**40.36**	18.51
	CoT	42.91	29.14	**20.47**
	OMC(Ours)	**51.30**	39.63	15.28

Table 4. Accuracy results(%) of ChatGPT, GPT-3, ChatGLM on PC subset in PH dataset. The best result for each task is shown in bold.

		ChatGPT	GPT-3	ChatGLM
Zero Shot	Answer-only	40.48	25.71	6.67
	CoT	34.76	25.24	**11.90**
	Self-Consistency	43.33	30.95	10.95
	OMC(Ours)	**47.14**	**31.90**	10.95
Few Shot	Answer-only	41.90	20.00	**20.00**
	CoT	40.48	30.95	16.19
	OMC(Ours)	**46.19**	**32.38**	12.86

The results of the additional 210 physics calculation problems (including numerical calculations and formula derivations) are shown in Table 4, which indicates that our OMC method achieves optimum on both ChatGPT and GPT-3 models. This suggests our approach significantly enhances the model's capability to solve problems involving extensive reasoning steps, whether in zero-shot or few-shot contexts.

Table 5 displays the performance of various models and prompt methods under the history dataset in our PH dataset. In nearly all tests with larger models and zero-shot setting, our OMC method outperforms Self-Consistency. When tackling history problems, using zero-shot CoT method results in a significant performance decline, a phenomenon potentially attributable to the misleading interpretations of CoT [21]. Our OMC method improves performance when zero-shot CoT method has significant side effects.

We observe that the OMC method outperforms Self-Consistency in models with larger parameter sets, regardless of whether physics or history. We surmise that when provided with diverse reasoning paths, the model can directly analyze the quality of answers, indicating that the model possesses not only "problem-

Table 5. Accuracy results(%) of ChatGPT, GPT-3, ChatGLM on history subset in PH dataset. For each model, the left and right part represents the accuracy of junior and senior high problems respectively. The best result for each task is shown in bold.

		ChatGPT		GPT-3		ChatGLM	
		JH	SH	JH	SH	JH	SH
Zero Shot	Answer-only	63.3	52.2	52.6	**47.8**	**45.2**	**29.9**
	CoT	51.9	41.4	52.7	46.5	29.5	20.5
	Self-Consistency	62.6	51.9	53.6	47.0	32.3	21.0
	OMC(Ours)	**64.8**	**52.8**	**53.7**	46.6	33.4	25.6
Few Shot	Answer-only	62.6	**54.6**	**55.7**	48.6	31.9	**30.3**
	CoT	62.6	45.7	53.6	**49.4**	**32.3**	23.2
	OMC(Ours)	**64.2**	52.4	53.5	46.9	25.0	18.1

solving ability" but also "answer-checking capability. The model undoubtedly discerns the better reasoning path, rather than making random or majority-based choices, as exemplified in Fig. 2.

5.3 Answer-Checking Ability Study

Upon obtaining the main results, we infer that the usage of LLM in question answering exhibits randomness and cannot guarantee the optimal response at the initial CoT. We postulate that LLM possesses two distinct capabilities: problem-solving ability and answer-checking ability, with the latter superseding the former. In previous prompt methods, LLM's answer-checking ability was scarcely employed, as it did not reassess answers post-generation. We conduct answer-checking ability experiments to substantiate this claim, the results of which are presented in Table 6.

Table 6. Accuracy result(%) of ChatGPT conducting answer-checking ability study on PH dataset. The best result for each task is shown in bold.

	PS	PC	JH	SH
Iter-1(CoT)	45.60	34.76	51.90	41.40
Iter-2	45.87	39.05	58.80	47.40
Iter-3	46.69	**41.43**	59.80	50.00
Iter-4	**47.61**	40.48	60.50	50.10
Iter-5	47.24	39.05	**60.70**	**50.60**

We conduct experiments on the PH dataset, using the responses from ChatGPT-CoT as the initial outputs. In each subsequent iteration, we add the previous response to the prompt and have the model evaluate its correctness before outputting a new reasoning path and answer. After repeating the process four times, we find that this iterative-check method consistently allows the model to evaluate the correctness of its previous reasoning path. Particularly in the history split, the LLM's answer-checking ability enables it to continually reject its own incorrect answers and improve itself without external assistance. We can hardly reach the model's peak performance even after the fifth test.

Previous methods have been insufficient to support this capability of answer verification. However, we believe that LLMs have the ability to continuously refine judgments, and prompting iterative corrections of answers by the model itself, without the use of external tools, can improve the performance.

5.4   Attending Actual History Examination

We collaborate with two Chinese secondary schools, No.2 Middle School Affiliated to Shanghai Jiao Tong University[5] and High School Affiliated to Shanghai

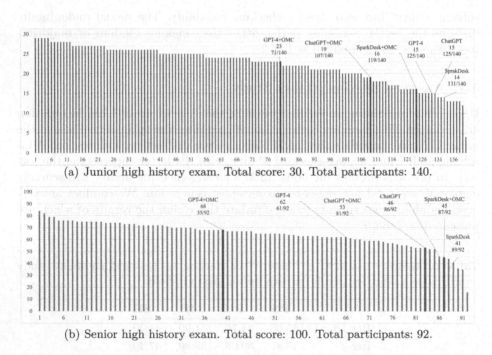

(a) Junior high history exam. Total score: 30. Total participants: 140.

(b) Senior high history exam. Total score: 100. Total participants: 92.

Fig. 3. Figures (a) and (b) respectively illustrate the scores and rankings of GPT-4, ChatGPT, and SparkDesk in junior and senior high school history exams. Blue bars represent the actual students' scores, while others correspond to different models. Bars enclosed in black borders represent the performance of models using OMC method. Specific models used, scores, and rankings are annotated within the labels.

---

[5] No.2 Middle School Affiliated to Shanghai Jiao Tong University: https://www.jd2fz.sjtu.edu.cn/.

Jiao Tong University[6], to have three public LLMs, including GPT-4[7], ChatGPT, and SparkDesk[8] by iFLYTEK, participate in two exams in junior and senior high schools, recording their rankings among students. In each exam, we select three classes of varying levels from the same grade, with the proportion of students at different levels assumed to be consistent.

Results of the exam[9] are shown in Fig. 3. In junior high exams, the scores of GPT-4, ChatGPT, and SparkDesk are closely matched, but GPT-4 significantly outperforms the others in high school exams. Unfortunately, all three lag in the overall student ranking, unable to surpass the top two-thirds. However, using our OMC method can further enhance each model's performance to varying degrees. GPT-4 surpasses half of the students after employing OMC method in junior and senior high exams.

The prominent capability of LLMs lies in their ability to generate comprehensive and fluent expressions, with grammatical and writing skills approaching or even surpassing the average human level. However, they still fall significantly below the average level of age-appropriate students in secondary school history knowledge tests.

6   Conclusion

In this paper, we establish a dataset, the PH dataset, containing more than 4,000 problems and targeting secondary school physics and history in the Chinese education sector. We propose a novel prompt method, One-More-Check (OMC), which leverages the Generator-Discriminator framework to enable the model to generate, evaluate, and optimize its own inference path and answers. We evaluate our OMC method on our PH dataset, outperforming mainstream methods such as Chain-of-Thought and Self-Consistency, achieving optimality in most tasks and improving accuracy by about 10% compared to Self-Consistency at most. Moreover, we identify two capabilities of LLMs in the educational field, problem-solving and answer-checking ability, and conduct an ability study. Finally, in collaboration with two secondary schools, we evaluate the real-world performance of three mainstream LLMs and our OMC method among age-appropriate students in secondary school history. The results suggest our OMC helps increase LLMs' scores in the actual exams at most 53.33% while there is still substantial room for improvement of LLMs in the educational field.

We hope to more comprehensively explore and integrate the problem-solving and answer-checking abilities of LLMs. And we will further optimize our methods to activate these two capabilities in future work, thereby enhancing LLMs' applicability.

---

[6] High School Affiliated to Shanghai Jiao Tong University: https://fz.sjtu.edu.cn/.

[7] GPT-4: https://openai.com/research/gpt-4. We use GPT-4 and ChatGPT for the exams at https://chat.openai.com/.

[8] SparkDesk by iFLYTEK can be accessed at https://xinghuo.xfyun.cn/.

[9] More detailed results can be found at: https://github.com/hcffffff/PH-dataset-OMC/tree/main/res/simulated-history-exam.

Acknowledgements. This work is supported by No.2 Middle School Affiliated to Shanghai Jiao Tong University and High School Affiliated to Shanghai Jiao Tong University.

References

1. Borji, A.: A categorical archive of ChatGPT failures (2023)
2. Brown, T., et al.: Language models are few-shot learners. Adv. Neural. Inf. Process. Syst. **33**, 1877–1901 (2020)
3. Creswell, A., White, T., Dumoulin, V., Arulkumaran, K., Sengupta, B., Bharath, A.A.: Generative adversarial networks: an overview. IEEE Signal Process. Mag **35**(1), 53–65 (2018). https://doi.org/10.1109/msp.2017.2765202 , https://doi.org/10.1109/2Fmsp.2017.2765202
4. Dhingra, S., Singh, M., SB, V., Malviya, N., Gill, S.S.: Mind meets machine: unravelling GPT-4's cognitive psychology (2023)
5. Dong, Q., et al: A survey on in-context learning (2023)
6. Du, Z., Qian, Y., Liu, X., Ding, M., Qiu, J., Yang, Z., Tang, J.: GLM: general language model pretraining with autoregressive blank infilling. In: Proceedings of the 60th Annual Meeting of the Association for Computational Linguistics (Volume 1: Long Papers), pp. 320–335 (2022)
7. Frieder, S., et al.: Mathematical capabilities of ChatGPT (2023)
8. Huang, Y., et al.: C-Eval: a multi-level multi-discipline Chinese evaluation suite for foundation models. arXiv preprint arXiv:2305.08322 (2023)
9. Inaba, T., Kiyomaru, H., Cheng, F., Kurohashi, S.: Multitool-cot: GPT-3 can use multiple external tools with chain of thought prompting (2023)
10. Kasai, J., Kasai, Y., Sakaguchi, K., Yamada, Y., Radev, D.: Evaluating GPT-4 and ChatGPT on Japanese medical licensing examinations (2023)
11. Kojima, T., Gu, S.S., Reid, M., Matsuo, Y., Iwasawa, Y.: Large language models are zero-shot reasoners (2023)
12. Li, X., et al.: Chain of knowledge: a framework for grounding large language models with structured knowledge bases (2023)
13. Min, S., et al.: Rethinking the role of demonstrations: what makes in-context learning work? In: Proceedings of the 2022 Conference on Empirical Methods in Natural Language Processing, pp. 11048–11064. Association for Computational Linguistics, Abu Dhabi, United Arab Emirates (2022). https://aclanthology.org/2022.emnlp-main.759
14. Mündler, N., He, J., Jenko, S., Vechev, M.: Self-contradictory hallucinations of large language models: evaluation, detection and mitigation (2023)
15. Nori, H., King, N., McKinney, S.M., Carignan, D., Horvitz, E.: Capabilities of GPT-4 on medical challenge problems (2023)
16. Nunes, D., Primi, R., Pires, R., Lotufo, R., Nogueira, R.: Evaluating GPT-3.5 and GPT-4 models on Brazilian university admission exams (2023)
17. OpenAI: GPT-4 technical report (2023)
18. Ouyang, L., et al.: Training language models to follow instructions with human feedback (2022)
19. Rae, J.W., et al.: Scaling language models: methods, analysis & insights from training gopher (2022)
20. Savelka, J., Agarwal, A., Bogart, C., Song, Y., Sakr, M.: Can generative pre-trained transformers (GPT) pass assessments in higher education programming courses? (2023)

21. Turpin, M., Michael, J., Perez, E., Bowman, S.R.: Language models don't always say what they think: unfaithful explanations in chain-of-thought prompting (2023)
22. Wang, B., Yue, X., Sun, H.: Can ChatGPT defend the truth? automatic dialectical evaluation elicits LLMs' deficiencies in reasoning (2023)
23. Wang, X., et al.: Self-consistency improves chain of thought reasoning in language models (2023)
24. Wei, J., et al.: Emergent abilities of large language models (2022)
25. Wei, J., et al.: Chain-of-thought prompting elicits reasoning in large language models (2023)
26. Yao, S., et al.: Tree of thoughts: deliberate problem solving with large language models (2023)
27. Yao, Y., Li, Z., Zhao, H.: Beyond chain-of-thought, effective graph-of-thought reasoning in large language models (2023)
28. Ye, X., Durrett, G.: The unreliability of explanations in few-shot prompting for textual reasoning (2022)
29. Yuan, Z., Yuan, H., Tan, C., Wang, W., Huang, S.: How well do large language models perform in arithmetic tasks? (2023)
30. Zeng, A., et al.: GLM-130B: an open bilingual pre-trained model. arXiv preprint arXiv:2210.02414 (2022)
31. Zhang, X., Li, C., Zong, Y., Ying, Z., He, L., Qiu, X.: Evaluating the performance of large language models on GAOKAO benchmark (2023)
32. Zhao, W.X., et al.: A survey of large language models (2023)
33. Zhu, W., Thomason, J., Jia, R.: Chain-of-questions training with latent answers for robust multistep question answering (2023)
34. Ziegler, D.M., et al.: Fine-tuning language models from human preferences (2020)

Sub-Instruction and Local Map Relationship Enhanced Model for Vision and Language Navigation

Yong Zhang[1,2], Yinlin Li[2(✉)], Jihe Bai[1,2], Yi Feng[3], and Mo Tao[3]

[1] School of Intelligence and Technology, University of Science and Technology
Beijing, Beijing 100083, China
[2] State Key Laboratory of Multimodal Artificial Intelligence Systems,
Institute of Automation, Chinese Academy of Sciences, Beijing 100190, China
yinlin.li@ia.ac.cn
[3] Science and Technology on Thermal Energy and Power Laboratory,
Wuhan 2nd Ship Design and Research Institute, Wuhan 430205, China

Abstract. In this paper, different from most methods in vision and language navigation, which primarily relies on vision-language cross-modal attention modeling and the agent's egocentric observations. We establish connections between sub-instructions and local maps to elaborately encode environment information and learn a path responsible for the whole instructions rather than the ultimate goal. We first obtain a local semantic map by ground projecting the RGB semantic segmentation map and the depth map. The segmented sub-instruction is passed through the sub-instruction attention module and then taken as input, together with the local map, to the cross-modal attention module. Finally, a set of waypoints are predicted by the navigation module until all sub-instructions of the long instruction are executed, which is the completion of an episode. Comparison experiments and ablation studies on the VLN-CE dataset show that our method outperforms most methods and has a good whole path predicting ability.

Keywords: Sub-instruction · Local map · VLN-CE

1 Introduction

An agent that can understand natural language instructions and perform corresponding actions in the visual world is one of the long-term challenges for artificial intelligence. Due to the complexity of instructions from humans, this requires the agent to be able to relate natural language to vision and action

This work is partly supported by the National Natural Science Foundation of China (grant no. 61702516), and the Open Fund of Science and Technology on Thermal Energy and Power Laboratory (No. TPL2020C02), Wuhan 2nd Ship Design and Research Institute, Wuhan, P.R. China.

B. Luo et al. (Eds.): ICONIP 2023, CCIS 1967, pp. 518–529, 2024.
https://doi.org/10.1007/978-981-99-8178-6_39

in unstructured and unseen environments. When the instruction provided by a human pertains to a navigation task, the challenge is referred to as Vision and Language Navigation(VLN) [1,12].

In vision and language navigation tasks, most existing work [4,11] uses panoramic RGB images as input, but they ignore environmental information, resulting in a trained agent that does not have an adequate perception of the 3D environment. The introduction of map information can integrate vision and environment information well and help the agent build better spatial representations. In addition, most of the traditional methods of VLN [6,12] nowadays ignore the large amount of information contained in language and have the problem that given a long natural language instruction. The instruction may contain landmarks outside the single view as references, leading to visual-language matching failure. The division of long instructions into sub-instructions can improve the short-term navigation ability of the agent and a similar problem was discussed in BabyWalk [18] at ACL2020, where the model can be better navigated by learning baby-steps for the shorter instructions.

Therefore, we propose a model with enhanced sub-instruction and local map relations for the continuous environment vision and language navigation tasks as illustrated in Fig. 1, where we introduce local maps and encourage the agent to proceed precisely on the described paths instead of just focusing on reaching the goal. We partition a long and complex natural language instruction into individual sub-instructions and feed the fused features into the Transformer model to predict the path on the egocentric map via a cross-modal attention module between the sub-instructions and the local map.

The contributions of this paper are as follows:

- We analyze the important role of the map for vision and language navigation tasks and propose a vision and language navigation model that enhances the relationship between sub-instructions and local maps.
- We consider the efficiency of sub-instructions for vision and language navigation tasks, and we combine the characteristics of navigation instructions and syntactic relations to effectively segment long instructions, which in turn functions as a supervisory process.
- Comparison experiments conducted on the VLN-CE dataset show that our results are better than most methods on val-seen, achieve competitive results on val-unseen, and our navigation error rate is better than the current state of the art.

2 Related Work

2.1 Sub-instruction

The current existing VLN methods rarely consider sub-instructions. Hong et al., [8] proposed the FGR2R dataset, by dividing long instructions and paths into sub-instructions and sub-paths based on R2R and using manual annotations to align sub-instructions and sub-paths. They used the shifting attention

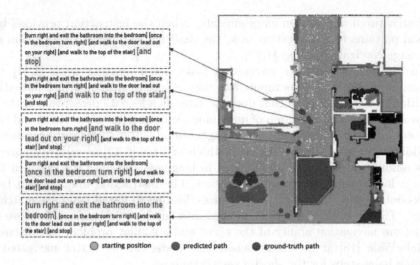

Fig. 1. The yellow waypoints indicate the starting position, the red waypoints are predicted by the model, and the blue are the waypoints obtained by sampling the ground-truth path. The sub-instruction marked in red indicates that the entire instruction has the greatest weight. (Color figure online)

module to infer whether a sub-instruction is completed and set a threshold to decide whether the agent should execute the next sub-instruction; Zhang et al., [17] distinguished and represented sub-instructions by indicator words and they used grounding between landmarks and objects to control the transition of sub-instructions; Zhu et al., [18] used memory buffers to record sub-instructions and convert past experiences into the context of future steps; He et al., [7] proposed to use multi-level attention to understand the low-level and high-level semantics of sub-instructions. Unlike them, we combined the characteristics and syntactic relations of navigation instructions to segment long instructions, and each step performs a self-attention mechanism on instruction, assigning a weight to each sub-instruction so that the agent focuses on the current sub-instruction while ensuring global awareness of the instruction.

2.2   Map in Navigation

In vision and language navigation, maps play a crucial role in navigation. A map is static information about the environment in which the agent is located, including the topology of the environment, location information, room layout, etc. In the navigation process, the map can help the agent better understand the instructions and comprehend the environment information and location information to plan the movement path more accurately. Chen et al., [2] built a global topological map by exploring the whole scene to learn the association between instructions and nodes; Wang et al., [15] constructed structured scene memory for persistent memory and topological environment representation; DUET

[3] constructs topological maps in real-time, dynamically combining fine-scale encoding of local observations and coarse-scale encoding of global maps through graph converters to balance large-action spatial reasoning and fine-grained linguistic grounding; SASRA [9] and CM2 [5] on the other hand, introduces egocentric semantic maps to perform cross-modal attention with language information to focus on structured, local top-down spatial information. We agree with this approach and we introduce local semantic maps to help the agent construct spatial feature representations and thus better spatial reasoning through cross-modal representations of local semantic maps and sub-instructions.

3 Method

In this section, we present our proposed model, called the sub-instruction and local map relationship enhanced model for vision and language navigation, as shown in Fig. 2, which consists of four main modules, the map module, the language module, the cross-modal attention module and the navigation module. We describe how to obtain the map, the sub-instruction, and their respective encoding processes. Then we present a method for feature fusion of sub-instructions and local maps. Finally, we present a navigation module based on the Transformer model to predict a set of waypoints.

3.1 Map Module

How to Get a Map. The single-view RGB observations of the agent are inputted into Swin Transformer [13] for semantic segmentation, resulting in the generation of a semantic segmentation map. Subsequently, the camera parameters and depth observation information are utilized to project the segmentation map into 3D space, creating a 3D semantic point cloud map. The 3D point cloud corresponding to each observation is then mapped to a grid of size h*w*c, where h represents the height, w represents the width, and c represents the number of semantic categories. A local 2D semantic map is obtained through this process.

Map Encoder. We use ResNet18 to encode the input egocentric local semantic map $s \in \mathbb{R}^{(h*w*c)}$ with Y=Enc(s), ResNet18 initially generates the feature representation $Y' \in \mathbb{R}^{(h'*w'*c')}(h' = h/16, w' = w/16, d = 128)$ and then reshapes it to $Y \in \mathbb{R}^{(N*d)}(N = h' * w')$.

3.2 Language Module

How to Get a Sub-instruction. We have devised chunking rules that utilize the syntactic relations provided by the Stanford NLP Parser. These rules enable us to partition long and complex instructions into shorter and more comprehensible sub-instructions. Each sub-instruction represents an independent navigation task, typically requiring the agent to perform one or two actions to complete. To

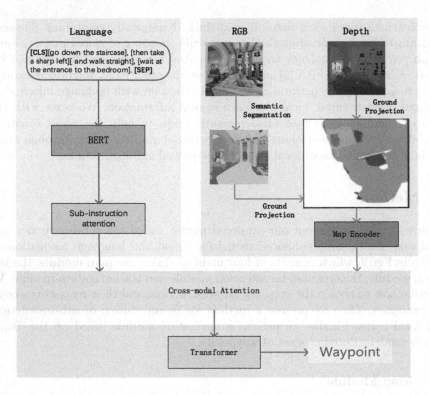

Fig. 2. We propose a sub-instruction and local map relationship enhanced model for vision and language navigation in continuous environments. At each step, the model assigns a weight to each sub-instruction to make the agent pay attention to the current sub-instruction. We use the local semantic map and sub-instructions as input to a cross-modal attention mechanism for feature fusion and then predict waypoints via the Transformer model.

achieve this, the complete instruction is initially processed through the Stanford NLP Parser, extracting the dependencies and governors for each word. With the aid of these properties, we establish two functions: a segmentation function and a checking function.

The segmentation function divides the long and complex instructions into temporary sub-instructions based on the characteristics of the navigation instructions of the R2R dataset [1]. The check function is then applied to the temporary sub-instructions to determine whether they should be included in the final sub-instruction list. Here, the check function checks whether the temporary sub-instruction satisfies three conditions: (1) the length of the sub-instruction should be longer than the minimum length of two words (2) it checks whether the last word of the previous sub-instruction is concatenated, and if it is, then it is divided into the current sub-instruction (3) the temporary sub-instruction should not contain only one action-related phrase, it should immediately follow the previous sub-instruction or lead to the next sub-instruction, and if this happens, then the temporary block should be appended to the previous sub-instruction or added to the next sub-instruction, respectively.

Here we provide an illustrative example. We split the given instruction "Go down the staircase, then take a sharp left and walk straight. Wait at the entrance to the bedroom." into 1 "Go down the staircase", 2 "then take a sharp left", 3 "and walk straight", 4 "wait at the entrance to the bedroom". Where the sub-instructions of the 2nd and 3rd satisfy condition (2).

Language Encoder. We use the pre-trained BERT [10] model as a language encoder to extract feature vectors for each word in the instruction, and the final feature representation is $X \in \mathbb{R}^{(l*M*d)}$, where l is the number of sub-instructions, M is the number of words in the sub-instruction and $d = 128$ is the feature dimension we set. These language representations will be used by the sub-instruction attention module to further help the agent in language comprehension.

Sub-instruction Attention. To encourage the agent can learn the relationship between map and language features in sub-instructions, we add the sub-instruction attention module, which makes the agent focus on a specific sub-instruction X_i at each time step. We design our sub-instruction attention module concerning the self-attention mechanism [14], and at each time step, the model computes the weight distribution of the sub-instruction.

$$q = X_i w_q, k = X_j w_k, v = X_j w_v$$
$$H_t^L = \mathrm{softmax}\left(\frac{qk^T}{\sqrt{d}}\right) v \qquad (1)$$

w_q, w_k, w_v are learning parameters, X_i and X_j represent sub-instructions. H_t^L is the language feature representation with weights at moment t. The sub-instruction attention module plays a crucial role in guiding the agent's attention towards the most relevant parts of the instruction. By focusing on the current sub-instruction and disregarding the completed or upcoming sub-instructions, the agent receives supervision during the navigation process. This attention mechanism aids the agent in better understanding the instruction and constructing more accurate spatial features. The selective attention to the relevant sub-instructions allows the agent to maintain a clear understanding of the immediate task at hand, ensuring that its actions align with the specific requirements of the instruction. By leveraging this attention mechanism, the agent can effectively process and interpret the instruction, leading to improved spatial feature construction.

3.3 Cross-Modal Attention Module

We designed our cross-modal attention module by continuing the design idea of the sub-instruction attention module. We use the egocentric local semantic map as the query, and the instruction output by the sub-instruction attention module as the key and value, using scaled dot product attention:

$$Q = Y_t^s W_q, K = H_t^L W_k, V = H_t^L W_v$$
$$H_t^s = \text{softmax}\left(\frac{QK^T}{\sqrt{d}}\right) V \tag{2}$$

Where Y_t^s is the feature representation of the egocentric local semantic map at time t, H_t^s is the representation of the features after the fusion of language and map, W_q, W_k, W_v are the learning parameters. The cross-modal attention module takes the sub-instruction and local map as input for feature fusion. By utilizing sub-instructions, the agent effectively avoids the interference caused by long instructions that may include landmarks not visible in the current field of view. Furthermore, sub-instructions prevent the occurrence of identical landmarks within long instructions, which can lead to confusion for the agent. This approach allows the model to align visual and language features more accurately. The integration of sub-instructions and maps, which contain visual and environmental information, assists the agent in constructing spatial representations more effectively. As a result, the agent gains a better understanding of the vision and language navigation tasks by using these spatial features.

3.4 Navigation Module

We define a set of 2d waypoints as paths on an egocentric local map, where the first and last waypoints are always the start and target positions. The remaining waypoints are obtained from real path sampling. The Transformer model is employed as a waypoint prediction model. The previous waypoint and the cross-modal feature representation of the sub-instruction and local map are used as input to the Transformer model to predict the next waypoint.

$$\text{waypoint}_t = T\left(H_t^s, \text{waypoint}_{t-1}\right) \tag{3}$$

Using each waypoint as a short-term goal, the off-the-shelf deep reinforcement learning model DD-PPO [16] trained for the PointNav task is used to output navigation operations to the agent, which can decide to stop the action when the agent is within a certain radius of the last predicted waypoint.

4   Experiments

4.1   Experiment Setup

We conducted experiments on the VLN-CE dataset [12], which uses the Habitat simulation platform to reconstruct a vision and language navigation task on the Matterport3D environment, with 16,844 path instruction pairs for over 90 visually realistic scenes available in the dataset. A STOP decision is considered successful if made within 3 m of the target location and the agent has a fixed time budget of 500 steps to complete an episode. The agent is granted access to egocentric RGB-D viewing with a horizontal field of view of 90 °. The local semantic map is configured with dimensions of 192*192, where each pixel represents an area of 5 cm*5 cm. We use discrete action space to evaluate our agent, including forward 0.25 m, left 15 °, right 15 °, and stop. We initialize our model with the weights of the CM2-GT [5] model.

The following metrics were used to evaluate our performance on the vision and language navigation tasks in continuous environments: Trajectory Length (TL), Navigation Error (NE), Oracle Success Rate (OS), Success Rate (SR), and Success weighted by inverse Path Length (SPL).

- TL: Trajectory Length, which means the total length of the actual path.
- NE: Navigation Error, which means the distance between the stop position and the target position.
- OS: Oracle Success Rate, the success rate under the oracle stopping rule.
- SR: Success Rate. When the agent gets within 3 m of the target, the trajectory is judged as success.
- SPL: Success weighted by inverse Path Length is the success rate divided by the trajectory length.

4.2   Experiment Results

Our method has been compared with eight models, as shown in the first eight rows of Table 1. Due to the consistency of the experimental setup, we put in the table the performance claimed in their articles, but there are discrepancies in the reproduction, which we analyzed in detail in the ablation experiments. Our experimental results are shown in Table 1, from which it can be seen that our proposed method outperforms most of the methods in val-seen, and the results in val-unseen are competitive. The navigation error of our method is lower than the current state of the art. It is worth noting that WPN-DD [11], R2R-CMTP [2] and AG-CMTP [2] use visual information in a panoramic view, while we use a single view. In contrast to the R2R-CMTP [2] and AG-CMTP [2] models, which allow the agent to first explore the environment and then construct a topological map of the agent's understanding of the environment, which combines language and topological maps in the cross-modal attention mechanism. Unlike their approach, we introduce a local semantic map, where we avoid previous exploration of the environment, which requires many computational resources, while the local semantic maps can better combine language information and the agent can better understand the language and the environment. Unlike the MLANet [7] model where they separately extract low-level semantics for each word of the instruction and high-level semantics for segmented sub-instructions and then combine them, we assign weights directly to each sub-instruction. This allows the agent to focus on the current sub-instruction while maintaining a global perception of the whole instruction. Compared with the CM2-GT [5] model, we incorporate sub-instructions and emphasize the cross-modal attention between sub-instructions and the local map. We contend that the local map is highly responsive to sub-instructions and plays a crucial role in aiding the agent's construction of spatial features.

Table 1. Evaluation on VLN-CE dataset. Our proposed method outperforms most of the methods in val-seen and the results in val-unseen are competitive. CM2-GT(reproduced)/Inst. is the result of our reproduction. /Inst. means it uses unsegmented instruction to fuse features with the map; /Sub-Inst. w/o Weights means sub-instruction without weights to fuse features with the map.

Methods	Val-seen					Val-unseen				
	TL↓	NE↓	OS↑	SR↑	SPL↑	TL↓	NE↓	OS↑	SR↑	SPL↑
AG-CMTP [2]	–	6.60	56.2	35.9	30.5	–	7.90	39.2	23.1	19.1
Seq2Seq+PM+DA+Aug [12]	9.37	7.02	46.0	33.0	31.0	9.32	7.77	37.0	25.0	22.0
R2R-CMTP [2]	–	7.10	45.4	36.1	31.2	–	7.90	38.0	26.4	22.7
WPN-DD [11]	9.11	6.57	44.0	35.0	32.0	8.23	7.48	35.0	28.0	26.0
SASRA [9]	8.89	7.71	–	36.0	34.0	7.89	8.32	–	24.0	22.0
CMA+PM+DA+Aug [12]	9.26	7.12	46.0	37.0	35.0	8.64	7.37	40.0	32.0	30.0
CM2-GT [5]	12.60	4.81	58.3	52.8	41.8	10.68	6.23	41.3	37.0	30.6
MLANet+DA+Aug+RL [7]	**8.10**	5.83	50.0	44.0	**42.0**	**7.21**	6.30	42.0	**38.0**	**35.0**
CM2-GT(reproduced)/Inst	12.57	4.06	51.7	45.6	36.3	12.67	4.85	36.3	28.0	20.3
Ours/Sub-Inst. w/o Weights	11.91	4.15	49.7	46.4	38.0	12.51	4.97	33.5	28.0	20.3
Ours	13.09	**3.57**	**59.6**	**54.4**	**42.0**	13.95	**4.49**	**42.3**	33.5	22.1

4.3   Ablation Study

Is Sub-Instruction Better Than Long Instruction? In order to compare the proposed feature fusion of sub-instruction and local map with the common instructions of the VLN-CE tasks, we reproduced the results of the CM2-GT [5] model, as shown in the CM2-GT(reproduced)/Inst. in Table 1. The results of our method are shown in the Ours in Table 1. The experimental results show that our method improved compared to them in both val-seen and val-unseen. Our SR improved by 19.3% on val-seen and 19.6% on val-unseen and our SPL improved by 15.7% on val-seen and 8.9% on val-unseen. It can be seen that the results of feature fusion with sub-instructions and local maps are better than those of feature fusion with unsegmented original instructions and local maps, which indicates that sub-instructions and local maps are indeed beneficial for agent performing vision language navigation tasks in continuous environments.

However, the results in val-unseen are not as good as our expectations. We have analyzed the possible reasons for this. The experimental results show that the navigation error rate of our method is better than the current state of the art, but our trajectory length is longer and the SPL is the success rate divided by the trajectory length, so the SPL is not very high. Our agent may prefer the path described by the language rather than the shortest path to the goal, as shown in Fig. 3, where the waypoints predicted by our method largely overlap with the real path, not just the last goal point. As can be seen from the Fig. 3, although both succeed in reaching the target point, our path is more in line with the path described. It will be a direction of our research in the future to enable the agent to plan optimal paths on the described paths.

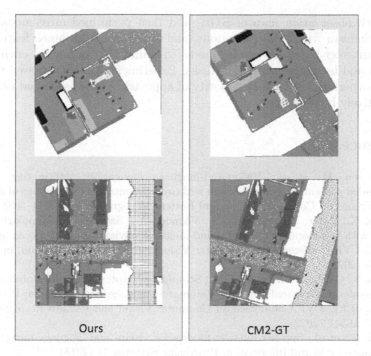

Fig. 3. Visualization of the model result where blue is the ground-truth path, red is the predicted path. (Color figure online)

Does Sub-Instruction Attention Module Improve Language Feature Learning? We aim to investigate the potential benefits of feature fusion with weighted sub-instructions and local maps for enhancing the agents' ability to construct spatial features. We experimented feature fusion with unweighted sub-instructions and local maps, and the experimental results are shown in the Ours/Sub-Inst. w/o Weights in Table 1. The experimental results show that Ours improved compared to Ours/Sub-Inst. w/o Weights in both val-seen and val-unseen, which SR improved by 17.2% on val-seen and 19.6% on val-unseen and SPL improved by 10.5% on val-seen and 8.9% on val-unseen. It can be seen that the results of feature fusion with weighted sub-instructions and local maps are better than those of feature fusion without weighted sub-instructions and local maps, which indicates that sub-instructions with weights can make the agent focus on the current sub-instruction and help the agent build global perception for better construction of spatial features.

5   Conclusion

In this work, we proposed a model for vision and language navigation tasks that enhanced the relationship between sub-instructions and local maps. First, we constructed a local semantic map from the semantic segmentation map

and depth observation map of RGB, and then performed cross-modal attention mechanism with sub-instructions, and inputted the obtained features into
the Transformer-based navigation model to realize the continuous environment
of vision and language navigation tasks. Experimental results showed that our
results outperformed the vast majority of approaches in val-seen and performed
competitively in val-unseen.

References

1. Anderson, P., et al.: Vision-and-language navigation: interpreting visually-grounded navigation instructions in real environments. In: Proceedings of the IEEE Conference on Computer Vision and Pattern Recognition, pp. 3674–3683 (2018)
2. Chen, K., Chen, J.K., Chuang, J., Vázquez, M., Savarese, S.: Topological planning with transformers for vision-and-language navigation. In: Proceedings of the IEEE/CVF Conference on Computer Vision and Pattern Recognition, pp. 11276–11286 (2021)
3. Chen, S., Guhur, P.L., Tapaswi, M., Schmid, C., Laptev, I.: Think global, act local: dual-scale graph transformer for vision-and-language navigation. In: Proceedings of the IEEE/CVF Conference on Computer Vision and Pattern Recognition, pp. 16537–16547 (2022)
4. Fried, D., et al.: Speaker-follower models for vision- and-language navigation. In: Advances in Neural Information Processing Systems 31 (2018)
5. Georgakis, G., et al.: Cross-modal map learning for vision and language navigation. In: Proceedings of the IEEE/CVF Conference on Computer Vision and Pattern Recognition, pp. 15460–15470 (2022)
6. Hao, W., Li, C., Li, X., Carin, L., Gao, J.: Towards learning a generic agent for vision-and-language navigation via pre-training. In: Proceedings of the IEEE/CVF Conference on Computer Vision and Pattern Recognition, pp. 13137–13146 (2020)
7. He, Z., Wang, L., Li, S., Yan, Q., Liu, C., Chen, Q.: MLANet: multi-level attention network with sub-instruction for continuous vision-and-language navigation. arXiv preprint arXiv:2303.01396 (2023)
8. Hong, Y., Rodriguez, C., Wu, Q., Gould, S.: Sub-instruction aware vision-and-language navigation. In: Proceedings of the 2020 Conference on Empirical Methods in Natural Language Processing (EMNLP), pp. 3360–3376 (2020)
9. Irshad, M.Z., Mithun, N.C., Seymour, Z., Chiu, H.P., Samarasekera, S., Kumar, R.: Semantically-aware spatio-temporal reasoning agent for vision-and-language navigation in continuous environments. In: 2022 26th International Conference on Pattern Recognition (ICPR), pp. 4065–4071. IEEE (2022)
10. Kenton, J.D.M.W.C., Toutanova, L.K.: BERT: pre-training of deep bidirectional transformers for language understanding. In: Proceedings of NAACL-HLT, pp. 4171–4186 (2019)
11. Krantz, J., Gokaslan, A., Batra, D., Lee, S., Maksymets, O.: Waypoint models for instruction-guided navigation in continuous environments. In: Proceedings of the IEEE/CVF International Conference on Computer Vision, pp. 15162–15171 (2021)
12. Krantz, J., Wijmans, E., Majumdar, A., Batra, D., Lee, S.: Beyond the Nav-Graph: vision-and-language navigation in continuous environments. In: Vedaldi, A., Bischof, H., Brox, T., Frahm, J.-M. (eds.) ECCV 2020. LNCS, vol. 12373, pp. 104–120. Springer, Cham (2020). https://doi.org/10.1007/978-3-030-58604-1_7

13. Liu, Z., et al.: Swin transformer: hierarchical vision transformer using shifted windows. In: Proceedings of the IEEE/CVF International Conference on Computer Vision, pp. 10012–10022 (2021)
14. Vaswani, A., et al.: Attention is all you need. In: Advances in Neural Information Processing Systems 30 (2017)
15. Wang, H., Wang, W., Liang, W., Xiong, C., Shen, J.: Structured scene memory for vision-language navigation. In: Proceedings of the IEEE/CVF Conference on Computer Vision and Pattern Recognition, pp. 8455–8464 (2021)
16. Wijmans, E., et al.: DD-PPO: learning near-perfect pointgoal navigators from 2.5 billion frames. arXiv preprint arXiv:1911.00357 (2019)
17. Zhang, Y., Guo, Q., Kordjamshidi, P.: Towards navigation by reasoning over spatial configurations. In: Proceedings of Second International Combined Workshop on Spatial Language Understanding and Grounded Communication for Robotics, pp. 42–52 (2021)
18. Zhu, W., et al.: BabyWalk: going farther in vision-and-language navigation by taking baby steps. In: Proceedings of the 58th Annual Meeting of the Association for Computational Linguistics, pp. 2539–2556 (2020)

STFormer: Cross-Level Feature Fusion in Object Detection

Shaobo Wang⊙, Renhai Chen^(✉)⊙, Tianze Guo⊙, and Zhiyong Feng⊙

College of Intelligence and Computing, Tianjin University, Tianjin 300350, China
{w_shinbow,renhai.chen,2020244244,zyfeng}@tju.edu.cn

Abstract. Object detection algorithms can benefit from multi-level features, which encompass both high-level semantic information and low-level location details. However, existing detection methods face numerous challenges in effectively utilizing these multi-level features. Most existing detection techniques utilize simplistic operations such as feature addition or concatenation to fuse multi-level features, thereby failing to effectively suppress redundant information. Consequently, the performance of these algorithms is significantly constrained in complex scenarios. To address these limitations, this paper presents a novel feature extraction network that incorporates joint modeling and multi-dimensional feature fusion. Specifically, the network partitions the features of each level into tiles and employs hybrid self-attention mechanisms to extract these grouped features more comprehensively. Additionally, a hybrid cross-attention-based approach is utilized to regulate the transmission proportion of each grouped feature, facilitating the seamless integration of high-level semantic features obtained from deep encoders and the low-level position details retained by the pipeline. Consequently, the network effectively suppresses noise and enhances performance. Experimental evaluation on the MS COCO dataset demonstrates the effectiveness of the proposed approach, achieving an impressive accuracy of 54.3%. Notably, the algorithm showcases exceptional performance in detecting small-scale targets, surpassing the capabilities of other state-of-the-art technologies.

Keywords: Feature extraction · Cross-attention · Object detection

1 Introduction

The integration of multi-level features, encompassing both high-level semantic information and low-level location details, is pivotal in bolstering the performance of object detection algorithms. However, existing detection methods encounter several challenges in effectively harnessing these multi-level features. Firstly, conventional object detection networks primarily rely on the final convolution features for detection while neglecting intermediate pooling operations that lead to a substantial reduction in feature size, often dwindling to just 1/32 of the input image size. This significant reduction presents a formidable

© The Author(s), under exclusive license to Springer Nature Singapore Pte Ltd. 2024
B. Luo et al. (Eds.): ICONIP 2023, CCIS 1967, pp. 530–542, 2024.
https://doi.org/10.1007/978-981-99-8178-6_40

obstacle to detection accuracy. Consequently, two categories of methods have been proposed to address this issue. The first category [1], focuses on leveraging high-resolution features to mitigate information loss. The second category [2], combines high-level features and low-level features to enhance the semantic information for detection. In this paper, our proposed method draws inspiration from both categories of networks. It not only constructs high-resolution end features but also progressively incorporates semantic knowledge from high-level features to enhance the localization accuracy of low-level features during the feature extraction process.

Secondly, in many detection methods [10, 19], high-level features and low-level features are often fused using simple cascading or addition operations, without effectively suppressing redundant information. Consequently, the performance of the network is significantly restricted, particularly in complex scenarios. To address this limitation, the approach presented in this paper employs feature grouping to organize multi-level feature information and utilizes cross-attention to regulate the transmission proportion within each feature group. This approach facilitates the effective transmission of information that is crucial for detection.

This paper presents a novel approach for feature extraction in which joint modeling and multi-dimensional feature fusion techniques are employed. The proposed network takes advantage of the attention mechanism for global modeling and the convolutional mechanism for local modeling, facilitating the effective learning of spatial and semantic information at different levels. By using self-attention, the network captures global context information and establishes long-range dependencies, enabling the extraction of valuable features. Additionally, cross-attention is utilized to adjust the contribution of each feature group, highlighting relevant information while suppressing redundant information. The network is composed of three key modules: the Hybrid Self-Attention Module, the Hybrid Cross-Attention Module, and the Encoder-Decoder Module. The hybrid self-attention and cross-attention modules play vital roles in the feature extraction and fusion reconstruction processes at each layer. The encoder module is designed based on hybrid self-attention, and the decoder module employs hybrid cross-attention. This architecture enables the network to effectively capture and combine relevant information from multiple dimensions, enhancing the overall feature extraction capabilities.

In summary, our contributions are as follows:

- We introduce cross-attention mechanisms in object detection that highlight relevant information and suppress redundant information by adjusting the contributions of each feature group. This multi-dimensional feature fusion method can make full use of multiple feature dimensions, improve the consistency and richness of feature expression.
- We propose Hybrid Self-Attention Module, the Hybrid Cross-Attention Module, and the Encoder-Decoder Module. This approach successfully integrates local information with global information and merges high-level features with low-level features.

– We design STFormer which achieves an impressive accuracy of 54.3% on the MS COCO dataset, demonstrates the effectiveness of our approach. This achievement serves as compelling evidence in support of the efficacy and effectiveness of our approach.

2   Related Work

CNN and Variants. CNNs serve as the standard network model throughout computer vision. Until the introduction of AlexNet [13] that CNN took off and became mainstream. Since then, deeper and more effective convolutional neural architectures [21] have been proposed to further propel the deep learning wave in computer vision, e.g. ResNet [11], Faster-RCNN [22], and Efficient-Net [24]. In addition to these architectural advances, there has also been much work on improving individual convolution layers, such as depthwise convolution and deformable convolution. While CNN and its variants are still the primary backbone architectures for computer vision applications, we highlight the strong potential of Transformer-like architectures for unified modeling between vision and language. Our work achieves strong performance on several basic visual recognition tasks, and we hope it will contribute to a modeling shift.

Self-attention for Object Detection. Also inspired by the success of self-attention layers and Transformer architectures in the NLP field, some works [4,5] employ self-attention layers to replace some or all of the spatial convolution layers in the popular ResNet. In these works, the self-attention is computed within a local window of each pixel to expedite optimization, and they achieve slightly better accuracy/FLOPs trade-offs than the counterpart ResNet architecture. However, their costly memory access causes their actual latency to be significantly larger than that of the convolutional networks. Instead of using sliding windows, we propose to shift windows between consecutive layers, which allows for a more efficient implementation in general hardware. Another work is to augment a standard CNN architecture with self-attention layers or Transformers. The self-attention layers can complement backbones or head networks by providing the capability to encode distant dependencies or heterogeneous interactions. More recently, the encoder-decoder design in Transformer has been applied for object detection and instance segmentation tasks. Our work explores the adaptation of Transformers for basic visual feature extraction and is complementary to these works.

Transformer Based Vision Backbones. Most related to our work is the Vision Transformer [27] (ViT) and its follow-ups. The pioneering work of ViT directly applies a Transformer architecture on nonoverlapping medium-sized image patches for image classification. It achieves an impressive speed-accuracy tradeoff in image classification compared to convolutional networks. While ViT requires large-scale training datasets (i.e., JFT-300M [23]) to perform well, DeiT

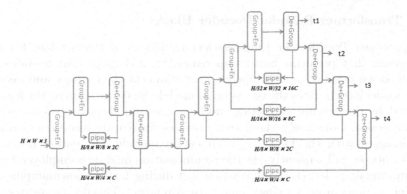

Fig. 1. The architecture of Transformer Encoder-Decoder Net.

introduces several training strategies that allow ViT to also be effective using the smaller ImageNet-1K dataset [6]. The results of ViT on image classification are encouraging, but its architecture is unsuitable for use as a general-purpose backbone network on dense vision tasks or when the input image resolution is high, due to its low-resolution feature maps and the quadratic increase in complexity with image size. Empirically, we find Swin Transformer [20] approach is both efficient and effective, achieving state-of-the-art accuracy on both MS COCO dataset [18] and ADE20K [29] semantic segmentation.

3 Method

On the basis of the above solutions, we propose STFormer framework. We format our framework into three parts: Hybrid Self-Attention Module, Hybrid Cross-Attention Module, and the Encoder-Decoder Module. In this section, we will introduce the main components of STFormer.

3.1 Overall Architecture

Figure 1 depicts the design framework of the proposed feature fusion network, STFormer. The RGB input image is divided into non-overlapping patches using patch partitioning. Each patch comprises 16 (4 × 4) adjacent pixels, encompassing the top, bottom, left, and right regions, resulting in a total of 48 (4 × 4 × 3) dimensional feature information. Subsequently, the original feature information of each patch is mapped to a predefined dimension, C, using Linear embedding.

To address the loss of spatial information during bottom-up reasoning, it becomes necessary to incorporate a mechanism that effectively processes and integrates cross-scale features during top-down reasoning. To achieve this, we adopt a single pipeline approach with skip layers, akin to the Hourglass network, which helps retain spatial information at each resolution. This ensures the preservation of crucial spatial details throughout the reasoning process.

3.2   Transformer Encoder-Decoder Blocks

The proposed Transformer Encoder-Decoder blocks in this section, featuring an encoder that performs bottom-up reasoning, a decoder that operates in a bottom-down manner, and a pipeline that connects the encoder and decoder. The encoder incorporates the tile merging module for downsampling the features, followed by the feature extraction module based on the hybrid self-attention module design. Conversely, the decoder utilizes the feature reconstruction module, designed with the hybrid cross-attention module, for feature fusion and reconstruction. Subsequently, the tile recombination module is employed for feature upsampling. To retain information lost during feature downsampling, the pipeline section employs a single pipe with skip layers. This design ensures effective feature processing and integration throughout the encoder-decoder structure.

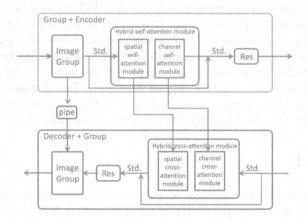

Fig. 2. A set of corresponding Transformer Encoder-Decoder blocks (notation presented with Eq. 1). SATT is the Hybrid self-attention module, XATT is the Hybrid cross-attention module.

As shown in Fig. 2, the strategy of window division in the first group of encoder modules divides the 16×16 feature map into 4×4 windows (each window consists of 4×4 patches). The corresponding last group of decoder modules uses the same window division strategy to divide the feature map of the same size into 4×4 windows.

A set of corresponding Transformer Encoder-Decoder blocks are calculated as follows:

$$q^{el}, k^{el}, v^{el} = IG(z^{el-1})$$
$$z^{el} = SATT(q^{el}, k^{el}, v^{el}) + z^{el-1} \qquad (1)$$
$$z^{el} = Res(Std(z^{el})) + z^{el}$$

$$q^{dl}, k^{dl}, v^{dl} = IG(z^{el+1})$$
$$z^{dl} = XATT(q^{dl}, k^{dl}, v^{dl}) + z^{dl-1} \qquad (2)$$
$$z^{dl} = Res(Std(z^{dl})) + z^{dl}$$

where z^{el} and z^{dl} mean a set of corresponding output feature maps of the Encoder module and the Decoder module. IG represents our image fusion module, SATT and XATT respectively represent our Hybrid self-attention module and Hybrid cross-attention module, Res represents residual blocks, and Std represents the standardization operation.

3.3 Patch Merge and Patch Demerge

In contrast to conventional convolutional neural networks that employ maximum pooling for downsampling and depooling for upsampling, this section introduces the Patch Merge module for downsampling and Patch Demerg module for upsampling. This choice is motivated by the fact that the fundamental unit of self-attention operation is the tile vector, and sampling involves rearranging and mapping the tile vectors. Similarly, Patch Merge and Patch Demerg modules serve the purpose of rearranging and mapping the tile vectors, aligning them with the underlying principles of self-attention.

The patch merge method will rearrange input feature maps of shape $h \times w \times C$ into output feature maps of shape $\frac{h}{r} \times \frac{w}{r} \times 2C$.

The patch demerge method and the patch merge method are inverse operations of each other and will rearrange the input feature map of shape $h \times w \times C$ into an output feature map of shape $rh \times rw \times \frac{C}{2}$.

The variables w, h, and c denote the dimensions of the input data T in the graph structure. T represents a matrix or tensor. h is the vector representation of the matrix T on the abscissa, w is the vector representation of the matrix T on the ordinate, c is the vector representation of the matrix T on the feature dimension.

3.4 Hybrid Self-attention Module

The hybrid self-attention module proposed in this section is a modular unit composed of a window division module, spatial self-attention module, and channel self-attention module. It consists of multiple sets of linear mapping, L2 regularization operations, and matrix operations. Unlike the self-attention module in the Swin-Transformer [20] network, we add channel attention to ensure equal attention to both spatial and channel dimensions. However, the introduction of the channel self-attention module doubles the computational complexity, slows down network prediction time, and increases training difficulty. To mitigate these challenges, we employ a window mechanism and the linear attention mechanism concurrently [15] to maintain computational complexity on a linear scale. The

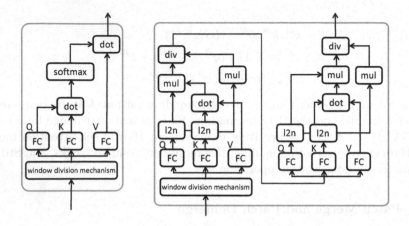

Fig. 3. Hybrid self-attention module

hybrid self-attention module proposed in this chapter exhibits stronger information processing capabilities compared to the self-attention module in the Swin-Transformer network, achieving a smaller computing scale while preserving consistent input and output. The structural differences between the self-attention module in the Swin-Transformer network and the hybrid self-attention module in this chapter can be observed in the comparison Fig. 3, illustrating the overall operation process of input information after passing through different modules.

Figure 3(left) shows the calculation process of the Swin-Transfromer network self-attention module, which splits the image into several non-overlapping independent windows through the window division mechanism, and divides a calculation problem into $(\frac{h}{M} \times \frac{w}{M})$ sub-problems, which greatly reduces the computational complexity. The formula (3) shows the computational complexity of the self-attention module, where h represents the input image length, w represents the input image width, C represents the number of channels of the input image, and M represents the window length and width (M is set to 7 in the experiment).

$$O((\frac{h}{M} \times \frac{w}{M}) \times (4M^2C^2 + 2M^4C)) = O(4hwC^2 + 2M^2hwC) \tag{3}$$

Figure 3(right) shows the calculation process of the mixed self-attention module proposed in this chapter, which relies on the window mechanism to reduce the complexity, and reduces the calculation scale of the attention inside the window from the original inner attention $O(4M^2C^2 + 2M^4C)$ to the current linear attention $O(4M^2C^2 + 2M^2C)$ by expanding the first-order Taylor formula to taking approximations and multiplexing intermediate results. After further adding the channel attention module, the computational complexity of the mixed attention module is described in the Eq. (4). Considering $M = 7$ in the experiment, the final hybrid self-attention module pays attention to the channel dimension information while the computational scale also decreases.

$$O((\frac{h}{M} * \frac{w}{M}) * (8M^2C^2 + 4M^2C)) = O(8hwC^2 + 4hwC) \tag{4}$$

The formula (5) shows the function expression of the spatial self-attention module, and the formula (6) shows the jth column of the attention function. Due to the intermediate result produced in the operation $\sum_{i=1}^{C} v_i(\frac{k_i}{|k_i|_2})^T$ and $\sum_{i=1}^{C}(\frac{k_i}{|k_i|_2})^T$ can be reused, so the two intermediate results are the same as $\frac{q_j}{|q_j|_2}$ together introduce the computational complexity of $O(2M^2C)$. where $Q \in R^{M^2C}$, $K \in R^{M^2C}$, and $V \in R^{M^2C}$, respectively, are query matrices, key matrices, and value matrices constructed by the spatial self-attention module, and $q_j \in R^{M^2}$, $k_i \in R^{M^2}$, and $v_i \in R^{M^2}$ are column vectors in the Q matrix, K matrix, and V matrix, respectively.

$$SpaceAttn(Q, K, V) = \frac{\sum_C V + (V(\frac{K}{\|K\|_2})^T)\frac{Q}{\|Q\|_2}}{C + (\sum_C(\frac{K}{\|K\|_2})^T)\frac{Q}{\|Q\|_2}} \tag{5}$$

$$SpaceAttn(Q, K, V)_j = \frac{\sum_{i=1}^{C} v_i + (\sum_{i=1}^{C} v_i(\frac{k_i}{\|k_i\|_2})^T)\frac{q_j}{\|q_j\|_2}}{C + (\sum_{i=1}^{C}(\frac{k_i}{\|k_i\|})^T)\frac{q_j}{\|q_j\|_2}} \tag{6}$$

The formula (7) shows the function expression of the channel from the attention module, and the formula (8) shows the ith line of the attention function. Due to the intermediate result produced in the operation $\sum_{j=1}^{M^2}(\frac{k_j}{|k_j|_2})v_j^T$ and $\sum_{j=1}^{M^2}(\frac{k_j}{|k_j|_2})$ can be reused, so the two intermediate results are the same as $(\frac{q_i}{|q_i|_2})^T$ together introduce the computational complexity of $O(2M^2C)$. where $q_i \in R^C$, $k_j \in R^C$, and $v_j \in R^C$, respectively, are column vectors in the Q^T matrix, the K^T matrix, and the V^T matrix.

$$ChanlAttn(Q, K, V) = \frac{\sum_{M^2} V + \frac{Q}{\|Q\|_2}((\frac{K}{\|K\|_2})^T V)}{M^2 + \frac{Q}{\|Q\|_2}\sum_{M^2}(\frac{K}{\|K\|_2})^T} \tag{7}$$

$$ChanlAttn(Q, K, V)_i = \frac{\sum_{j=1}^{M^2} v_j + (\frac{q_i}{\|q_i\|_2})^T \sum_{j=1}^{M^2}(\frac{k_j}{\|k_j\|_2})v_j^T}{M^2 + (\frac{q_i}{\|q_i\|_2})^T \sum_{j=1}^{M^2}(\frac{k_j}{\|k_j\|_2})} \tag{8}$$

3.5   Hybrid Cross-Attention Module

The proposed hybrid cross-attention module in this section is a modular unit that combines a window division module, spatial cross-attention module, spatial self-attention module, channel cross-attention module, and channel self-attention module. It aims to jointly decode and align the semantic information obtained from network forward propagation with the supplementary information saved from the residual structure. The output features of the module are then transmitted to the deep network for further processing.

The spatial cross-attention component facilitates the exchange of spatial information, establishing internal connections between semantic information and supplementary information from shallow networks, thereby achieving joint encoding in the spatial dimension. By fusing the semantic feature X and the supplementary feature F_r, a new feature I is obtained through spatial cross-attention coding, representing the joint encoding of the two types of information. Specifically, the semantic feature X and the supplementary feature F_r are initially mapped to two distinct feature spaces using fully connected layers, resulting in Q, K, V, Q_r, K_r, and V_r as defined in formula (9).

$$\begin{aligned} Q, K, V &= Linear(X) \\ Q_r, K_r, V_r &= Linear(F_r) \end{aligned} \tag{9}$$

Using Q, K, V, and Q_r,K_r,V_r, the whole calculation process is shown in the formula (10), where $lambda_r$ is a custom constant.

$$I = \frac{\sum_C V + (V(\frac{K}{\|K\|_2})^T)\lambda_r \frac{Q_r}{\|Q_r\|_2}}{C + (\sum_C (\frac{K}{\|K\|_2})^T)\lambda_r \frac{Q_r}{\|Q_r\|_2}} \tag{10}$$

I_s is obtained by going through the fully connected layer and normalization, and the calculation process is shown in the formula (11), where A_1 and B_1 are the parameters to be learned.

$$I_s = Normalization(A_1 I + B_1) \tag{11}$$

To enhance the retrieval of contextual information, the output I_s from the spatial cross-attention module is fed into the spatial self-attention module. Initially, the input feature I_s undergoes linear mapping to obtain the feature vectors Q_s, K_s, and V_s. These vectors are then used to calculate the self-attention weights. Subsequently, the resulting feature vector F_s from the spatial self-attention module is further processed by the channel cross-attention module and the channel self-attention module, undergoing a similar fusion process. The entire procedure is represented by formula (12), with A_2 and B_2 denoting the learnable parameters.

$$\begin{aligned} Q_s, K_s, V_s &= Linear(I_s) \\ F &= SpaceAttention(Q_s, K_s, V_s) \\ F_s &= Normalization(A_2 F + B_2) \end{aligned} \tag{12}$$

$$egH_o = F_{C\_3\times3}\left(\sum_{i=x}^{n} F_{Adp\_dn}(C_i')\right) \tag{13}$$

4 Experiments

Our experimental evaluation conducted on the widely recognized MS COCO dataset [18] forms the foundation of our study. These findings confirm that our STFormer architecture aligns with, and even surpasses, the performance levels achieved by prevailing and widely adopted backbone structures.

4.1 Object Detection on MS COCO Dataset

In order to verify the effectiveness of the backbone network STFormer, the proposed backbone is combined with CenterNet [8], SR-CenterNet, and MR-CenterNet [9] detection methods, and the hyperparameter settings of all experimental models use the hyperparameters provided by MR-CenterNet. Training uses the adam optimizer with an epochs of 300, a batch size of 16 (on both GPUs), a base learning rate of 0.0005, and a learning rate decay strategy using cosine with a weight decay of 0.05, dropout of 0.1, and warmup epochs of 5. The image was randomly cropped and scaled to a size of 511×511, while some common data enhancement methods such as color dithering and brightness dithering were applied. During the test, the resolution of the original input image is kept unchanged and filled to a fixed size with zeros before feeding into the network. The detection results are given in the Table 1, and compared with other object detection models, mainly involving DeNet [25], Faster R-CNN [22], Faster R-CNN w/FPN and other models.

Table 1. Performance comparison (%) with state-of-the-art methods on the MS COCO dataset. Our approach demonstrates significant superiority over existing one-stage detectors and achieves top-ranking performance among state-of-the-art two-stage detectors.

Method	Backbone	Train input	AP	AP_{50}	AP_{75}	AP_S	AP_M	AP_L
two-stage:								
Faster R-CNN [22]	Resnet101 [11]	1000×600	34.9	55.7	37.4	15.6	38.7	50.9
Cascade R-CNN [3]	Resnet101	1300×800	42.8	62.1	46.3	26.6	23.7	45.5
Cascade Mask R-CNN	ResNeXt101 [26]	1000×600	48.3	66.4	52.3	–	–	–
GFLV2 [16]	Res2Net101-DCN [12]	2000×1200	53.3	70.9	59.2	35.7	56.1	65.5
LSNet [30]	Res2Net101-DCN	1400×840	53.5	71.1	59.2	35.7	**56.4**	65.8
one-stage:								
RetinaNet [17]	ResNet101	800×800	39.1	54.9	42.3	21.8	42.7	50.2
RefineDet [28]	ResNet101	512×512	41.8	62.9	45.7	25.6	45.1	54.1
CornerNet [14]	Hourglass104 [31]	511×511	42.1	57.8	45.3	20.8	44.8	56.7
ExtremeNet [32]	Hourglass104	511×511	43.2	59.8	46.4	24.1	46.0	57.1
CenterNet [8]	Hourglass104	511×511	45.1	63.9	49.3	26.6	47.1	57.7
SR-CenterNet [9]	Hourglass104	511×511	47.0	64.5	50.7	28.9	49.9	58.9
CentripetalNet [7]	Hourglass104	511×511	48.0	65.1	51.8	29.0	50.4	59.9
MR-CenterNet [9]	Res2Net101-DCN	511×511	53.7	70.9	59.7	35.1	56.0	66.7
CenterNet	STFormer	511×511	48.9	65.7	53.7	28.9	52.6	64.2
SR-CenterNet	STFormer	511×511	50.8	69.4	55.2	33.1	53.7	65.1
MR-CenterNet	STFormer	511×511	**54.3**	**71.6**	**62.4**	**37.3**	56.4	**66.9**

The two-stage detection method typically employs a significantly higher number of anchor boxes compared to the one-stage method, leading to more precise detection results at the cost of lower efficiency. While STFormer-SR-CenterNet serves as a one-stage detection method, it exhibits a notable advantage over certain two-stage detection methods such as Faster R-CNN [22], Cascade R-CNN [3], and Cascade Mask R-CNN. However, it still lags behind other two-stage detection methods like Cascade Mask R-CNN+Swin Transformer, GFLV2 [16], and LSNet [30]. It is noteworthy that these methods utilize deep residual networks as backbone networks and employ feature pyramid structures for detecting targets of varying scales. For instance, the GFLV2 network achieves a detection accuracy of 53.3% using the Res2Net-101-DCN backbone, while the LSNet network achieves a detection accuracy of 53.5% using the same backbone. As a one-stage detection method, STFormer-MR-CenterNet leverages a pyramid inference structure, enabling it to outperform Cascade Mask R-CNN - Swin Transformer, GFLV2, and LSNet networks, particularly in detecting small targets. We also found that STFormer has a stronger ability to detect small objects compared to conventional networks. In the detection based on the MS COCO dataset, we compared the performance difference between Hourglass and STFormer. When compared to various one-stage detection methods including RetinaNet [17], RefineDet [28], CornerNet [14], ExtremeNet [32], CenterNet [8], SR-CenterNet [9], CentripetalNet [7], and MR-CenterNet, the STFormer method exhibits clear advantages across all evaluation criteria. The table highlights these advantages, showing that when the STFormer backbone network proposed in this chapter is applied, it achieves a performance improvement of nearly 4% compared to methods that detect key points on a single-resolution feature map, such as CenterNet (Objects as Points) and SR-CenterNet. Additionally, when compared to the method of detecting key points on a feature pyramid structure map, such as MR-CenterNet, the STFormer backbone network proposed in this chapter outperforms Res2Net-101-DCN, which also employs a feature pyramid structure, by 0.6%. These results emphasize the superior performance of the STFormer method in enhancing detection accuracy.

5   Conclusion

This paper introduces STFormer, a novel feature extraction network that leverages joint modeling and multi-dimensional feature fusion technology to integrate low-level and high-level features, as well as global and local information. This approach significantly enhances the accuracy of detection algorithms, particularly in small target detection scenarios. By employing encoders to process high semantic features and preserving high-resolution target details through pipelines, followed by feature fusion using decoders at different levels, the model facilitates information exchange. Furthermore, attention and convolution mechanisms are utilized to extract comprehensive global and local information, resulting in a comprehensive model that improves detection accuracy and efficiency. Compared to mainstream detection networks, the combination of MR CenterNet and STFormer achieves superior results on both the PASCAL VOC and MS COCO datasets. Ablation experiments clearly demonstrate the performance of the three modules, supported by graphs and data visualization. The results illustrate the significant improvement in detection accuracy achieved by the proposed modules.

Acknowledgment. Project supported by the National Natural Science Foundation of China (No. 62072333).

References

1. Bao, H., et al.: UniLMv2: pseudo-masked language models for unified language model pre-training. In: International Conference on Machine Learning, pp. 642–652. PMLR (2020)
2. Bochkovskiy, A., Wang, C.Y., Liao, H.Y.M.: YOLOv4: optimal speed and accuracy of object detection. arXiv preprint arXiv:2004.10934 (2020)
3. Cai, Z., Vasconcelos, N.: Cascade R-CNN: delving into high quality object detection. In: Proceedings of the IEEE Conference on Computer Vision and Pattern Recognition, pp. 6154–6162 (2018)
4. Carion, N., Massa, F., Synnaeve, G., Usunier, N., Kirillov, A., Zagoruyko, S.: End-to-end object detection with transformers. arXiv preprint arXiv:2005.12872 (2020)
5. Chi, C., Wei, F., Hu, H.: RelationNet++: bridging visual representations for object detection via transformer decoder. Adv. Neural. Inf. Process. Syst. **33**, 13564–13574 (2020)
6. Deng, J., Dong, W., Socher, R., Li, L.J., Li, K., Fei-Fei, L.: ImageNet: a large-scale hierarchical image database. In: 2009 IEEE Conference on Computer Vision and Pattern Recognition, pp. 248–255. IEEE (2009)
7. Dong, Z., Li, G., Liao, Y., Wang, F., Ren, P., Qian, C.: CentripetalNet: pursuing high-quality keypoint pairs for object detection. In: Proceedings of the IEEE/CVF Conference on Computer Vision and Pattern Recognition, pp. 10519–10528 (2020)
8. Duan, K., Bai, S., Xie, L., Qi, H., Huang, Q., Tian, Q.: CenterNet: keypoint triplets for object detection. In: Proceedings of the IEEE/CVF International Conference on Computer Vision, pp. 6569–6578 (2019)
9. Duan, K., Bai, S., Xie, L., Qi, H., Huang, Q., Tian, Q.: CenterNet++ for object detection. arXiv preprint arXiv:2204.08394 (2022)
10. Ghiasi, G., Lin, T.Y., Le, Q.V.: NAS-FPN: learning scalable feature pyramid architecture for object detection. In: Proceedings of the IEEE/CVF Conference on Computer Vision and Pattern Recognition, pp. 7036–7045 (2019)
11. He, K., Zhang, X., Ren, S., Sun, J.: Deep residual learning for image recognition. In: Proceedings of the IEEE Conference on Computer Vision and Pattern Recognition, pp. 770–778 (2016)
12. Huang, G., Liu, Z., Van Der Maaten, L., Weinberger, K.Q.: Densely connected convolutional networks. In: Proceedings of the IEEE Conference on Computer Vision and Pattern Recognition, pp. 4700–4708 (2017)
13. Krizhevsky, A., Sutskever, I., Hinton, G.E.: ImageNet classification with deep convolutional neural networks. Commun. ACM **60**, 84–90 (2012)
14. Law, H., Deng, J.: CornerNet: detecting objects as paired keypoints. In: Proceedings of the European Conference on Computer Vision (ECCV), pp. 734–750 (2018)
15. Li, R., Zheng, S., Duan, C., Su, J., Zhang, C.: Multistage attention ResU-Net for semantic segmentation of fine-resolution remote sensing images. IEEE Geosci. Remote Sens. Lett. **19**, 1–5 (2021)

16. Li, X., Wang, W., Hu, X., Li, J., Tang, J., Yang, J.: Generalized focal loss V2: learning reliable localization quality estimation for dense object detection. In: Proceedings of the IEEE/CVF Conference on Computer Vision and Pattern Recognition, pp. 11632–11641 (2021)
17. Lin, T.Y., Goyal, P., Girshick, R., He, K., Dollár, P.: Focal loss for dense object detection. In: Proceedings of the IEEE International Conference on Computer Vision, pp. 2980–2988 (2017)
18. Lin, T.-Y., et al.: Microsoft COCO: common objects in context. In: Fleet, D., Pajdla, T., Schiele, B., Tuytelaars, T. (eds.) ECCV 2014. LNCS, vol. 8693, pp. 740–755. Springer, Cham (2014). https://doi.org/10.1007/978-3-319-10602-1_48
19. Liu, W., et al.: SSD: single shot multibox detector. In: Leibe, B., Matas, J., Sebe, N., Welling, M. (eds.) ECCV 2016. LNCS, vol. 9905, pp. 21–37. Springer, Cham (2016). https://doi.org/10.1007/978-3-319-46448-0_2
20. Liu, Z., et al.: Swin transformer: hierarchical vision transformer using shifted windows. In: Proceedings of the IEEE/CVF International Conference on Computer Vision, pp. 10012–10022 (2021)
21. Redmon, J., Divvala, S., Girshick, R., Farhadi, A.: You only look once: unified, real-time object detection. In: Proceedings of the IEEE Conference on Computer Vision and Pattern Recognition, pp. 779–788 (2016)
22. Ren, S., He, K., Girshick, R., Sun, J.: Faster R-CNN: towards real-time object detection with region proposal networks. In: Advances in Neural Information Processing Systems, pp. 91–99 (2015)
23. Sun, C., Shrivastava, A., Singh, S., Gupta, A.: Revisiting unreasonable effectiveness of data in deep learning era. In: Proceedings of the IEEE International Conference on Computer Vision, pp. 843–852 (2017)
24. Tan, M., Pang, R., Le, Q.V.: EfficientDet: scalable and efficient object detection. In: Proceedings of the IEEE/CVF Conference on Computer Vision and Pattern Recognition, pp. 10781–10790 (2020)
25. Tychsen-Smith, L., Petersson, L.: DeNet: scalable real-time object detection with directed sparse sampling. In: Proceedings of the IEEE International Conference on Computer Vision, pp. 428–436 (2017)
26. Xie, S., Girshick, R., Dollár, P., Tu, Z., He, K.: Aggregated residual transformations for deep neural networks. In: Proceedings of the IEEE Conference on Computer Vision and Pattern Recognition, pp. 1492–1500 (2017)
27. Yuan, L., et al.: Tokens-to-token ViT: training vision transformers from scratch on ImageNet. In: Proceedings of the IEEE/CVF International Conference on Computer Vision, pp. 558–567 (2021)
28. Zhang, S., Wen, L., Bian, X., Lei, Z., Li, S.Z.: Single-shot refinement neural network for object detection. In: Proceedings of the IEEE Conference on Computer Vision and Pattern Recognition, pp. 4203–4212 (2018)
29. Zhou, B., Zhao, H., Puig, X., Fidler, S., Barriuso, A., Torralba, A.: Scene parsing through ADE20K dataset. In: Proceedings of the IEEE Conference on Computer Vision and Pattern Recognition, pp. 633–641 (2017)
30. Zhou, W., Zhu, Y., Lei, J., Yang, R., Yu, L.: LSNet: lightweight spatial boosting network for detecting salient objects in RGB-thermal images. IEEE Trans. Image Process. **32**, 1329–1340 (2023)
31. Zhou, X., Wang, D., Krähenbühl, P.: Objects as points. arXiv preprint arXiv:1904.07850 (2019)
32. Zhou, X., Zhuo, J., Krahenbuhl, P.: Bottom-up object detection by grouping extreme and center points. In: Proceedings of the IEEE/CVF Conference on Computer Vision and Pattern Recognition, pp. 850–859 (2019)

Improving Handwritten Mathematical Expression Recognition via an Attention Refinement Network

Jiayi Liu[1], Qiufeng Wang[1(✉)], Wei Liao[1], Jianghan Chen[1], and Kaizhu Huang[2]

[1] School of Advanced Technology, Xi'an Jiaotong-Liverpool University,
Suzhou, China
Qiufeng.Wang@xjtlu.edu.cn

[2] Data Science Research Center, Duke Kunshan University, Suzhou, China

Abstract. Handwritten mathematical expression recognition (HMER), typically regarding as a sequence-to-sequence problem, has made great progress in recent years, where RNN based models have been widely adopted. Although Transformer based model has demonstrated success in many areas, its performance is not satisfied due to the issue of standard attention mechanism in HMER. Therefore, we propose to improve the performance via an attention refinement network in the Transformer framework for HMER. We firstly adopt a shift window attention (SWA) from Swin Transformer to capture spatial contexts of the whole image for HMER. Moreover, we propose a refined coverage attention (RCA) to overcome the issue of lack of converge in the standard attention mechanism, where we utilize a convolutional kernel with a gating function to obtain coverage features. With the proposed RCA, we refine coverage attentions to attenuate the repeating issue of focused areas in the long-sequence. In addition, we utilize a pyramid data augmentation method to generate mathematical expression images with multiple resolutions to enhance the model generalization. We evaluate the proposed attention refinement network on the HMER benchmark datasets of CROHME2014/2016/2019, and extensive experiments demonstrate its effectiveness.

Keywords: Handwritten mathematical expression recognition · Shift window attention · Refined coverage attention · Pyramid data augmentation

1 Introduction

Handwritten Mathematical Expression Recognition (HMER) has attracted much attention due to its potential application in the digital learning and AI education, making mathematical computations and education more accessible and interactive. Since the rise of deep learning [5], Recurrent Neural Networks (RNN)-based models [11,12,16] have been widely adopted and demonstrated effective in HMER, however, it is unsatisfied especially when dealing with mathematical

B. Luo et al. (Eds.): ICONIP 2023, CCIS 1967, pp. 543–555, 2024.
https://doi.org/10.1007/978-981-99-8178-6_41

formulas with long length due to the vanishing gradient during the training of RNN models [18].

Recently, Transformer-based models have demonstrated more powerful recognition ability than RNN-based models in most of computer vision and natural language processing (NLP) tasks [18] due to three advantages: (1) the Transformer's self-attention mechanism allows it to capture long-range dependencies more effectively, which is vital for understanding complex mathematical expressions with nested structures, (2) the parallel computation capabilities of the Transformer enable faster training and inference, making it more suitable for large-scale applications, (3) the Transformer's multi-head attention mechanism allows the model to learn diverse representations of the input data, which is beneficial for capturing various writing styles and spatial relationships between symbols in handwritten mathematical expressions. These factors contribute to the Transformer's enhanced performance in HMER tasks [10], making it a promising approach for tackling the challenges in the HMER task.

However, it is still challenging to recognize the handwritten mathematical expressions. In this paper, we aim to improve Transformer-based model in HMER via an attention refinement network. Specifically, we propose to modify the attention mechanisms by shift window attention (SWA) and coverage attention mechanisms, which effectively capture local and global contextual information in handwritten mathematical expressions. Shift window attention (SWA) [8] calculates the attention weights of elements within a fixed-sized window, and therefore reduces the computational complexity of model significantly. Besides, SWA can better capture local dependencies between adjacent symbols in expressions by focusing on the immediate neighbours of the current symbol. In addition, when humans read handwritten mathematical expressions, we typically focus on a part of the expression at each moment, rather than the entire expression. SWA's sliding window mechanism simulates this process well and enables the model to better understand mathematical expressions.

Meanwhile, the coverage attention [9] is widely adopted in the HMER, which calculates a coverage matrix to address the difficulty of the over-parsing (some symbols are decoded more than once) and under-parsing (some symbols are never decoded) in the standard attention mechanisms. However, such calculation is usually step-by-step, then may repeat to focus on some areas in the long sequence. To overcome this issue, we propose a Refined Coverage Attention (RCA) method to adopt a convolutional operation with a gating function, thus attenuating the repeating issue of the normal coverage attention.

Furthermore, we propose a pyramid data augmentation method to generate multiple expression images based on the original input image, and each generated image is recognized by the proposed Transformer model, then aggregate all output s to predict the final recognition results. Through such augmentation strategy, we can improve the generalization ability of the proposed model.

The main contributions of this work can be summarized as follows:

- We propose to incorporate shift window attention (SWA) in the handwritten mathematical expression recognition task.

- We propose a refined coverage attention (RCA) in the transformer model to improve the attention coverage efficiently by a convolutional operation with gating mechanism.
- We propose a pyramid data augmentation for processing handwritten mathematical expressions images with multiple scales.
- We evaluate the proposed method on the HMER benchmark datasets of CROHME2014/2016/2019, and the extensive experimental results demonstrate its effectiveness.

2 Related Works

We can roughly divide the research of handwritten mathematical expression recognition (HMER) into two categories: (1) Traditional HMER method, and (2) Deep learning based HMER model. The traditional HMER method basically contains two steps: symbol recognition and structure analysis [18]. However, this isolation of each symbol in a mathematical formula can lead to some recognition results that are completely incompatible with the principles of mathematical syntax such as the "C2" may be misidentified as "(2". To address this issue, manually pre-defined grammar rules such as context-free grammars need to be added to solve the lack of association between individual symbols and alleviate the problem of poor generalization. Besides, in some cases that the pre-defined grammar rules have not covered, the relationship between some unusual symbols will not be identified successfully. With the rise of deep learning, HMER has made great progress. Recurrent Neural Networks (RNNs) have become particularly influential due to their ability to model temporal dependencies, and several RNN based models like WAP (Watch, Attend and Parse) [16], Dense-WAP [12], TAP (Track, Attend, and Parse) [13], ABM [1], and CAN [6] have been developed specifically for HMER tasks. These models demonstrate significant improvements over traditional methods. However, they face limitations in processing long sequences and lack parallel computation capabilities, impacting the computational efficiency.

Recently, Transformers-based model [10] have emerged as promising alternatives in HMER task. The strength of Transformers lies in their ability to model long-range dependencies and provide parallel computation, potentially addressing the issues associated with RNNs. However, the current Transformer-based models of HMER, such as BTTR [18], have yet to yield competitive performance, potentially due to some regions in mathematical expressions remaining unparsed. By introducing a coverage attention mechanism [9] into the Transformer architecture, it could remedy this issue by ensuring each region of the sequence is adequately parsed. Concurrently, considering the spatial variability in handwriting, the SWIN Transformer model [8] introduces a shift window attention mechanism that could cater to this characteristic. In this work, we explore the incorporation of both shift window attention and coverage attention within the Transformer model, aiming to enhance its performance in the HMER task.

3   Methodology

In this section, we will describe the proposed model under the Transformer framework for Handwritten mathematical expression recognition (HMER), and Fig. 1 shows an overview of our model. We firstly adopt a pyramid data augmentation method to generate samples with multiple scales. Secondly, we utilize the encoder and positional encoding to extract the features from the augmented images. Thirdly, we employ the shift window attention (SWA) to enhance the spatial contexts. Then, we propose a refined coverage attention (RCA) to capture long-range historical contexts by adjusting alignment without repeating issue. Finally, we merge the multiple recognition results of each augmented image to obtain the final predicted results. In the following, we will describe each component in details.

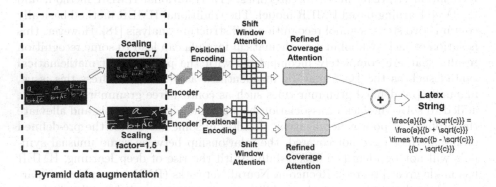

Fig. 1. Overview of proposed attention refined network for HMER.

3.1   Pyramid Data Augmentation

As the size, shape and structure of expressions may vary over a wide range in the task of handwritten mathematical expression recognition, we propose a pyramid data augmentation method in this paper to perform multi-scale processing of handwritten mathematical expression images. Specifically, a large scale (i.e., low-resolution, for example, the top branch in the Fig. 1) helps the model to understand the overall structure of the mathematical expression, while a small scale allows the model to better capture details such as specific mathematical symbols. Given the original image I, we can create an image pyramid P by applying a series of scaling operations to it. Suppose we choose n different scaling factors $\{k_1, k_2, \ldots, k_n\}$, we can calculate P by:

$$P_i = Resize(k_i, I), \quad i = 1, 2, \ldots, n, \tag{1}$$

where $Resize(\cdot)$ function denotes the scaling operation of the image and k_i is the i-th scaling factor. Specifically, we call the torch.nn.functional.interpolate()

from Pytorch directly in our work. In this way, we obtain an image pyramid consisting of n-scale images with different resolutions. We can then apply our model to each scale of this image pyramid and finally fuse these results by a weighted average to obtain the final recognition result. The Fig. 1 shows how images are processed by our model with two different scales in pyramid data augmentation when the scaling factor is 0.7 and 1.5, and how they are finally combined by weighted average. To be noted, we use this strategy in both training and test stage to improve the model generalization ability.

3.2   Encoder and Positional Encoding

In this work, we adopt the widely used model DenseNet [4] as the backbone in the Encoder to extract features of the inputted handwritten mathematical expression image. The key idea of DenseNet is to use concatenation operation to improve information flow between convolutional layers. In details, assuming H_l represent the convolution function of the l-th layer, then the output feature of layer l can be computed by:

$$x_l = H_l([x_0; x_1; \ldots; x_{l-1}]), \tag{2}$$

where x_l represents the feature map generated in the l-th convolution layer and the symbol ";" represents the concatenation operation.

To efficiently process the image in the Transformer encoder, we downsample the feature map from DenseNet. Each pixel in the downsampled image now represents an $8*8$ patch of the original input image, thereby reducing the computational complexity while maintaining essential features. To record positional information for each patch in the Transformer encoder, we add positional encoding [2], which are calculated by the sinusoidal functions of different frequencies. Specifically, the absolute positional encoding can be represented by:

$$P^w_{(p,d)}[2i] = \sin\left(\frac{p}{10000} \cdot \frac{2i}{d}\right), P^w_{(p,d)}[2i+1] = \cos\left(\frac{p}{10000} \cdot \frac{2i}{d}\right), \tag{3}$$

where d denotes the size of encoding dimension, p denotes position information (x, y), and i denotes the index of feature dimension. Then, a 2D coordinates tuple (x,y) will be used to record the position of each pixel in the input image, which is then calculated by the following equation:

$$P^I_{(\frac{x}{h_0}, \frac{y}{\omega_0}, d)} = \left[P^w_{(\frac{x}{h_0}, \frac{d}{2})}; P^w_{(\frac{y}{\omega_0}, \frac{d}{2})}\right], \tag{4}$$

where h_0 and ω_0 represent the shape of feature, and the symbol ";" represents the concatenation operation.

3.3 Shift Window Attention

In most of handwriting recognition tasks, the character string is assumed to be linearly written from left to right (or right to left). However, this phenomenon is usually not right in the handwritten mathematical expressions. Symbols in mathematical expressions are arranged not just linearly (left to right), but also vertically (top to bottom), with symbol sizes varying to represent superscripts, subscripts, or other elements. Thereby, recognition of mathematical expressions cannot merely rely on sequential analysis while ignoring the spatial contexts, because spatial relations between symbols significantly impact their semantic meaning. Therefore, shift window attention has been adopted in this work to addresses such issue.

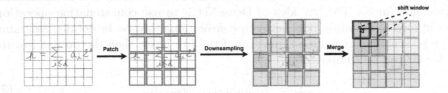

Fig. 2. The workflow of shift window attention (SWA).

Figure 2 shows a general workflow of the shift window attention strategy in our work. By regarding the image of a mathematical expression as a 2D grid of tokens and applying self-attention within local windows, it can capture the local spatial relations between symbols. Furthermore, by shifting the windows across different layers, it also encourages long-range interactions between symbols that might be far apart in the image but closely related in the mathematical semantics.

In details, given an input sequence of tokens $X \in \mathbb{R}^{n \times d}$, we first reshape it into a 2D grid of size H × W. Then, we divide the grid into non-overlapping windows of size M × M. As the standard self-attention mechanism is applied within each window, the original multi-head self-attention [10] is revised by

$$
\begin{aligned}
A_i &= \frac{Q_i K_i^{\mathsf{T}}}{\sqrt{d_k}} + E, \ \text{Attention}(Q, K, V) = A_i V_i, \\
\text{head}_i &= \text{Attention}(QW_Q^i, KW_K^i, VW_V^i), \\
\text{MultiHead}(Q, K, V) &= \text{Concat}(\text{head}_1, \dots, \text{head}_h) W_o,
\end{aligned}
\tag{5}
$$

where E is a matrix whose value $E_{i,j}$ indicates the i-th and j-th elements in the whole image are in the same window if $E_{i,j} = 0$; otherwise, $E_{i,j} = $ -∞. In this way, when softmax is applied, the attention score of elements that are not within the same window will approach 0, achieving the effect of window restriction. In the next layer, the windows are shifted by M/2 in both vertical and horizontal directions to encourage interactions between neighboring windows.

3.4   Refined Coverage Attention

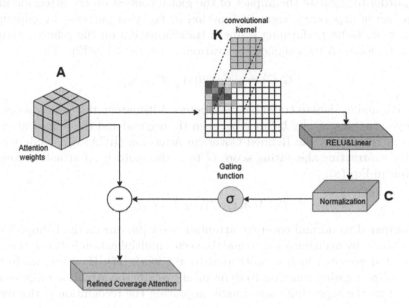

Fig. 3. The workflow of proposed refined coverage attention (RCA).

We utilize shift window attention mechanism to capture spatial dependencies between mathematical symbols, however, it still has limitation in handling long sequences due to the sequential nature of its attention window. To overcome this limitation, the converge attention has been widely adopted for HMER [9, 17]. Although it has improved the HMER performance, it sometimes repeats high-attention areas excessively ignoring the normal attention weights. To refine the coverage attention mechanism, we propose a Refined Coverage Attention (RCA) to adaptively attenuate such repeating issue, creating a more balanced and effective approach to processing long sequences. By introducing a sigmoid gating mechanism, RCA dynamically adjusts the global and local contexts, thus improving the recognition accuracy for complex mathematical expressions.

Figure 3 shows an overall workflow of the proposed RCA, which basically can be divided into five steps: (1) Convolutionalization, (2) RELU & Linear, (3) Batch-Normalization, (4) Gating mechanism, and (5) Subtraction. In the following, we will describe more details. Consider an input sequence $X \in \mathbb{R}^{N \times D}$ (N is the sequence length and D is the embedding dimension), we can represent the first three steps by

$$C = \mathrm{norm}\left(\max\left(0, \mathbf{K} * A + b_c\right) W_c\right), \tag{6}$$

where $*$ means the convolutional operation, K refers to the 5×5 convolutional kernel, $b_c \in \mathbb{R}^h$ is a bias, and $W_c \in \mathbb{R}^h \times \mathrm{d}$ is a projection matrix. In the Eq. (6), we actually reshape the attention matrix A from the sequence to image pattern for the following convolutional operation.

In order to regulate the impact of the global context on the attention mechanism, we incorporate a sigmoid function σ. For this purpose, we compute a gating score G by performing a linear transformation on the refined attention matrix C, followed by a sigmoid activation, as expressed in Eq. (7):

$$G(C) = \text{sigmoid}(W_g \cdot C + b_g). \tag{7}$$

Here, W_g and b_g denote trainable parameters. Ultimately, this gating score G is employed to dynamically balance between the original and the refined attention matrices. Therefore, the Refined Coverage Attention (RCA) can be derived by directly subtracting the gating score G from the multi-head attention weights, as given in Eq. (8):

$$RCA(A, G(C)) = A - G(C). \tag{8}$$

Compared to normal coverage attention work [9], our method simplifies the computation by utilizing a single matrix-vector multiplication between the input matrix and vector, which is motivated by the work [17]. However, we furthermore utilize a gating function to dynamically adapt the attention alignment by attenuating the repeating issue, finally improving the recognition performance.

3.5 Loss Function

In general, we adopt the widely used Cross Entropy loss in our work:

$$CE = - \sum (y \log(p) + (1 - y) \log(1 - p)). \tag{9}$$

Furthermore, we consider to balance the attention weights for all symbols in the expression, particularly in scenarios with longer expression. To this end, we add one regularization term in the loss function, which is calculated by:

$$\text{Regu}(RCA) = \frac{1}{T \times T \times N} \sum_{i=1}^{T} \sum_{j=1}^{T} \sum_{k=1}^{N} |RCA_{ik} - RCA_{jk}|, \tag{10}$$

where RCA_{ik} and RCA_{jk} represent the attention weights in Eq. (8) of the ith and jth target positions at the kth position of the input, respectively. N is the length of the input sequence and T is the length of the target sequence.

In summary, the final loss function is defined by:

$$Total = CE + \lambda \cdot Regu, \tag{11}$$

where λ is a weight parameter to balance the cross entropy loss and regularization term.

4   Experiments

4.1   Dataset and Evaluation Metrics

The proposed model was trained on the CROHME [16] 2014 training dataset, which contains 8,836 handwritten mathematical expressions images. In the test stage, we evaluate the proposed model on three HMER benchmark datasets including CROHME2014/2016/2019 test set, which contains 986/1,147/1,199 handwritten mathematical expression images.

We adopt the widely used expression recognition rates (ExpRate) [16] as the evaluation metric in our experiments, which refers to the percentage of predicted mathematical expressions completely same with the ground truth. In addition to this strict evaluation metric, we also report two relaxation metrics, namely $R_{\leq 1}$ and $R_{\leq 2}$ with maximum of 1 and 2 symbol errors tolerated, respectively. Moreover, we report the Character Error Rate (CER) in our experiments, which is derived from the Levenshtein distance between the predicted expressions and the ground truth at the character level. Specifically, the CER is calculated by $CER = \frac{S+D+I}{N}$, where N denotes the number of symbols (i.e., characters) in the ground truth, and S, D, I evaluates all errors in the prediction expressions including the number of substitutions (S), deletions (D) and insertions (I).

4.2   Implementation Details

In our work, we utilize BTTR [18] as our baseline model and all experiments are conducted with the PyTorch framework on four Tencent Cloud NVIDIA 3090Ti GPUs. We adopt the same DenseNet architecture used in the baseline, consisting of three DenseNet blocks in the encoder, each with 16 bottleneck layers. In the decoder part, the model dimension size is set to 256 with 8 attention heads, following the transformer specification. The feed-forward layer dimension size is set to 1024. We employ three stacked decoder layers with a dropout rate of 0.3. For our proposed Refined Coverage Attention (RCA), the kernel size is set to 5. During the training, the learning rate is set as 0.8, the weight decay and momentum of SGD are set as 10^{-8} and 0.9 respectively. The scaling factor in pyramid data augmentation is $X \in [0.7, 2]$.

4.3   Experimental Results

In this part, we compare the proposed method with other HMER methods on the three benchmark datasets, and the results are shown in Table 1. We can see that the proposed method improves the ExpRate from 53.02% to 61.32% by comparing to the baseline model (i.e., BTTR [18]) on the CROHME 2019 test set, and is supervisor to other works. In detail, the ExpRate of our method outperforms WAP [16] by 18.12 %, DWAP-TD [15] by 11.82%, SRD [14] by 6.62% and ABM [1] by 6.3% and CAN [6] by 4.8% on CRHOME 2014, 2016 and 2019, averagely. All of these demonstrate the effectiveness of the proposed method in the HMER.

Table 1. Comparison with previous HMER works on ExpRate (%), $R_{\leq 1}$ (%), and $R_{\leq 2}$ (%). The bold and underline numbers indicate the best and second-best respectively. (*) means that this BTTR is also our adopted baseline model.

Method	CROHME 2014			CROHME 2016			CROHME 2019		
	ExpRate ↑	$R_{\leq 1}$ ↑	$R_{\leq 2}$ ↑	**ExpRate** ↑	$R_{\leq 1}$ ↑	$R_{\leq 2}$ ↑	**ExpRate** ↑	$R_{\leq 1}$ ↑	$R_{\leq 2}$ ↑
WAP [16]	40.49	26.11	25.43	44.62	57.81	61.60	–	–	–
DWAP [12]	50.18	–	–	47.55	–	–	–	–	–
DWAP-TD [15]	49.16	64.24	67.83	48.56	62.37	65.34	51.44	66.15	69.16
SRD [14]	53.56	62.43	66.42	–	–	–	–	–	–
BTTR (*) [18]	54.07	66.04	70.32	52.32	63.95	68.66	53.02	66.05	69.13
Li et al. [7]	56.62	69.15	75.31	54.64	69.35	73.84	–	–	–
Ding et al. [3]	<u>58.75</u>	–	–	<u>57.74</u>	70.03	76.41	<u>61.32</u>	<u>75.25</u>	<u>80.26</u>
ABM [1]	56.85	73.74	80.29	52.99	69.69	78.73	53.96	71.12	78.78
CAN [6]	57.04	<u>74.23</u>	<u>80.62</u>	56.10	<u>71.53</u>	<u>79.51</u>	54.96	71.92	79.42
Ours	**60.22**	**76.83**	**81.29**	**61.02**	**75.44**	**82.91**	**61.47**	**78.76**	**82.17**

4.4 Ablation Study

In this section, we will conduct a series of ablation studies to explore the impact of the proposed attention refinement network, including the shift window attention (SWA), refined coverage attention (RCA) and pyramid data augmentation (PDA), and the results are shown in Table 2.

From the Table 2, we can see that each component contributes positively to the model performance while the RCA plays the most important role. When RCA is removed (i.e., Ours W/O RCA), the ExpRate dropped by roughly 4% and the CER increased by 1.4%. Meanwhile, the removal of SWA (i.e., Ours W/O SWA) also results in a reduction of the recognition performance including ExpRate reduction by approximately 1.5% and CER increment by around 0.8% on average. Albeit less impacted than the removal of RCA, SWA still contributes positively to the overall model performance. Lastly, the absence of PDA (i.e., Ours W/O PDA) also leads to a performance reduction although the reduction is the smallest. But the model with such removal still performs better than the baseline model without any proposed component. Finally, by incorporating all three proposed components, our model demonstrates the best performance across all datasets, showing their collective importance in achieving optimal results.

Furthermore, we test the performance with respect to the input expression length when employing different components of the model on the CROHME2014 dataset, aiming to identify which component exhibits the most advantage in handling longer sequences. Figure 4 shows the ExpRate of different models with various expression lengths, and we can see when the length exceeds 20, the RCA can significantly improve the ExpRate. Concurrently, employing both RCA and SWA help the model obtain a satisfactory ExpRate even with longer expression.

Table 2. Ablation Study on the ExpRate (%) and CER (%) for the proposed three improvements.

Method	CROHME 2014		CROHME 2016		CROHME 2019	
	ExpRate↑	CER↓	ExpRate↑	CER↓	ExpRate↑	CER↓
Baseline	54.07	11.9	52.32	12.3	53.02	12.1
Ours W/O RCA	56.24	11.8	57.42	11.4	60.38	10.9
Ours W/O SWA	58.76	11.2	58.79	11.3	60.72	10.6
Ours W/O PDA	59.81	10.7	59.23	11.0	61.01	10.3
Ours	60.22	10.4	61.02	10.2	61.32	10.1

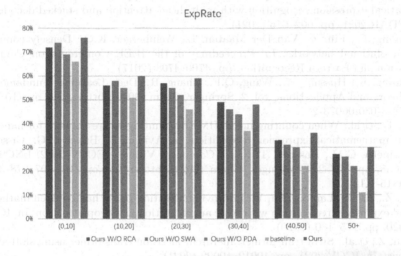

Fig. 4. ExpRate of different models for expressions with various lengths.

5   Conclusion

In this paper, we aim to improve the transformer-based model for handwritten mathematical expression recognition (HMER) by an attention refinement network. Specifically, we propose to utilize shift window attention to improve the spatial contexts in the whole image, and adopt a refined coverage attention to attenuate the repeating issue in the attention alignment in the historical attentions. Furthermore, we conduct a pyramid data augmentation strategy in both training and test stage to enhance the recognition generalization. Through the extensive experiments on the benchmark datasets CROHME2014/2016/2019, we can see the effectiveness of the proposed method.

Acknowledgements. This research was funded by National Natural Science Foundation of China (NSFC) no. 62276258, and Jiangsu Science and Technology Programme no. BE2020006-4.

References

1. Bian, X., Qin, B., Xin, X., Li, J., Su, X., Wang, Y.: Handwritten mathematical expression recognition via attention aggregation based bi-directional mutual learning. In: AAAI 2022, vol. 36, pp. 113–121 (2022)
2. Chu, X., et al.: Conditional positional encodings for vision transformers. arXiv preprint arXiv:2102.10882 (2021)
3. Ding, H., Chen, K., Huo, Q.: An encoder-decoder approach to handwritten mathematical expression recognition with multi-head attention and stacked decoder. In: ICDAR 2021, pp. 602–616 (2021)
4. Huang, G., Liu, Z., Van Der Maaten, L., Weinberger, K.Q.: Densely connected convolutional networks. In: Proceedings of the IEEE Conference on Computer Vision and Pattern Recognition, pp. 4700–4708 (2017)
5. Huang, K., Hussain, A., Wang, Q.F., Zhang, R.: Deep Learning: Fundamentals, Theory and Applications, vol. 2. Springer, Cham (2019). https://doi.org/10.1007/978-3-030-06073-2
6. Li, B., et al.: When counting meets HMER: counting-aware network for handwritten mathematical expression recognition. In: Avidan, S., Brostow, G., Cissé, M., Farinella, G.M., Hassner, T. (eds.) Computer Vision – ECCV 2022. LNCS, vol. 13688, pp. 197–214. Springer, Cham (2022). https://doi.org/10.1007/978-3-031-19815-1_12
7. Li, Z., Jin, L., Lai, S., Zhu, Y.: Improving attention-based handwritten mathematical expression recognition with scale augmentation and drop attention. In: ICFHR 2020, pp. 175–180 (2020)
8. Liu, Z., et al.: Swin transformer: hierarchical vision transformer using shifted windows. In: ICCV 2021, pp. 10012–10022 (2021)
9. Mi, H., Sankaran, B., Wang, Z., Ittycheriah, A.: Coverage embedding models for neural machine translation. arXiv preprint arXiv:1605.03148 (2016)
10. Vaswani, A., et al.: Attention is all you need. In: Advances in Neural Information Processing Systems, vol. 30 (2017)
11. Wang, J., Du, J., Zhang, J., Wang, B., Ren, B.: Stroke constrained attention network for online handwritten mathematical expression recognition. Pattern Recogn. **119**, 108047 (2021)
12. Zhang, J., Du, J., Dai, L.: Multi-scale attention with dense encoder for handwritten mathematical expression recognition. In: ICPR 2018, pp. 2245–2250 (2018)
13. Zhang, J., Du, J., Dai, L.: Track, Attend, and Parse (TAP): an end-to-end framework for online handwritten mathematical expression recognition. IEEE Trans. Multimedia **21**(1), 221–233 (2018)
14. Zhang, J., Du, J., Yang, Y., Song, Y.Z., Dai, L.: SRD: a tree structure based decoder for online handwritten mathematical expression recognition. IEEE Trans. Multimedia **23**, 2471–2480 (2020)
15. Zhang, J., Du, J., Yang, Y., Song, Y.Z., Wei, S., Dai, L.: A tree-structured decoder for image-to-markup generation. In: ICML 2020, pp. 11076–11085 (2020)
16. Zhang, J., et al.: Watch, attend and parse: an end-to-end neural network based approach to handwritten mathematical expression recognition. Pattern Recogn. **71**, 196–206 (2017)

17. Zhao, W., Gao, L.: CoMER: modeling coverage for transformer-based handwritten mathematical expression recognition. In: Avidan, S., Brostow, G., Ciss, M., Farinella, G.M., Hassner, T. (eds.) Computer Vision – ECCV 2022. LNCS, vol. 13688, pp. 392–408. Springer, Cham (2022). https://doi.org/10.1007/978-3-031-19815-1_23
18. Zhao, W., Gao, L., Yan, Z., Peng, S., Du, L., Zhang, Z.: Handwritten mathematical expression recognition with bidirectionally trained transformer. In: Lladós, J., Lopresti, D., Uchida, S. (eds.) ICDAR 2021. LNCS, vol. 12822, pp. 570–584. Springer, Cham (2021). https://doi.org/10.1007/978-3-030-86331-9_37

Dual-Domain Learning for JPEG Artifacts Removal

Guang Yang[1], Lu Lin[3], Chen Wu[1(✉)], and Feng Wang[2]

[1] University of Science and Technology of China, Hefei, China
{guangyang,wuchen5X}@mail.ustc.edu.cn
[2] Hefei Institute of Physical Science, Chinese Academy of Sciences, Hefei, China
[3] Central China Normal University, Wuhan, China

Abstract. JPEG compression brings artifacts into the compressed image, which not only degrades visual quality but also affects the performance of other image processing tasks. Many learning-based compression artifacts removal methods have been developed to address this issue in recent years, with remarkable success. However, existing learning-based methods generally only exploit spatial information and lack exploration of frequency domain information. Exploring frequency domain information is critical because JPEG compression is actually performed in the frequency domain using the Discrete Cosine Transform (DCT). To effectively leverage information from both the spatial and frequency domains, we propose a novel Dual-Domain Learning Network for JPEG artifacts removal (D2LNet). Our approach first transforms the spatial domain image to the frequency domain by the fast Fourier transform (FFT). We then introduce two core modules, Amplitude Correction Module (ACM) and Phase Correction Module (PCM), which facilitate interactive learning of spatial and frequency domain information. Extensive experimental results performed on color and grayscale images have clearly demonstrated that our method achieves better results than the previous state-of-the-art methods. Code will be available at https://github.com/YeunkSuzy/Dual_Domain_Learning.

Keywords: JPEG Artifacts Removal · Dual-Domain Learning · Fourier Transform

1 Introduction

JPEG [26], based on the Discrete Cosine Transform (DCT) [2], is one of the most widely used image compression algorithms due to its extremely high compression ratio. In JPEG, the image is divided into 8×8 blocks and each block is encoded separately. The DCT is then applied to each block, followed by quantization and

B. Luo et al. (Eds.): ICONIP 2023, CCIS 1967, pp. 556–568, 2024.
https://doi.org/10.1007/978-981-99-8178-6_42

entropy coding. After these steps, we can obtain a good quality of the coded images with a small size. However, this comes with the loss of information, and complex artifacts inevitably appear in the compressed images. This image degradation not only causes visual discomfort, but also affects the performance of other image processing tasks, such as object detection, image super-resolution, and so on (Fig. 1).

Fig. 1. Visual comparisons. The left side of the image is the JPEG compressed image and the right side is the image that has been reconstructed by our method.

In order to reduce the impact of JPEG compression artifacts, many methods have been proposed. Before deep learning was widely used in computer vision tasks, most methods solved the problem by designing a specific filter [10,23], but they were usually limited to solving specific artifacts. In recent years, with the rapid development of deep learning, JPEG artifacts removal methods based on convolutional neural networks (CNNs) [4,5,7,14,16,19,28,31,39] have prevailed and achieved better performance. However, most of the existing CNN-based methods primarily exploit the spatial information and neglect the distinguished frequency information. As we all know, JPEG compression actually occurs in the frequency domain by the DCT. Thus, exploring the effective solutions for the JPEG artifacts removal in the frequency domain is necessary.

In this paper, we explore the manifestation of JPEG compression artifacts in the frequency domain. We can transform the spatial domain image to the frequency domain by the fast Fourier transform (see Fig. 2 (a)). Then, we restore the phase spectrum and amplitude spectrum separately and reconstruct the image by applying the inverse fast Fourier transform [32]. From Fig. 2 (b), we can see that: (1) the Fourier phase spectrum preserves important visual structures, while the amplitude spectrum contains low-level features. (2) After JPEG compression, the Fourier phase spectrum loses some high-frequency information, while the Fourier amplitude spectrum becomes slightly blurred.

Based on the above observation, we propose a novel Dual-Domain Learning Network (D2LNet) for JPEG artifacts removal. We address JPEG artifacts removal by jointly exploring the information in the spatial and frequency domains. In order to utilize the information in both frequency domain and spatial domain effectively, we propose two core modules, namely Amplitude Correction Module (ACM) and Phase Correction Module (PCM), which are composed of multiple Amplitude Correction Blocks (ACB) and Phase Correction Blocks (PCB), respectively. Specifically, the ACM restores the amplitude spectrum of

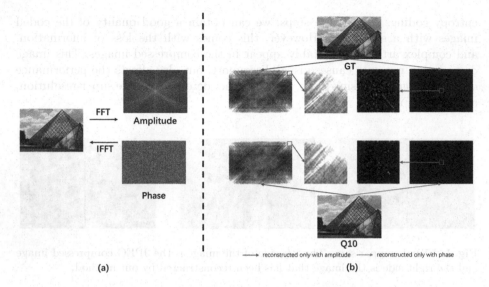

Fig. 2. Fourier transform and Fourier reconstructions. From (b), we can see that: (1) the Fourier phase spectrum preserves the important visual structures, while the amplitude spectrum contains low-level features. (2) After JPEG compression, the Fourier phase spectrum loses some high-frequency information, while the Fourier amplitude spectrum becomes slightly blurred.

degraded images to remove JPEG artifacts, and the PCM restores the phase spectrum information to refine the high-frequency information. The qualitative and quantitative experimental results on the benchmarks show that the proposed method is effective compared to state-of-the-art methods.

2   Related Work

2.1   JPEG Artifacts Removal

JPEG compression can be represented by a formula:

$$\mathbf{Y} = \mathcal{D}(\hat{\mathbf{X}}; QF), \tag{1}$$

where $\hat{\mathbf{X}}$ and $\mathbf{Y}$ denote the original uncompressed image and the compressed image respectively, and $\mathcal{D}$ stands for the compression algorithm, and QF represents the *quality factor* determined and used for adjusting the degree of compression. Removing unwanted image artifacts which might appear on the compressed image $\mathbf{Y}$ is what we want to do. Hopefully, the restored image has a much-improved image quality, as close to $\hat{\mathbf{X}}$ as possible; that is,

$$\mathbf{X} = \mathcal{N}(\mathbf{Y}) \approx \hat{\mathbf{X}}, \tag{2}$$

where $\mathcal{N}$ means the neural network on $\mathbf{Y}$ to reconstruct or restore a very high-quality image $\mathbf{X}$ that is close to the ground-truth image $\hat{\mathbf{X}}$.

Significant progress has recently been made in reducing JPEG artifacts through the application of deep convolutional neural networks. ARCNN [7] is a relatively shallow network that first uses CNN to solve this problem. RED-Net [21] designs a deep encoding-decoding structure to exploit the rich dependencies of deep features. RNAN [37] incorporates both local and non-local attention mechanisms into its learning process, thereby enhancing its ability to represent complex relationships within images. This approach has demonstrated promising results in various image restoration tasks, such as image denoising, reducing compression artifacts, and improving image super-resolution. FBCNN [16] is a flexible blind CNN which can predict the adjustable quality factor to control the trade-off between artifacts removal and details preservation. Some GAN-based JPEG artifacts reduction works [12,13] also have good performance because they are able to produce more realistic details than MSE or SSIM [30] based networks.

2.2 Spatial-Frequency Interaction

There are several frequency domain learning methods [6,9,15,18,20,22] have achieved good results in different tasks such as image classification. For JPEG artifacts removal, learning in the frequency domain is critical because the JPEG compression is actually performed in the frequency domain using the DCT. MWCNN [19] uses wavelet to expand the receptive field to achieve image restoration. D3 [29] introduces a DCT domain prior to facilitating the JPEG artifacts removal. DWCNN [36] also removes the JPEG artifacts in the DCT domain. The DCT is a special case of the FFT, and the FFT can produce an accurate representation of the frequency domain. Although there is currently no existing work that utilizes FFT to remove JPEG artifacts, it is worth exploring the use of FFT for this purpose.

3 Method

In this section, we begin by presenting the fundamental characteristics of the Fourier transform, which hold significant relevance in comprehending our research. Subsequently, we provide a comprehensive elaboration on the proposed model and its corresponding loss function.

3.1 Fourier Transform

The Fourier transform operation, denoted by $\mathcal{F}$, allows us to convert an image $\mathbf{X}$ from the spatial domain to the frequency domain. In our work, we independently apply the Fourier transform to each channel of the image.

For a given image $\mathbf{X}$ with dimensions $H \times W$, where H represents the height and W represents the width, the Fourier transform $\mathcal{F}(\mathbf{X})(u,v)$ at frequency domain coordinates u and v is computed according to the following equation:

$$\mathcal{F}(\mathbf{X})(u,v) = \frac{1}{\sqrt{HW}} \sum_{h=0}^{H-1} \sum_{w=0}^{W-1} \mathbf{X}(h,w) e^{-j2\pi(\frac{h}{H}u + \frac{w}{W}v)}, \tag{3}$$

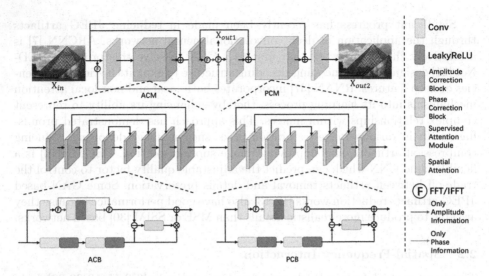

Fig. 3. The architecture of our network, which consists of two main modules: the Amplitude Correction Module (ACM) and the Phase Correction Module (PCM). Specifically, the ACM restores the amplitude spectrum of degraded images to remove JPEG artifacts, and the PCM restores the phase spectrum information to refine the high-frequency information.

It is noteworthy that the Fourier transform can be efficiently implemented using the Fast Fourier Transform (FFT) algorithm.

After performing the Fourier transform on an image $\mathbf{X}$, we can define the amplitude and frequency components as follows:

$$\mathcal{A}(\mathbf{X})(u,v) = \left[R^2(\mathbf{X})(u,v) + I^2(\mathbf{X})(u,v)\right]^{\frac{1}{2}}, \tag{4}$$

$$\mathcal{P}(\mathbf{X})(u,v) = \arctan\left[\frac{I(\mathbf{X})(u,v)}{R(\mathbf{X})(u,v)}\right]. \tag{5}$$

Here, $R(\mathbf{X})$ and $I(\mathbf{X})$ represent the real and imaginary parts of the Fourier transform of image $\mathbf{X}$. The components $\mathcal{A}(\mathbf{X})$ and $\mathcal{P}(\mathbf{X})$ correspond to the amplitude and phase of the image, respectively.

In general, the phase component of an image captures its fundamental structure and semantic information, while the amplitude component represents low-level details such as style. In the context of JPEG compression, the spatial structure of the image is compromised, and high-frequency information is significantly lost during the compression process. As a result, both the phase and amplitude suffer varying degrees of degradation.

Consequently, restoration in the frequency domain offers the advantage of separating the restoration of spatial structure and high-frequency information. Additionally, leveraging the spectral convolution theorem, image processing in the frequency domain inherently provides a global receptive field. This characteristic facilitates the capture of global information in the image.

3.2 Network Framework

Our network architecture, depicted in Fig. 3, consists of two main modules: the amplitude correction module and the phase correction module, which yield two outputs, $\mathbf{X_{out1}}$ and $\mathbf{X_{out2}}$ respectively. The purpose of the amplitude correction module is to mitigate the effects of JPEG compression, while the phase correction module focuses on restoring the fundamental image details, bringing them closer to the ground truth image. To define an input image $\mathbf{X_{in}}$, we initially apply convolution to project it into the feature space. The resulting feature map is then passed through the amplitude correction module, resulting in the reconstructed feature map. Subsequently, we employ SAM [33] to derive $\mathbf{X_{out1}}$ and $\mathbf{X_{amp}}$, where $\mathbf{X_{amp}}$ is the output of the phase correction module. By replacing the amplitude of $\mathbf{X_{in}}$ with that of $\mathbf{X_{out1}}$, we obtain $\mathbf{X_{inv}}$. We concatenate $\mathbf{X_{inv}}$ and $\mathbf{X_{amp}}$, transmitting them to the phase correction module for further refining the high-frequency details of the image. Finally, we obtain the output $\mathbf{X_{out2}}$ through channel modulation. This restoration approach effectively separates the reduction of JPEG blocking artifacts from the restoration of image details.

Amplitude Correction Module. As shown in Fig. 3, the amplitude correction module comprises **n** amplitude correction blocks, which aim to alleviate the influence of JPEG blocking artifacts by rectifying the amplitude. Various amplitude reconstruction units establish direct connections, known as skip links, between the lower and upper layers of the network, thereby preserving essential details. Each amplitude correction unit is composed of two branches, enabling both spatial and frequency domain learning. These branches interact in a dual domain manner, leveraging the spatial convolution's local characteristics and the frequency domain's global features. Specifically, given the feature $\mathbf{F_{Xi}}$, the amplitude correction unit can be defined as follows:

$$\mathbf{F_{spa}} = Conv(\mathbf{F_{Xi}}), \tag{6}$$

$$\mathcal{A}(F_{Xi}), \mathcal{P}(F_{Xi}) = \mathcal{F}(F_{Xi}), \tag{7}$$

$$\mathbf{F_{fre}} = \mathcal{F}^{-1}(Conv(\mathcal{A}(F_{Xi})), \mathcal{P}(F_{Xi})), \tag{8}$$

$$\mathbf{F_{dif}} = \mathbf{F_{spa}} - \mathbf{F_{fre}}, \tag{9}$$

$$\mathbf{F_{out}} = SA(F_{dif}) \cdot F_{spa} + F_{fre}. \tag{10}$$

Here, $conv(.)$ denotes a sequence of convolutional operations followed by rectified linear units (ReLU). The Fourier transform and inverse Fourier transform are represented by $\mathcal{F}(.)$ and $\mathcal{F}^{-1}(.)$, respectively. The term $SA(.)$ denotes the spatial attention mechanism [27]. In this process, the input feature $\mathbf{F_{Xi}}$ is convolved to obtain the refined feature $\mathbf{F_{spa}}$. Additionally, the amplitude and phase components, $\mathcal{A}(F_{Xi})$ and $\mathcal{P}(F_{Xi})$, are obtained by applying Fourier transform to $\mathbf{F_{Xi}}$. While keeping the phase constant, the amplitude features are reconstructed, resulting in $\mathbf{F_{fre}}$ through inverse Fourier transform. To facilitate dual domain interaction, we subtract $\mathbf{F_{fre}}$ from $\mathbf{F_{spa}}$, apply spatial attention to obtain significant weights, and finally combine the weighted fusion of $\mathbf{F_{spa}}$ and $\mathbf{F_{fre}}$ to obtain the output.

Phase Correction Module. Upon passing through the Amplitude Correction Module (ACM), $\mathbf{X_{in}}$ yields the outputs $\mathbf{X_{amp}}$ and $\mathbf{X_{out1}}$ through the Supervised Attention Module (SAM) [33]. The phase of $\mathbf{X_{amp}}$ is then substituted with the phase of $\mathbf{X_{in}}$, resulting in $\mathbf{X_{inv}}$. Subsequently, $\mathbf{X_{inv}}$ and $\mathbf{X_{amp}}$ are concatenated along the channel dimension and transmitted to the Phase Correction Module.

The Phase Correction Module comprises n phase correction blocks, aiming to enhance the high-frequency details of the image by restoring the phase component. The implementation methodology of each phase correction block is identical to that of the amplitude correction block, with the sole distinction being the preservation of the amplitude and solely addressing the phase component. Given the outputs $\mathbf{X_{amp}}$ and $\mathbf{X_{out1}}$ from the SAM, this stage can be defined by the following process:

$$A(X_{out1}), P(X_{out1}) = \mathcal{F}(X_{out1}), \tag{11}$$

$$A(X_{in}), P(X_{in}) = \mathcal{F}(X_{in}), \tag{12}$$

$$\mathbf{X_{inv}} = \mathcal{F}^{-1}(A(X_{out1}), P(X_{in})), \tag{13}$$

$$\mathbf{X_{out2}} = PCM(concat(\mathbf{X_{inv}}, \mathbf{X_{amp}})). \tag{14}$$

3.3   Loss Function

Our loss function consists of three components: spatial loss, amplitude loss, and phase loss, which collectively guide the reconstruction process.

Given the output X_{out2} and the ground truth image GT, the spatial loss $\mathcal{L}spa$ is defined as the L1 distance between the two:

$$\mathcal{L}spa = ||X_{out2} - GT||_1, \tag{15}$$

To guide the amplitude reconstruction module, we utilize the amplitude loss. Specifically, for X_{out1} and the ground truth image GT, the amplitude loss $\mathcal{L}amp$ is computed as:

$$A(X_{out1}), P(X_{out1}) = \mathcal{F}(X_{out1}), \tag{16}$$

$$A(GT), P(GT) = \mathcal{F}(GT), \tag{17}$$

$$\mathcal{L}amp = ||A(X_{out1}) - A(GT)||_1. \tag{18}$$

Similarly, the phase loss is employed to guide the reconstruction of spatial details in the image. Given X_{out2} and the ground truth image GT, the phase loss $\mathcal{L}pha$ is computed as follows:

$$A(X_{out2}), P(X_{out2}) = \mathcal{F}(X_{out2}), \tag{19}$$

$$A(GT), P(GT) = \mathcal{F}(GT), \tag{20}$$

$$\mathcal{L}pha = ||P(X_{out2}) - P(GT)||_1. \tag{21}$$

The total loss $\mathcal{L}$ is the weighted sum of each component:

$$\mathcal{L} = \mathcal{L}spa + \alpha * (\mathcal{L}amp + \mathcal{L}_{pha}), \tag{22}$$

Here, α is set to 0.05 based on empirical observations.

Table 1. PSNR/SSIM/PSNR-B results of our method compared to other nine methods in three grayscale datasets, with the best outcomes being highlighted in red.

Dataset	QF	JPEG			ARCNN			DNCNN			MWCNN			DCSC		
LIVE1	10	27.77	0.773	25.33	28.96	0.808	28.68	29.19	0.812	28.90	29.69	0.825	29.32	29.34	0.818	29.01
	20	30.07	0.851	27.57	31.29	0.873	30.76	31.59	0.880	31.07	32.04	0.889	31.51	31.70	0.883	31.18
	30	31.41	0.885	28.92	32.67	0.904	32.14	32.98	0.909	32.34	33.45	0.915	32.80	33.07	0.911	32.43
BSD500	10	27.80	0.768	25.10	29.10	0.804	28.73	29.21	0.809	28.80	29.61	0.820	29.14	29.32	0.813	28.91
	20	30.05	0.849	27.22	31.28	0.870	30.55	31.53	0.878	30.79	31.92	0.885	31.15	31.63	0.880	30.92
	30	31.37	0.884	28.53	32.67	0.902	31.94	32.90	0.907	31.97	33.30	0.912	32.34	32.99	0.908	32.08
Classic5	10	27.82	0.760	25.21	29.03	0.793	28.76	29.40	0.803	29.13	30.01	0.820	29.59	29.62	0.810	29.30
	20	30.12	0.834	27.50	31.15	0.852	30.59	31.63	0.861	31.19	32.16	0.870	31.52	31.81	0.864	31.34
	30	31.48	0.867	28.94	32.51	0.881	31.98	32.91	0.886	32.38	33.43	0.893	32.63	33.06	0.888	32.49

Dataset	QF	RNAN			RDN			QGAC			FBCNN			Ours		
LIVE1	10	29.63	0.824	29.13	29.70	0.825	29.37	29.51	0.825	29.13	29.75	0.827	29.40	30.08	0.840	30.06
	20	32.03	0.888	31.12	32.10	0.889	31.29	31.83	0.888	31.25	32.13	0.889	31.57	32.47	0.898	32.42
	30	33.45	0.915	32.22	33.54	0.916	32.62	33.20	0.914	32.47	33.54	0.916	32.83	33.91	0.924	33.85
BSD500	10	29.08	0.805	28.48	29.24	0.808	28.71	29.46	0.821	28.97	29.67	0.821	29.22	29.61	0.810	29.53
	20	31.25	0.875	30.27	31.48	0.879	30.45	31.73	0.884	30.93	32.00	0.885	31.19	32.04	0.890	31.67
	30	32.70	0.907	31.33	32.83	0.908	31.60	33.07	0.912	32.04	33.37	0.913	32.39	33.41	0.916	32.97
Classic5	10	29.96	0.819	29.42	30.03	0.819	29.59	29.84	0.812	29.43	30.12	0.822	29.80	31.18	0.838	31.11
	20	32.11	0.869	31.26	32.19	0.870	31.53	31.98	0.869	31.37	32.31	0.872	31.74	33.41	0.886	33.27
	30	33.38	0.892	32.35	33.46	0.893	32.59	33.22	0.892	32.42	33.54	0.894	32.78	34.71	0.907	34.49

4 Experiments

4.1 Experimental Datasets and Implementation Details

In our experiments, we employ DIV2K [1] and Flickr2K [25] as our training data. During training, we randomly crop 256×256 patches from the images. In addition, we have compressed them with JPEG with different quality factors Q = 10, 20 and 30. To optimize the parameters of D2LNet, we adopt the Adam optimizer [17] with $\beta_1 = 0.9$ and $\beta_2 = 0.999$. We train our model on one NVIDIA GeForce GTX 3060 GPU by using PyTorch.

During testing, we evaluate the performance of our model on Classic5 [34], LIVE1 [24], and the test set of BSDS500 [3] for grayscale images. For color images, we do not use the Classic5 but the ICB [8] instead. We use PSNR, SSIM (structural similarity) [30], and PSNR-B (specially designed for JPEG artifacts removal) to quantitatively assess the performance of our JPEG artifacts removal model.

4.2 Results

To evaluate the effectiveness of our model, we conducted experiments on both grayscale and color images. We use the Y channel of YCbCr space for grayscale image comparison, and the RGB channels for color image comparison.

Grayscale JPEG Image Restoration. We first evaluate the effect of our model on the Y-channel JPEG compressed images. For benchmarking purposes, we chose ARCNN [7], DNCNN [35], MWCNN [19], DCSC [11], RNAN [37], RDN [38], QGAC [8], and the powerful FBCNN [16] as reference methods. Table 1 presents the comparison results, with the superior outcomes highlighted in red. Our method consistently achieved the best performance across multiple datasets, as evaluated using three assessment metrics. This observation underscores the considerable potential of incorporating frequency domain information in JPEG restoration (Fig. 4).

Color JPEG Image Restoration. To further showcase the efficacy of our model, we conducted restoration experiments on color datasets. Considering the increased complexity of color image restoration, we selected QGAC [8] and FBCNN [16] methods for comparison. As shown in Table 2, the results clearly demonstrate the superiority of our approach in color image restoration, reinforcing the robustness of our model and the significance of frequency domain information.

Table 2. PSNR/SSIM/PSNR-B results of different methods on the three color datasets, with the best outcomes being highlighted in red.

Dataset	QF	JPEG			QGAC			FBCNN			Ours		
LIVE1	10	25.69	0.743	24.20	27.62	0.804	27.43	27.77	0.803	27.51	27.82	0.805	27.80
	20	28.06	0.826	26.49	29.88	0.868	29.56	30.11	0.868	29.70	30.14	0.871	30.11
	30	29.37	0.861	27.84	31.17	0.896	30.77	31.43	0.897	30.92	31.49	0.899	31.33
BSD500	10	25.84	0.741	24.13	27.74	0.802	27.47	27.85	0.799	27.52	27.66	0.778	27.61
	20	28.21	0.827	26.37	30.01	0.869	29.53	30.14	0.867	29.56	30.15	0.869	29.75
	30	29.57	0.865	27.72	31.33	0.898	30.70	31.45	0.897	30.72	31.52	0.897	30.97
ICB	10	29.44	0.757	28.53	32.06	0.816	32.04	32.18	0.815	32.15	33.02	0.829	32.97
	20	32.01	0.806	31.11	34.13	0.843	34.10	34.38	0.844	34.34	34.57	0.847	34.49
	30	33.95	0.831	32.35	35.07	0.857	35.02	35.41	0.857	35.35	35.55	0.856	35.44

Fig. 4. Visual comparisons of JPEG image "BSD: 3096" with $QF = 10$.

4.3 Ablation Studies

To provide additional insights into the functionality of our proposed module, we performed ablation experiments on three color image datasets with a quantization factor of Q10. The ablation experiments were designed to evaluate the impact of the Amplitude Correction Block and the Phase Correction Block. Two groups of experiments were conducted, wherein we replaced these units with res-blocks having similar parameter settings. This allowed us to assess the effects of eliminating the amplitude reconstruction and phase reconstruction functionalities.

Amplitude Correction Block. The Amplitude Correction Block primarily focuses on the image's amplitude, aiming to mitigate the effects of JPEG compression and alleviate block artifacts. In the first set of experiments, we substituted the amplitude reconstruction unit with a resblock, as indicated in the first row of the Table 3. It is evident from the results that various metrics exhibit a certain degree of decline, providing evidence that processing the amplitude effectively restores the compression-induced degradation in the image.

Phase Correction Block. The Phase Correction Block is designed to restore fine details in the image, bringing the texture edges closer to the ground truth. In the second set of experiments, we substituted the phase reconstruction unit with a resblock, as presented in the second row of the Table 3. Upon removing the phase reconstruction unit, the evaluation metrics exhibited a decrease across all three datasets. This observation underscores the significance of phase information in the image restoration task.

Table 3. The results of the ablation experiments conducted on the three datasets.

CONFIG	ACB	PCB	BSD500			LIVE1			ICB		
			PSNR	SSIM	PSNR-B	PSNR	SSIM	PSNR-B	PSNR	SSIM	PSNR-B
(I)	✗	✓	27.54	0.776	27.49	27.72	0.801	27.69	31.74	0.813	31.73
(II)	✓	✗	27.47	0.773	27.44	27.71	0.799	27.63	31.75	0.813	31.74
Ours	✓	✓	27.66	0.778	27.61	27.82	0.805	27.80	33.02	0.829	32.97

5   Conclusions

In this paper, we propose a Dual-Domain Learning Network for JPEG artifacts removal (D2LNet). In contrast to previous JPEG artifacts removal methods performed directly in the spatial domain, we combined the information from the frequency domain. The Amplitude Correction Module (ACM) and Phase Correction Module (PCM) have effectively achieved information interaction between the spatial and frequency domains. Extensive experiments on the grayscale JPEG images and the color JPEG images demonstrate the effectiveness and generalizability of our proposed D2LNet. Nevertheless, our approach is sensitive to the quality factor (QF), which is clearly a drawback. We should delve into the research on utilizing frequency domain information to achieve flexible blind JPEG artifacts removal.

Acknowledgements. This work is supported by National MCF Energy R&D Program of China (Grant No: 2018YFE0302100).

Author contributions. Guang Yang and Lu Lin are contributed equally to this work.

References

1. Agustsson, E., Timofte, R.: NTIRE 2017 challenge on single image super-resolution: dataset and study. In: Proceedings of the IEEE Conference on Computer Vision and Pattern Recognition Workshops, pp. 126–135 (2017)
2. Ahmed, N., Natarajan, T., Rao, K.R.: Discrete cosine transform. IEEE Trans. Comput. **100**(1), 90–93 (1974)
3. Arbelaez, P., Maire, M., Fowlkes, C., Malik, J.: Contour detection and hierarchical image segmentation. IEEE Trans. Pattern Anal. Mach. Intell. **33**(5), 898–916 (2010)
4. Cavigelli, L., Hager, P., Benini, L.: CAS-CNN: a deep convolutional neural network for image compression artifact suppression. In: 2017 International Joint Conference on Neural Networks (IJCNN), pp. 752–759. IEEE (2017)
5. Chen, Y., Pock, T.: Trainable nonlinear reaction diffusion: a flexible framework for fast and effective image restoration. IEEE Trans. Pattern Anal. Mach. Intell. **39**(6), 1256–1272 (2016)
6. Chi, L., Jiang, B., Mu, Y.: Fast Fourier convolution. In: Advances in Neural Information Processing Systems, vol. 33, pp. 4479–4488 (2020)
7. Dong, C., Deng, Y., Loy, C.C., Tang, X.: Compression artifacts reduction by a deep convolutional network. In: Proceedings of the IEEE International Conference on Computer Vision, pp. 576–584 (2015)
8. Ehrlich, M., Davis, L., Lim, S.-N., Shrivastava, A.: Quantization guided JPEG artifact correction. In: Vedaldi, A., Bischof, H., Brox, T., Frahm, J.-M. (eds.) ECCV 2020. LNCS, vol. 12353, pp. 293–309. Springer, Cham (2020). https://doi.org/10.1007/978-3-030-58598-3_18
9. Ehrlich, M., Davis, L.S.: Deep residual learning in the JPEG transform domain. In: Proceedings of the IEEE/CVF International Conference on Computer Vision, pp. 3484–3493 (2019)

10. Foi, A., Katkovnik, V., Egiazarian, K.: Pointwise shape-adaptive DCT for high-quality denoising and deblocking of grayscale and color images. IEEE Trans. Image Process. **16**(5), 1395–1411 (2007)
11. Fu, X., Zha, Z.J., Wu, F., Ding, X., Paisley, J.: JPEG artifacts reduction via deep convolutional sparse coding. In: Proceedings of the IEEE/CVF International Conference on Computer Vision, pp. 2501–2510 (2019)
12. Galteri, L., Seidenari, L., Bertini, M., Del Bimbo, A.: Deep generative adversarial compression artifact removal. In: Proceedings of the IEEE International Conference on Computer Vision, pp. 4826–4835 (2017)
13. Galteri, L., Seidenari, L., Bertini, M., Del Bimbo, A.: Deep universal generative adversarial compression artifact removal. IEEE Trans. Multimed. **21**(8), 2131–2145 (2019)
14. Guo, J., Chao, H.: Building Dual-domain representations for compression artifacts reduction. In: Leibe, B., Matas, J., Sebe, N., Welling, M. (eds.) ECCV 2016. LNCS, vol. 9905, pp. 628–644. Springer, Cham (2016). https://doi.org/10.1007/978-3-319-46448-0_38
15. Huang, J., et al.: Deep Fourier-based exposure correction network with spatial-frequency interaction. In: Avidan, S., Brostow, G., Cissé, M., Farinella, G.M., Hassner, T. (eds.) ECCV 2022. LNCS, vol. 13679, pp. 163–180. Springer, Cham (2022). https://doi.org/10.1007/978-3-031-19800-7_10
16. Jiang, J., Zhang, K., Timofte, R.: Towards flexible blind jpeg artifacts removal. In: Proceedings of the IEEE/CVF International Conference on Computer Vision, pp. 4997–5006 (2021)
17. Kingma, D.P., Ba, J.: Adam: a method for stochastic optimization. arXiv preprint arXiv:1412.6980 (2014)
18. Li, Z., et al.: Fourier neural operator for parametric partial differential equations. arXiv preprint arXiv:2010.08895 (2020)
19. Liu, P., Zhang, H., Zhang, K., Lin, L., Zuo, W.: Multi-level wavelet-CNN for image restoration. In: Proceedings of the IEEE Conference on Computer Vision and Pattern Recognition Workshops, pp. 773–782 (2018)
20. Liu, T., Cheng, J., Tan, S.: Spectral Bayesian uncertainty for image super-resolution. In: Proceedings of the IEEE/CVF Conference on Computer Vision and Pattern Recognition, pp. 18166–18175 (2023)
21. Mao, X., Shen, C., Yang, Y.B.: Image restoration using very deep convolutional encoder-decoder networks with symmetric skip connections. In: Advances in Neural Information Processing Systems, vol. 29 (2016)
22. Mao, X., Liu, Y., Shen, W., Li, Q., Wang, Y.: Deep residual Fourier transformation for single image deblurring. arXiv preprint arXiv:2111.11745 (2021)
23. Ren, J., Liu, J., Li, M., Bai, W., Guo, Z.: Image blocking artifacts reduction via patch clustering and low-rank minimization. In: 2013 Data Compression Conference, pp. 516–516. IEEE (2013)
24. Sheikh, H.: Live image quality assessment database release 2 (2005). http://live.ece.utexas.edu/research/quality
25. Timofte, R., Agustsson, E., Van Gool, L., Yang, M.H., Zhang, L.: NTIRE 2017 challenge on single image super-resolution: methods and results. In: Proceedings of the IEEE Conference on Computer Vision And Pattern Recognition Workshops, pp. 114–125 (2017)
26. Wallace, G.K.: The JPEG still picture compression standard. IEEE Trans. Consum. Electr. **38**(1), xviii–xxxiv (1992)
27. Wang, H., Fan, Y., Wang, Z., Jiao, L., Schiele, B.: Parameter-free spatial attention network for person re-identification. arXiv preprint arXiv:1811.12150 (2018)

28. Wang, X., Fu, X., Zhu, Y., Zha, Z.J.: JPEG artifacts removal via contrastive representation learning. In: Avidan, S., Brostow, G., Cissé, M., Farinella, G.M., Hassner, T. (eds.) ECCV 2022. LNCS, vol. 13677, pp. 615–631. Springer, Cham (2022). https://doi.org/10.1007/978-3-031-19790-1_37

29. Wang, Z., Liu, D., Chang, S., Ling, Q., Yang, Y., Huang, T.S.: D3: deep dual-domain based fast restoration of JPEG-compressed images. In: Proceedings of the IEEE Conference on Computer Vision and Pattern Recognition, pp. 2764–2772 (2016)

30. Wang, Z., Bovik, A.C., Sheikh, H.R., Simoncelli, E.P.: Image quality assessment: from error visibility to structural similarity. IEEE Trans. Image Process. 13(4), 600–612 (2004)

31. Xu, L., Ren, J.S., Liu, C., Jia, J.: Deep convolutional neural network for image deconvolution. In: Advances in Neural Information Processing Systems, vol. 27 (2014)

32. Xu, Q., Zhang, R., Zhang, Y., Wang, Y., Tian, Q.: A Fourier-based framework for domain generalization. In: Proceedings of the IEEE/CVF Conference on Computer Vision and Pattern Recognition, pp. 14383–14392 (2021)

33. Zamir, S.W., et al.: Multi-stage progressive image restoration. In: Proceedings of the IEEE/CVF Conference on Computer Vision and Pattern Recognition, pp. 14821–14831 (2021)

34. Zeyde, R., Elad, M., Protter, M.: On single image scale-up using sparse-representations. In: Boissonnat, J.-D., et al. (eds.) Curves and Surfaces 2010. LNCS, vol. 6920, pp. 711–730. Springer, Heidelberg (2012). https://doi.org/10.1007/978-3-642-27413-8_47

35. Zhang, K., Zuo, W., Chen, Y., Meng, D., Zhang, L.: Beyond a gaussian denoiser: residual learning of deep CNN for image denoising. IEEE Trans. Image Process. 26(7), 3142–3155 (2017)

36. Zhang, X., Yang, W., Hu, Y., Liu, J.: DMCNN: dual-domain multi-scale convolutional neural network for compression artifacts removal. In: 2018 25th IEEE International Conference on Image Processing (ICIP), pp. 390–394. IEEE (2018)

37. Zhang, Y., Li, K., Li, K., Zhong, B., Fu, Y.: Residual non-local attention networks for image restoration. arXiv preprint arXiv:1903.10082 (2019)

38. Zhang, Y., Tian, Y., Kong, Y., Zhong, B., Fu, Y.: Residual dense network for image restoration. IEEE Trans. Pattern Anal. Mach. Intell. 43(7), 2480–2495 (2020)

39. Zini, S., Bianco, S., Schettini, R.: Deep residual autoencoder for blind universal JPEG restoration. IEEE Access 8, 63283–63294 (2020)

Graph-Based Vehicle Keypoint Attention Model for Vehicle Re-identification

Yunlong Li, Zhihao Wu, Youfang Lin, and Kai Lv[✉]

Beijing Key Laboratory of Traffic Data Analysis and Mining, School of Computer
and Information Technology, Beijing Jiaotong University, Beijing 100044, China
lvkai@bjtu.edu.cn

Abstract. Vehicle re-identification is the task of locating a particular
vehicle image among a set of images of vehicles captured from different cameras. In recent years, many methods focus on learning distinctive global features by incorporating keypoint details to improve re-
identification accuracy. However, these methods do not take into account
the relation between different keypoints and the relation between keypoints and the overall vehicle. To address this limitation, we propose
the Graph-based Vehicle Keypoint Attention (GVKA) model that integrates keypoint features and two relation components to yield robust
and discriminative representations of vehicle images. The model extracts
keypoint features using a pre-trained model, models the relation among
keypoint features using a Graph Convolutional Network, and employs
cross-attention to highlight important areas of the vehicle and establish the relation between keypoint features and the overall vehicle. Our
experimental results on three large-scale datasets demonstrate the effectiveness of our proposed method.

Keywords: Vehicle re-identification · Keypoint features · Graph ·
Cross-attention

1 Introduction

Vehicle Re-Identification (ReID) is a task that involves locating a particular
vehicle image within a set of images of vehicles captured from various cameras. This area of study has gained significant attention due to its wide range
of applications in the field of intelligent transportation, such as public security
and smart cities [19,35–37]. In recent years, much attention has been given to
developing methods for vehicle ReID that focus on learning distinctive global
features [12,20–22]. Yan *et al.* [23] propose a multi-task learning framework that
leverages multi-dimensional information about vehicles to perform both vehicle
classification and similarity ranking [23]. Liu *et al.* [13] fuse global features with
license plate and contextual information. Several studies in vehicle ReID [3,20]
have explored the use of keypoint information as a complement to global features
in improving the performance of their models. The keypoint provide locally significant information as well as orientation information, which can significantly

B. Luo et al. (Eds.): ICONIP 2023, CCIS 1967, pp. 569–580, 2024.
https://doi.org/10.1007/978-981-99-8178-6_43

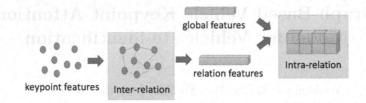

Fig. 1. The example of the relation between different keypoint and the relation between keypoint and the entire vehicle.

enhance the results. Khorramshahi *et al.* [3] propose a novel dual-path adaptive attention model to extract keypoint features from a vehicle image and then concatenate the keypoint features with the global feature to obtain the final features. Wang *et al.* [20] extract the keypoint from four different views and then concatenate keypoint features to global features to improve the result of their model. Those approaches get successful outcomes in vehicle re-identification tasks. The above methods primarily emphasize the extraction of keypoint and the simple concatenation of these features with global features. However, these methods do not adequately consider the interconnections between differnet keypoint, as well as the relation between keypoint and the overall vehicle, which can undermine their robustness. In order to enhance the robustness of the model, as illustrated in Fig. 1, we seek to develop a framework that can effectively capture the inter-relation and the intra-relation to yield more robust features. In light of the significance of keypoint features and the need to capture their relation with the overall vehicle, we propose the Graph-based Vehicle Keypoint Attention (GVKA) model. This model integrates keypoint features and two relation components to yield robust and discriminative representations of vehicle images. Firstly, keypoint features are extracted from vehicle images through the use of a pre-trained model. Subsequently, a Graph Convolutional Network (GCN) is applied to model the relation among different keypoint features, utilizing a relation matrix to depict the graph topology. Finally, cross-attention between the keypoint features and the global features is employed to emphasize important areas of the vehicle and establish the relation between keypoint features and the overall vehicle. Experimental results on three large-scale datasets demonstrate the effectiveness of our proposed method. The contributions of our method can be summarised as:

- We establish a topological relation matrix to model the graph structure and employ cross-attention to establish the relation between keypoint and the overall vehicle.
- To derive more representative features, we introduce a Graph-based Vehicle Keypoint Attention model that captures the relation between different keypoint features extracted by a pre-trained model.
- Our method demonstrates superior performance compared to other existing methods through comprehensive experiments conducted on three widely used vehicle re-identification benchmarks and ablation studies.

2 Related Works

Deep learning methods have dominated the vehicle ReID task [32–34]. In this section, we review two mainstream approaches related to our method: local region feature approaches and attention-based approaches.

Early deep learning methods directly integrated vehicle basic information such as color and spatiotemporal information to enhance global features [29,30]. However, color and spatiotemporal information are not always exits so those methods are easy to over-fitting and not general. In recent years, some approaches utilize some local features to improve their methods. [13] used multi-modal information, such as license plate, camera locations, to enhance vision features, which achieving good performance. [20] provides a dataset with 20 vehicle keypoint and the local region features extracted using these keypoint are divided into four different branches according to the different view perspective, and then integrating them into global features to get promising results. Furthermore, [3] also use a model to extract keypoint on a vehicle image, and then simply concatenate these keypoint features to enhance global features.

The attention mechanism has also attracted more and more people's interest in ReID, and many methods have begun to use attention-based methods to improve their performance. [25] proposed a two-branch adaptive attention network, the combination of non-local attention and channel attention, which can effectively identify features and eliminate the negative influence brought by the background of vehicle images to obtain promising results. [31] proposed an attention model guided by local regions. First, the target detection model is used to extract the key parts of vehicle image such as brand, car lights, logo, and use them as candidate search regions in the learning process, finally conducting competitive results.

However, those methods do not consider the relation between different keypoint features and the relation between keypoint features and the global features. To solve these, our method first extract local keypoint features and then using a graph convolutional network to establish the relation between different keypoint features. Furthermore, our network employs self-attention and cross-attention blocks to highlight important areas of the vehicle and establish the relation between keypoint features and the global features, which can obtain more robust and generative features.

3 Proposed Method

3.1 Overall Framework

To enhance the robustness of the vehicle re-identification model, we propose the Graph-based Vehicle Keypoint Attention (GVKA) model, as illustrated in Fig. 2. Our approach consists of two branches, one for learning global features and the other for extracting keypoint features and establishing the relation between them. The GVKA incorporates three components: Relation Matrix, Vehicle Keypoint Graph model, and Cross-Attention. Our method not only establishes the relation between keypoint, but also the relation between keypoint and the entire vehicle, thus allowing the model to learn more robust features. The GVKA is optimized through the Center Loss and Triplet Loss.

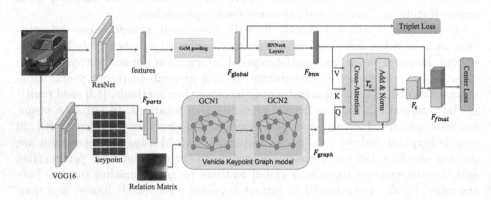

Fig. 2. An overview of the Graph-based Vehicle Keypoint Attention (GVKA) model. It consists of two branches: (1) The global branch which extracts the global features of a vehicle, and (2) the keypoint branch which first extracts the keypoint of a vehicle and then masks these keypoint with shallow features obtained through a pre-trained model [3].

3.2   Relation Matrix

We denote the proposed Graph Convolutional Network (GCN) as $G = \{V, E\}$, where $V = \{v_1, v_2, ..., v_n\}$ represents the nodes of the network with a total of n nodes and E denotes the relation between the nodes. In our approach, each node corresponds to a keypoint feature extracted by a pre-trained keypoint detection model. The topology of the graph is represented by an adjacency matrix A, where the weights on each edge $(V_i, V_j) \in E$ denote the relation between the nodes. The relation matrix A can be formed as:

$$A = \begin{bmatrix} A_{ii} & A_{ij} \\ A_{ji} & A_{jj}, \end{bmatrix}. \tag{1}$$

where $A_{ij} \in R^{n \times n}$ is the keypoint features correlation matrix. Following the description in [26], we decided to construct the relation matrix A. The probability of occurrence of keypoint feature j when the feature i occurs is represented by A_{ij}. The formula is as follows:

$$A_{ij} = \frac{C_{ij}}{L_i}, \tag{2}$$

where L_i represents the frequency of the keypoint i appears in the training set, and C_{ij} stands for the co-occurrence times of keypoint i and j in the training set.

3.3   Keypoint Feature Extractor

Firstly, we utilze a pre-trained keypoint detection model [3] to obtain vehicle keypoint coordinates. Then, we take those obtained coordinate as the center point to get a circle area $P = \{P_i\}_{i=1}^{20}$ of radius r, where P_i is a binary mask with the same size as input image. Then, we multiply the mask P_i with shallow features $F_{shallow}$ from the pre-trained model to get the keypoint feature $F_{parts} \in R^{C \times W \times H}$. And each location features F_{parts_i} can be computed as below:

$$F_{parts_i} = P_i(F_{shallow}). \tag{3}$$

3.4   Vehicle Keypoint Graph Model

To obtain vehicle keypoint features F_{parts}, we firstly utilize a pre-trained vehicle keypoint detection model [3] to get the keypoint and then mask the keypoint with shallow features from the pre-trained model. Then to encoding these keypoint features into more representativeness and discriminative, we decide to choose GCN to establish the relation between each keypoint features. Formally, GCN can be thought of as network that propagate in multiple layers on a graph G. Generally, each layer of GCN can be defined as a function $f = (I, A)$ which can update nodes features $I \in R^{N \times d}$ by propagating information between nodes under the constraints of the relation Matrix A. After the input I passing through the k-th layers of GCN, we get $H^{(k)} \in R^{N \times d}$ which denotes the feature matrix. Then following the formulation which proposed in [26], we take feature matrix $H^{(k)}$ and relation matrix A as input to pass through the $(k + 1)$-th layers of GCN. After that, we can get $H^{(k+1)} \in R^{N \times d}$. Hence, according to [26], we can formulate every GCN layer as follows:

$$H^{(k+1)} = \sigma(\hat{A}H^{(k)}W^{(k)}), \tag{4}$$

where σ is the activation function, in our method we use LeakyRelu. $W^{(k)} \in R^{d \times d}$ is a learnable linear transformation and $\hat{A}$ is the normalization of the A matrix, which can be formulated as:

$$\hat{A} = \tilde{D}^{-\frac{1}{2}}(A + I)\tilde{D}^{-\frac{1}{2}}, \tag{5}$$

$$\tilde{D} = D + I, \tag{6}$$

where $I \in R^{N \times N}$ is the identity matrix and D is the diagonal of the degree matrix A. We add I to D, similar to a residual connection. Hence, our aim is to learn these learnable parameters $W = \{W_1, W_2, ..., W_n\}$.

3.5   Cross-Attention

Inspired by the success of attention and non-local feature learning mechanism in many areas of computer vision and to build the relation between keypoint features and global features. During the training phase, we get F_{global} after we pass the features which captured from Resnet to the generalized-mean (GeM) pooling layer and get F_{bnn} which obtained by passing the global features F_{global} to the BNNeck Layers. As shown in Fig. 2, when we obtained the features F_{bnn}, we first flatten both width and height dimensions into a sequence $\in R^{HW \times C}$. Then by using different FC layers, F_{bnn} is transformed to Keys ($K \in R^{HW \times C}$) and Values ($V \in R^{HW \times C}$) and the output of GCN F_{graph} is mapped to Queries ($Q \in R^{L \times C}$), respectively. After obtaining Q, K, V, following the processes mentioned in [28], we first calculate the similarity between the F_{graph} and the global features F_{bnn}. The calculation formula is as follows:

$$S(Q, K) = s(\frac{QK^T}{\sqrt{C}}), \tag{7}$$

$$T_c = SV, \tag{8}$$

$$F_c = F_{graph} + N(T_c), \tag{9}$$

where, s represents softmax function and N represents LayerNorm. C denotes the channel dimensions. The $S \in R^{L \times HW}$. Here, the similarity S_{ij} means how important the j-th local feature to global features F_{bnn}. In order to obtain different semantic information, we compute S with Mutli-Head Attention. After obtaining the similarity matrix S, we multiply it by V to obtain the T_c. Furthermore, we add T_c to the F_{graph} to get the output F_c. Then, we concatenate the output F_c with the F_{bnn} to get the final features F_{final}. In training process, the final loss function of our method is as follows:

$$L = L_{triplet} + L_{center}. \tag{10}$$

4   Experiment

4.1   Datasets and Settings

We conduct experiments on three public benchmarks for vehicle re-identification: VeRi-776, VehicleID, and VERI-Wild. VeRi-776 is a dataset containing more than 50,000 vehicle images captured by 20 cameras, including 8 different perspectives. It includes 37,778 images of 576 different vehicles in the training set and 11,579 images of 200 different vehicles in the test set. VehicleID is another public dataset for vehicle re-identification, containing 221,567 images of 26,328 different vehicles from 2 different perspectives. The VehicleID dataset includes 110,178 images for training and 111,585 images for testing, and the test set is split into three sizes (small, medium, and large). VERI-Wild is the largest dataset among the three benchmarks, containing 416,314 images of 40,671 vehicles. The training set includes 277,797 images of 30,671 different vehicles, and the test set is also split into three sizes (small, medium, and large).

Table 1. Performance on VeRi-776 and VehicleID.

Method	VeRi-776		VehicleID					
			Small		Medium		Large	
	mAP	R1	R1	R5	R1	R5	R1	R5
FDA-Net [2]	55.5	84.3	59.8	77.1	55.5	74.7	-	-
GSTE [18]	59.5	96.2	75.9	84.2	74.8	83.6	74.0	82.7
AAVER [3]	58.52	88.68	74.7	93.8	68.6	90.0	63.5	85.6
MAVN [14]	72.53	92.59	72.58	83.07	-	-	-	-
RPN [5]	74.3	94.3	78.4	92.3	75.0	88.3	74.2	86.4
VSCR [6]	75.53	94.11	74.58	87.12	-	-	-	-
PVEN [17]	79.5	95.6	84.7	97.0	80.6	94.5	77.8	92.0
Baseline [8]	80.73	96.66	83.87	97.05	80.93	94.21	77.56	91.81
DFNet [9]	80.97	97.08	84.76	96.22	80.61	94.10	**79.15**	**92.86**
PANet+PMNet [10]	81.60	96.50	85.30	97.30	80.50	94.50	77.60	92.20
GVKA(ours)	**82.06**	**97.30**	**85.32**	**97.79**	**81.76**	**94.67**	78.48	92.84

Table 2. Performance on VERI-Wild.

Methods	Small		Medium		Large	
	mAP	Rank-1	mAP	Rank-1	mAP	Rank-1
DRDL [12]	22.5	57.0	19.3	51.9	14.8	44.6
GoogLeNet [16]	24.3	57.2	24.2	53.2	21.5	44.6
GSTE [18]	31.42	60.5	52.1	26.2	45.4	19.5
FDA-Net [2]	35.1	64.0	29.8	57.8	22.8	49.4
VARID [7]	75.4	75.3	70.8	68.8	64.2	63.2
VSCR [6]	75.79	93.13	70.47	89.68	64.19	86.29
PVEN [17]	82.5	95.7	77.0	94.8	69.7	92.3
DFNet [9]	83.09	94.79	77.27	93.22	69.85	89.38
Baseline [8]	85.57	96.12	81.20	94.86	74.42	92.2
GVKA(ours)	**87.55**	**96.58**	**82.29**	**95.24**	**75.54**	**93.38**

4.2   Implement Details and Evaluation Protocols

In the training of our model, the size of the vehicle images utilized is set to 256×256. Additionally, it is important to note that the pre-trained keypoint detection model is fixed during the training process. The learning rate is initially initialized at 0.01 and is subject to a cosine learning rate decay, ultimately reaching a final value of 1.04e-4. The training methodology closely follows the approach presented in the fastreid [8] baseline.

In testing, we utilize mean Average Precision (mAP) and rank-k accuracy as our evaluation metrics. The mAP metric is appropriate for n *vs.* n scenarios where each probe image has several corresponding gallery images. The rank-k accuracy measures the proportion of probe images that are correctly matched with one of the top k images in the gallery set.

4.3 Ablation Study

To demonstrate the effectiveness of our proposed method, which incorporates Graph Convolutional Network (GCN) and cross-attention mechanisms, we conduct ablation experiments on the VeRi-776 dataset. The results are presented in Table 3. The baseline model is based on fastreid [8]. The terms "baseline + GCN" and "baseline + cross-attention" denote the implementation of the two main components, respectively. The full proposed method, GVKA, incorporates both components. The results indicate that the best performance is achieved when both GCN and cross-attention are utilized together. Each component individually enhances the accuracy of mAP, rank-1, and rank-5. Moreover, it can be observed that GCN has a greater impact on performance compared to cross-attention, suggesting the significance of inter-keypoint feature relation.

Table 3. Ablation Study on VeRi-776.

Methods	mAP	Rank-1	Rank-5
baseline	80.73	96.66	98.57
baseline + GCN	81.78	97.08	98.69
baseline + cross-attention	81.42	96.72	98.33
GVKA(ours)	**82.06**	**97.3**	**98.95**

Furthermore, in order to verify the effect of the size of the mask area, we conduct some experiments on the radius r of the mask area on VeRi-776 dataset. As shown in Table 4, at the begining as r increases, mAP continues to increase and then decrease until $r=16$ to achieve the best results. This is due to the fact that the obtained keypoint feature information increases continuously with the

Table 4. mAP comparison of radius size on VeRi-776.

radius size	mAP	Rank-1	Rank-5
baseline	80.73	96.66	98.57
$r = 5$	81.32	96.08	97.58
$r = 7$	81.41	96.32	98.29
$r = 9$	81.72	97.12	98.31
$r = 16$	**82.06**	**97.3**	**98.95**
$r = 25$	81.65	97.10	98.24

increase of the radius r. This is due to the lack of capability to provide keypoint features when r is small. However, when the radius size become large, the local features will contain more noise and make the model more difficult to converge.

4.4 Parameter Analysis

We also aim to determine the optimal values of the model hyper-parameters. To this end, we perform additional experiments to examine the influence of the hyper-parameters K_{gcn} (the number of GCN layers) and $K_{cross-attn}$ (the number of cross-attention blocks). The results, as shown in Tables 5, indicate that the best overall performance in terms of mAP, rank-1, and rank-5 is achieved when $K_{gcn} = 2$ and $K_{cross-attn} = 3$.

Table 5. Analysis of K_{gcn} and $K_{cross-attn}$ on VeRi-776.

hyper parameter	setting	mAP	Rank-1	Rank-5
K_{gcn}	1	81.24	96.90	98.45
	2	**82.06**	**97.3**	**98.95**
	3	81.38	96.60	98.75
$K_{cross-attn}$	1	81.70	97.08	98.39
	2	81.83	97.18	98.69
	3	**82.06**	**97.3**	**98.95**
	4	80.17	96.72	98.29

4.5 Comparison Results

This section presents a comparative evaluation of our proposed approach against other existing methods on three publicly available benchmarks. The results in Table 1 and Table 2 indicate that our proposed model yields highly competitive performance when compared to other state-of-the-art methods on the VeRi-776 and VERI-Wild datasets. These results confirm that incorporating the relation between keypoint features, as well as the relation between keypoint features and global features, is more effective in deriving robust features as opposed to simply concatenating keypoint features with global features.

From the Table 1, on VehicleID, our model is outperformed by DFNet [9] on the large test set. This may be attributed to the fact that our method is better suited for datasets with richer viewpoint information. However, our method still achieves superior accuracy in the other test sets than other state-of-the-arts. Hence, it can also illustrate that our method is robustness and generative.

4.6   Visualization of the Results

In this section, we present some query examples with Top 5 retrieval results in Fig. 3. Our model obtains better results than the baseline because our method can pay more attention to keypoint information. For example, in the first row, the baseline gets false result due to the similarity of vehicle logo. But our model extracts features are more robust to the keypoint information, leading to better retrieval results.

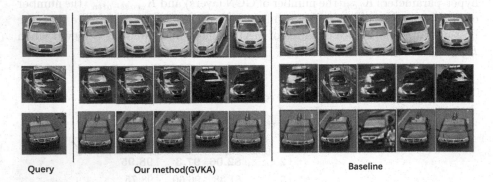

Query   Our method(GVKA)   Baseline

Fig. 3. Top 5 retrieval results of some queries on VeRi-776. Note that the red/blue boxes denote true/false retrieval results, respectively. (Color figure online)

5   Conclusion

This paper presents a novel Graph-based Vehicle keypoint Attention model (GVKA), which contains three essential components: Relation Matrix, Vehicle Keypoint Graph model, and Cross-Attention. Firstly, we utilize a Relation Matrix and Vehicle Keypoint Graph model to establish the relation between distinct keypoint features. Then, we use cross-attention to establish the connection between keypoint features and global features. To demonstrate the efficacy of our proposed model, we conduct experiments on three public benchmarks. The results of our experiments are promising and serve to support the effectiveness of our model.

Acknowledgements. This work was supported by the National Natural Science Foundation of China under Grant 62206013.

References

1. Shen, Y., Xiao, T., Li, H., Yi, S., Wang, X.: Learning deep neural networks for vehicle re-id with visual-spatio-temporal path proposals. In: Proceedings of the IEEE International Conference on Computer Vision, pp. 1900–1909 (2017)

2. Lou, Y., Bai, Y., Liu, J., Wang, S., Duan, L.: Veri-wild: a large dataset and a new method for vehicle re-identification in the wild. In: Proceedings of the IEEE/CVF Conference on Computer Vision and Pattern Recognition, pp. 3235–3243 (2019)
3. Khorramshahi, P., Kumar, A., Peri, N., Rambhatla, S.S., Chen, J.C., Chellappa, R.: A dual-path model with adaptive attention for vehicle re-identification. In: Proceedings of the IEEE/CVF International Conference on Computer Vision, pp. 6132–6141 (2019)
4. Tang, Z., et al.: Pamtri: pose-aware multi-task learning for vehicle re-identification using highly randomized synthetic data. In: Proceedings of the IEEE/CVF International Conference on Computer Vision, pp. 211–220 (2019)
5. He, B., Li, J., Zhao, Y., Tian, Y.: Part-regularized near-duplicate vehicle re-identification. In: Proceedings of the IEEE/CVF Conference on Computer Vision and Pattern Recognition, pp. 3997–4005 (2019)
6. Teng, S., Zhang, S., Huang, Q., Sebe, N.: Viewpoint and scale consistency reinforcement for UAV vehicle re-identification. Int. J. Comput. Vision 129(3), 719–735 (2021)
7. Li, Y., Liu, K., Jin, Y., Wang, T., Lin, W.: Varid: viewpoint-aware re-identification of vehicle based on triplet loss. IEEE Trans. Intell. Transp. Syst. 23(2), 1381–1390 (2020)
8. He, L., Liao, X., Liu, W., Liu, X., Cheng, P., Mei, T.: Fastreid: a pytorch toolbox for general instance re-identification, arXiv preprint arXiv:2006.02631 (2020)
9. Bai, Y., Liu, J., Lou, Y., Wang, C., Duan, L.-Y.: Disentangled feature learning network and a comprehensive benchmark for vehicle re-identification. IEEE Trans. Pattern Anal. Mach. Intell. 44(10), 6854–6871 (2021)
10. Tang, L., Wang, Y., Chau, L.-P.: Weakly-supervised part-attention and mentored networks for vehicle re-identification. IEEE Trans. Circuits Syst. Video Technol. 32(12), 8887–8898 (2022)
11. He, S., Luo, H., Wang, P., Wang, F., Li, H., Jiang, W.: Transreid: transformer-based object re-identification. In: Proceedings of the IEEE/CVF International Conference on Computer Vision, pp. 15013–15022 (2021)
12. Liu, H., Tian, Y., Yang, Y., Pang, L., Huang, T.: Deep relative distance learning: tell the difference between similar vehicles. In: Proceedings of the IEEE Conference on Computer Vision and Pattern Recognition, pp. 2167–2175 (2016)
13. Liu, X., Liu, W., Mei, T., Ma, H.: Provid: progressive and multimodal vehicle reidentification for large-scale urban surveillance. IEEE Trans. Multimedia 20(3), 645–658 (2017)
14. Teng, S., Zhang, S., Huang, Q., Sebe, N.: Multi-view spatial attention embedding for vehicle re-identification. IEEE Trans. Circuits Syst. Video Technol. 31(2), 816–827 (2020)
15. Zhou, Y., Shao, L.: Aware attentive multi-view inference for vehicle re-identification. In: Proceedings of the IEEE Conference on Computer Vision and Pattern Recognition, pp. 6489–6498 (2018)
16. Yang, L., Luo, P., Change Loy, C., Tang, X.: A large-scale car dataset for fine-grained categorization and verification. In: Proceedings of the IEEE Conference on Computer Vision and Pattern Recognition, pp. 3973–3981 (2015)
17. Meng, D., et al.: Parsing-based view-aware embedding network for vehicle re-identification. In: Proceedings of the IEEE/CVF Conference on Computer Vision and Pattern Recognition, pp. 7103–7112 (2020)
18. Bai, Y., Lou, Y., Gao, F., Wang, S., Wu, Y., Duan, L.-Y.: Group-sensitive triplet embedding for vehicle reidentification. IEEE Trans. Multimedia 20(9), 2385–2399 (2018)

19. Sheng, H., et al.: Near-online tracking with co-occurrence constraints in blockchain-based edge computing. IEEE Internet Things J. **8**(4), 2193–2207 (2020)
20. Wang, Z., et al.: Orientation invariant feature embedding and spatial temporal regularization for vehicle re-identification. In: Proceedings of the IEEE International Conference on Computer Vision, pp. 379–387 (2017)
21. Jin, Y., Li, C., Li, Y., Peng, P., Giannopoulos, G.A.: Model latent views with multi-center metric learning for vehicle re-identification. IEEE Trans. Intell. Transp. Syst. **22**(3), 1919–1931 (2021)
22. Sheng, H., et al.: Combining pose invariant and discriminative features for vehicle reidentification. IEEE Internet Things J. **8**(5), 3189–3200 (2020)
23. Yan, K., Tian, Y., Wang, Y., Zeng, W., Huang, T.: Exploiting multi-grain ranking constraints for precisely searching visually-similar vehicles. In: Proceedings of the IEEE International Conference on Computer Vision, pp. 562–570 (2017)
24. Hu, J., Shen, L., Sun, G.: Squeeze-and-excitation networks. In: Proceedings of the IEEE Conference on Computer Vision and Pattern Recognition, pp. 7132–7141 (2018)
25. Liu, K., Xu, Z., Hou, Z., Zhao, Z., Su, F.: Further non-local and channel attention networks for vehicle re-identification. In: Proceedings of the IEEE/CVF Conference on Computer Vision and Pattern Recognition Workshops, pp. 584–585 (2020)
26. Kipf, T.N., Welling, M.: Semi-supervised classification with graph convolutional networks, arXiv preprint arXiv:1609.02907 (2016)
27. Weinberger, K.Q., Saul, L.K.: Distance metric learning for large margin nearest neighbor classification. J. Mach. Learn. Res. **10**(2) (2009)
28. Vaswani, A., et al.: Attention is all you need. In: Advances in Neural Information Processing Systems, vol. 30 (2017)
29. Guo, H., Zhao, C., Liu, Z., Wang, J., Lu, H.: Learning coarse-to-fine structured feature embedding for vehicle re-identification. In: Proceedings of the AAAI Conference on Artificial Intelligence, vol. 32, no. 1 (2018)
30. Huang, Y., et al.: Dual domain multi-task model for vehicle re-identification. IEEE Trans. Intell. Transp. Syst. **23**(4), 2991–2999 (2020)
31. Zhang, X., Zhang, R., Cao, J., Gong, D., You, M., Shen, C.: Part-guided attention learning for vehicle instance retrieval. IEEE Trans. Intell. Transp. Syst. **23**(4), 3048–3060 (2020)
32. Lv, K., Sheng, H., Xiong, Z., et al.: Pose-based view synthesis for vehicles: a perspective aware method. IEEE Trans. Image Process. **29**, 5163–5174 (2020)
33. Lv, K., Sheng, H., Xiong, Z., et al.: Improving driver gaze prediction with reinforced attention. IEEE Trans. Multimedia **23**, 4198–4207 (2020)
34. Sheng, H., Lv, K., Liu, Y., et al.: Combining pose invariant and discriminative features for vehicle reidentification. IEEE Internet Things J. **8**(5), 3189–3200 (2020)
35. Hu, X., Lin, Y., Wang, S., et al.: Agent-centric relation graph for object visual navigation. IEEE Trans. Circuits Syst. Video Technol. (2023)
36. Zhang, H., Lin, Y., Han, S., et al.: Lexicographic actor-critic deep reinforcement learning for urban autonomous driving. IEEE Trans. Veh. Technol. (2022)
37. Wang, S., Wu, Z., Hu, X., et al.: Skill-based hierarchical reinforcement learning for target visual navigation. IEEE Trans. Multimedia (2023)

POI Recommendation Based on Double-Level Spatio-Temporal Relationship in Locations and Categories

Jianfu Li[✉] and Xu Li

Civil Aviation University of China, Tianjin 300300, China
jfli@cauc.edu.cn

Abstract. The sparsity of user check-in trajectory data is a great challenge faced by point of interest (POI) recommendation. To alleviate the data sparsity, existing research often utilizes the geographic and time information in check-in trajectory data to discover the hidden spatio-temporal relations. However, existing models only consider the spatio-temporal relationship between locations, ignoring that between POI categories. To further reduce the negative impact of data sparsity, motivated by the method to integrate the spatio-temporal relationship by attention mechanism in LSTPM, this paper proposes a POI recommendation model based on double-level spatio-temporal relationship in locations and categories-(POI2TS). POI2TS integrates the spatio-temporal relationship between locations and that between categories through attention mechanism to more accurately capture users' preferences. The test results on the NYC and TKY datasets show that POI2TS is more accurate compared with the state-of-the-art models, which verifies that integrating the spatio-temporal relationship between locations and that between categories can effectively improve POI recommendation models.

Keywords: Attention Mechanism · Categories · Graph Convolutional Network · POI Recommendation · Spatio-temporal Information

1 Introduction

In recent years, the rapid development and widespread application of mobile internet technologies offer users a chance to share their Point of Interest (POI) in the form of check-in anytime and anywhere, resulting in a massive of check-in data. POI recommendation aims to capture users' preferences from the check-in trajectory data and recommend the next POI suitable to them.

POI recommendation is essentially a time series prediction problem and earlier works in POI recommendation mainly apply Markov chains to model sequential transitions.

This work is supported by the State Key Laboratories Development Program of China Development Fund for Key Laboratory of Energy and Electric Power Knowledge Calculation in Hebei Province (HBKCEP202202) and the Open Fund of Information Security Evaluation Center of Civil Aviation University of China (ISECCA-202002).

With the rapid development of deep learning, many deep learning models have been exploited for POI recommendation, such as RNNs, attention networks, GNNs [1, 2]. Deep learning based POI recommendation models have shown promising performance comparing with conventional methods.

Due to the large number of POIs and the limited activity of a user, the user check-in data usually are sparse. Data sparsity prevents POI recommendation from accurately capturing user preferences, which in turn affects the accuracy of recommendations. The data sparsity is currently a great challenge faced by POI recommendation [2]. There has been a lot of researches dedicated to alleviate data sparsity. A general idea is to introduce the contexts hidden in the check-in trajectory data. Geographic, time and category information are commonly used contexts. Some studies use geographic and time information to discover the spatio-temporal relations between POIs, such as ST-RNN [3] and LSTPM [4], while others exploit the transition or hierarchical relationship between categories.

We argue that existing models only consider the spatio-temporal relations between POIs and the dependency between categories, ignoring the temporal relationship between categories. In fact, there is dependency between categories and time. Figure 1 shows the visits of three categories of POIs in the NYC dataset [1] at different time slots, including coffee shops, homes and parks, where the vertical axis indicates the percentage of visits to a category of POI in total visits and the horizontal axis indicates 48 time slots (0–23 corresponds to 24 slots for hours on weekdays and 24–47 corresponds to slots for hours on weekends). It can be seen that the number of visits to different categories shows different distributions over time. For example, the visits to cafes from 11 am to 2 pm on weekdays is much higher than other time slots, and the visits on weekdays is higher than that on weekends. On the other hand, the visits to parks in different time slots is more relatively uniform. In addition, the visit frequency to a certain category of POI by different users at different time slots is determined by the users' status and preferences. For example, the visit time of a restaurant employee to the restaurant is different from that of the diners to the restaurant. So, to study the time distribution of users' visit to different categories of POIs has a positive impact on capturing their preferences. However, how to simultaneously fuse the spatio-temporal dependency between categories and that between POI locations into current POI recommendation models is also a challenge.

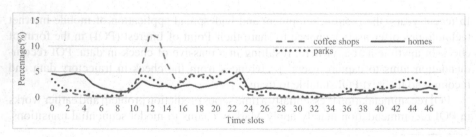

Fig. 1. The percentage of visits to categories of POIs in total visits at different time slots

In order to address the above challenge, motivated by the method to integrate the spatio-temporal information by attention mechanism in LSTPM, this paper proposes a new POI recommendation model based on Double-level Spatio-Temporal relationship in locations and categories (POI2TS). POI2TS adopts the basic framework of LSTPM, which includes data embedding, user preference learning and prediction. The main difference from LSTPM is that POI2TS exploits not only the spatial-temporal relationship between POI locations like LSTPM but that between categories when learning user preferences. In order to avoid confusion between the spatial-temporal relationship between categories and that between POI locations, if there is no specific explanation in the following text, the spatial-temporal relationship of POIs refers to that between POI locations.

The main contributions of this paper can be summarized as follows:

(1) We have proposed for the first time the dependency between categories and time and design a method to fuse the dependency between categories and time into POI recommendation models.
(2) We propose a method for capturing the dependency between categories based on GCN (a variant of GNN) and further fuse the dependency between categories into POI recommendation models.
(3) A new POI recommendation model is proposed by integrating the spatial-temporal relationship between POIs and that between categories.

2 Related Work

2.1 POI Recommendation with Spatio-Temporal Contexts

Context-aware POI recommendation models usually introduce the spatio-temporal contexts by the following two ways:

(1) To change the structure of backbone networks to incorporate the geographic and visit time of POIs. For example, ST-RNN [3] changes the input gates of RNN, and takes the spatio-temporal weighted sum of all inputs within a range before the current time as the current input. STGN [5] adds two time gates and two distance gates to LSTM to capture the spatio-temporal relationship between POIs.
(2) To convert the spatio-temporal relationship between POIs into the weights of check-in nodes by an attention mechanism. For example, Graph-Flashback [6] represents a check-in node as the weighted sum of other nodes, and the weights are calculated based on the spatio-temporal similarity between nodes by attention mechanism. STAN [7] designs a two-layer self-attention network to capture the spatio-temporal relationship between POIs. LSTPM [4] fuses the long-term and short-term preference by the spatio-temporal relationship between POIs. Due to that attention mechanism is simple and easy to implement, it is more popular to use attention mechanism to fuse geographic and time into POI recommendation.

2.2 POI Recommendation with Category Contexts

For category information, the most primary approach is to encode it as a part of a check-in representation, such as CatDM [8], PLSPL [9]. Recently, some works further reduce data sparsity by mining the complex relationship hidden in the category information by the following two ways:

(1) To exploit the transition or hierarchical relationship between categories. For example, Ref [10] computes the representation of check-ins by combining the basic embeddings of POIs and its parent category nodes via an attention mechanism on the basis of a hierarchical relationship between categories. SGRec [11] utilizes graph attention network to learn transition relations between categories.

(2) To regard the occurrence frequencies of categories in historical check-in data as the user preference, and then use a generative model to fit the preference to obtain the user preference model. Both STARec [12] and CSNS [13] follow the above ideas.

3   Problem Formulation and LSTPM

3.1   Problem Formulation

Let denote each POI q_i as (g_i, c_i), where $g_i = (lat_i, lon_i)$ represents the geographic location, lat_i and lon_i are the latitude and longitude respectively; c_i is the category. Each check-in b_i is made of a three-tuple (u_i, q_i, t_i), which means that user u_i visited POI q_i at past time t_i. The check-in trajectory $S = \{b_1, b_2,...,b_{|S|}\}$ is an ordered sequence of check-ins, where $|S|$ is the length of the trajectory. Given a user u, historical trajectory set $\{S_1, S_2,..., S_{n-1}\}$ and current trajectory $S_n = \{b_1, b_2,..., b_{t-1}\}$ where b_{t-1} is the most recent check-in generated by u, POI recommendation aims to recommend the most preferable POI to user u at the next time slot t.

3.2   LSTPM

As shown in Fig. 2, LSTPM includes three modules: data embedding, user preference learning and prediction. The user preference learning module is further divided into short-term user preference learning, long-term user preference learning based on temporal attention, and long-term user preference learning based on spatial attention. Due to we only draws on the approach to learn long-term user preference by temporal and spatial attention in LSTPM, this section only describes the user long-term preference learning module. In addition, the proposed model uses the prediction module of LSTPM. This section presents the prediction module.

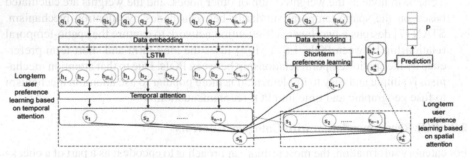

Fig. 2. The architecture of LSTPM

Long-Term User Preference Learning Based on Temporal Attention. This part mainly uses the temporal relationship between POIs to learn the representation of each historical trajectory $S_h = \{b_1, b_2, \ldots, b_{|S_h|}\}$, and then learns long-term user preference by fusing the representations of each S_h and the current trajectory S_n.

First, given a user u, to learn the latent representation h_i of each b_i in each historical trajectory S_h by LSTM, that is, $h_i = LSTM(x_i, h_{i-1})$ for $i \in \{1,2,\ldots\ldots,|S_h|\}$, where x_i the one-hot encoding of the location id of b_i.

Next, LSTMP further takes the weighted sum of each h_i in S_h as the representation of S_h, that is, $s_h = \sum_{i=1}^{|S_h|} w_i h_i$ and the weight w_i is generated by:

$$w_i = \frac{exp(T_{t_i,t_c})}{\sum_{j=1}^{|S_h|} exp(T_{t_j,t_c})}, \quad T_{t_i,t_j} = \frac{|G_{t_i} \cap G_{t_j}|}{|G_{t_i} \cup G_{t_j}|} \tag{1}$$

where t_c is visit time of $q_{|S_h|}$, t_i is the visit time of q_i, G_{t_i} is the POIs visited by user u at time t_i. Finally, to calculate the similarity $f(s_n, s_h)$ between the representation s_n of S_n and each s_h, and further obtain the long-term preference s_n^* by information aggregation:

$$s_n^* = \frac{1}{C(S)} \sum_{h=1}^{n-1} f(s_n, s_h) W_h^1 s_h \tag{2}$$

where $C(S) = \sum_{h=1}^{n-1} f(s_n, s_h)$ is the normalization factor; W_h^1 is weight matrix.

Long-Term Preference Learning Based on Spatial Attention. To improve the accuracy of long-term preference representation, this part further aggregates each s_h into s_n^* via the spatial attention:

$$s_n^+ = \frac{1}{\sum_h^{n-1} exp(\frac{1}{d_{n,h}}(s_n'^T s_h))} \sum_{h=1}^{n-1} exp(\frac{1}{d_{n,h}}(s_n'^T s_h)) W_h^2 s_h \tag{3}$$

where $s_n' = s_n^* + h_{t-1}$ and h_{t-1} is the hidden representation of b_{t-1} in S_n; W_h^2 is weight matrix; $d_{n,h}$ is the spatial distance between q_{t-1} and S_h and it is computed by:

$$d_{n,h} = \sqrt{(lon_{t-1} - lon_{S_h})^2 + (lat_{t-1} - lat_{S_h})^2} \tag{4}$$

where lon_{t-1} and lat_{t-1} represents the longitude and latitude of q_{t-1} respectively, and lon_{S_h} and lat_{S_h} represent the average latitude and longitude of S_h respectively.

Prection and Optimization. This prediction module is to predict the probability distribution P over the all POIs as follow:

$$P = soft - max(W_p^3 \cdot (s_n^+ \oplus h_{t-1}^+)) \tag{5}$$

where h_{t-1}^+ is the user's short-term preference got by the short-term user preference learning module. Finally, the model is optimized by minimizing the following objective function:

$$L = -\sum_{k=1}^{N} log(P_k) \tag{6}$$

where N is the total number of training samples, P_k represent the probability of the ground truth POI generated by the model regarding the k-th training samples.

4   The Proposed Model

We present our proposed model POI2TS in detail here. As shown in Fig. 3, POI2TS adopts the basic framework of LSTPM, which includes data embedding, short-term user preference learning, long-term user preference learning based on temporal attention, long-term user preference learning based on spatial attention and prediction. POI2TS differs from LSTPM in all four modules except for prediction.

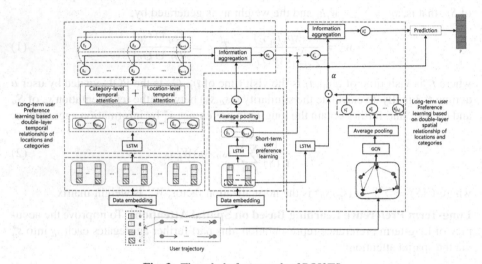

Fig. 3. The whole framework of POI2TS

4.1   Data Embedding

In order to reflect the relations among users and geographical location, category and visit time, the data embedding module jointly encodes the u_i, g_i, c_i and t_i into the embedding vector p_i of b_i, that is, $p_i = e_i \oplus x_i \oplus y_i \oplus z_i$, where e_i, x_i, y_i, and z_i are the initial embedding vector of u_i, g_i, c_i and t_i, respectively.

4.2   Short-Term User Preference Learning

First POI2TS learns the hidden representation of each b_i in S_n by LSTM, that is, $\tilde{h}_i = $ LSTM $(p_i, \tilde{h}_{i-1})$ for $i = 1,2,\ldots t-1$. The latent representation $\tilde{h}_{t-1}$ of b_{t-1} is taken as the short-term user preference.

In addition, use an average pooling to get the representation $\tilde{s}_n$ of S_n as follow:

$$\tilde{s}_n = \frac{1}{t-1} \sum_{i=1}^{t-1} \tilde{h}_i \tag{7}$$

4.3 Long-Term User Preference Learning Based on Double-Layer Temporal Relationship in Locations and Categories

Like in LSTMP, this part learns the representation of each historical trajectory $S_h = \{b_1, b_2, \ldots, b_{|S_h|}\}$ by a temporal attention, and then the long-term user preference by fusing the representations of each S_h and the current trajectory S_n. Different from LSTMP, when learning the representation of each S_h, POI2TS utilizes not only the temporal relationship between POI locations like LSTMP but also the dependency between categories and time.

In order to obtain a user's preferences for categories at different time slots, we construct a time-category matrix $M_{48 \times m}$ based on all historical trajectories of user u, where $M_{i,j}$ represents the average visit probability to POIs belonging to category c_j within time slot t_i, and m is the total number of categories, 48 corresponds to the 48 time slots. Next we design a time-attention operation to integrate the user's preferences for different categories at different time slots into the representation of S_h. That is, the category weight w_i^c of the check-in b_i is calculated as follow:

$$w_i^c = \frac{exp(M_{t_i,c_i})}{\sum_{j=1}^{|S_h|} exp(M_{t_j,c_i})} \tag{8}$$

The larger M_{t_i,c_i} is, this check-in node is given a higher weight.

Then, combing the weight w_i in Formula (1) and the category weights w_i^c in Formula (8), we obtain the representation of S_h as $\tilde{s}_h = \sum_{i=1}^{|S_h|} (w_i \tilde{h}_i + w_i^c \tilde{h}_i)$. Finally, according to the similarity $f(\tilde{s}_n, \tilde{s}_h)$ between $\tilde{s}_n$ and each $\tilde{s}_h$, we use information aggregation to obtain the long-term user preference $\tilde{s}_n^*$ as follow:

$$\tilde{s}_n^* = \frac{1}{C(S)} \sum_{h=1}^{n-1} f(\tilde{s}_n, \tilde{s}_h) \cdot g(\tilde{s}_h) \tag{9}$$

where $g(\tilde{s}_h) = w_h^4 \tilde{s}_h$, $w_h^4 \tilde{s}_h$ is weight matrix.

4.4 Long-Term User Preference Learning Based on Double-Layer Spatial Relationship in Locations and Categories

Like in LSTPM, this part reintegrates the representation of each historical trajectory into the long-term user preference representation by an attention operation. Different from LSTPM, POI2TS not only considers the spatial relationship between POI locations but also that between categories in the attention operation.

To capture the dependency among categories, we first build a directed graph $G = <V, E>$ according to all historical trajectories, where each node $v \in V$ corresponds to a category, E is the set of edges. If a user visited the POI belonging to c_j after the POI belonging to c_i, a directed edge is established from node c_i to node c_j. Then we use GCN to learn the dependency between categories in G to iteratively update the representation of the category nodes by $Y^l = \sigma (D^{-1} A^{-1} Y^{l-1} W^l)$, where σ is the activation function, Y^{l-1} is the node representation matrix in the l-th layer, and $Y_i^0 = y_i$; A is the adjacency matrix of G, D is the degree diagonal matrix of A; W^l represents the weight matrix in the l-th layer. For convenience, denote the representation of each c_i after GCN as $\tilde{y}_i$.

Then, we perform average pooling on the category representation of each check-in in S_h to obtain the main intention of S_h as follow:

$$s_h^c = \frac{1}{|S_h|} \sum\nolimits_{i=1}^{|S_h|} \tilde{y}_i \tag{10}$$

In order to get the similarity between the intention of S_h and the user's current intention, we sort the category of each check-in in S_n in order of visit time to form a category sequence. Then we use LSTM to get the latent representation $\tilde{h}_{t-1}^c$ of the last category in the category sequence, and use it as the user's current intention. Then, we calculate the similarity α between $\tilde{h}_{t-1}^c$ and s_h^c as follow:

$$\alpha = \frac{exp(s_h^{c^T} \tilde{h}_{t-1}^c)}{\sum_{h=1}^{n-1} exp(s_h^{c^T} \tilde{h}_{t-1}^c)} \tag{11}$$

Finally, we use the attention mechanism to re-aggregate each $\tilde{s}_h$ into long-term user preference representation based on the spatial distance $d_{n,h}$ between q_{t-1} and S_h and the similarity α between $\tilde{h}_{t-1}^c$ and s_h^c:

$$\tilde{s}_n^+ = \frac{1}{\sum_{h=1}^{n-1} exp(\frac{\alpha}{d_{n,h}}(\tilde{s}_n^{'T} \tilde{s}_h))} \sum\nolimits_{h=1}^{n-1} exp(\frac{\alpha}{d_{n,h}}(\tilde{s}_n^{'T} \tilde{s}_h)) W_h^5 \tilde{s}_h \tag{12}$$

where $\tilde{s}_{n'} = \tilde{s}_n^* + \tilde{h}_{t-1}$, w_h^5 is trainable projection weight matrix.

5 Experiments

5.1 Experiment Settings

Data Sets. To test the performance of POI2TS, we conduct experiments on two public real datasets: NYC and TKY [4]. Before the test, the following preprocessings were performed: (1) POIs visited by less than 10 users were deleted. (2) All check-ins of a user within a day were regarded as a trajectory. Both trajectories with less than 3 check-ins and users with less than 5 trajectories are filtered out. The final datasets is shown in Table 1, where the sparsity is calculated as 1-Check-ins/(Users × POIs). From Table 1, we can seen that both NYC and TKY have heavy sparsity.

Table 1. The statistical properties of datasets

Datasets	Users	POIs	Categories	Check-ins	Sparsity
NYC	1020	14085	374	227428	98.41%
TKY	2163	19216	348	500000	98.79%

Evaluation Metrics: We evaluate the models on two widely used evaluation metrics: Recall@K and NDCG@K. Recall@K measures the presence of the correct POI among the top K recommended POIs and NDCG@K measures the quality top-K ranking list. The larger the two metrics, the more accurate the models are. Here we set the popular K = {1, 5, 10}for evaluation.

5.2 Performance Comparison with Baseline Models

We compare POI2TS with the following six POI recommendation models:

(1) STRNN [3]: changes the input gates of RNN to fuse the spatio-temporal information of POI locations.
(2) STGN [5]: fuses the spatio-temporal information of POI locations by adding two spatio-temporal gating mechanisms to LSTM.
(3) Deep-Move [14]: introduces the short-term and long-term preference learning for the first time, but it only considers the sequence in trajectories and does not exploit any contexts.
(4) PLSPL [9]: embeds the location, category, and time into the embedding of check-ins, and uses an attention mechanism to learn weights for each node.
(5) STAN [7]: utilizes a self-attention network to model temporal and spatial correlations between non-adjacent locations.
 6) LSTPM [4]: fuses the long-term and short-term preference by the spatio-temporal relationship between POI locations.

During experiments, we set the number of hidden layer units of POI2TS and its variants (in Sect. 5.3) to 100, and all parameters are optimized using Adam, with a batch size of 32 and a learning rate of 0.0001.

The test results of POI2TS and six baseline models on NYC and TKY are respectively listed in Table 2, where the best results for each column are highlighted in bold, the second best results are highlighted in underline; Row "Percentage1" and Row "Percentage2" show the percentage improvement of POI2TS compared to LSTPM and the second best model respectively.

From Table 2, the following consistent conclusions can be obtained in terms of all the 6 evaluation metrics on both NYC and TKY:

(1) Our proposed POI2TS is the best and outperforms all baselines. Specifically, on NYC, compared with the second-best model, POI2TS improve the Recall@K indexes by 7.1%, 14.5% and 1.1%, and the NDCG@K indexes by 7.1%, 14.3% and 15.3%, and the average improvement is 9.9%. On the other hand, on TKY, compared with the second-best model, POI2TS improve the Recall@K indexes by 6.1%, 5.4%, and 1.1%, and the NDCG@K indexes by 6.1%, 9.8% and 9.8%, respectively and the average improvement is 13.0%. The above data verifies the effectiveness of POI2TS.
(2) POI2TS outperforms its basic model- LSTPM.On NYC, compared with LSTPM, POI2TS improve the Recall@K indexes by 7.1%, 16.6% and 17.4%, and the NDCG@K indexes by 7.1%, 14.3% and 15.3%, and the average improvement is 13.0%. On TKY, compared with LSTPM, POI2TS improve the Recall@K indexes

by 6.1%, 5.7% and 6.4%, and the NDCG@K indexes by 6.1%, 9.8% and 7.5% respectively, and the average improvement was 6.9%. The above data show that POI2TS can effectively improve the accuracy by integrating POI spatio-temporal between POIs and that between categories.

(3) Among the baselines, STRNN and STGN are worse than the other 5 models. This is because STRNN and STGN only learn the short-term preference according to the current trajectory, while the other 5 models such as Deep-Move not only learn the short-term preference according to the current trajectory, but also learn the long-term preference contained in historical trajectories. It can be seen that using the historical trajectory as a reference can improve the accuracy.

(4) PLSPL outperforms Deep-move. This is because PLSPL embeds the location, category and time of POIs into node representations, while Deep-move only considers the sequential information in the trajectory, without any contexts. This proves that contexts information is helpful to improve the recommendation performance.

(5) LSTPM and STAN are better than PLSPL. The main reason is that although PLSPL considers the location, category and time information of POIs, it only embeds these information into the node representation, and does not fully exploit the complex relationship in these contexts. Although LSTPM and STAN do not use category information, they make fuller use of spatio-temporal information. It can be seen that the appropriate use of spatio-temporal information also has a positive effect on the accuracy of the model.

5.3 Ablation Study

In order to compare the contribution of different modules in POI2TS to the model performance gain, this section uses the method of gradually adding modules to construct the following three variants.

(1) POI2TS-I: The variant only considers the spatio-temporal relationship between POI locations.
(2) POI2TS-II: This variant adds the temporal attention module based on the dependency between categories and time on the basis of POI2TS-I.
(3) POI2TS-III: This variant adds the spatio attention module based on the relationship between categories on the basis of POI2TS-I.

The above three variants were tested on NYC and TKY. Table 3 list the test results of three variants and POI2TS on NYC and TKY.

From Table 3, we can get the following consistent conclusions on NYC and TKY under the six metrics:

(1) POI2TS-II outperforms POI2TS-I. The average improvement of POI2TS-II over POI2TS-I is 5.2% and 6.0% on NYC and TKY, respectively. This clearly demonstrates the benefit of adding the attention module based on the category-time relationship.
(2) POI2TS-III outperforms POI2TS-I. The average improvement of POI2TS-III over POI2TS-I is 6.0% and 7.5% on NYC and TKY, respectively. This demonstrates the adding the attention module based on relationship between categories is effective.

Table 2. Test results of POI2TS and 6 baselines on NYC and TKY

Datasets	Model	Rec@1	Rec@5	Rec@10	NDCG@1	NDCG@5	NDCG@10
NYC	STRNN	0.093	0.193	0.229	0.093	0.159	0.173
	STGN	0.087	0.149	0.176	0.087	0.120	0.129
	Deep-Move	0.113	0.258	0.308	0.113	0.197	0.213
	PLSPL	0.129	0.301	0.367	0.129	0.217	0.234
	STAN	0.132	0.337	0.473	0.132	0.224	0.259
	LSTPM	0.155	0.331	0.407	0.155	0.245	0.268
	POI2TS	**0.166**	**0.386**	**0.478**	**0.166**	**0.280**	**0.309**
	Percentage1	7.1%	16.6%	17.4%	7.1%	14.3%	15.3%
	Percentage2	7.1%	14.5%	1.1%	7.1%	14.3%	15.3%
TKY	STRNN	0.120	0.262	0.323	0.120	0.197	0.216
	STGN	0.122	0.216	0.269	0.122	0.175	0.190
	Deep-Move	0.114	0.280	0.348	0.114	0.213	0.234
	PLSPL	0.123	0.310	0.381	0.123	0.233	0.246
	STAN	0.126	0.336	0.441	0.126	0.223	0.258
	LSTPM	0.148	0.335	0.419	0.148	0.246	0.279
	POI2TS	**0.157**	**0.354**	**0.446**	**0.157**	**0.270**	**0.300**
	Percentage1	6.1%	5.7%	6.4%	6.1%	9.8%	7.5%
	Percentage2	6.1%	5.4%	1.1%	6.1%	9.8%	7.5%

Table 3. Test results of 3 variants and POI2TS on NYC and TKY

Datasets	Model	Rec@1	Rec@5	Rec@10	NDCG@1	NDCG@5	NDCG@10
NYC	POI2TS-I	0.143	0.368	0.452	0.143	0.265	0.276
	POI2TS-II	0.156	0.375	0.461	0.156	0.279	0.287
	POI2TS-III	0.159	0.377	0.471	0.159	0.271	0.289
	POI2TS	0.166	0.386	0.478	0.166	0.280	0.309
TKY	POI2TS-I	0.142	0.323	0.391	0.142	0.245	0.267
	POI2TS-II	0.151	0.346	0.414	0.151	0.251	0.289
	POI2TS-III	0.153	0.339	0.430	0.153	0.260	0.289
	POI2TS	0.157	0.354	0.446	0.157	0.270	0.300

(3) POI2TS outperforms POI2TS-II and POI2TS-III. POI2TS improves by 4.6% and 3.8% on NYC compared to POI2TS-II and POI2TS-III, respectively; it improves by 4.8% and 3.5% on TKY compared to POI2TS-II and POI2TS-III, respectively. The above data shows that fusing the spatial-temporal relationship between locations and that between categories has a positive impact the recommendation performance, and also proves that the approach to fuse the two kinds of spatial-temporal relationship in POI2TS is effective.

6 Conclusion

To alleviate the impact of data sparsity, we propose a POI recommendation model based on double-level spatio-temporal relationship between locations and categories--POI2TS. POI2TS presents for the first time that the spatio-temporal relationship between categories can be helpful for recommendations performance, and proposes an approach to integrate the spatial-temporal relationship between POIs and that between categories. Experimental results on NYC and TKY datasets show that POI2TS is more accurate compared with the state-of-the-art models.

References

1. Islam, M.A., Mohammad, M.M., Das, S.S.S., Ali, M.E.: A survey on deep learning based Point-of-Interest (POI) recommendations. Neurocomputing **472**, 306–325 (2022)
2. Da'u, A., Salim, N.: Recommendation system based on deep learning methods: a systematic review and new directions. Artif. Intell. Rev. **53**(4), 2709–2748 (2019). https://doi.org/10.1007/s10462-019-09744-1
3. Liu, Q., Wu, S., Wang, L., Tan, T.: Predicting the next location: a recurrent model with spatial and temporal contexts. In: Proceedings of the Thirtieth AAAI Conference on Artificial Intelligence, pp. 194–200 (2016)
4. Sun, K., Qian, T., Chen, T., Liang, Y., Nguyen, Q.V.H., Yin, H.: Where to go next: modeling long-and short-term user preferences for point-of-interest recommendation. In: Proceedings of the AAAI Conference on Artificial Intelligence, pp.214–221 (2020)
5. Zhao, P., Luo, A., Liu, Y., Xu, J., Li, Z., Zhuang, F., et al.: Where to go next: a spatio-temporal gated network for next POI recommendation. IEEE Trans. Knowl. Data Eng. **34**(5), 2512–2524 (2020)
6. Rao, X., Chen, L., Liu, Y., Shang, S., Yao, B., Han, P.: Graph-flashback network for next location recommendation. In: Proceedings of the 28th ACM SIGKDD Conference on Knowledge Discovery and Data Mining, pp. 1463–1471(2022)
7. Luo, Y., Liu, Q., Liu, Z.: Stan: spatio-temporal attention network for next location recommendation. In: Proceedings of the Web Conference 2021, pp. 2177–2185 (2021)
8. Yu, F., Cui, L., Guo, W., Lu, X., Li, Q., Lu, H.: A category-aware deep model for successive POI recommendation on sparse check-in data. In: Proceedings of the Web Conference 2020, pp.1264–1274 (2020)
9. Wu, Y., Li, K., Zhao, G., Qian, X.: Personalized long-and short-term preference learning for next POI recommendation. IEEE Trans. Knowl. Data Eng. **34**(4), 1944–1957 (2020)
10. Zang, H., Han, D., Li, X.: Cha: categorical hierarchy-based attention for next POI recommendation. ACM Trans. Inf. Syst. **40**(1), 1–22 (2021)

11. Li, Y., Chen, T., Luo, Y., Yin, H., Huang, Z.: Discovering collaborative signals for next POI recommendation with iterative Seq2Graph augmentation. In: Proceedings of the Thirtieth International Joint Conference on Artificial Intelligence, pp. 1491–1497 (2021)
12. Ji, W., Meng, X., Zhang, Y.: STARec: adaptive learning with spatiotemporal and activity influence for POI recommendation. ACM Trans. Inf. Syst. **40**(4), 1–40 (2021)
13. Dong, Z., Meng, X., Zhang, Y.: Exploiting category-level multiple characteristics for POI recommendation. IEEE Trans. Knowl. Data Eng. **35**(02), 1488–1501 (2023)
14. Feng, J., et al.: DeepMove: predicting human mobility with attentional recurrent networks. In: Proceedings of the 2018 World Wide Web Conference, pp. 1459–1468 (2018)

Multi-Feature Integration Neural Network with Two-Stage Training for Short-Term Load Forecasting

Chuyuan Wei[✉] and Dechang Pi

Nanjing University of Aeronautics and Astronautics, Nanjing, China
884237467@qq.com

Abstract. Accurate short-term load forecasting (STLF) helps the power sector conduct generation and transmission efficiently, maintain stable grid operation while reducing energy waste, and thus achieve sustainable development. However, short-term load forecasting suffers from complex temporal dynamics and many environment variables, which causes considerable difficulties for the power sector. Therefore, short-term load forecasting is an essential yet challenging task. In this paper, we propose a short-term load forecasting model that integrates historical load, environment variables and temporal information, named TCN-GRU-TEmb. Our method utilizes temporal convolutional network (TCN) to capture the regularity of historical loads and gated recurrent unit (GRU) to extract useful features from environmental variables. As to temporal information, we propose a temporal embedding (TEmb) self-learning module, which can automatically capture the power consumption patterns of different time periods. We further propose a two-stage training algorithm to facilitate model convergence. Comparison experiments show that our model outperforms all the baselines, exhibiting an average reduction in MAE, MAPE, and RMSE of 8.24%, 9.23%, and 7.48%, respectively. Another experiment proves the effectiveness of the proposed temporal embedding method and two-stage training algorithm.

Keywords: Short-term load forecasting · Multi-feature integration · Temporal embeddings · Two-stage model training

1 Introduction

Electric load forecasting is one of the most significant tasks of the power sector. Accurate load forecasting ensures the stable operation of the power grid, thus promoting social and economic development [1]. Moreover, the power sector can cut unnecessary generation with the help of load forecasting, thus reducing carbon emissions, which is of great significance for energy conservation and sustainable development.

Depending on the prediction horizons, electric load forecasting can be categorized into short-term load forecasting (predict the load after a few minutes to a few hours), medium-term load forecasting (predict the load after a month to a year) and long-term load forecasting (predict the load after several years) [2]. With the development of smart

© The Author(s), under exclusive license to Springer Nature Singapore Pte Ltd. 2024
B. Luo et al. (Eds.): ICONIP 2023, CCIS 1967, pp. 594–606, 2024.
https://doi.org/10.1007/978-981-99-8178-6_45

grid technologies, short-term load forecasting has gained a lot of attention in recent years. Short-term load forecasting (STLF) can reflect the supply and demand changes in the grid on time, thus assisting the power sector to develop more economical and efficient daily generation plans. Therefore, STLF is of great research value [3].

The mainstream STLF methods can be divided into two categories: machine learning methods and deep learning methods. Classical machine learning methods such as ARIMA [4], SVM [5], and XGBoost [6] are widely used. However, these methods are incapable of fitting the non-linear relationships and thus tend to have low accuracy while dealing with complex data.

In recent years, the rapid development of deep learning techniques provides a new course for STLF. Many researchers proposed a lot of advanced STLF methods. Kim et al. [8] proposed a recursive inception convolutional neural network, which utilizes a 1-D CNN to adjust the hidden state of RNN. Wang et al. [9] proposed an STLF method based on a bidirectional LSTM where weights are adjusted using attention mechanism. Wu et al. [10] proposed a hybrid GRU-CNN model where features are extracted by GRU and CNN separately and then fed into fully connected layers for prediction, achieving higher forecast accuracy on real-world datasets than single models. Li et al. [11] proposed an STLF method that decomposes the load data into low-frequency and high-frequency components. The low-frequency and high-frequency parts are predicted by multivariable linear regression (MRLR) and LSTM respectively. Bohara et al. [12] proposed a CNN-BiLSTM model which utilizes a single-layer CNN followed by a pooling layer to extract key features in historical load. Results are obtained by feeding the extracted features into a bi-directional LSTM.

However, most existing neural network-based STLF methods only utilize CNNs for early-stage feature extraction; the actual forecasting module is usually composed of RNN and its variants. In our opinion, CNN is significantly underestimated in prediction tasks. A significant advantage of CNN is parallel computing, which can considerably improve time efficiency while dealing with long sequences. To further explore the potential of CNN in short-term load forecasting, we employ the temporal convolutional network (TCN) [13] in our model.

In addition, temporal information also plays an important role in short-term load forecasting. Commonly used temporal information includes year, month, day of the week, hour, minute, second, etc. Eskandari et al. [14] encoded season, weekend, weekday, and holiday by one-hot coding. Deng et al. [15] proposed a trigonometric function-based periodic coding strategy for temporal information. In this paper, we propose a temporal embedding self-learning module, which can automatically capture the electricity consumption patterns of different time periods (e.g., the difference between on-peak periods and off-peak periods), and contains richer information than the commonly used coding strategies (e.g., one-hot coding, label coding).

The main contributions of this paper are summarized as follows:

- Propose a short-term load forecasting method which integrates historical load, environment variables and temporal information.
- Propose a temporal embedding self-learning module which can automatically capture powers consumption patterns at different time periods.
- Propose a two-stage training algorithm to facilitate model convergence.

2   Framework of TCN-GRU-TEmb

The TCN-GRU-TEmb model mainly contains a load feature extraction module, an environment variables feature extraction module, a temporal embedding self-learning module, and an output module. The overall structure of TCN-GRU-TEmb is shown in Fig. 1.

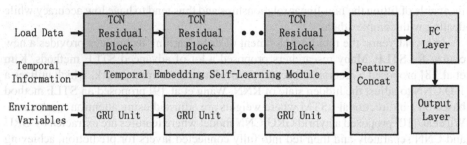

Fig. 1. The architecture of TCN-GRU-TEmb.

2.1   Load Feature Extraction Module

The load feature extraction module consists of a temporal convolutional network (TCN), which is composed of several 1D fully convolutional networks (FCNs). The structure of TCN is organized according to two core principles: causal convolution and dilated convolution. Additional residual connections are employed to stabilize the gradient.

Causal convolution is the key component of TCN. As shown in Fig. 2(a), the i-th output y_i depends only on the first i input x_1, x_2, …, x_i, thus avoiding the leakage of future information. However, causal convolution still has an obvious drawback in that the receptive field of the network only increases linearly as the depth of the network increases.

To address the problem of restricted receptive field, TCN introduces dilated convolution. Figure 2(b) shows the structure of dilated convolution with a kernel size of 2 and a dilated factor of $\{1, 2, 4\}$. After employing dilated convolution, the receptive field of TCN increases exponentially as the depth of the network increases. For an input sequence of length n, only a TCN network of depth $\log(n)$ is required to make the receptive field cover the entire input sequence.

In addition to increasing the kernel size and dilated factor, TCN can further expand the receptive field by adding more convolution layers. Therefore, TCN introduces the residual connection to stabilize the gradient of the deep network. Figure 2(c) shows the detail of the residual block.

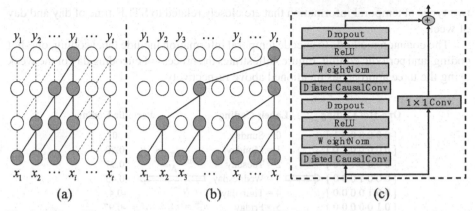

Fig. 2. The architecture of three key components in TCN. (a) shows the architecture of causal convolution. (b) shows the architecture of dilated convolution. (c) shows the detail of a residual block.

2.2 Environment Variables Feature Extraction Module

Considering that historical environment data is usually time series data, we adopt gated recurrent unit (GRU) to extract the key feature. The formula of GRU is shown in Eq. (1).

$$
\begin{aligned}
r_t &= \sigma\left(W_r \cdot [h_{t-1}, x_t]\right) \\
z_t &= \sigma\left(W_z \cdot [h_{t-1}, x_t]\right) \\
\tilde{h}_t &= \tanh\left(W_h \cdot [r_t * h_{t-1}, x_t]\right) \\
h_t &= (1 - z_t) * h_{t-1} + z_t * \tilde{h}_t
\end{aligned}
\tag{1}
$$

where r_t denotes the reset gate, z_t denotes the update gate, h_t denotes the hidden state, W_r, W_z, W_h is the trainable parameters, and σ represents the sigmoid activation function.

2.3 Temporal Embedding Self-Learning Module

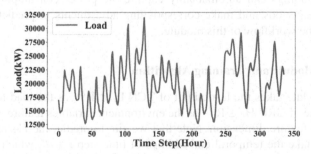

Fig. 3. A two-week load profile from a real-world dataset.

As shown in Fig. 3, periodicity is a common characteristic in modern power systems. Therefore, temporal information is vital to load forecasting. In this paper, we focus on two types of temporal information that are closely related to STLF, time of day and day of week.

The common methods to encode temporal information include label coding, one-hot coding, and periodic coding. Figure 4 illustrates the process of encoding day of the week using the three approaches mentioned above respectively.

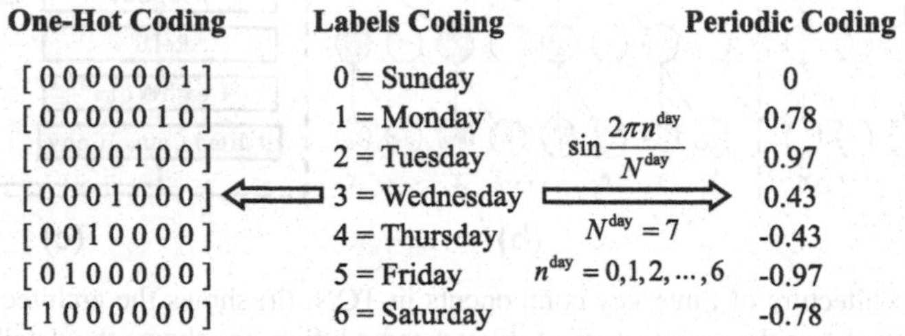

One-Hot Coding	Labels Coding		Periodic Coding
[0 0 0 0 0 0 1]	0 = Sunday		0
[0 0 0 0 0 1 0]	1 = Monday	$\sin \dfrac{2\pi n^{\text{day}}}{N^{\text{day}}}$	0.78
[0 0 0 0 1 0 0]	2 = Tuesday		0.97
[0 0 0 1 0 0 0]	3 = Wednesday		0.43
[0 0 1 0 0 0 0]	4 = Thursday	$N^{\text{day}} = 7$	-0.43
[0 1 0 0 0 0 0]	5 = Friday	$n^{\text{day}} = 0,1,2,\dots,6$	-0.97
[1 0 0 0 0 0 0]	6 = Saturday		-0.78

Fig. 4. The illustration of label coding, periodic coding, and one-hot coding.

However, each coding strategy mentioned above has certain shortcomings. For example, in label coding and periodic coding, Friday and Saturday are in adjacent positions, thus they tend to have similar coding values. However, the electricity consumption patterns are likely to vary significantly due to Saturday being a rest day. Although one-hot coding does not have similar problems, it contains limited information, which may cause low prediction accuracy.

We propose a temporal embedding self-learning module to address these problems. We divide a day into $N_{\text{time}} = 24$ h and a week into $N_{\text{day}} = 7$ days, two sets of trainable temporal embeddings $T^{\text{time}} \in R^{N_{\text{time}} \times D_{\text{T}}}$ and $T^{\text{day}} \in R^{N_{\text{day}} \times D_{\text{T}}}$ are utilized to capture the electricity consumption patterns for different time in a day and different days in a week, respectively, where D_{T} denotes the dimension of temporal embeddings. We denote the temporal embeddings of time step t as $T_t^{\text{time}} \in R^{D_{\text{T}}}$ and $T_t^{\text{day}} \in R^{D_{\text{T}}}$.

For example, residential electricity consumption is usually much lower in the early hours of the working day than in the evening on the rest day. With model training, temporal embeddings can automatically capture the power consumption patterns of these two time periods and make corresponding adjustments in prediction. Figure 5 demonstrates the workflow of this module.

2.4 Output Module and Training Algorithm

The output module takes the last output of TCN $L_t \in R^{D_L}$ as the load feature and the last hidden state of GRU $H_t \in R^{D_H}$ as the environment variables feature, where D_L and D_H denote the feature dimension. Note that we aim to predict the F-step-ahead value y_{t+F}, thus we take the temporal embeddings of time step $t + F$, which are T_{t+F}^{time} and T_{t+F}^{day}. We get Z_t by concatenating L_t, H_t, T_{t+F}^{time} and T_{t+F}^{day}. Subsequently, Z_t is fed into two fully connected layers to obtain the final result y_{t+F}, shown in Eq. (2).

$$\hat{y}_{t+F} = \text{FC}_2(\text{LeakyReLU}(\text{FC}_1(Z_t))) \tag{2}$$

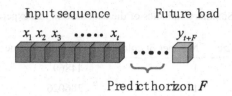

Predict horizon F

Get the temporal information of time step $t+F$

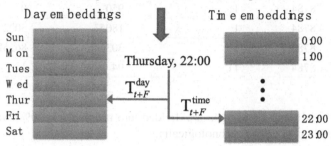

Fig. 5. The workflow of temporal embedding self-learning module.

We choose Mean Square Error (MSE) as the loss function of our model and Adam optimizer as the model optimizer.

To exploit the full potential of temporal embeddings, we further propose a novel two-stage model training algorithm. In stage one, we train a naïve model to pre-optimize the randomly initialized temporal embeddings. As illustrated in Fig. 6, this naïve model consists of just the temporal embeddings and a fully connected layer. In stage two, we initialize the main model (TCN-GRU-TEmb) with pre-optimized temporal embeddings, then we train the main model. It should be noted that during the main model training, temporal embeddings are not fixed and will be further optimized.

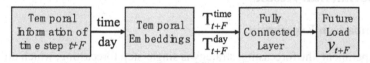

Fig. 6. The architecture of naïve model.

3 Experimental Setup

3.1 Data Preparation

We select the load dataset in GEFCOM2012 [16] as the experimental dataset. This dataset contains hourly load data for 20 zones in the US, and hourly temperature data from 11 weather stations, from January 2004 to June 2008. For this experiment, load data from zones 1–6 and all temperature data in 2004 are selected. The summary statistics of the selected data are presented in Table 1.

Table 1. Summary statistics of the data used in this experiment.

Dataset	Sample size	Number of columns	Maximum value	Minimum value
Zone 1	8784	1	44869	8688
Zone 2	8784	1	286026	82672
Zone 3	8784	1	308623	89204
Zone 4	8784	1	950	0
Zone 5	8784	1	18053	3022
Zone 6	8784	1	303646	86652
Temperature	8784	11	94	1

Both load and temperature data are divided into training set (60%), validation set (20%), and testing set (20%) chronologically.

3.2 Baselines and Evaluation Metrics

In this paper, a total of six baseline methods are selected for comparison experiments. TCN [13], LSTM [7], and CNN-BiLSTM [12] are non-GNN load forecasting methods. Considering that there is a strong dependence between environment variables and load, we choose three GNN-based load forecasting methods, which are IGMTF [17], MTGNN [18], and MTGODE [19].

Mean Absolute Error (MAE), Mean Absolute Percentage Error (MAPE), and Root Mean Square Error (RMSE) are selected as evaluation metrics to assess the model performance.

4 Experimental Results

4.1 Comparison Experiment

In this experiment, the forecasting strategy is single-step forecasting, which utilizes load and temperature data from the last 48 h to predict the load after 3 h, 6 h, 12 h, and 24 h, respectively. To reduce the impact of randomness, we repeat the experiment 10 times and record the average value of each evaluation metric.

Table 2 records the performance of TCN-GRU-TEmb and all non-GNN baselines on zone 1, 2, and 3. From Table 2, we can observe that TCN-GRU-TEmb outperforms all baselines on each dataset. Compared with the best baseline (TCN), our model exhibits an average reduction in MAE, MAPE, and RMSE of 12.37%, 12.55%, and 10.93%.

Table 3 records the performance of TCN-GRU-TEmb and all GNN-based baselines on zone 4, 5, and 6. From Table 3, It can be seen that our model still outperforms all baselines. Compared with the best baseline (MTGNN), our model exhibits an average reduction in MAE, MAPE, and RMSE of 4.08%, 5.89%, and 4.04%.

Figure 7 shows the forecasting results under the horizon of 24 h on each dataset.

Table 2. Comparison results with non-GNN baselines.

Dataset		Zone1				Zone2				Zone3			
		horizon				horizon				horizon			
Method	Metrics	3 h	6 h	12 h	24 h	3 h	6 h	12 h	24 h	3 h	6 h	12 h	24 h
TCN [13]	MAE	1060	1544	1936	2153	6269	8417	10613	13187	6813	9330	11388	14035
	MAPE	0.0577	0.0839	0.1006	0.1111	0.0379	0.0516	0.0635	0.0779	0.0383	0.053	0.0633	0.0772
	RMSE	1413	2128	2791	3208	8275	10906	14139	17911	8989	12010	15291	19055
LSTM [7]	MAE	2483	2766	2671	2291	9646	14472	14817	13212	10405	15615	15987	14256
	MAPE	0.138	0.158	0.1468	0.1232	0.0585	0.0888	0.0893	0.0787	0.0585	0.0888	0.0893	0.0787
	RMSE	3139	3658	3589	3257	12169	18013	18766	18017	13128	19436	20249	19439
CNN-BiLSTM [12]	MAE	1262	1592	2022	2280	7121	9222	11625	14577	7683	9949	12544	15728
	MAPE	0.0694	0.0853	0.1061	0.119	0.0433	0.0561	0.07	0.087	0.0433	0.0561	0.07	0.087
	RMSE	1720	2254	2917	3348	9297	11951	15145	19352	10032	12894	16342	20879
TCN-GRU-TEmb	**MAE**	**931**	**1302**	**1756**	**2051**	**5381**	**7200**	**9404**	**11541**	**5846**	**7660**	**10157**	**12418**
	MAPE	**0.0503**	**0.0706**	**0.0922**	**0.1061**	**0.0323**	**0.0435**	**0.0568**	**0.0684**	**0.0325**	**0.0429**	**0.0568**	**0.0683**
	RMSE	**1291**	**1782**	**2535**	**3055**	**7401**	**9562**	**12553**	**15904**	**8019**	**10221**	**13572**	**17108**

Table 3. Comparison results with GNN-based baselines.

Dataset		Zone4				Zone5				Zone6			
		horizon				horizon				horizon			
Method	Metrics	3 h	6 h	12 h	24 h	3 h	6 h	12 h	24 h	3 h	6 h	12 h	24 h
IGMTF [17]	MAE	33.35	45.32	45.39	53.51	555	916.8	936.7	970.7	6926	9794	11375	12457
	MAPE	0.0624	0.0855	0.0844	0.0978	0.0699	0.1199	0.116	0.1184	0.04	0.0568	0.0655	0.0702
	RMSE	49.04	60.37	60.89	71.14	745.9	1170	1273	1347	9344	12939	15719	17688
MTGNN [18]	MAE	32.02	39.3	44.28	52.59	516.7	720.2	866.8	953	5884	8181	11096	13146
	MAPE	0.0597	0.0762	0.0819	0.0957	0.0657	0.0929	0.1106	0.1167	0.0341	0.0475	0.0626	0.0733
	RMSE	46.58	53.67	59.19	70.47	684.7	943.1	1147	1315	8096	10969	15470	18443
MTGODE [19]	MAE	34.93	41.12	44.41	52.87	522.9	719.2	869	977.0	7234	9429	12505	13006
	MAPE	0.065	0.078	0.0824	0.0968	0.0652	0.0916	0.1102	0.1239	0.0416	0.055	0.0749	0.0737
	RMSE	50.2825	54.77	58.84	71.37	703.3	970.4	1187	1317	9581	12369	16363	18307
TCN-GRU-TEmb	MAE	**29.40**	**35.91**	**42.34**	**50.45**	**495.8**	**680**	**823.8**	**940.6**	**5756**	**7695**	**10017**	**12346**
	MAPE	**0.0536**	**0.0661**	**0.0771**	**0.0913**	**0.0604**	**0.0846**	**0.1009**	**0.1150**	**0.0328**	**0.0444**	**0.0576**	**0.0697**
	RMSE	**43.75**	**49.24**	**57.04**	**67.96**	**678.3**	**893.7**	**1123**	**1314**	**7879**	**10215**	**13301**	**16931**

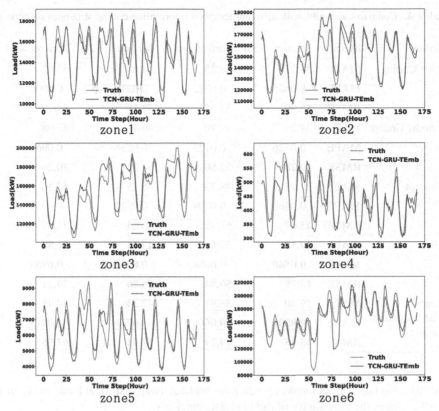

Fig. 7. Forecasting results of TCN-GRU-TEmb model under the horizon of 24h.

In summary, TCN-GRU-TEmb model achieves the highest accuracy in all six datasets and exhibits a considerable improvement compared with the baselines. In addition, we found that it is more challenging to make accurate predictions when the prediction horizon gets longer.

4.2 Comparison of Different Temporal Information Coding Strategies

To evaluate the effectiveness of the proposed temporal embedding self-learning module, we compare our method with some common coding strategies, including label coding, periodic coding, and one-hot coding. For fairness, all models are implemented based on TCN-GRU-TEmb, only the temporal information module is replaced accordingly. We choose zone 4 as experimental data.

From Table 4, we can observe that just TEmb alone outperforms three commonly used temporal information coding strategies. When TEmb and two-stage training are used in combination, the accuracy is further improved. Such results prove the effectiveness of both TEmb and two-stage training algorithm. To demonstrate the power of temporal embeddings more intuitively, we plot the forecasting results of the naïve model in Fig. 8.

Table 4. Comparison results with different temporal information coding strategies on zone 4.

Method	Metrics	horizon = 3 h	horizon = 6 h	horizon = 12 h	horizon = 24 h
Label Coding	MAE	34.17	39.25	44.68	55.62
	MAPE	0.0625	0.0722	0.0819	0.1000
	RMSE	48.99	52.65	59.10	73.79
Periodic Coding	MAE	34.29	39.20	46.17	53.08
	MAPE	0.0626	0.0723	0.0836	0.0969
	RMSE	49.28	52.50	61.31	70.26
One-Hot Coding	MAE	30.95	38.73	45.25	53.36
	MAPE	0.0562	0.0709	0.0822	0.0959
	RMSE	45.67	51.95	60.66	71.57
TEmb	MAE	**29.53**	**37.61**	**43.84**	**51.88**
	MAPE	**0.0540**	**0.0688**	**0.0792**	**0.0938**
	RMSE	**43.79**	**50.84**	**58.80**	**70.21**
TEmb (with two-stage training)	MAE	**29.40**	**35.91**	**42.34**	**50.45**
	MAPE	**0.0536**	**0.0661**	**0.0771**	**0.0913**
	RMSE	**43.75**	**49.24**	**57.04**	**67.96**

Note that this naïve model makes predictions without using historical load, and even so, it still captures the periodicity of the load data precisely.

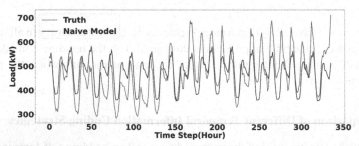

Fig. 8. Forecasting result of the naïve model in zone 4. Because this naïve model makes predictions without using historical load, different prediction horizons make no difference to it. Therefore, we only show one image here.

5 Conclusion

In this paper, we propose an STLF model named TCN-GRU-TEmb which integrates historical load, environment variables and temporal information. We compare our model with both GNN-based and non-GNN baselines. Results show that our model achieves the highest accuracy. We also observe that the overall performance of GNN-based baselines is much better than that of non-GNN baselines. This might be because GNN is a powerful tool to handle the dependencies between load and environment variables. In addition, we conduct another experiment to explore the effectiveness of the proposed TEmb method and two-stage training algorithm. Results show that just TEmb alone outperforms three commonly used temporal information coding strategies and the performance is further improved after applying two-stage training.

Our plans for future work are as follows. First, we would like to test our model on more datasets, especially those containing richer environment variables. Second, if we aim to predict the power load at a future moment, whether we should predict the future environment variables first? This is a question worthy of further study.

References

1. Feinberg, E.A., Genethliou, D.: Load forecasting. In: Applied Mathematics for Restructured Electric Power Systems: Optimization, Control, and Computational Intelligence, pp. 269–285 (2005)
2. Alfares, H.K., Nazeeruddin, M.: Electric load forecasting: literature survey and classification of methods. Int. J. Syst. Sci. **33**(1), 23–34 (2002)
3. Gross, G., Galiana, F.D.: Short-term load forecasting. Proc. IEEE **75**(12), 1558–1573 (1987)
4. Lee, C.M., Ko, C.N.: Short-term load forecasting using lifting scheme and ARIMA models. Expert Syst. Appl. **38**(5), 5902–5911 (2011)
5. Chen, B.J., Chang, M.W.: Load forecasting using support vector machines: a study on EUNITE competition 2001. IEEE Trans. Power Syst. **19**(4), 1821–1830 (2004)
6. Abbasi, R.A., Javaid, N., Ghuman, M.N.J., et al.: Short term load forecasting using XGBoost. In: Barolli, L., Takizawa, M., Xhafa, F., Enokido, T. (eds.) Web, Artificial Intelligence and Network Applications, pp. 1120–1131. Springer, Cham (2019). https://doi.org/10.1007/978-3-030-15035-8_108
7. Kong, W., Dong, Z.Y., Jia, Y., et al.: Short-term residential load forecasting based on LSTM recurrent neural network. IEEE Trans. Smart Grid **10**(1), 841–851 (2017)
8. Kim, J., Moon, J., Hwang, E., et al.: Recurrent inception convolution neural network for multi short-term load forecasting. Energy Build. **194**, 328–341 (2019)
9. Wang, S., Wang, X., Wang, S., et al.: Bi-directional long short-term memory method based on attention mechanism and rolling update for short-term load forecasting. Int. J. Electr. Power Energy Syst. **109**, 470–479 (2019)
10. Wu, L., Kong, C., Hao, X., et al.: A short-term load forecasting method based on GRU-CNN hybrid neural network model. Math. Probl. Eng. **2020**, 1–10 (2020)
11. Li, J., Deng, D., Zhao, J., et al.: A novel hybrid short-term load forecasting method of smart grid using MLR and LSTM neural network. IEEE Trans. Industr. Inf. **17**(4), 2443–2452 (2020)
12. Bohara, B., Fernandez, R.I., Gollapudi, V., et al.: Short-term aggregated residential load forecasting using BiLSTM and CNN-BiLSTM. In: 2022 International Conference on Innovation and Intelligence for Informatics, Computing, and Technologies (3ICT), pp. 37–43. IEEE (2022)
13. Bai, S., Kolter, J.Z., Koltun, V.: An empirical evaluation of generic convolutional and recurrent networks for sequence modeling. arXiv preprint arXiv:1803.01271 (2018)
14. Eskandari, H., Imani, M., Moghaddam, M.P.: Convolutional and recurrent neural network based model for short-term load forecasting. Electr. Power Syst. Res. **195**, 107173 (2021)

15. Deng, Z., Wang, B., Xu, Y., et al.: Multi-scale convolutional neural network with time-cognition for multi-step short-term load forecasting. IEEE Access **7**, 88058–88071 (2019)
16. Hong, T., Pinson, P., Fan, S.: Global energy forecasting competition 2012. Int. J. Forecast. **30**(2), 357–363 (2014)
17. Xu, W., Liu, W., Bian, J., et al.: Instance-wise graph-based framework for multivariate time series forecasting. arXiv preprint arXiv:2109.06489 (2021)
18. Wu, Z., Pan, S., Long, G., et al.: Connecting the dots: Multivariate time series forecasting with graph neural networks. In: Proceedings of the 26th ACM SIGKDD International Conference on Knowledge Discovery & Data Mining, pp. 753–763 (2020)
19. Jin, M., Zheng, Y., Li, Y.F., et al.: Multivariate time series forecasting with dynamic graph neural ODEs. IEEE Trans. Knowl. Data Eng. **01**, 1–14 (2022)

Author Index